T0180184

Theory and Technology of Drilling Engineering

Zhichuan Guan · Tinggen Chen ·
Hualin Liao

Theory and Technology of Drilling Engineering

Zhichuan Guan
School of Petroleum Engineering
China University of Petroleum (East China)
Qingdao, Shandong, China

Tinggen Chen
School of Petroleum Engineering
China University of Petroleum (East China)
Qingdao, Shandong, China

Hualin Liao
School of Petroleum Engineering
China University of Petroleum (East China)
Qingdao, Shandong, China

ISBN 978-981-15-9329-1   ISBN 978-981-15-9327-7   (eBook)
https://doi.org/10.1007/978-981-15-9327-7

Jointly published with China University of Petroleum Press
The print edition is not for sale in China (Mainland). Customers from China (Mainland) please order the
print book from: China University of Petroleum Press.

This Springer imprint is published by the registered company Springer Nature Singapore Pte Ltd.
The registered company address is: 152 Beach Road, #21-01/04 Gateway East, Singapore 189721,
Singapore

Preface

The textbook *Theory and Technology of Drilling Engineering* (Chinese version) is a professional course textbook according to the teaching plan and talent training requirements of petroleum engineering. The first edition of the book was formally printed and published in August 2000. Over the past 20 years since its publication, the book has been revised and reprinted for 12 times. It has been used by more than dozens of undergraduate students majoring in petroleum engineering, as well as a large number of adult education students and online education students.

By continuously enriching the teaching contents and soliciting opinions and suggestions from field engineers and previous graduates, the content of the textbook has been enhanced. The content of the book is based on the combination of theory and practice in terms, and it is compacted and refined, covering a wide range, reflecting as much as possible the new technology and new achievements in drilling technologies.

The principal teachers of the department of oil and gas well engineering, School of Petroleum Engineering, China University of Petroleum (East China) jointly wrote this book in the Chinese language. Chief Editors Prof. Zhichuan Guan and Prof. Tinggen Chen organized the compilation of the book. Prof. Tinggen Chen and Prof. Ruihe Wang wrote and modified the introduction and Chap. 8, Prof. Yuanfang Chen modified Chap. 1, Prof. Deyong Zou and associate Prof. Yequan Jin wrote and modified Chap. 2, Prof. Zhengsong Qiu modified Chap. 3, Prof. Zhichuan Guan wrote and modified Chap. 4. Professor Zhiyong Han wrote and modified Chap. 5. Professor Gang Liu modified Chap. 6. Professor Zhichuan Guan and Prof. Yuhuan Bu wrote and modified Chap. 7. Associate Editor Prof. Hualin Liao conducted translation, correction, and the overall revision of the book. During the phase of translation and correction, many teachers and experts provided various kinds of help, the editors representing the whole writing group would like to express their thanks to all of them, who are Chichen Xu, Xiaochun Jin in the USA, Chunyuan Zeng in Canada, Lei Wang in Kazakhstan, associate professors Yucai Shi, Xunchen Song, Jiafang Xu, Genlu Huang, Ph.D. candidates Alaa Dandash, Alex William Mwang'ande in UPC. Also, we would like to express our sincere gratitude to Ms. Lina Mu of China University of Petroleum Press for her excellent work in publishing this book.

This textbook presents the theories and technologies of drilling operations. It covers the gamut of the formulas and calculations for petroleum engineers. Some of these formulas and calculations have been used for decades, while others are meant to help guide engineers through some of the more recent breakthroughs in the industry's technology. However, there are inevitably improprieties and mistakes, and readers are kindly asked to criticize and correct them.

July 2020   Editors

About This Book

This book mainly describes the fundamental technological theories in the process of oil and gas well drilling and modern major drilling technologies. This book consists of eight chapters as well as tutorials for readers to self-learn and to master the necessary knowledge. The main contents include engineering geological conditions of drilling, rigs and drilling tools, drilling fluids, optimization of drilling parameters, well trajectory design and inclination control, well control, cementing and completion, and other drilling technologies and operations.

This book has a complete system, clear levels, and appropriate depth and breadth, which can be used as a textbook for petroleum engineering majors in ordinary colleges and universities, as well as a reference book for engineers and researchers engaged in oil and gas well drilling and production.

Introduction

I

Drilling plays a significant role in various tasks of oil and gas exploration development, such as finding and confirming oil and gas bearing structures, obtaining industrial oil or gas flow, proving geological and development data about oil, and finally taking crude oil to the ground, etc., all of which are completed by drilling. Drilling is an important link in the exploration and exploitation of petroleum and natural gas resources, and it is an essential means of exploration and exploitation of oil and gas resources as well.

The process of oil and gas exploration and development consists of many phases with different requirements and missions. The purpose and task of drilling are different at different phases. Some are for proving reserves structures, and others are for oil field development and crude oil exploitation. Therefore, oil and gas wells are generally defined as cylindrical holes with a certain depth drilled in the formation to explore and exploit underground resources such as oil and natural gas, and obtaining underground information. To realize these above objectives, the types of drilling could be categorized as follows:

Datum Well

The well is drilled during a regionally general investigation to understand the sedimentary characteristics of the formation and the oil and gas bearing situation, and verify the results of geophysical exploration, and provide geophysical parameters. The datum well is generally drilled to the bedrock and requires full wellbore coring.

Profile Well

The well is drilled along a large regional profile. The purpose of drilling is to expose local geological sections, to study stratigraphic lithology and lithofacies, and to find structures. Profiles are mainly used during the regionally general investigation.

Parameter Well

The well is drilled in petroliferous basins to understand regional structures and provide rock physical parameters. Parameter wells are mainly used in the comprehensive investigation stage.

Tectonic Well

The well is drilled to make the structural map of a standard layer in the formation, to understand its geological structural characteristics, and to verify the results of geophysical exploration.

Exploration Well

The well is drilled to determine the existence of oil and gas reservoirs, to delineate the boundaries of oil and gas reservoirs, to perform a technical evaluation of oil and gas reservoirs, and to obtain the geological data required for oil and gas development in the favorable oil and gas collection structures or oil and gas fields. The wells drilled in each exploration stage can be divided into pre-exploration wells, preliminary exploration wells, and detailed exploration wells.

Information Well

The well is drilled to make oil and gas field development plans, or to obtain data for particular thematic research during the development process.

Production Well

The well is drilled for oil and gas exploitation during oil and gas field development. Production wells can be divided into oil-producing and gas-producing wells.

Water or Gas Injection Well

The well is drilled for injecting water or gas to supply and make reasonable use of formation energy to improve oil and gas recovery and development process. Wells drilled specifically for injecting water or gas are called water injection or gas injection wells, respectively. Sometimes they are collectively referred to as injection wells.

Inspection Well

The well is drilled to clarify the pressure and distribution of oil, gas, and water in various oil and gas layers, and the distributions and changes of remaining oil saturation in oil and gas fields to a particular water-cut stage, to understand the effect of the adjustments of various submerged excavation measures.

Observation Well

The well is drilled during the development of oil and gas fields, which are used to understand the dynamics of underground oil or gas fluids, for observing the pressures variations of oil or gas at different layers, water cuts, and flooding as examples. Observation wells are generally not used for production.

Adjustment Well

The well is drilled to adjust the well network to improve the development effect and recovery rate in the middle and later stages of oil and gas field development, which includes production wells, injection wells, observation wells, etc. The production layer pressure of such wells may be low due to the later stage of oil production or be high pressure due to the energy held by the injection well.

The whole oilfield development process is divided into several stages of exploration, construction, and production, and each stage is interrelated, where a large amount of drilling work is required. High-quality, fast and efficient drilling is an important approach to develop oil and gas fields.

II

Drilling is not only applied in the petroleum industry but also in national economic construction. For example, drilling methods are often used to obtain relevant information. They are used in engineering construction in the areas of prospecting, hydrogeology, railway, hydraulics, and various types of infrastructure construction. In ancient times, humans began to dig wells to obtain underground resources. The development of drilling technology can generally be divided into four stages: manual digging, manual percussion drilling, and mechanical percussion drilling, rotary drilling. China has a long history of using drilling to develop underground resources. According to records, salt wells were drilled in Sichuan area more than 2000 years ago using the method similar to the modern percussive drilling method, and the basic principle is still used by people today. In the Northern Song Dynasty, the manual rope drilling method was developed. In 1521, oil and fire wells (natural gas wells) were drilled. In 1835, a fire well with a depth of 1200 m was drilled in Sichuan, which was the deepest well in the world at that time. It is generally believed that mechanical drilling (1859) was the beginning of modern oil drilling. Later, in 1901, a rotary drilling method was developed, using a turning table to drive the drill string, and a drill bit to break the rock at the bottom of the hole, as well as to circulate drilling fluid to clean the wellbore. Former Soviet engineers developed turbine drilling tools in 1923, which have been widely used since the 1940s. Later, electric drilling tools and screw drilling tools appeared, collectively known as downhole power drilling tools, which have particular advantages in drilling directional wells.

So far, the rotary drilling method is still the primary method of oil and gas drilling. With the development of modern science and technology, the rotary drilling technology has also been rapidly developed. Its characteristics are: ① From empirical drilling to scientific drilling; ② From shallow wells to deep wells or ultra-deep wells; ③ Development from vertical and directional wells to highly deviated directional wells, cluster wells, and horizontal wells; ④ Development from land drilling to offshore and deep-water drilling.

Drilling researchers have divided the development of rotary drilling technology into the following four periods:

- Concept period (1901–1920): This period began to combine drilling and washing wells, and the techniques of roller bit and cemented casing were used.
- Development period (1920–1948): During this period, technologies such as roller cone bit, cementing technology, drilling fluid were further developed. High-power drilling equipment appeared at the same time.

- Scientific drilling period (1948–1969): During this period, a lot of research work was carried out to study the natural laws in the drilling process, which led to the rapid development of drilling technologies. Its main technical achievements were: full use of water power (jet drilling); inserted teeth, sliding sealed bearing drill bits; low solid phase, non-solid-phase non-dispersion system of drilling fluid, and solid-phase control; optimized drilling parameters; well control technologies and balanced pressure drilling technologies.
- Automated drilling period (1969 to present): During this period, automatic measurement of drilling parameters, integrated logging, and measurement while drilling technology (MWD) were developed. Computers were widely employed in drilling operations. New techniques and equipment such as optimized drilling, automated drilling, wellhead mechanization, automation tools, remote control of wellbore trajectory, and automatic closed-loop control also came into practice.

In recent years, slim wells, extended reach wells, multi-branch wells, underbalanced pressure drilling, and coiled tubing drilling have been developed. The development of these technologies is conducive to improving drilling efficiency and enhancing oilfield production and recovery.

III

In oil and gas well drilling, although the purpose of drilling is different and the depth of the well is different, the rotary drilling method is still dominant, including rotary table rotary drilling, downhole power rotary drilling, and top drive rotary drilling, wherever in the land or in the sea.

The process of well construction, from determining the well location to the final test and production, includes a series of operations. According to the sequence, it can be divided into three stages, namely pre-drill preparation, drilling, cementing, and completion. Each stage also includes many practical techniques and operations.

Pre-drill Preparation

After determining the well location and completing the design of the well, the first procedure in the well construction by the pre-drilling engineer mainly includes the following four steps:

- Road construction: Construction of a transportation road to the well site to transport drilling equipment.
- Well site and equipment foundation preparation: Leveling the site according to the depth of the well, the type of equipment, and the design requirements, and carrying out equipment foundation construction (including the foundation of the rig, derrick, drilling pumps, etc.).

- Transportation and installation of drilling equipment: Placing, adjusting, and fixing drilling equipment, and installation of drilling circulation pipelines and oil, gas, water thermal insulation pipelines and tanks, insulation boilers, etc.
- Wellhead equipment preparation: Including digging a cellar (or not appliable), running conductor and cementing, drilling a mouse hole and rat hole, etc.

Drilling

Drilling process is to apply weight on bit to break the rock at the bottom of the hole. The cuttings generated by bit are carried to the surface by circulating drilling fluid. Weight on bit is accomplished by the gravity of a part of the drill string (drill collar). The rotation of the drill bit is achieved by the rotary table or the top drive device. When using downhole motors, the rotary table may not rotate. During the drilling process, as long as drilling tools are in the wellbore, the drilling fluid should be continuously circulated to avoid downhole problems.

During drilling, the drill bit continues to break the rock, and the wellbore will be gradually deepened. The drill string also needs to be lengthened, so it is necessary to connect drill pipes (make-up single joint) intermittently.

Since the drill bit breaks the rock at the bottom of the hole, the drill bit will gradually wear out, and the rate of penetration will decrease correspondingly. When the drill bit wears to a certain extent, it needs to be replaced with a new one. For this purpose, all the drill strings need to be taken out of the well. After a new drill bit is replaced, the new drill bit and all the drill strings are again lowered into the well. This process is called trip-out and trip-in. Sometimes it is necessary to carry out a tripping operation to deal with accidents and logging.

During drilling, the borehole shall be deepened continuously and the wellbore wall formed shall be stable without complications to ensure continued drilling. It is necessary to drill through a variety of formations, but the characteristics of each layer are different. Some formations have high or low rock strength; some contain high-pressure water, oil, gas and other fluids; some contain salt, gypsum, sodium sulfate and other compositions; all of which have adverse effects on the drilling fluid. Low-strength formations may collapse or be fractured by dense drilling fluid, preventing drilling from continuing. This requires running casing and cementing to seal the hole, and then a new section is drilled with a smaller bit. The process of changing the bit size (hole size) to start a new section drilling is called spud-in. In general, there should be several spud-ins in the course of drilling a well. The number of spud-in is different depending on the depth and formation. The basic process of drilling a well includes:

First spud-in: drill a large hole from the ground to a certain depth, and then running surface casing to this extent.

Second spud-in: Continue drilling from the surface casing with a smaller bit. If the formation is not complicated, it can directly drill to the depth objective and running production casing for completion; if formations are complex and it is difficult to control well safety with drilling fluid, intermediate casings might be required.

Third spud-in: Drill down from the intermediate casing with a smaller drill bit. Depending on situations, it may drill to the target layer, or the second and third intermediate casings will be needed, and the fourth and fifth spud-in will be carried out until the final depth of the destination reach, and production casing will be run for cementing and completion operations.

Cementing and Completion

Cementing is a process in which casing and formation are boned together by pumping cement slurry into the annular space between casing and wellbore (low part or full of annulus space). Cementing prevents complications occurring to ensure safe drilling of the next hole (for the surface casing and intermediate casing) or to ensure the successful production of oil and gas (for production casing). The upper part of the casing string is fixed on the ground with a casing head. The completion engineering includes drilling out oil and gas bearing layers, determining the connection mode between the reservoir and the wellbore (i.e., the bottom hole structure of completion), determining the wellhead device of the completion, and related technical measures. Completion bottom hole structures can be divided into four types, namely closed bottom holes, open bottom holes, mixed bottom holes, and sand control completions, etc., which are respectively adapted to different reservoir conditions. Completion operations also include running tubing, installing tubing heads, and X-tree, followed by replacing injection and inducing the flow to bring oil and gas into the wellbore for production.

In addition, cuttings logging, electrical logging, gas logging, and other logging operations need to be performed during the construction of an entire well. If it is necessary, coring is required as well. When the exploration well is drilled into the oil-bearing layer, drill pipe testing is needed. The process of oil well construction is shown in Fig. 1.

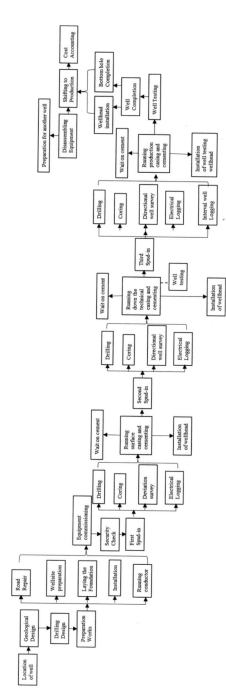

Fig. 1 The construction process of an oil and gas well

Contents

Engineering Geological Conditions for Drilling Operations

1

Abstract

Fundamental concepts related on subsurface pressures are firstly introduced, including hydrostatic pressure, in situ stress, formation pore pressure, and formation fracture pressure. Based on them, the origin of abnormal pressures is analyzed. After that, prediction and detection of abnormal formation pore pressures, formation fracture pressure, and formation collapse pressure are discussed in detail. In the second section, the rock engineering mechanical characteristics are presented with respect to uniaxial strengths and triaxial strengths, rock brittleness, rock hardness, rock drillability, rock abrasiveness. Finally, the influencing factors of confining pressure on rock mechanical properties are analyzed.

Engineering geological conditions for drilling operations refer to a combination of geological factors associated with drilling engineering, including engineering mechanical properties and types of rocks and soils, geological framework, formation fluids, and formation pressures. As is known to drilling engineers, drilling operations are performed by continuously crushing bottomhole rocks and it is necessary for them to fully understand engineering mechanical properties of rocks which provide significant evidence for selecting bit types and optimizing drilling parameters.

As the borehole comes into shape, the formation is exposed to the borehole face, which is closely related to borehole stability and pressure balance between the borehole and the formation. Improper solutions to these problems will lead to some downhole troublesome situations or even severe accidents such as well kick, well blowout, borehole collapse, and lost circulation, resulting in the obstruction in drilling operations or even wells abandonment. Accordingly, drilling engineers are supposed to comprehend engineering geological factors (such as engineering mechanical properties of rocks, characteristics of the formation pressure) of the target region before drilling processes, which lays a significant foundation for efficient well planning and successful drilling operations.

1.1 Characteristics of Subsurface Pressures

Theories and evaluation techniques on the detection of subsurface pressures play a guiding role in the exploration and development of oil and gas fields. Based on the formation pore pressure, the formation collapse pressure, and the formation fracture pressure in the drilling engineering, drilling engineers can scientifically perform well planning and drilling operations, and it is firmly necessary to make accurate evaluations for these pressures. This section will give specific introductions to concepts and evaluations for subsurface pressures mentioned above.

1.1.1 Concepts of Various Subsurface Pressures

1.1.1.1 Hydrostatic Pressure

Produced by the gravitation of the hydrostatic column itself, the hydrostatic pressure can be expressed in terms of the fluid density and vertical depth of the hydrostatic column.

$$p_h = 0.009\,81 \rho h \tag{1.1}$$

where

p_h Hydrostatic pressure, MPa;
ρ Fluid density, g/cm^3;
h Vertical depth of the fluid column, m.

As illustrated in Fig. 1.1, hydrostatic pressure shows a linear proportion to vertical depth. Hydrostatic pressure per unit vertical depth, or hydrostatic pressure gradient in terms of G_h, represents the changing hydrostatic pressure with the vertical depth.

$$G_h = p_h/h = 0.009\,81 \rho \tag{1.2}$$

where

G_h Hydrostatic pressure gradient, MPa/m.

The hydrostatic gradient is a function of the concentration of minerals and gases dissolved in the drilling fluid. The hydrostatic pressure gradient varies significantly because of the different types of drilling fluids used in oil and gas well drilling. For example, the hydrostatic gradient is generally below 0.0034 MPa/m for gas drilling fluids; 0.005–0.009 MPa/m for foam drilling fluids, and 0.01–0.025 MPa/m for regular water-based or oil-based drilling fluids.

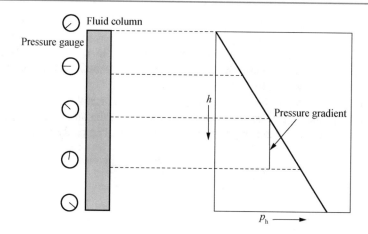

Fig. 1.1 Schematic diagram of hydrostatic pressure versus vertical depth of fluid column

1.1.2   Underground Stress

The underground stress defined as the resistance of the formation matrix to compaction refers to the state of in situ stress at some point in the formation before the drilling process, including overburden pressure (vertical compressive stress), the maximum horizontal stress, and the minimum horizontal stress, respectively.

Overburden pressure

Overburden pressure at some formation point is caused by total gravitation of rock matrix and fluids in porous rocks above the point.

$$p_o = \frac{matrix\ gravitation + fluid\ gravitation}{Corresponding\ Area}$$

$$p_o = 0.00981D\rho_b = 0.00981D\left[(1 - \phi)\rho_m + \phi\rho_f\right] \tag{1.3}$$

where

p_o Overburden pressure, MPa;
D Vertical depth above the calculation point, m;
ϕ Formation porosity, %;
ρ_b Bulk density, g/cm^3;
ρ_m Matrix density, g/cm^3;
ρ_f Fluid density, g/cm^3.

   The overburden pressure increases with depth due to compaction effects. Specifically, the overburden pressure gradient is 0.0245 MPa/m if the average bulk density of sedimentary rocks is 2.5 g/cm^3. During drilling processes, the drill floor is generally regarded as the base level for computing the overburden pressure. Therefore, in the offshore drilling, the overburden pressure gradient is greatly influenced by the distance from sea level to the drill floor, sea water depth and unconsolidated submarine sediments. Technically, the actual overburden pressure gradient is less than 0.0245 MPa/m.

As the bulk density of rocks increases along with the increase of burial depth, the overburden pressure gradient needs to be calculated in the stratified sections, and the section with similar density and lithology is used as an analysis section, i.e.,

$$G_o = \frac{\sum p_{oi}}{\sum D_i} = \frac{\sum 0.00981 \rho_{bi} D_i}{\sum D_i} \tag{1.4}$$

where

G_o Overburden pressure gradient, MPa/m;

p_{oi} Overburden pressure at the ith layer of the target formation, MPa;

D_i Vertical depth of the ith layer of the target formation, m;

ρ_{bi} Average bulk density at the ith layer of the target formation, g/cm^3.

Horizontal stress

Horizontal stress is the result of lateral stress and tectonic stress caused by tectonic movement. It is characterized by maximum and minimum horizontal in situ stresses. As a result of the tectonic movement, these two horizontal stress components are generally unequal

1.1.2.1   Formation Pore Pressures

Formation pore pressure refers to the pressure of the fluid in the pores of the rock, also known as formation pressure expressed by p_p. In all kinds of geological deposits, formation pressure is divided into normal pressure and abnormal pressure.

The term normal formation pressure (in terms of p_n) is equal to the hydrostatic pressure of continuous formation water from the surface to somewhere underground. The value is related to the sedimentary environment and depends on the density and environmental temperature of the fluid in the pore. If the formation water is fresh water (density less than 1.02g/cm^3), then the average formation pressure gradient (expressed as G_p) is 0.01 MPa /m. If the formation water is saline, the normal formation pressure gradient varies with the salinity of the formation water. The typical brine concentration is 80000 PPM, the density is 1.07g/cm^3, and the pressure gradient is 0.0105 MPa/m. Formation water encountered in oil & gas drilling is mostly brine.

The term abnormal formation pressure refers to the cases that the formation pressure is higher or lower than the normal pressure. Abnormally high pressure or super pressure is used to describe the formation pressure greater than the normal value, while abnormally low pressure or subnormal pressure describes the formation pressure lower than the normal value. The result of statistical data is illustrated in Fig. 1.2 for 100 wells located in the formation with abnormally high pressures in the USA.

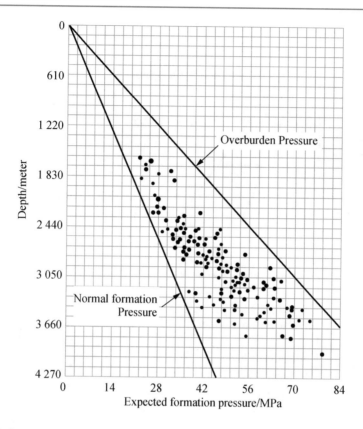

Fig. 1.2 Supper pressure formation statistics of 100 wells in the USA

1.1.2.2 Matrix Stress

Matrix stress refers to the part of the overburden load sustained by the rock matrix, symbolized by σ. In terms of overburden pressure, matrix stress can be also defined as effective overburden pressure or rock framework stress.

The relationship among stresses discussed above is illustrated in Fig. 1.3 and is formulated in the Eq. 1.5.

$$p_o = p_p + \sigma \tag{1.5}$$

The overburden load is collectively supported by both the rock matrix and fluids in porous rocks. Therefore, any factor causing decreased matrix stress will definitely lead to increased formation pressure.

1.1.2.3 Mechanisms for Abnormal Formation Pressure

Abnormally high/low formation pressures collectively refer to as *abnormal formation pressure*. For the case of abnormally low formation pressure, the gradient is

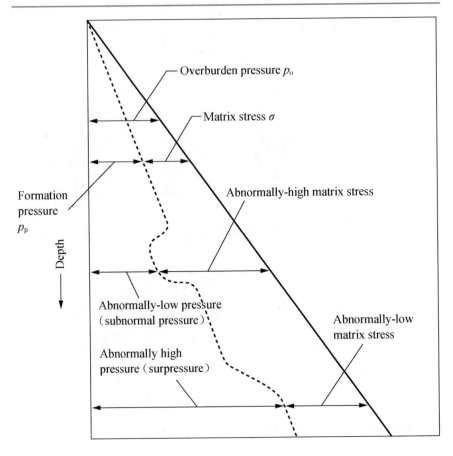

Fig. 1.3 Schematic diagram for the relationship among p_o, p_p and σ

usually under 0.01 MPa/m (or 0.0105 MPa/m), or even half of the hydrostatic gradient. As demonstrated in the worldwide drilling operations, the cases for formations with abnormally low pressure are less than those with abnormally high pressure. It is generally believed that the reservoir developed for many years will produce abnormal low pressure without sufficient energy supplement. Abnormal low pressure also occurs in areas with low underground water levels. In such an area, the normal hydrostatic pressure gradient begins at the surface of the ground. Abnormal high-pressure strata exist widely in the world, from Cenozoic Pleistocene to Paleozoic Cambrian and Sinian.

The mechanism of abnormal pressure formation is complex and may be different in different regions. For sedimentary formations, the normal fluid pressure system can be thought of as a hydraulic "open" system. This permeable, fluidable formation allows for the establishment or re-establishment of hydrostatic pressure conditions. In contrast, the pressure system in abnormal high-pressure formations is essentially "closed." There is a barrier between the abnormal pressure and the normal pressure that prevents or at least greatly restricts the flow of fluid. In this way, the upper rock gravity is partly supported by the fluid in the pores of the rock, resulting in under compaction. It is generally believed that under compaction is the most critical mechanism for the formation of abnormally high pressure. Figure 1.4 gives a simple model to simulate the compaction process. The container is filled with fluid and spring. The fluid represents the pore fluid, the spring represents the rock skeleton, and the force on the piston represents the overburden pressure. The overburden pressure is assumed jointly by spring force and fluid pressure. Therefore, the relationship between overburden pressure (p_o), matrix stress (σ), and formation pressure (p_p) satisfies Eq. 1.5.

It is obvious to observe from the model that the increased overburden pressure will result in both increased matrix stress and formation pressure. If the fluids cannot escape from pore spaces (the valve shuts shown in Fig. 1.4b), the formation pressure will exceed the normal value, and the system of abnormally high pressure will form, showing that the additional overburden load will fall on the pore fluid rather than the rock matrix due to the incompressibility of the fluid. If the pore fluid can freely flow out of pore spaces (the valve opens showed in Fig. 1.4c), the rock matrix will sustain the additional overburden load, and the formation pressure will remain at the hydrostatic pressure, forming a system of normal formation pressure we discussed above.

It is generally considered that the upper limit of abnormal high pressure is overburden pressure. Although the abnormal pressure by no means exceeds the overburden pressure on the basis of stability principle, formations with abnormal pressure greater than the overburden pressure (even 1.4 times greater) are also encountered in some countries and regions, including Pakistan, Iran, Barbier, and the South margin of Junggar Basin in Xinjiang, the Uygur Autonomous Region in China. This situation can be explained by the theory of *pressure bridge* (Fig. 1.5). The tensile strength of overburden formations may help to offset part of the upward huge force produced by pore fluids.

A system of abnormal pressures results from a variety of combined factors that are closely associated with geological processes, tectonic movements, and sedimentation rates. It is generally acknowledged at present that the abnormally high pressure is associated with the sedimentary under compaction, hydrothermal pressurization, permeation effects, and tectonic movements. Further descriptions for the sedimentary under compaction will be given as follows on account of the fact that the mechanism for the sedimentary under compaction lays a theoretical foundation for subsurface pressures evaluation for the present.

The sedimentary compaction is a result of the overburden gravitation. With continuous sedimentation, overburden sediments increase, and lower formations are compacted gradually. As long as the pore fluid escapes from pore spaces as required by the relatively low compaction rate, rock grains at sedimentary formations can rearrange for the decreased porosity. In the case of an open geological environment, the fluid squeezed out of pore spaces will flow along the direction with low resistance or with low pressure and high permeability. In this way, the system for

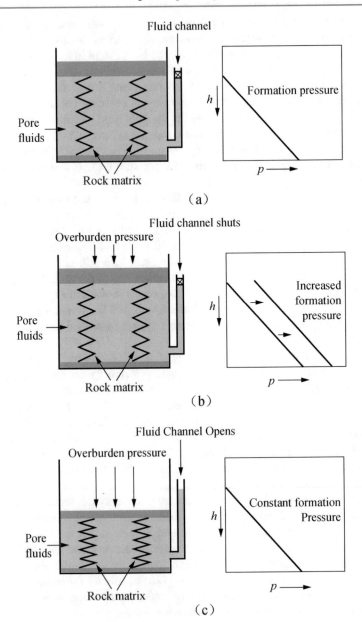

Fig. 1.4 The sediment compaction model for the formation pressure system

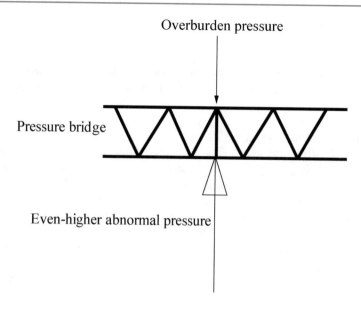

Fig. 1.5 Pressure bridge

normal hydrostatic pressure is established where increased burial depth will lead to increased rock density and reduced formation porosity. Four elements are pivotal for balanced formation compaction.

- Overburden sedimentation rates;
- Formation permeability;
- Declining rates of formation porosity;
- Capability of squeezing out fluids from pores.

If the rate of sedimentation is relatively lower than that of other processes, the process of normal compaction is formed, and the system for the normal hydrostatic pressure is also maintained.

In the process of stable deposition, normal sedimentary balance will be destroyed if any condition of equilibrium is affected. If the deposition rate is very fast, the rock particles do not have enough time to arrange, the fluid discharge in the pores is limited, and the matrix stress cannot be increased, that is, its supporting capacity to the overlying strata cannot be increased. As the overburden continues to accumulate and the support capacity of the underlying bedrock does not increase, the fluid in the pores must begin to partially support the overburden that should be supported by the rock skeleton, leading to the abnormal high pressure.

In a certain geological environment, a sealing medium is strongly necessary for a trap of abnormal pressures. In sequentially deposited basins, strata with low permeability, such as pure shale, are the most common media. Shale formations can efficiently restrict pore fluid dissipation, resulting in the under compaction and the abnormal formation pressure. Compared with normally compacted formations, under compacted formations show smaller rock density and larger formation porosity.

Rapid sedimentation is likely to happen in continental margins, particularly delta regions where the sedimentation rate tends to exceed the value required by balance conditions. Therefore, abnormally pressured formations are usually encountered in such regions.

1.1.2.4 Pressure Transition Zone

Figure 1.6 illustrates the relationship between well depths and formation pressure profile of a well in the Gulf of Mexico, USA. It is obvious that the upper formation shows normal pressure system with hydrostatic gradient, while abnormally high pressure happens at the lower formation featured by slightly less than the overburden gradient. The well section between normal and abnormal pressure systems is typically defined as *pressure transition zone*, or transition zone for simplicity.

The pressure in the pressure transition zone and the abnormal high-pressure stratum is significantly higher than that in the normal pressure stratum. The pressure

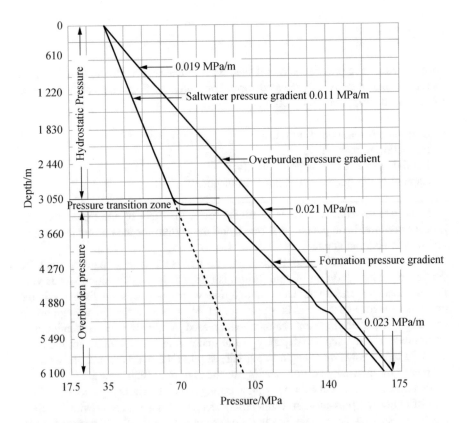

Fig. 1.6 Well depth versus pressure profile of a well in the Gulf of Mexico, USA

transition zone is the cover layer of the abnormal high-pressure stratum. Therefore, the pressure transition zone in Fig. 1.6 represents the thick shale. The shale has very low porosity, which makes the fluid in the pore space over pressurized. Effective traps are formed because the permeability of the shale is so low that the over-pressure in and below the shale does not flow upward through the shale. Therefore, the cap layer of a reservoir is not completely impervious, but in general its permeability is extremely low.

Specifically, if the cap is a thick shale layer, pressure of the cap formation will gradually increase which provide a way for formation pressure detection; if the cap is impermeable crystalline halite, the pressure transition zone will no longer exist, and the pressure change cannot be detected either.

If drill wells in high pressure areas, the drilling crew tends to detect formation pressure at the transition zone by performance parameters of drilling fluids and cuttings, which makes it easier to drill into the super pressure formation for them. It should be noted that despite the high pore pressure in the pressure transition zone, formation fluids cannot flow into the wellbore. In other words, the permeability of cap layer is extremely low, and the transition zone of pressure will not produce overflow, so other methods must be used to detect the overpressure.

1.1.2.5 Drilling Problems Associated with Abnormal Formation Pressures

In the conventional drilling process, the hydrostatic pressure of the drilling fluid should be maintained to prevent the borehole collapsing, and to prevent influx of formation fluid. To meet these two requirements, the drilling fluid pressure is kept slightly higher than formation pressure. This is known as overbalance. If, however, the overbalance is too great, this may lead to:

- Reduction of rate of penetration (ROP) due to chip hold down effect.
- Fracturing the formation and causing well loss.
- Leading to differential pressure pipe sticking.

Formation pressure profile provides significant evidence for casing programs. If there is a high pressure section above a low-pressure section, it is not possible to drill through both sections with the same drilling fluid density; otherwise the low-pressure section may be fractured. Therefore, the upper high pressure section must be cased, and then the lower pressure section could be drilled safely with lower density of drilling fluid. However, a common problem is that the surface casing is run to a relatively shallow depth. When overflow occurs during drilling lower sections with high formation pressure, it is impossible to circulate the overflow out of the hole with high-density drilling fluid without fracturing the upper formations. This phenomenon is particularly prominent for offshore drilling. Therefore, each casing must be run deeper than the required cased point so that there will not be fractured risk during killing operations. If this requirement is not met, an additional casing level is required, which not only increases drilling costs, but also results in a smaller hole diameter. After completion, the choice of production string size is limited.

Accordingly, accurate comprehension of formation pressures before drilling processes can help optimize casing programs, avoid or reduce well kick, well loss, and other drilling risks.

1.1.3 Evaluation Methods for Formation Pressures

In the long-term practice, petroleum engineers have summed up a variety of methods to evaluate formation pressure. However, each method has certain limitations, so it is difficult to accurately assess the formation pressure in an area by using only one method at present, and multiple ways should be used for comprehensive analysis and interpretation. The methods of formation pressure assessment can be divided into two categories. One is to use the data of adjacent wells to predict pressure and establish a formation pressure profile, which is often used in the design of new wells. The other is to conduct pressure monitoring according to the real-time data of the drilling, to grasp the actual variation law of formation pressure and decide the current drilling measures accordingly. These two types of methods require detailed and truthful recording of relevant data during logging and drilling, and then scientific judgments will be obtained after systematic analyses for relevant data.

Due to diverse mechanisms for abnormally high pressure, on the mudstone or sandstone profile may exist several cap formations (strata comprising a couple of tight barrier layers) with different thicknesses on top of formations with abnormally high pressure. Quite a few pressure transition zones may also exist in these cap formations. And faults would make the situation much more complicated. In addition, changes in lithological properties, like intra-mud calcic or silt may result in inaccurate evaluation for formation pressures. In general, comprehensive analyses for relevant data collected from specific situations are significant for reasonable assessments for formation pressures in the target region.

1.1.3.1 Prediction of Formation Pressure

Before drilling processes, formation pressures prediction and profile build-ups are significant for reasonable well planning and feasible drilling operations. Estimates of formation pressures are primarily based on the seismic data, the interval transit time data, and the shale resistivity data. This section will give a specific introduction to how to utilize the *acoustic time* data as follows.

A commonly used and effective way is to make use of geophysical well logging data for formation pressures prediction. Data for acoustic velocities are usually utilized since acoustic velocities through different formations are efficient for lithologic identification, reservoirs recognition, and calculations for the formation porosity and the formation pressure.

Acoustic waves through porous rocks are specifically divided into P-wave and S-wave. In the same rock, P-wave is about twice faster than the S-wave in wave velocity and is first captured by the receiver. For convenience, the regular acoustic logging primarily focuses on the propagation of P-wave through porous rocks.

Accordingly, the propagation speed of the acoustic wave through porous rocks, or *acoustic time*, is typically measured in terms of the time per unit distance the acoustic wave travels through.

$$\Delta t = \sqrt{\frac{\rho(1+\mu)}{3E(1-\mu)}} \qquad (1.6)$$

where

Δt Acoustic time, ms/m;
ρ Rock density, g/cm^3;
μ Poisson's ratio of the rock, dimensionless;
E Elastic modulus of the rock, MPa.

From the above formula, it can be seen that the speed of sound wave propagation in the formation is related to the density and elasticity coefficient of the rock, which in turn depends on the nature, structure, porosity, and burial depth of the formation. Different stratum and different lithology have different propagation velocity. Therefore, the formation characteristics can be studied and identified by measuring the propagation velocity of sound waves in the formation.

In well logging, the acoustic time refers to the interval transit time that the acoustic wave reaches two points at different depths in a borehole of well. The velocity of the acoustic wave declines with the increase of formation porosity in the formation with the same lithologic properties. For mudstones and shale with sedimentary compaction, the correlation between the interval transit time and the formation porosity can be expressed as follows.

$$\phi = \frac{\Delta t - \Delta t_m}{\Delta t_f - \Delta t_m} \qquad (1.7)$$

where

ϕ Formation porosity, %;
Δt Interval transit time of the formation, μs /m;
Δt_m Interval transit time of the rock matrix, μs /m;
Δt_f Interval transit time of the pore fluid, μs/m.

Interval transit times of the rock matrix and the pore fluid can be measured in the laboratory, which would be constants in terms of invariant properties of the rock and the pore fluid.

During normal processes of sedimentation, the relationship between the porosity of mudstones or shale and well depth may satisfy the following expression.

$$\phi = \phi_o e^{-cD} \qquad (1.8)$$

where

ϕ Porosity of mudstones or shale, %;

ϕ_o Surface porosity of mudstones or shale, %;

c Constant;

D Well depth, m.

The expression below is then obtained by studying the correlation of surface porosity and the interval transit time,

$$\phi_o = \frac{\Delta t_o - \Delta t_m}{\Delta t_f - \Delta t_m} \tag{1.9}$$

where

Δt_o refers to the initial interval transit time at the surface (D=0), which could be considered as constant in certain regions.

Substituting Eqs. 1.7 and 1.9 into Eq. 1.8, it yields,

$$\Delta t - \Delta t_m = (\Delta t_o - \Delta t_m)e^{-cD} \tag{1.10}$$

Δt_m can be also considered as constant in terms of the same lithologic properties of mudstones or shale. If Δt_m equals to 0, the Eq. 1.10 would become,

$$\Delta t = \Delta t_o e^{-cD} \tag{1.11}$$

Therefore, a linear relationship between the interval transit time and the well depth is found in a semi-logarithmic plot, where the well depth D is in the linear ordinate and the interval transit time Δt is in the semi-logarithmic abscissa.

In the normal formation pressure section, with the increase of well depth, the porosity of rock decreases, acoustic velocity increases, and acoustic time difference decreases. According to the data of acoustic time difference, the curve can be drawn on the semi-logarithmic coordinates, as shown in Fig. 1.7. In the normal pressure formation, the curve is a straight line, known as the normal trend line of acoustic time. After entering the abnormal high-pressure stratum, the porosity of the rock increases, the acoustic velocity decreases, and the acoustic time difference increases, and the rock deviates from the normal trend line. The point at which it begins to deviate is the top of formation with t abnormal high pressure.

For a region, there is a good correlation between the formation pressure gradient G_p expressed by equivalent density (g/cm^3) and the difference between the measured interval transit time in the formation with abnormal high pressure Δt and normal interval transit time at the corresponding depth Δt_n. As shown in Fig. 1.8, such a curve can be used to calculate the formation pressure quantitatively.

The steps to calculate formation pressure using acoustic logging data of mud and shale are as follows:

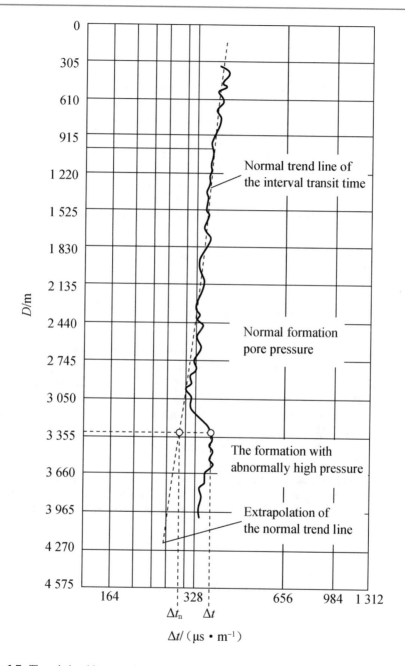

Fig. 1.7 The relationship curve for Δt versus D

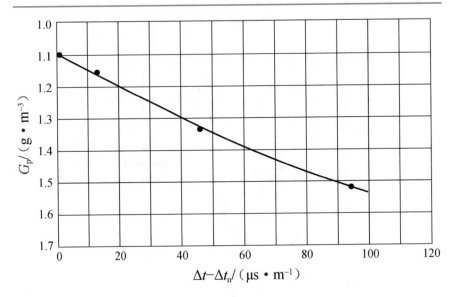

Fig. 1.8 The relationship curve for $\Delta t - \Delta t_n$ versus G_p

- Select pure mudstone or shale strata from standard well logging data, read out the interval transit time and the corresponding well depth on the logging curve at the interval of about 5 meters, and trace the points on the semi-logarithm coordinates.
- Select as many reliable data points as possible in the formation with normal pressure to build a normal trend line of the interval transit time, and extrapolate it to the region where abnormally high pressure exists.
- Record the measured interval transit time Δt at a certain depth and t_n on the normal trend line at the corresponding depth, and then calculate $\Delta t - t_n$.
- Record the corresponding formation pressure gradient G_p from the relationship curve for $\Delta t - t_n$ vs. G_p, and calculate the formation pressure using the following equation.

$$p_p = 0.009\,81 G_p D \tag{1.12}$$

As the relationship curve for $\Delta t - \Delta t_n$ vs. G_p varies in different regions, it is necessary to draw a suitable relation curve between the deviation value of interval transit time and formation pressure according to a large number of statistical data in this region.

1.1.3.2 Detection of Formation Pressures

There may be some errors in the predicted value of formation pressure before drilling, so the drilling data are used to monitor the formation pressure in real-time drilling, to correct the predicted value of formation pressure. The commonly used detecting methods of formation pressure include the d_c exponent method, standardized ROP method, and shale density method. The following mainly introduces the exponential method.

The d_c exponent method is a mechanical ROP method, monitoring of the formation pressure by the ROP produced from compaction laws of mudstones or shale and the pressure difference between hydrostatic pressure of drilling fluids and the formation pressure both at downhole. ROP is closely associated with weight on bit, rotary speed, bit types and sizes, hydraulic parameters, the performance of drilling fluids, and lithologic properties of the formation. If other factors remain constant and only the influence of pressure difference is considered, ROP will increase with the decrease of pressure difference.

In the scenario of normal formation pressure, ROP declines with the increase of well depths if lithologic properties of the formation and drilling conditions keep unchanged. When drilling into the pressure transition zone where the porosity increases gradually, the increasing pore pressure leads to the decreasing pressure difference and increasing ROP. Therefore, this characteristic can be used to predict abnormal high pressure formation. Since ROP is also influenced by the weight on bit, rotary speed, and hydraulic parameters which can hardly remain unchanged, ROP can't be affected only by the formation compaction law and pressure difference. Therefore, it is difficult to accurately predict and quantitatively calculate the formation pressure with only the variation of the ROP. Accordingly, the exponential method is developed.

The d_c exponent method is based on Bingham drilling rate equation. Without considering the influence of hydraulic factors, Bingham put forward the drilling rate equation as follows.

$$v_{pc} = Kn^e (W/d_b)^d \tag{1.13}$$

where

v_{pc} ROP (rate of penetration);
K Formation drillability;
n Rotary speed;
e Rotary speed exponent;
W Weight on bit;
d_b Bit size;
d The exponent of weight on bit.

Bingham, based on his experience in the Gulf, found that exponents of rotary speed in soft rocks were so close together that he treated them as the constant integer ($e=1$). Assuming that the drilling conditions and lithology remain unchanged, the Eq. 1.13 is simplified as follows.

$$v_{pc} = Kn(W/d_b)^d \tag{1.14}$$

Take the logarithm for Eq. 1.14 on both sides and then rearrange, it yields,

$$d = \frac{\lg(v_{pc}/n)}{\lg(W/d_b)} \tag{1.15}$$

After rearranging the expression above in metric units, it becomes,

$$d = \frac{\lg \frac{0.0547 v_{pc}}{n}}{\lg \frac{0.0684 W}{d_b}} \tag{1.16}$$

where

v_{pc} ROP (rate of penetration), m/h;
n Rotary speed, r/min;
W Weight on bit, kN;
d_b Bit size, mm.

According to the range of relevant parameters applied in the most of oil fields, both numerator and denominator in the Eq. 1.16 are negative as values for both $0.0547 v_{pc}/n$ and $0.068 W/d_b$ are less than 1.0. Meanwhile, it is easily observed that the absolute value of $\lg (0.0547 v_{pc}/n)$ is inversely proportional to the ROP v_{pc}, and thereby, the d exponent is also inversely proportional to the ROP v_{pc}. Therefore, d exponent can be applied for the detection of abnormally high pressure in the target formation since it is associated with the pressure difference. Specifically, in the formation with normal pressure, the ROP v_{pc} tends to reduce and the d exponent is inclined to increase with increasing well depths. The d exponent tends to deviate from the normal trend line and lean toward a smaller value after encountering the pressure transition zone and abnormally pressured formation.

The assumption of constant density of drilling fluids yields a correct derivation of d exponential method, which could be hardly achieved in practice. When drilling into the pressure transition zone, it is often necessary to increase the drilling fluid density, which deviates d exponent from the normal trend line. To eliminate this influence, Rehm and MeClendon put forward a corrected d exponential method, or d_c exponent method in 1971. The corrected expression for d_c is given as follows.

$$d_c = d \frac{\rho_n}{\rho_d} \tag{1.17}$$

where

d_c Corrected d exponent;

ρ_b Equivalent density in the formation with normal pressure (equal to the density of formation water), g/cm^3;

ρ_d Drilling fluid density, g/cm^3.

The following steps should be utilized to estimate the formation pressure by d_c the exponential method:

- Select data points at certain intervals along the well section of pure mudstones or shale at least 300 m above the top of the pressured formation (or select data points at mudstone or shale layers for sand–mud cross strata). Specifically, select one data point in every 1.5 m or 3 m along the well section for the ideal case; in every 5m, 10m or even larger intervals for the case of higher ROP; in every one meter for the key well section. Meanwhile, record the corresponding ROP, bit weight, rotary speed, and diameter of the bit, formation water density, and drilling fluid density.
- Calculate d and d_c exponents using the recorded data.
- Plot the curve for d_c exponent (on the abscissa axis) and corresponding well depth (on the ordinate axis) on semi-logarithm coordinates.
- Mark out the normal trend line of d_c exponent in terms of data for the formation with normal pressure, illustrated in Fig. 1.9.
- Calculate the formation pressure. After the d_c-D curve and the normal trend line are plotted; it is obvious to find where the formation with abnormal high pressure occurs and what the deviation value of d_c is in this formation. It is also easily observed that greater deviation indicates a larger formation pressure. Accordingly, the corresponding equivalent density of the formation pressure can be expressed as follows in terms of the deviation value of d_c.

$$\rho_p = \rho_n \frac{d_{cn}}{d_{ca}} \tag{1.18}$$

where

ρ_p Equivalent density of actual formation pressure at given depth, g/cm^3;

ρ_n Equivalent density of normal formation pressure at given depth, g/cm^3;

d_{cn} Normal value of d_c at given depth;

d_{ca} Measured value of d_c at given depth.

The above formula (i.e., the formation water density of normal formation pressure) varies from region to region and should be determined according to the statistical data of different areas. The density of formation water depends on the salinity of the water (namely salinity). Sampling analysis should be conducted at different formation to determine the salinity (PPM) and convert to density during the calculation.

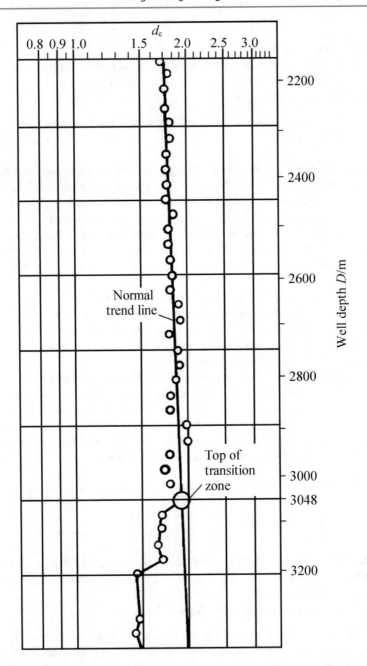

Fig. 1.9 The curve for d_c versus well depth D

Additionally, the method of equivalent well depth can also be applied for calculating the formation pressure. As discussed above, d_c exponent largely reflects how much mudstone or shale is compacted and we conclude that the same d_c exponent for formations implies the same matrix stress of rocks. As the overburden pressure is equal to the addition of matrix stress of rocks and the formation pressure, the matrix stress of rocks is known in the formation with normal pressure. Therefore, according to the conclusion, it is necessary to find equivalent depth D_e (as illustrated in Fig. 1.10) in the formation with normal pressure which shares the

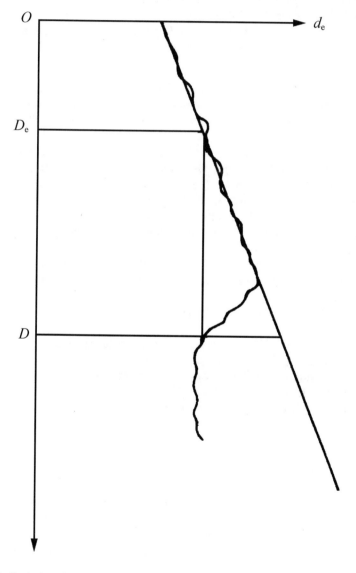

Fig. 1.10 Equivalent depth of d_c exponent

same d_c exponent of the well depth D in the formation with abnormal pressure and to calculate the pressure value in the abnormally pressured formation by the following equation.

$$p_p = G_o D - (G_o - G_{pn}) D_e \tag{1.19}$$

where

p_p Formation pressure at given depth, MPa;
G_o Overburden gradient, MPa/m;
D Corresponding well depth of formation pressure p_p at given depth, m;
G_{pn} Normal formation pressure gradient at equivalent depth D_e, MPa/m;
D_e Equivalent depth, m.

Divergent behavior may occur for values of d_c exponent and the points of divergence are undesirable when plotting the normal trend line with calculated values of d_c exponent. Reasons for this divergence are mainly as follows.

- Changes of lithologic properties. The d_c exponent closely depends on rock matrix strength, and lithologic properties vary for different strengths of rock skeleton. The regular trend for d_c exponent may be disturbed due to changes of lithologic properties, such as sand–shale staggered formation.
- Hydraulic parameters. Great changes in hydraulic parameters will cause different effects on drill cuttings cleaning by jetting flow, resulting in changes in the regular trend of d_c exponent.
- Bit types. Different types of drill bits have different rock-breaking mechanisms, so changes in drill bit will lead to deviation of the d_c exponent from the normal trend line.
  Additionally, the value of d_c exponent is undesirable under the following conditions.
- Deviation control with low WOB in drilling vertical section;
- Drilling with drag bits or core bits;
- During running wear period of bits and later period of bit ;
- Insufficient cleaning at the bottom of hole;
- Encountering faults or fractures while drilling.

1.1.4 Formation Fracture Pressure

After the borehole forms, the wellbore is filled with drilling fluid maintaining the borehole stability. Drilling practice indicates that when the hydrostatic pressure of the drilling fluid is high enough, the borehole wall will break. The pressure of the drilling fluid column is defined as *formation fracture pressure*. As one of the significant stimulations for oil wells, hydraulic fracturing for formations has been applied since the 1940s. However, for drilling engineering, it is not desirable for the formation to be fractured, because it is easy to cause well leakage and a series of complex downhole problems. Therefore, it is very important for drilling optimization design and construction to understand formation fracture pressure.

In order to have a better understanding on the formation fracture pressure, many scholars have put forward different methods to detect and calculate the formation fracture pressure. Most of them are based on field exprience and semi-theoretical formulas, which have their limitations and need to be further developed. The following are some common methods for calculating formation fracture pressure.

1.1.4.1 The Hubbert and Willis Method

In 1957, Hubert and Willis deduced from the mechanism and experiment of hydraulic fracturing of rocks that the underground stress state in the geological area where normal faulting occurs is characterized by the three-dimensional non-uniform principal stress state, and the three principal stresses are perpendicular to each other. It is also indicated that the vertical maximum principal stress σ_1 is numerically equal to the effective overburden pressure (namely matrix stress), and the minimum principal stress σ_3, one-third to half of σ_1, is perpendicular to the principal stress σ_2 between σ_1 and σ_3 on a horizontal plane.

Formation failure occurs when wellbore pressure reaches to the addition of the formation pressure and the minimum principal stress σ_3, which can be expressed by,

$$P_f = p_p + \sigma_3 = p_p + (1/3-1/2)\sigma_1 \tag{1.20}$$

Since the matrix stress σ_1 is given by $\sigma_1 = p_o - p_p.$
The formation fracture pressure can then be expressed by

$$P_f = p_p + (1/3-1/2)(p_o - p_p) \tag{1.21}$$

According to Eq. 1.21, the formation fracture gradient is expressed by

$$G_f = p_p/D + (1/3-1/2)(p_o - p_p)/D \tag{1.22}$$

where

G_f Formation fracture gradient at well depth D, MPa/m;
p_f Formation fracture pressure at well depth D, MPa;
p_p Formation pressure at well depth D, MPa;
p_o Overburden pressure at well depth D, MPa;
D Well depth, m.

Hubert and Willis laid the foundation for the measurement of formation fracture pressure theoretically and technically. However, Hubert and Willis's theory was limited in its industrial application because drilling in normal fault areas was rare.

1.1.4.2 The Mathews and Kelly Method

In 1967, Matthews and Kelly proposed a method to detect the fracture pressure of sandstone reservoirs in the Gulf region based on some empirical data. They assumed the minimum fracture pressure equal to the formation pressure and the maximum fracture pressure equal to the overburden pressure. If the actual fracture pressure is greater than the formation pressure, it is considered to be due to overcoming the matrix stress. Accordingly, they replaced the assumption that horizontal stress was one-third to half of σ_1 and were prone to believe that the matrix stress was in line with formation compaction. Specifically, greater formation compaction would lead to larger horizontal matrix stress. Therefore, they came up with a new correlation for the formation fracture pressure which can be expressed in terms of the formation pressure and the matrix stress.

$$G_f = \frac{p_p}{D} + K_i \frac{\sigma}{D} \tag{1.23}$$

where

K_i Matrix stress coefficient;
σ Matrix stress, MPa.

Matrix stress coefficient K_i is typically determined by substituting empirical data points from different regions into Eq. 1.23. K_i is a function of the well depth and is related to lithologic properties. Basically, K_i is relatively larger in sandstones with higher shale content. In the formation with normal pressure, K_i increases with increasing well depth. In abnormally pressured formations, K_i declines with the increasing formation pressure resulting from less compacted formations.

1.1.4.3 The Eaton Method

Eaton published in 1969 a more suitable method for calculating the fracture pressure of a formation. In this method, the overburden pressure gradient is considered as a variable and Poisson's ratio is also introduced into the calculation of fracture pressure gradient. Generally, within the elastic limit of an elastomer, vertical load on it will result in both transverse and longitudinal strains, and Poisson's ratio is typically defined as the ratio of the transverse strain to the longitudinal strain. In this way, Poisson's ratio can reflect the nature of rocks that are regarded as elastomers. In the Eaton correlation, however, Poisson's ratio was considered as a function of regional stress field rather than a function of rock natures. Accordingly, Eaton defined a new Poisson's ratio as the ratio of the horizontal stress to the vertical stress.

Considering the overburden pressure as a source of pressure, Eaton derived a correlation between horizontal and vertical stresses since horizontal restraints on rocks produced no horizontal stresses.

$$\sigma_h = \mu \sigma_v / (1 - \mu) \tag{1.24}$$

where

σ_h Horizontal stress, MPa;
σ_v Vertical stress ($\sigma_v = \sigma_1$), MPa;
μ Poisson's ratio of rocks.

The above formula is introduced into the calculation formula of formation fracture pressure gradient Eq. 1.23, thus expanding Matthews and Kelly's theory, namely,

$$G_f = \frac{p_p}{D} + \frac{\mu}{1-\mu}\frac{\sigma}{D} \qquad (1.25)$$

Eaton put forward the concept that the overburden gradient was changeable. It was found that Poisson's ratio varies with depth due to the change of the overburden pressure gradient. After calculating Poisson's ratio for the Gulf of Mexico, Eaton plotted the empirical curves of Poisson's ratio and depth, as illustrated in Fig. 1.11. In the calculation of fracture pressure, the overburden pressure plays an important role. If the exact increment of the overburden pressure gradient can be obtained, the calculation accuracy of the fracture pressure can be improved.

If Poisson's ratio curve of a region is known, then the Eaton method can be applied where Poisson's ratio is inversely calculated from Eq. 1.25.

1.1.4.4 The Rongzun Huang's Method

Rongzun Huang, a professor at the China University of Petroleum, found that certain errors may exist in previous methods on account of neglecting effects on the formation fracture pressure produced by the tensile strength of rocks and tectonic stresses. On the basis of a group of methods proposed by foreign researchers, professor Huang considered various factors that may influence the formation fracture pressure and came up with a new correlation for computing the formation fracture pressure by performing rigorous mathematical derivation and a number of laboratory experiments. The correlation is expressed as Eq. 1.26.

$$p_f = p_p + \left(\frac{2\mu}{1-\mu} - K_{ss}\right)(p_o - p_p) + S_{rt} \qquad (1.26)$$

where

K_{ss} Tectonic stress coefficient;
S_{rt} Tensile strength of rocks, MPa.

Compared with the first three methods mentioned above, two notable characteristics in Huang's method are shown as follows:

- Generally, the in-situ stress is inhomogeneous, consisting of three principal stresses, the vertical stress (overburden pressure), the maximum horizontal stress, and the minimum horizontal stress. Horizontal stresses are composed of

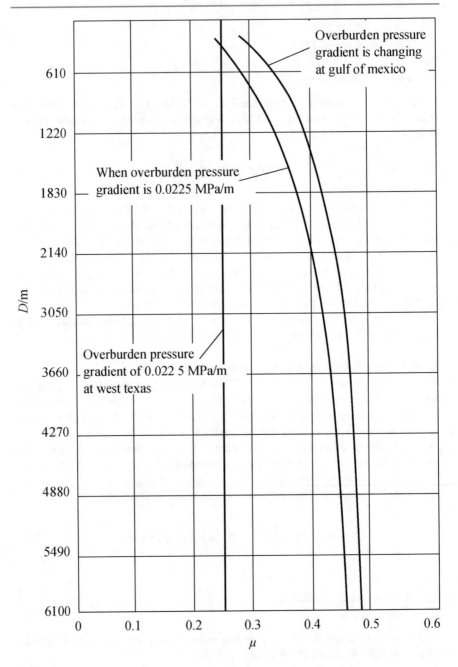

Fig. 1.11 The profile of μ-D correlation in the region of Gulf of Mexico

two components. One is the lateral pressure caused by the overburden pressure. The lateral pressure is a function of Poisson's ratio of rocks. Another is the tectonic stress produced by tectonic movements. They are independent of Poisson's ratio of rocks and not equal in two directions.

– Formation failure depends on the stress distribution on the borehole wall. Specifically, deep formations failure caused by hydraulic pressure occurs when the effective tangential stress on the borehole wall reaches or even exceeds the tensile strength of rocks.

The tensile strength of rock (S_{rt}) is measured by using drilling cores in the Brazilian test.

The tectonic stress coefficient (K_{ss}) varies for different geologic structure. However, it is a constant in the same tectonic block. The tectonic stress coefficient K_{ss} is measured by the field test for the formation fracture pressure combined with the laboratory experiment on Poisson's ratio of rocks. An accurate value for K_{ss} requires a better knowledge of Poisson's ratio μ, the formation fracture pressure p_f, and the tensile strength S_{rt}.

The methods mentioned above have certain limitations. Some errors may exist between calculated values and actual ones, even if the conditions are suitable. An accurate and useful field measurement method is presented below.

1.1.4.5   Leak-off Test Method

The leak-off test is performed after a cementing operation and drilling through the cement plug. When the pressure reaches a certain value by pumping drilling fluid with mud pumps or cementing pumps, the first sandstone layer (if there is one) at the casing shoe breaks first. At this time, the borehole pressure is called the fracture pressure of this layer. It is generally believed that the formation fracture pressure increases gradually from top to bottom, so as long as the formation at casing shoe is not broken, the possibility of well leakage in the lower well section will be greatly reduced.

Steps for leak-off test are:

– Drill through the cement plug and stop after drilling 5–10 meters in the new formation or encountering the first sandstone formation.
– Circulate and control drilling fluid properties to keep it stable, lift the bit into the casing shoe, and shut off the blowout preventer.
– Inject the drilling fluid at a relatively low flow rate between 0.66 L/s to 1.32 L/s and record the injection volume and the stand pipe pressure at each time interval.
– Plot a curve for the relationship between stand pipe pressure and cumulative pumping volume as illustrated in Fig. 1.12.
– Confirm the value for each pressure on the figure. The leakage pressure p_L is at the point where the deviation from the straight line occurs; the pressure continues to increase and reaches the maximum which is defined as the fracture pressure p_f; after the maximum value, the pressure decreases and tends to be flat. The flat pressure is called propagation pressure p_r.

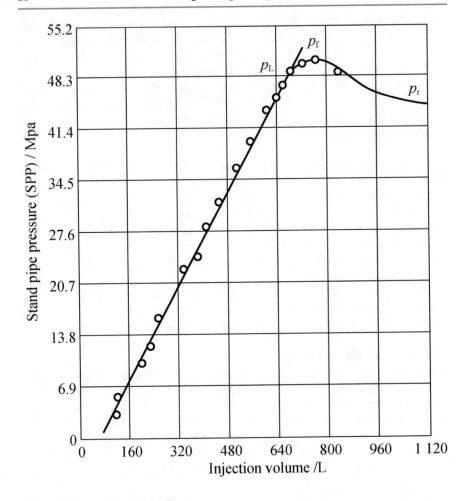

Fig. 1.12 The profile of leak-off test

- Calculate the equivalent density of the formation fracture pressure by the following expression

$$\rho_f = \rho_d + p_L/(0.00981D) \tag{1.27}$$

where

ρ_d Density of test drilling fluid, g/cm^3;
p_L Leakage pressure, MPa;
D Test well depth, m.

The test pressure should not exceed the carrying capacity of the ground equipment and casing; otherwise it will increase the drilling fluid density for the test.

1.1.5 Formation Collapse Pressure

Before drilling, subsurface formations remain balanced due to effects from a combination of overburden pressure, maximum horizontal stress, minimum horizontal stress, and pore pressure. With drilling processes going on, drilling fluid in the wellbore will sustain the borehole wall instead of rocks removed from the wellbore. That must result in stress redistribution of surrounding rocks at the borehole and high stress concentration near the borehole. Low strength of rocks near the borehole will lead to borehole instability. As stress concentration on the borehole wall is associated with the hydrostatic pressure of drilling fluid, borehole stability can be maintained by changing the density of drilling fluid to alter stress distribution near the borehole.

In general, on account of greater wall stresses than the shear strength of rocks, low density of drilling fluid will result in shear failure featured by borehole enlargement resulting from borehole collapse or borehole shrinkage due to borehole yield. When the failure occurs, this hydrostatic pressure is typically defined as the formation collapse pressure. Specifically, there are two common types of shear failure.

– Brittle failure mainly encountered in brittle rocks, results in borehole size enlargement which has a negative influence on subsequent operations such as well cementing and logging. For weakly consolidated formations, washing action may cause borehole size enlargement as well.
– Borehole shrinkage may occur in soft shale, sandstones, and halites, and some limestone strata may reproduce this phenomenon. In engineering, if this phenomenon encountered, we have to keep reaming; otherwise there will be stuck phenomenon.

Essentially, borehole stability depends on the comparison of stress distribution of surrounding rocks at the borehole with rock failure criterion selected. If the wellbore stress exceeds the formation strength, the wellbore will be destroyed. Otherwise, the wall is stable. However, there are many factors that affect the stress state and formation strength of surrounding rock, which make the problem very complicated.

– Geomechanics. Some factors are invariant, including initial in-situ stress, formation pore pressure, in-situ temperature, tectonic structures. Precise determination of these values is strongly necessary.
– Comprehensive properties of rocks, consisting of rock strength, deformation characteristics, formation porosity, water content, clay content, rock composition, and formation compaction.

- Comprehensive properties of drilling fluid, chemical compositions, properties of continuous phases, compositions and types of internal phases, additive associated with continuous phases and maintenance for drilling fluid system. It is notable that drilling fluids have a great influence on the physical and mechanical properties of mud shale and argillaceous sandstones.
- Other engineering factors include drilling time, the length of open-hole section, parameters for casing designs such as well depth, inclination, azimuth angle, pressure regulation, and swabbing effect.

These factors and parameters interact and influence each other, which makes the problem of wellbore stability very complicated.

1.1.5.1   Stress Distribution of Near Borehole in Vertical Wells

Since the borehole size is far less than the well depth, the borehole model of vertical wells can be largely simplified into a model of plane stress. As illustrated in Fig. 1.13, this mechanical borehole model of a vertical well shows the maximum horizontal stress σ_H at infinity along the x-direction, the minimum horizontal stress σ_h at infinity along the y-direction, hydrostatic pressure of drilling fluid in the borehole and pore pressure in the formation. If the formation is considered as a linear elastic structure, stress distribution of rocks near borehole of vertical wells can be expressed by,

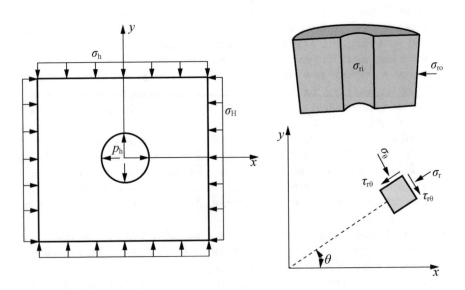

Fig. 1.13 The mechanical borehole model of a vertical well

$$\sigma_r = \frac{\sigma_H + \sigma_h}{2}\left(1 - \frac{a^2}{r^2}\right) + \frac{\sigma_H - \sigma_h}{2}\left(1 - 4\frac{a^2}{r^2} + 3\frac{a^4}{r^4}\right)\cos 2\theta + \frac{a^2}{r^2}p_h - \alpha p_p$$

$$\sigma_\theta = \frac{\sigma_H + \sigma_h}{2}\left(1 + \frac{a^2}{r^2}\right) - \frac{\sigma_H - \sigma_h}{2}\left(1 + 3\frac{a^4}{r^4}\right)\cos 2\theta - \frac{a^2}{r^2}p_h - \alpha p_p$$

$$\sigma_z = \sigma_v - \mu\left[2(\sigma_H - \sigma_h)\frac{a^2}{r^2}\cos 2\theta\right] - \alpha p_p$$

$$\tau_{r\theta} = \frac{\sigma_H - \sigma_h}{2}\left(1 - 3\frac{a^4}{r^4} + 2\frac{a^2}{r^2}\right)\sin 2\theta$$

$$(1.28)$$

where

σ_r Radial stress of the study point, MPa;

σ_θ Circumferential stress of the study point, MPa;

σ_z Vertical stress of the study point, MPa;

$\tau_{r\theta}$ Shear stress of the study point, MPa;

σ_H Maximum horizontal stress, MPa;

σ_h Minimum horizontal stress, MPa;

a Borehole radius, m;

r Radius of the study point, m;

θ Angle at the circumference, °;

α Effective stress coefficient;

p_p Formation pressure, MPa;

p_h Wellbore hydrostatic pressure, MPa.

Figure 1.14 illustrates the variation of each stress component in surrounding rocks with the distance r from the borehole under conditions of homogeneous horizontal stresses and two kinds of borehole pressures. It is obvious that stresses mainly concentrate on the borehole wall and farther distance from the borehole results in less stress concentration and gets much closer to the initial in stress. It is also easily observed that changes of borehole pressure can significantly alter the stress distribution of surrounding rocks. Accordingly, wellbore collapse starts from the borehole wall, and increasing borehole pressure can avoid this collapse. Each wallbore stress component can be expressed by,

$$\sigma_r = p_h - \alpha p_p$$
$$\sigma_\theta = \sigma_H + \sigma_h - 2(\sigma_H - \sigma_h)\cos 2\theta - p_h - \alpha p_p$$
$$\sigma_z = \sigma_v - \mu[2(\sigma_H - \sigma_h)\cos 2\theta] - \alpha p_p$$
$$\tau_{r\theta} = 0$$

$$(1.29)$$

Figure 1.15 illustrates the profiles of radial stress σ_r, circumferential stress σ_θ, vertical stress σ_z, and circumferential angle θ on the borehole wall. It is easily

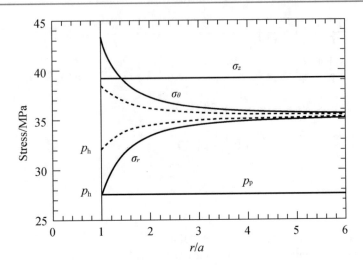

Fig. 1.14 Stress distribution of near wellbore under different borehole pressures

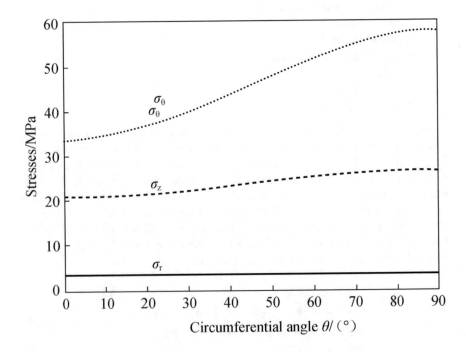

Fig. 1.15 Change rule of borehole stress component with angle

observed that the maximum stress difference occurs at the point ($\theta = 90°$ along with the direction where horizontal stress is minimum) where the first point of collapse happens.

1.1.5.2 Formula for the Calculation of Formation Collapse Pressure in Vertical Wells

Based on the Mohr–Coulomb failure criterion, the calculation formula for the formation collapse pressure of vertical wells can be expressed as follows.

$$p_c = \frac{(3\sigma_H - \sigma_h)f_1 - (f_1 - f_2)\alpha p_p - 2\tau_0}{f_1 + f_2} \tag{1.30}$$

where

$$f_1 = \sqrt{f^2 + 1} - f$$
$$f_2 = \sqrt{f^2 + 1} + f$$

In this equation:

p_c Collapse pressure of vertical well formation, MPa;
f Internal friction coefficient, i.e., tangent value of internal friction angle;
τ_0 Cohesion, MPa.

f, τ_0 are shear strength parameters, which are determined by the test.

Therefore, if in-situ stresses and shear strength are known, the minimum drilling fluid column for maintaining wellbore stability can be obtained, that is, formation collapse pressure.

1.2 Rock Engineering Mechanical Properties

Rock is a primary working object of drilling. In the process of drilling and completion, on the one hand, it needs to improve rock-breaking efficiencies; on the other hand, to ensure the stability of the wellbore, which depends on recognition and understanding of rock engineering mechanical properties. This section combines the drilling engineering and illuminates rock engineering mechanical properties and the related factors which affect these properties. This provides a necessary foundation for mastering main theories and techniques of drilling engineering.

1.2.1 Mechanical Properties of Rock

1.2.1.1 Sedimentary Rock

The rocks encountered during oil and gas well drilling are mainly sedimentary rocks.

Rock is the combination of rock-forming mineral particles, and the main rock-forming minerals are divided into eight categories, with more than 20 species, as shown in Table 1.1. To a great extent, the property of rock depends on the properties of rock-forming minerals, and the structure and tectonics of rock have an important influence on the mechanical properties of rock.

Table 1.1 Major diagenetic minerals

Number	Mineral name	Density g/cm^3	Mohs' hardness	Crystalline form
I. Aluminosilicate (Feldspar groups)				
1	Orthoclase	2.57	6	Triclinic
2	Potassium microcline fledspar	2.54	6–6.5	Triclinic
3	Albite	2.62–2.65	6–6.5	Triclinic
4	Oligocllase	2.65–2.67	5.5–6	Triclinic
5	Andesine	2.68–2.69	5–6	Triclinic
6	Labradorite	2.70–2.73	5–6	Triclinic
7	Anorthite	2.74–2.76	6–6.5	Triclinic
Analogous feldspar groups				
8	Nelpheline	2.55–2.65	5.5–6	Hexagonal system
9	Leucite	2.45–.50	5.5–6	Isometric system
II. Mica (Phyllosilicate)				
10	White mica	2.76–3.0	2–2.5	Monoclinic system
11	Black mica	2.70–3.1	2.5–3	Monoclinic system
III. Ferromagnesium silicate				
12	Pyroxene	3.3	5–6	Monoclinic system
13	Augite	3.26–3.43	5–6	Monoclinic system
14	Hornblende	3.05–3.47	5–6	Monoclinic system
15	Olivine	3.27–3.37	6.5–7	Orthorhombic system
IV. Oxide class				
16	Quartz	2.60–2.66	7	Hexagonal system

(continued)

Table 1.1 (continued)

Number	Mineral name	Density g/cm³	Mohs' hardness	Crystalline form
17	Chalcedony	–	7	Cryptocrystalline
18	Opal	1.9–2.3	5.5–6.5	Non-crystal
19	Magnetite, Hematite and others	–	–	–
V. Carbonate minerals				
20	Calcite	2.71–2.72	3	Triclinic
21	Aragonite	2.93–2.95	3.5–4	Orthorhombic system
22	Dolomite	2.8–2.9	3.5–4	Triclinic
VI. Sulfate minerals				
23	Anhydrous gypsum	2.9–2.99	3–3.5	Orthorhombic system
24	Gypsum	2.37–2.33	1.5–2	Monoclinic system
VII. Haloid				
25	Halite	2.13	2–2.5	Isometric system
VIII. Clay minerals (Phyllosilicate)				
26	Kaolinite	2.6–2.63	1–2.5	Monoclinic system
27	Microcrystalline kaolinite	–	–	

The structure of rock illustrates the microstructural characteristic of small rocks, which mainly refers to the structure of rock crystal and cementitious material. From this point of view, sedimentary rocks can be divided into crystalline sedimentary rocks and clastic sedimentary rocks. Crystalline sedimentary rocks are formed by the precipitation of salts from aqueous solution or chemical reactions in the Earth's crust, including limestone, dolomite, gypsum, and so on. Clastic sedimentary rocks are formed by the cementing action of cementitious material of rock detritus deposited, compressed in the solution of sediments, including sandstone, mudstone, and conglomerate, and cementitious materials usually have several kinds of siliceous, calcareous, iron, and clayey materials.

Rock tectonics refers to structural characteristics of rock in a large range, and sedimentary rock structure mainly includes bedding and lamination. Bedding refers to variation of rock composition and structure in vertical direction, which is mainly manifested as rock particles with different compositions changing alternately in vertical direction, rock particle size changes regularly in vertical direction, and some rock particles which are arranged in a certain direction, and so forth. Lamination is the ability of rocks to split into thin slices along a parallel plane, which is related to rock microstructure. The lamination surface is usually inconsistent with the bedding surface.

The rock petrophysical properties related to drilling engineering are also porosity and density. The porosity of rock Φ is the ratio of pore volume to rock volume.

1.2.1.2   Rock elasticity

When an object deforms under the action of external forces, the deformation disappears after the external force is removed, and then the object reverts to its original shape and volume; this character is called elastic deformation. When the external force is removed, and deformation cannot disappear, it is called plastic deformation. The relationship between stress and strain follows to Hooke's law in the deformation stage of an object that produces elastic deformation:

$$\sigma = E\varepsilon \tag{1.31}$$

where

σ Stress, MPa;
ε Strain;
E Elasticity modulus, MPa.

In elastic deformation stage of an object, the stress in one direction will not only produce strain in that direction, but also cause strain in other directions perpendicular to this direction. For example, when material exerts a stress σ_z in direction of Z-axis, expect for strain in Z-direction, and it will also cause strains ε_x and ε_y in horizontal direction (X and Y directions), and if material is isotropic, then:

$$\mu = -\frac{\varepsilon_x}{\varepsilon_z} = -\frac{\varepsilon_y}{\varepsilon_z} \tag{1.32}$$

$$\varepsilon_x = \varepsilon_y = -\mu\frac{\sigma_z}{E} \tag{1.33}$$

where

μ Poisson's ratio.

In elastic deformation stage, shearing deformation is subjected to Hooke's law, i.e.,

$$\tau = G\gamma \tag{1.34}$$

where

τ Shear stress, MPa;
γ Shear strain;
G Shear modulus (or shear modulus), MPa.

For the same material, three elastic constants E, G, and μ have the following relationship:

$$G = \frac{E}{2(1+\mu)} \tag{1.35}$$

Table 1.2 Rock elastic modulus and Possion's ratio

Rock name	(GPa)	μ	Rock name	$\dfrac{E}{(\text{GPa})}$	μ
Clay	0.3	0.38–0.45	Granite	26–60	0.26–0.29
Compact mudstone	–	0.25–0.35	Basalt	60–100	0.25
Shale	15–25	0.10–0.20	Quartzite	75–100	–
Sandstone	33–78	0.30–0.35	Orthoclase	68	0.25
Limestone	13–85	0.28–0.33	Diorite	70–100	0.25
Marble	39–92	–	Diabase	70–110	0.25
Dolomite	21–165	–	Halite	05–10	0.44

Table 1.3 Mineral elastic modulus

Mineral name	$\dfrac{E}{(\text{GPa})}$
Corundum	520
Topaz	300
Quartz	78.5–100
Feldspar	≤ 80
Calcite	58–90
Gypsum	12–15
Halite	≤ 40

The rocks show great differences with ideal elastic materials, due to the characteristics of mineral composition and structure, especially for sedimentary rocks. The relevant rock elastic constants can be measured to meet the demands of engineering and construction. The force-deformation behavior of the minerals that make up the rock when they exist alone is generally subjected to Hooke's law. The elastic constants of some minerals and rocks are listed separately in Tables 1.2 and 1.3.

1.2.1.3 Rock Strength

Rock strength refers to the maximum stress that rock can resist without failure, and in rock mechanics, the failure stress is often defined as the rock strength in the unit of MPa. The strength characteristics of rocks are different under different stress conditions, such as uniaxial compressive strength, uniaxial tensile strength, shear strength, triaxial strength, and so on.

Rock strength under simple stress condition

Rock strength under simple stress condition refers to rock strength under uniaxial external load, including uniaxial compressive strength, uniaxial tensile strength, shear strength, and flexural strength. Table 1.4 lists the strength of some rocks under simple stress conditions.

Table 1.4 Rock compressive, tensile, shear, flexural strength

Rock name	Compressive strength σ_c (MPa)	Tensile strength σ_t (MPa)	Shear strength σ_s (MPa)	Flexural strength σ_s (MPa)
Coarse sandstone	142	5.14	–	10.3
Medium sandstone	151	5.2	–	13.1
Fine sandstone	185	7.95	–	24.9
Shale	14–61	1.7–8	–	36
Clay	18	3.2	–	3.5
Gypsum	17	1.9	–	6
Calciferous limestone	42	2.4	–	6.5
Andesite	98.6	5.8	98	–
Dolomite	162	6.9	118	–
Limestone	138	9.1	145	–
Marble	166	12	198	–
Orthoclase	215.2	14.3	221	–
Gabbro	230	13.5	244	–
Quartzite	305	14.4	316	–
Diabase	343	13.4	347	–

A large number of experimental results show that the strength of rocks under simple stress conditions has the following laws:

- Under the conditions of simple stress, the strength of the same rock varies with different loading modes. In general, the strength of rocks has the following sequential relationship:

Tensile strength < Flexural strength ≤ Shear strength < Compressive strength
    If compressive strength is 1, the relationship between the strength of other loading modes and compression ratio is shown in Table 1.5.

Table 1.5 The proportional relationship between the strength of rocks

Rock name	Compressive strength	Tensile strength	Shear strength	Flexural strength
Marble	1	0.09	0.03	0.02–0.04
Sandstone	1	0.10–0.12	0.06–0.20	0.02–0.05
Limestone	1	0.15	0.08–0.10	0.04–0.10

Table 1.6 The anisotropy of some certain sedimentary strength

Rock name	Tensile strength (MPa)		Shear strength (MPa)		Flexural strength (MPa)		Compressive Strength (MPa)	
	//	⊥	//	⊥	//	⊥	//	⊥
Coarse sandstone	4.43	5.1–5.3	11.1–17.2	10.3	48.3	47	118.5–157.5	142.3–176.0
Medium sandstone	7.7	5.2	16.2–22.6	13.1–19.4	33.6–59.4	48.2–61.8	117.0–210.0	147.0–200.0
Fine sandstone	8.1–10.2	6–8	20.9–26.5	17.75	45.2–59.5	52.4–64.8	137.8–241.0	133.5–220.5
Siltstone	–	–	2.3–16.6	4.3	4.8–11.3	12.9–19.8	34.4–104.3	55.4–114.7

– Sedimentary rocks have different strengths in different directions due to the influence of bedding Table 1.6 shows the result of four kinds of strength measured in parallel to bedding direction (Denote as//) and perpendicular (Denote as ⊥).

Although rock compressive strength cannot be directly used in downhole conditions of drilling, it still can be used as a reference for bit selection in many cases.

Rock strength under complex stress conditions

Under actual geological conditions, rock is buried underground, which is subjected to compression in each direction. Rock is in a complicated three-direction stress state, and it is of great practical significance to study rock strength under the action of the complex stress.

- **The test method under triaxial stress conditions**

The triaxial stress test is a reliable method to quantitatively test rock mechanical properties quantitatively under complex stress states. Figure 1.16 shows several schemes for triaxial tests, of which Scheme (a) is the most common one, known as

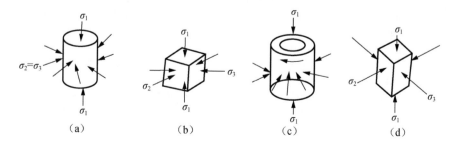

Fig. 1.16 Diagram of rock triaxial stress experiments

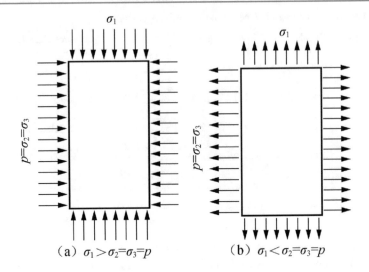

Fig. 1.17 Conventional triaxial experiment (compressive and tensile)

the conventional triaxial test. It is to place a cylindrical rock sample in a high-pressure vessel. Firstly, it is put in a stress state of uniform compression around it by hydraulic pressure. Then, the pressure is kept unchanged and the rock sample is subjected to longitudinal loading until it is damaged. Vertical stress and strain is recorded during experimental process, and stress–strain relation curve is plotted. Triaxial compression experiments can be performed as well as triaxial tensile tests. The former's loading scheme is $\sigma_1 > \sigma_2 = \sigma_3 = p$, and the latter's loading scheme is $\sigma_1 < \sigma_2 = \sigma_3 = p$, where σ_2 or σ_3 are usually called confining pressures, as shown in Fig. 1.17.

- **Characteristics of rock strength variation under triaxial stress condition**

Rock strength increases obviously under triaxial stress condition. Figure 1.18 presents rock triaxial compressive test results obtained by Kalman at room temperature; Figure 1.19 shows the variation of triaxial compressive strength of some rocks with confining pressures according to the experimental data compiled by Handing and Hager.

For all rocks, strength increases when confining pressure increases, but the increased magnitude is different for various types of rocks. Generally speaking, the pressure effect on the strength of sandstone and granite is larger than that of limestone and marble. In addition, the extent to which pressure affects strength is not the same within all pressure ranges. When confining pressure begins to increase, rock strength increases more obviously and then continues to increase, corresponding strength increment becomes smaller; finally, when confining pressure is very high, some rock strength, such as limestone, tends to be constant.

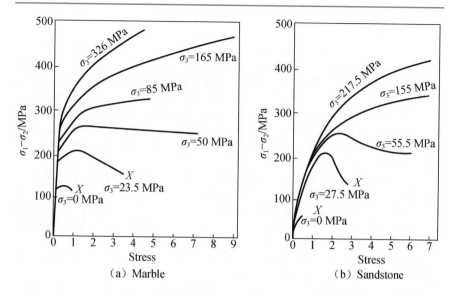

Fig. 1.18 Stress–strain curve for triaxial compressive test (X is brittle damage)

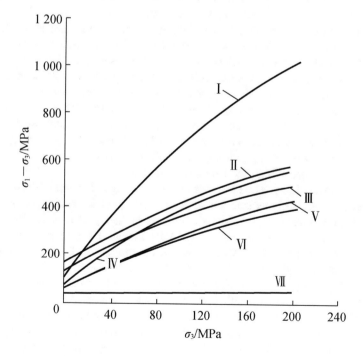

Fig. 1.19 Impact of confining pressure on rock strength (room temperature 24 °C) I—Oil Creek quartz sandstone; II—Hasmark dolomite; III—Blain anhydrock; IV—Yule marble; V—Barns sandstone and Marianna limestone (curve overlapping); VI—Muddy shale; VII—Halite salt rock

Factors influencing rock strength

Rock strength is influenced and restricted by many factors which can be divided into two aspects; on the one hand, the factors of rock itself, such as rock mineral composition, particle size, cementitious material, formation conditions, bedding structure, bulk density, porosity and water content, and on the other hand, the effects of experimental methods and physical environment, such as specimen size and shape, specimen processing, friction between press head and specimen, loading rate, and surrounding physical environment (such as temperature, confining pressure, and so forth.).

Rock with different mineral compositions has different compressive strength. Generally speaking, if the rock contains more quartz, feldspar, and other minerals, then rock compressive strength is relatively increased. Conversely, if the rock contains more mica, kaolin, chlorite, talc, pyrophyllite and so on, rock compressive strength is relatively reduced. But even rocks with the same mineral composition, the compressive strength is also very far due to the influence of particle size, agglutination, and formation conditions.

Quartz is a strong mineral in rock-forming minerals, and if quartz grains are connected mutually into matrix, the increase of quartz content can increase rock strength. Price (1934) investigated the relationship between quartz content and uniaxial compressive strength. He studied a series of sandstones containing calcite impurities and mudstones containing clay mineral impurities, and concluded that the compressive strength of these two types of rocks increased when the quartz content increased. However, rock strength is not always enhanced with the increase of quartz contents. For example, in granite, if quartz grains are dispersed (not into matrix), the increase of quartz content will not enhance rock strength.

In granite, mica-like schistose minerals and feldspar with a well-developed cleavage surface in two directions makes granite form a concealed weak surface, thus reducing rock strength. Therefore, this kind of mineral content in granite is more and the particle is larger; it has a bad effect on granite strength and becomes the main factor determining the strength of granite.

Particle size also plays a role in controlling rock strength. In general, the strength of fine-grained rocks is higher than that of coarse-grained rocks with the same mineral composition. For example, the strength of coarse-grained granite is 120 MPa, while for fine-grained granite, it can reach to 260 MPa. Likewise, the compressive strength of compacted limestone is 140 MPa, for oolitic limestone with diameter 0.5 mm, 118 MPa in drying condition, 97 MPa in saturated water condition.

For sedimentary rocks, cementation has a great effect on strength. Siliceous cements have the highest strength, followed by iron and calcareous cements, and the lowest strength of clay cements. Siliceous cemented rock has a high strength, for instance, compressive strength of siliceous cemented sandstone is above 200 MPa; the limestone cemented rock is lower, such as compressive strength of gray cementing sandstone is 20 to 100 MPa, and the strength of muddy cemented rock is the lowest, and the weak rock often belongs to this category. In terms of clay particles, the strength of siliceous cemented shale can be up to 200 MPa or higher, while that of clay cemented shale will not exceed the maximum strength of 100 MPa.

In addition, rock density often affects its strength, such as limestone density increases from 1.5 g/cm³ to 2.7 g/cm³, its compressive strength increases from 4.9 MPa to 176.6 MPa; sandstone density increases from 1.87 g/cm³ to 2.57 g/cm³, its compressive strength increases from 14.7 MPa to 88.3 MPa. Rock porosity has a great effect on compressive strength, and compressive strength decreases with the increase of rock porosity. If water infiltration causes cementation softening, rock strength can be significantly reduced.

Thus, even for the same name rock, due to different origin, rock internal structure, particle size, cementation, bulk density, porosity and water content, and so on, the strength is very different; data from Table 1.4 can only provide a reference.

1.2.1.4 Rock Brittleness and Plasticity

Rock-crushing experiment is carried out on the device shown in Fig. 1.20. The test is to load and press into the rock with cemented carbide flat press head (Fig. 1.21), and during the pressing process, the correlation curve between load and penetration is recorded (Fig. 1.22). All rock experimental curves can be divided into three typical shapes shown in Fig. 1.22. According to these three typical shapes, rocks can be divided into three categories: brittle rock, plastic rock, and plastic brittle rock.

It can be seen from Fig. 1.22 that the deformation and failure processes of rocks under the action of external forces are different. In one case, rock only changes its shape and size without breaking its continuity, which is called plasticity. The other case is when rock is subjected to external forces until it is broken without appreciable change in shape, which is called brittleness. Rock plasticity is the characteristic of mechanical energy which reflects rock absorbing residual deformation or

Fig. 1.20 Experimental device for testing rock hardness 1—Hydraulic cylinder block, 2—Cylinder plunger;3—Rock sample; 4—Press head; 5—Upper press board;6—Dial gauge; 7—Plunger guide rod

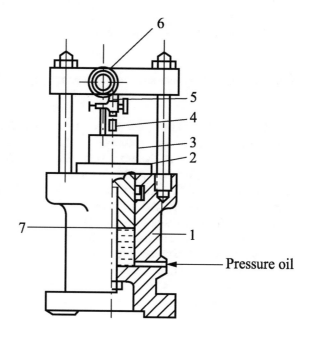

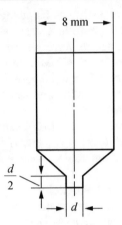

Fig. 1.21 Carbide flat press head

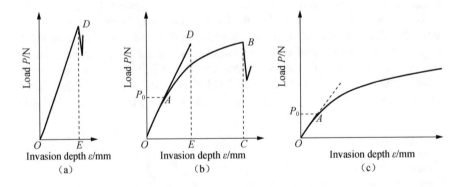

Fig. 1.22 Deformation curve when press head penetrating into rock

absorbing rock irreversible deformation before it is broken. The rock brittleness reflects that the irreversible deformation of the rock does not absorb mechanical energy obviously, that is, there is no obvious plastic deformation.

Fig. 1.22 (a) is a brittle rock, OD section is elastic deformation stage, after reaching D point brittle breakage occur; (b) for plastic brittle rock, OA section is elastic deformation stage, AB section is plastic deformation zone, reaching point B means brittle fracture has occurred; (c) for plastic rock, a small load is produced by plastic deformation, and then deformation increases with prolonging of deformation time, and there is no obvious brittle failure phenomenon.

Rock plasticity coefficient is used as a quantitative parameter to characterize rock plasticity and brittleness. Plasticity coefficient is the ratio of total work A_F to the elastic deformation work A_E before rock crushing. The calculation is based on

Table 1.7 The classification of rock plasticity

Category	Brittleness	Plastic brittle				Plasticity
		Low plasticity→High plasticity				
Grade	1	2	3	4	5	6
Plastic coefficient	1	(1,2]	(2,3]	(3,4]	(4,6]	(6,∞)

the curve of load-penetration depth in the process of rock crushing as shown in Fig. 1.22. For brittle rock, the total work before crushing A_F is equal to elastic deformation work A_E, that is, plasticity coefficient $K_p = 1$; for plastic brittle rock:

$$K_P = \frac{A_F}{A_E} = \frac{\text{OABC's Area}}{\text{ODE's Area}} \qquad (1.36)$$

For plastic rock, $K_p = \infty$.

According to rock plasticity coefficient, rocks can be divided into three categories, six grades as in Table 1.7.

Under triaxial stress conditions, a remarkable change of rock mechanics properties is that rock exhibits a transition from brittle to plastic with confining pressure increases, and greater plasticity before rock failure with greater confining pressure.

Rock plastic properties under high confining pressure can be judged from stress–strain curves. It is generally believed that when rock total strain reaches 3–5%, it can be said that rock has already begun to reflect plastic properties or has realized the transformation from brittle to plastic. For example, confining pressure of marble and sandstone in Fig. 1.18 is over 23.5 MPa and 27.5 MPa, respectively, these two kinds of rock begin to show plastic state. Table 1.8 shows several rocks stain in Fig. 1.19 before they are destroyed at room temperature. As can be seen from

Table 1.8 Rock plastic deformation under confining pressure

Rock name		Strain amount before damage under different confining pressure %	
		$p=100$ MPa	$p=200$ MPa
Oil Creek	Quartz sandstone	2.9	3.8
Hasmark	Dolomite	7.3	13.0
Blain	Anhydrite	7.0	22.3
Yule	Marble	22.0	28.8
Barns	Sandstone	25.8	25.9
Marianna	Limestone	29.1	27.2
Muddy	Shale	15.0	25.0
–	Halite	28.8	27.5

Table, in addition to Oil Creek quartz sandstone in 200 MPa confining pressure remaining brittle damage, the other rocks in 100 MPa confining pressure above all have clear plastic properties, but the extent of plasticity is different.

In 1978, Black A.D. and Green S.J. published the results of drilling tests on a full-scale simulated deep well rig in Salt Lake City, USA, which determined the pressures of the Bonne Terre. Dolomite, Colton sandstone, and Mancos shale in their brittle to plastic transition from 100 to 150MPa, 40 to 70MPa, and 20 to 40 MPa, respectively.

For deep well drilling, it will be of great practical significance to recognize and understand transition pressure (or threshold pressure) of rock from brittle to plastic. Brittle and plastic damage are two fundamentally different methods of failure; therefore, to destroy these two types of rocks by applying different crushing tools (different types of drill bit), different crushing methods (impact, crushing, extrusion, shearing or cutting, grinding, etc.) and reasonable combination of different parameters, this can achieve better rock-breaking effect.

Therefore, it is indicated that the comprehension of rock plasticity, brittleness properties, and threshold pressure is an important basis for designing, selecting, and using drill bits.

1.2.1.5   Rock Hardness

Rock hardness is the ability of a rock to resist penetration and abrasion of other objects surface.

Hardness is associated with compressive strength, but there is a big difference. Hardness is the resistance of a partial solid surface to the penetration or press of another object while compressive strength is the resistance of a solid to the failure of the whole solid. Therefore, rock compressive strength cannot be used as hardness index. A distinction should be made between the hardness of the mineral particles that make up the rock, which has a significant effect on tool wear during drilling, and the combined hardness of the rock, which has a significant effect on the rate of rock breaking during drilling.

There are many kinds of methods for measuring and expressing rock and mineral hardness, only two common methods in oil drilling will be introduced here.

Moh's hardness

This is a popular, simple method that represents rock or other materials relative hardness. The method of this measurement is to use two kinds of materials to carve each other, and less hardness is to leave scratches on the surface. It is represented by ten minerals, as a standard of hardness, successively: talc(1), gypsum(2), calcite (3), fluorite(4), apatite(5), feldspar(6), quartz(7), topaz(8), corundum(9), diamond (10).

In the field, it is often easier method, that is, to use a fingernail (2.5), iron knife (3.5), ordinary steel knife(5), glass(5.5), saw blades(6), file(7), hard alloy(9), and other carved minerals or rock to identify its hardness.

Moh's hardness in rocks minerals is an important reference for rock tool selection, and if minerals Mohs's hardness in a certain proportion of rock is reached or near the working part of rock tool, tools wear quickly.

Table 1.9 The classification of rock hardness

Type	Soft		Medium soft		Medium hard		Hard		Rigidity	
Grade	1	2	3	4	5	6	7	8	9	10
Hardness MPa	≤1	1–2.5	2.5–5	5–10	10–15	15–20	20–30	30–40	40–50	50–60

Rock indentation hardness

Rock indentation hardness was made by Srinel in former Soviet Union, also known as Shi's hardness. Shi's hardness test device is shown in Fig. 1.20. Figure 1.22 shows calculation method of three types of rock indentation hardness.

For brittle rocks and plastic brittle rocks, they eventually produce obvious breaking pits, and rock compressive strength is:

$$p_Y = \frac{P}{S} \tag{1.37}$$

where

P_Y Rock indentation hardness, MPa;
P The load on press head when brittle crushing occurs, N;
S Bottom area of press head, mm^2.

For plastic rocks, load P_0 is used to replace P when yielding (i.e., from elastic deformation to plastic transformation), rock indentation hardness is:

$$p_Y = \frac{P_0}{S} \tag{1.38}$$

During drilling process, rock-breaking tool exerts loads on the surface of rock at the bottom of the hole, causing local fracture on the surface of rock. Rock indentation hardness is a certain representative in rock breaking in drilling process, to a certain extent, and it can reflect the ability of rock crush resistance during drilling. In China, the rocks are divided into six categories and 12 grades (Table 1.9) according to rock indentation hardness, which is one of the main bases for selecting drill bits.

1.2.2 Rock Mechanical Properties Under Bottomhole Pressure Conditions and Its Influencing Factors

In the process of drilling, especially in deep wells, the mechanical properties of rocks change greatly under the conditions of high-pressure and multidirectional compression. It is of great significance to study the mechanical properties and influencing factors of rocks under such conditions for guiding the drilling engineering practice.

1.2.2.1 Stress Status of Strata Rocks Around Wellbore

The stress on the formation rocks around the borehole is shown in Figure 1.23, which includes the following aspects:

– Overburden pressure σ_1. It comes from the gravity of the upper rock, which covers the pressure above wellbore rocks. And the difference between σ_1 and pore fluid pressure in rock (σ_1-p_p) is called the effective overburden pressure.
– Pore fluid in rock is p_p.
– Horizontal in-situ stress. Horizontal in-situ stress σ_2 and σ_3 are composed of overburden lateral pressure and geological tectonic stress. Overburden pressure is the source of partial horizontal stress; if strata in horizontal direction are isotropic, then this part of horizontal stress is evenly distributed in horizontal direction; it can be considered only related to Poisson's ratio μ; this part of effective horizontal stress value is $\mu(\sigma_1-p_p)/(1-\mu)$.

The other part of horizontal stress originates from geological tectonic stress, which is not equal in two main horizontal directions, but it increases linearly with burial depth increasing, that is, it is proportional to effective overburden pressure.

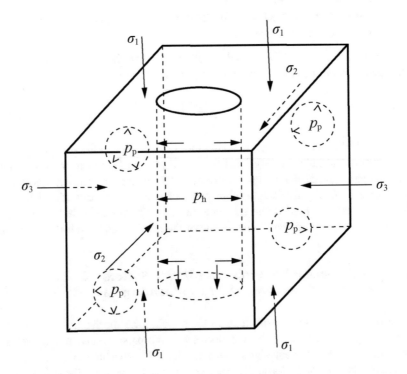

Fig. 1.23 Diagram of stress near wellbore

Therefore, effective horizontal stress in two horizontal directions at a certain depth can be expressed as:

$$\sigma_2 = (\frac{\mu}{1 - \mu} + \alpha)(\sigma_1 - P_P)$$
$$\sigma_3 = (\frac{\mu}{1 - \mu} + \beta)(\sigma_1 - P_P)$$

(1.39)

where

α, β—Tectonic stress coefficients of 2 and 3 horizontal principal direction respectively, $\alpha > \beta$.

σ_1, σ_2 and σ_3 are stresses due to interaction between underground rocks, therefore, which are generally called underground stresses.

– Hydrostatic pressure of drilling fluid column pressure p_h.

1.2.2.2 Influence of Various Pressure on Rock Performance at Bottom Hole
The effect of in-situ stress

The full-scale bit drilling test in Salt Lake City, USA, adopted the loading method as shown in Fig. 1.24 to simulate the influence of borehole stress and other pressure conditions on drilling and rock breaking, where σ_3 represent the uniform horizontal in situ stress.

Fig. 1.24 Diagram of bottomhole pressure

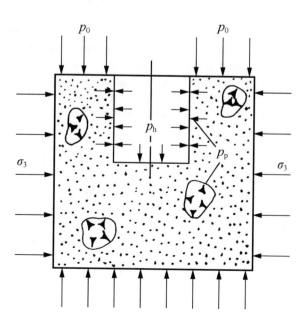

During the test, pressure gradient was: overburden pressure gradient $G_0 = 0.023$ MPa/m, horizontal in situ stress gradient $G_3 = 0.016$ MPa/m, and formation pore pressure gradient $G_P = 0.0104$ MPa/m. The test results show that test pressure is small (Both p_o and σ_3 are 0 to 35 MPa, whether overlying strata pressure p_o or horizontal ground stress has no obvious effect on drilling speed.

Theoretical analysis shows that both vertical overburden pressure and horizontal stresses (uniform or non-uniform) will affect the stress state of borehole, thus affecting wellbore stability. The greater difference between the maximum and minimum principal stress, the more serious problem could be. If drilling fluid density in the well is too small, some weak formation will shear and collapse, or plastic flow can cause wellbore to shrink; if drilling fluid density is too large, it will cause some formations to fracture. Formation fracture pressure depends on the stress state on the borehole wall, and the stress state is closely related to the magnitude of the in situ stress.

The effect of hydrostatic pressure and pore pressure effects

- Conventional triaxial test results considering pore pressure

In the conventional triaxial experiment as shown in Fig. 1.17, if the rock is dry or impermeable, or porosity is small, and there is no liquid or gas in the pores, the increase of confining pressure will increase rock strength as well as the plasticity these two aspects are collectively referred to as "anisotropic compression effect."

When rock pore contains fluid and has certain pore pressure, J. Handin et al. believed that the strength and plasticity of the porous rock depends on compression effect, but when pore fluid is chemically inert, rock permeability is sufficient to ensure that fluid circulates in pores and forms the same pressure, and the shape of pore space can pass all the pore pressure to rock matrix, the "anisotropic compression effect" equals to the difference between external pressure and internal pressure. In triaxial tests, external pressure refers to confining pressure σ_3, and internal pressure is pore pressure p_p. In other words, pore pressure reduces the anisotropic compression effect of rock.

The difference between external pressure and internal pressure is called effective stress (σ'), that is:

$$\sigma' = \sigma_3 - p_p \tag{1.40}$$

The effect of effective stress can be clearly got from the Ardrey test results (Table 1.10). The test is a triaxial stress conducted with Berea sandstone rock samples at room temperature and confining pressure of 0 to 69 MPa.

It can be seen from the table that under above experimental conditions, the strength of only rock only depends on the value of effective stress, that is, under different matching schemes of confining pressure and pore pressure, as long as σ' is identical, rock strength value is almost the same.

Table 1.10 Effect of effective stress on rock strength

Confining pressure σ_3 MPa	Pore pressure p_p MPa	Effective stress (σ_3-p_p) MPa	Rock fracture stress (Ultimate strength) $(\sigma_1-\sigma_3)$ MPa
0	0	0	59.8
34.5	34.5	0	58.4
13.8	6.9	6.9	106.0
20.7	0	20.7	171.0
34.5	13.8	20.7	169.1
44.8	24.1	20.7	166.7
69.0	48.3	20.7	167.1
34.5	0	34.5	211.0
48.3	13.8	34.5	211.0
69.0	34.5	34.5	212.8
55.2	0	55.2	253.8
69.0	13.8	55.2	250.4

Robinson studied the effects of pore pressure in limestone, sandstone, and shale samples by triaxial stress experiments, as shown in Fig. 1.25.

As can be seen from Fig. 1.25, the yield strength of rock increases with pore pressure decreasing. When the confining pressure is constant, the rock presents only plastic failure when pore pressure is relatively small. Increasing pore pressure will transform rock from plastic failure to brittle failure. Therefore, we should pay enough attention to pore pressure when considering shale borehole stability. In contrast, pore pressure contributes to rock fragmentation in drilling, which increases drilling speed.

- Effect of hydrostatic pressure

If the rock at the bottom of hole is impermeable and has no pore liquid, then hydrostatic pressure of drilling fluid will increase the directional rock compression effect on the rock. As a result, rock compressive strength (or hardness) and plasticity will increase, and rock will change from failure to plastic failure under certain hydrostatic pressure. This transition pressure is called the threshold pressure of brittle-plastic transition.

B.B.Булатов studied the effect of drilling fluid column pressure on rock mechanical properties by press method; he observed that compressive strength (or hardness) of rock significantly increased with the increase of liquid hydrostatic pressure p_h increasing, as shown in Table 1.11. At the same time, it was found that the smaller rock hardness is, the more significant influence of p_h effect on its hardness (the higher the multiple of hardness increased by a higher multiple at the same value).

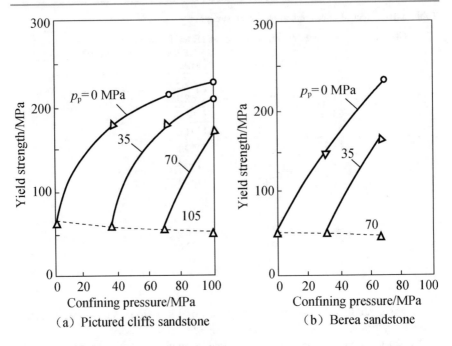

Fig. 1.25 Confining pressure and pore pressure effect on rock yield strength △—Brittle fracture O—Plastic fracture

Table 1.11 The effect of hydrostatic pressure p_h on rock hardness

Hydrostatic pressure/MPa	Muddy limestone		Marble		Dolomite	
	Hardness (MPa)	Relative value (%)	Hardness (MPa)	Relative value (%)	Hardness (MPa)	Relative value (%)
0	498	100.0	803	100.0	–	–
20	602	121.0	–	–	3670	100.0
35	633	127.2	980	122.0	4136	112.7
65	773	155.4	1083	135.0	4220	115.0
85	856	172.0	1180	147	4504	124.0
95	1301	261.3	–	–	4626	126.2
100	1626	306.0	1490	185.5	4940	134.7

In addition to rock strengthening, hydrostatic pressure will increase rock plasticity coefficient. At a certain pressure value (for different rocks, the value is different), its fragmentation features from brittleness to plasticity (plastic coefficient $K_p \rightarrow \infty$) are characterized by the plastic coefficient, and fragmentation crater area is close to the bottom area of press head; specific data are shown in Table 1.12.

Table 1.12 The brittle-plastic transition pressure of some rocks

Rocks	Atmospheric pressure		Liquid column pressure when $K_p \to \infty$ MPa
	Hardness MPa	Plastic coefficient K_p	
Dolomite	3610	1.30	50–60
Sandstone	514	1.65	20–30
Siltstone	895	2.68	15

Table 1.13 The brittle-plastic transition pressure of some rocks

Hydrostatic pressure 0→35 MPa

Rock name		V/W decrease %	Rock name		Microbit speed reduce %
Soft ↓ Hard	Indianna limestone	93	Soft ↓ Hard	Rifle shale	78
	Berea sandstone	91		Spraberry shale	76
	Virgina greenstone	90		Wyoming red bed	63
	Danby marble	83		Pennsylvanian limestone	50
	Carthage marble	71		Rush Spring sandstone	33
	Hasmark dolomite	49		Ellenberger dolomite	22

Therefore, as well deepens or drilling fluid density increases, the rate of penetration decrease is related not only to the increase of rock hardness, but also to the increase of rock plasticity, especially due to the reason that broken rock volume decreases which is affected by bit teeth interaction with rocks every time.

Many researchers have studied the problem of single tooth and microdrill rock breaking in high pressure; they believe that hydrostatic pressure of drilling fluid has a significant effect on drilling speed, showing that with the hydrostatic pressure increasing, broken rock volume (V/W) per unit energy required for breaking rock (V/W) decreases, the influence of hydrostatic pressure on soft and easy drilling formation is greater. The experimental results are shown in Table 1.13 and Fig. 1.26.

1.2.3 Rock Drillability and Abrasiveness

1.2.3.1 Rock Drillability

Rock drillability is the ability of rock to resist breaking. It can be understood as the ability of rock to resist bit breakage under certain bit specifications, types, and drilling technology conditions. The concept of drillability has led the more general

Fig. 1.26 Effects of hydrostatic pressure on the rate of penetration (Microbit test) I—Rifle shale; II—Spraberry shale; III—Wyoming red bed; IV—Pennsylvanian limestone; V—Rush Spring sandstone; VI—Ellenberger dolomite

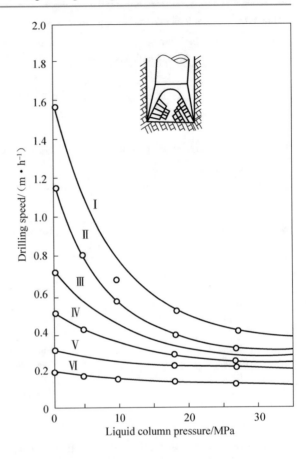

concept of rock properties such as strength and hardness to the concept related to drilling, which occupies an important position in practical application. Rock drillability is usually the basis of bit selection, production quota, bit working parameters, and bit working index prediction.

Rock drillability is a comprehensive index shown in drilling process. It depends on many factors, including physical and mechanical properties of rock itself and technological measures of rock breaking. Rock physical and mechanical properties mainly include rock hardness (or strength), elasticity, brittleness, plasticity, granularity, and bond properties between particles; technical measures include structural characteristics of rock-breaking tools, the action mode of tools on the rock, the nature of the load or force, the scale of rock-breaking energy, and cleaning efficiency at the bottom of the hole. Therefore, the rock drillability is related to many factors. It is complicated and difficult to find out the sensitive relationship between rock drillability and influencing factors, and rock drillability can only be determined by tests under such specific breaking methods and technological procedures.

At present, the measurement and classification of rock drillability is not uniform. Different kinds of drilling methods used by different departments, test methods for the determination of rock drillability are not the same. Measurements, conditions, and classification methods are also different in countries and regions. Based on the research results of professor Yin Hongjin and other researchers from the University of Petroleum, the former Ministry of Petroleum Industry determined the measurement and classification method of the rock drillability in the petroleum industry in China. This classification method is to use a microdrill bit to drill a rock sample, to determine the drillability of rock sample through practical drilling time (i.e., drilling rate). The specific method is to use rock drillability tester (that is, microdrill bit test device) to drill three hole in the sample with bit size of $\Phi 31.75$ mm (1.25 in), weight on bit of 889.66 N, rotational speed of 55 r/min, and then take the average drilling time of the three holes with depth of 2.4 mm as rock sample drilling time (t_d), use Eq. 1.26 to calculate rock drillability k_d, usually take an integer value.

$$k_d = \log_2 t_d$$

In China, formation drillability is divided into ten grades by integral value of k_d, and the proportion of drillability in major oil fields is shown in Table 1.14.

1.2.3.2 Rock Abrasiveness

In the process of rock breaking with a mechanical method, drill bit and rock have continuous or intermittent contact and friction, while the tool itself is gradually blunted by rock abrasiveness, until it is damaged, and the material that drill bit touches partial rock is generally steel, cemented carbide, or diamond. The ability of a rock to wear these materials is called rock abrasiveness. The study of rock abrasiveness is of great significance for the design and selection of drill bits, the extension of bit life, the improvement of bit footage, and the improvement of drilling speed. It is of great significance to study the grinding property of rock to correctly design and select the drill bit, prolong the bit life, and improve drill bit and drilling speed.

Table 1.14 The proportion of formation drillability grades in various oilfields in China

Oilfield or areas	Drillability grade									
	I	II	III	IV	V	VI	VII	VIII	IX	X
Dagang	6.4	16.8	29.2	27.9	14.8	4.24	0.1			
Shengli	8.0	11.9	18.9	22.1	18.9	12.0	5.0	2.0	0.8	0.2
Subei	6.5	7.7	11.9	15.5	17.1	15.5	11.7	7.6	3.9	2.6
Northwest	0.01	0.3	4.5	15.3	47.5	26.0	5.8	0.4	0.1	
Jianghan	2.1	5.1	11.4	18.4	21.7	19.5	12.7	6.2	2.3	
Northern China	4.2	6.3	11.2	16.1	18.5	17.2	12.9	7.7	3.8	2.1
Daqing	0.6	2.2	6.7	14.0	21.3	22.9	17.6	9.5	3.9	1.3
Sichuan	0.07	0.43	3.9	14.2	28.2	29.9	17.1	6.0	0.9	0.01

For drilling, the abrasiveness of the rock is reflected in the wear on the edge surface of the bit, i.e., abrasive wear. It is caused by microcutting, scratching, and so on during the contact between the working edge of the bit and the rock. This abrasive wear is not only related to the properties of friction accessory material, but also depends on the type and characteristics of friction, the shape and size of friction surface (such as surface roughness), friction surface temperature, relative motion speed of friction body, contact stress between friction bodies, wear debris and cleaning condition, and the participating friction medium. Therefore, abrasive wear is a very complex problem, which is worth of further investigation.

There is a variety of calculating methods for rock abrasiveness, but there is not a unified method of measurement, and many results are difficult to compare.

Questions and Exercises

1. Provide a brief description of basic concepts of various subsurface pressures, and the relationship between overburden pressure, formation pore pressure and matrix stress.
2. Describe the mechanism of abnormal high pressure with under compaction of formation deposits.
3. Describe the variation laws of rock density, strength, porosity, interval transit time, and d_c index with well depth in normally compacted formation.
4. Explain the concept of formation fracturing pressure. How to determine the formation fracture pressure according to leak-off test curve?
5. A well has 2000 m in depth, 25 MPa formation pressure, calculate the formation pressure equivalent density.
6. A well has 2500 m in vertical depth, and the drilling fluid density in well is 1.18 g/cm^3. If formation pressure is 27.5 MPa, calculate bottomhole pressure difference.
7. The depth of a well is 3200 m, and the pressure of the formation is 23.1 MPa, calculate the formation pressure gradient.
8. A well is drilled to the depth of 2500 m with bit size of 215 mm, weight on bit of 160 kN, rotary speed of 110 r/min, rate of penetration in 7.3 m/h, drilling fluid density of 1.28 g/cm^3. If drilling fluid density is 1.07 g/cm^3 under normal pressure conditon, calculate d and d_c index.
9. What is the difference between rock hardness and rock compressive strength?
10. What is the definition of rock plastic coefficient? Provide a brief description of brittle, plastic-brittle and plastic rock characteristics during fracturing process.
11. What is the difference between rock strength in parallel and vertical bedding direction? What is this rock character called?
12. When rock is subject to confining pressure, how does its strength and brittleness change?

13. What are factors that affect rock strength?
14. What is rock drillability? What method does Chinese petroleum industry use to evaluate rock drillability? How many grades can be classified according to rock drillability?
15. What forces are acting on the rock formation at the bottom of the hole and near around wellbore wall?
16. How does horizontal stress form? What is the relationship between horizontal stress and overburden pressure?
17. What are effective stress, effective overburden pressure, and anisotropic compression effect?

Drilling Rigs and Tools

2

Abstract

This chapter discusses the components and functions of a drilling rig and drilling tools. A drilling rig usually has six necessary subsystems classified as hoisting system, rotary system, circulating system, well control system, power and transmission system, and monitoring systems. Drilling tools are used to describe drill strings and drill bits for rock breakage in a wellbore. A drill string usually consists of kelly, drill pipes, drill collars, and other tools such as stabilizers and reamers, which are included in the drill string just above the drill bit.

2.1 Drilling Rig

A drilling rig is a complete set of drilling equipment for oil and gas exploration and development. It is a joint multi-functional working unit consisting of many kinds of machines. To satisfy the demand of drilling technology, the whole set of drilling rig has six basic systems classified as power system, hoisting system, circulating system, rotary system, well control system, and monitoring system. And the abilities of tripping, rotary drilling, and circulating the drilling fluid must be possessed.

To adapt to various geographical environment and geological conditions, various special drilling rigs have appeared in recent years, such as desert drilling rig, cluster drilling rig, inclined well-drilling rig, top drive drilling rig, slim-hole drilling rig, and coiled tubing drilling rig and so forth, which are called individual drilling rigs.

Since the 1990s, China has independently developed a series of different types of specialized drilling rigs, forming a diversified system of oil drilling rigs. After producing the first new electric drive drilling rig ZJ70D, some new oil drilling equipment has been built, including the world's first artificial island 7000 m circular orbit mobile module drilling rig, and the ultra-deep drilling rig with the designed drilling depth of up to 12,000 m.

© China University of Petroleum Press and Springer Nature Singapore Pte Ltd. 2021
Z. Guan et al., *Theory and Technology of Drilling Engineering*,
https://doi.org/10.1007/978-981-15-9327-7_2

2.1.1 Requirements of Drilling Technology to a Drilling Rig

The configuration of rig equipment is closely related to the drilling methods. Currently, the drilling methods widely used in the world are the rotary drilling method, that is, using drilling string hook, traveling block, crown block, top drive system (TDS), winch to trip the drill string; using drill pipe to deliver the drill bit to the bottom; using the rotary table or top drive system to make drill string and drill bit rotate, or using a downhole motor to make drill bit spin directly; using bit turning to crush the rock-forming borehole; using mud pump to circulating high-pressure drilling fluid and bring out the debris from the bottom.

Rotary drilling method requires the drilling machinery and equipment to have the following three necessary abilities.

2.1.1.1 The Capability of Rotary Drilling

Drilling technology requires the drilling machinery and equipment to provide a specific torque and speed for drilling tools(drill string and drill bit) and maintain a certain weight on bit (the gravity of drill string acting on drill bit).

2.1.1.2 The Capability of Tripping

Drilling technology requires the drilling machinery and equipment to have a specific lifting capacity and lifting speed, which can trip-in and out all drill strings and casing strings.

2.1.1.3 The Capability of Wellbore Cleaning

Drilling technology requires the drilling machinery and equipment to have the capacity of cleaning the bottom and carrying rock debris. The system could provide relative high pump pressure to make drilling fluid through drilling strings flow to bit nozzle, wash the bottom of the well, return to the wellhead through borehole annular space, and bring the cuttings out.

Besides, considering the flexibility of drilling operations, the drilling equipment should be easily installed, disassembled, and transported. The application and maintenance of the drilling rig must be simple and easy, and the vulnerable parts of the drilling rig should be easy to replace.

The configuration of the drilling rig device and the working ability and the technical index of various equipment are determined by the above three basic requirements of drilling technology to drilling rigs. The basic parameters of the drilling rig put forward the conditions about the torque and power of the rotary table, the lifting weight, and control of the hook and the permissible pump pressure and power of the mud pump. In these three sets of parameters, the torque of the rotary table, the lifting weight of the hook, and the permissible pump pressure of the mud pump are all limited by the strength of the machine parts.

Under the condition of the strength satisfying the requirements of operation, the rotary table should provide a specific rotating speed, the hook should give an appropriate lifting speed, and the mud pump should provide a certain displacement and pump pressure.

Otherwise, the drilling operation cannot go smoothly. The joint requirements of the torque and rotating speed of the rotary table, the lifting weight and lifting speed, and the pump pressure and displacement are the ones for the power and strength of the working machine. To ensure a specific rotating speed, lifting speed, and displacement, the engine should supply a certain required amount of power to the drilling machinery.

2.1.2 The Primary Working System of a Drilling Rig

The working system of the drilling rig is vast, and the working conditions and working characteristics of each unit are different. According to the requirements of rotary drilling method to a drilling rig, drilling rig mainly includes hoisting system, rotary system, circulation system, well control system, power and transmission system and control system, in which hoisting system, rotary system, and circulation system are three primary systems directly serving the drilling process and are the essential working systems of the drilling rig. The hoisting system, rotary table, (or top drive system) and mud pump are known as the three primary working machines of the drilling rig.

2.1.2.1 Hoisting System

To trip the drilling tools, lower the casing, control the drilling load, feed into a bit, and so on, drilling rigs are equipped with hoisting system to assist operations of drilling and completion. This set of equipment mainly is composed of drawworks, auxiliary brake, traveling system (including wireline, crown block, traveling block), hook, and derrick, besides, wellhead tools and mechanized equipment for tripping operation(such as rings, elevator, slip, power tong, stand transfer mechanism.).

Derrick

- Major functions of derrick

The derrick is an essential part of the oil drilling rig. It is a metal truss structure with a certain height and space, whose primary functions include

- Set the crown block, hang the traveling block, hook and specialized tools (such as tongs), and trip the drilling tools and lower the casing during the drilling.
- Used for depositing the stand during the process of tripping. The total length for depositing stands is called stacking stand capacity.

- Compositions of derrick

The derrick is mainly composed of the following six parts.

- The main body of derrick is mostly the space truss structure composed of section bar.
- Crown platform used for setting the crown block and gin pole.

- Crown frame used for installing and maintaining the crown block.
- Monkey board includes a platform for derrickman conducting tripping operation and fingerboard for depositing the stands.
- Stands platform assembles and disassembles a fire-hose operating platform.
- Working ladder.

• Properties of the derrick

The derrick should possess the following three properties:

- Enough bearing capacity, guarantee to trip a drilling string, or a casing string a certain depth. So-called enough means to be suited to hook nominal lifting weight (maximum drilling string weight) and hook maximum lifting weight equipped with the derrick.
- Enough dimensional space. The higher the height of derrick, the longer the length of the stand, which can save time; the bottom and the top of derrick should have necessary size to install the crown block and ensure the traveling system device run smoothly when tripping; guarantee that drilling floor has sufficient area to arrange equipment and install the tools, making workers operate safely and driller a good vision.
- Guarantee assembly and disassembly conveniently, and transfer quickly.

• Types of derricks

According to the main characteristics of basic structure form, drilling derricks could be divided into tower derrick, cantilever derrick, A-type derrick, and mast derrick, etc.

- Tower derrick.

As Fig. 2.1 shows, tower derrick is a space structure of four towers, whose cross section is usually square. The derrick itself is divided into four parallel trusses, every truss is divided into some, and tetrahedral truss at the same height in the space constitutes a layer of the derrick, so the derrick itself can be seen to consist of many layers space trusses.

The main characteristics of the integral structure form of tower derrick are

a. The derrick itself is a closed integral structure. Overall stability is excellent, and bearing capacity is large.
b. The whole derrick is a detachable structure made up of single components connected by bolts.
c. The size of derrick is not restricted by transportation conditions, allowing the vast inner space of derrick, tripping conveniently and safely. Still, assembly and disassembly work of a single piece is significant, working high above the ground, unsafe.

Fig. 2.1 Tower derrick

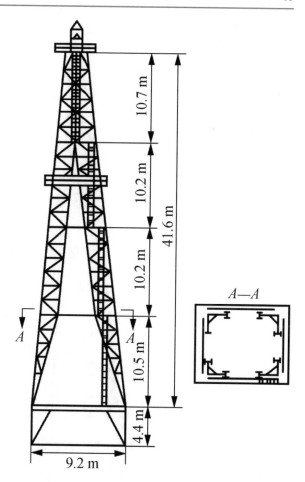

– Cantilever derrick (П-type derrick)

As shown in Fig. 2.2, the main features of П-type derrick are as follows:

a. The whole derrick body is divided into four–five sections, and each section is generally welded as an entire structure, the sections are positioned by tapered pins and connected by bolts, assembled on the ground or near the ground level, the whole derrick is lifted and placed, and transported in sections.

b. Because of the restriction of transportation size, the cross-sectional size of derrick itself is smaller than tower derrick. To facilitate traveling system devices moving up and down smoothly and depositing the stand, the derrick is made of a non-enclosed spatial structure of a front fan opening and П-type cross section. Some of the upper parts of П-type derrick is made into a four-sided closed structure to enhance overall stability.

Fig. 2.2 Π-type derrick

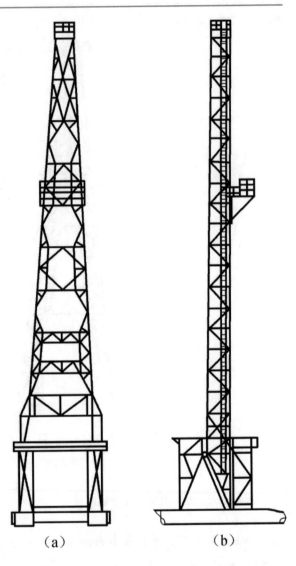

(a)   (b)

c. The structure form of two sections of fan truss in each part of the derrick is the same. To ensure the driller has a good vision, back fan adopts different belly bar arrangement, such as rhombus and so on. Back fan crosswise bars of some Π-type derrick are composed of taper pin and detachable structure connected by the left and right side segments for the convenience of segmented transportation.

– A-type derrick

The main features of A-type derrick are as follows:

a. Two legs connect into A-type through a crown platform, racking platform, and additional rod. In front of the thigh or behind the herringbone support, it constitutes a complete spatial structure. Entire derrick is installed horizontally on the ground or near the ground, erected up and down, transported separately.
b. Legs can be spatial bar structure, divided into three–five segments. According to the different profiles, sections of legs generally are divided into rectangular and triangular. Using pipes as leg chords mostly adopt triangularly and using angle steel mainly takes square for the convenience of manufacturing. Bracing rods have a rod string structure and rectangular section welded column structure or pipe column structure.
c. Each leg of A-type derrick is enclosed integral structure, and bearing capacity and stability are relatively good. But because there are only two legs and the connection between legs is relatively weak, the overall balance of the derrick is not ideal.

Figure 2.3 indicates a transformation of A-type derrick. The upper part is made into an enclosed integral structure to enhance the overall stability of the derrick.

Fig. 2.3 Upper closed A-type derrick

– Mast derrick

Mast derrick is a single column derrick composed of a section or several sections rod structure or column structure, holistic, and retractable two kinds. Mast derrick generally uses a hydraulic cylinder or hoist system to erect up and lay down, which could be transported as a whole or several disassemble parts.

Mast derrick at working is tilted to the wellhead, which needs to utilize guy-wires to keep the stability of the structure to give full play to its bearing capacity. This is the critical characteristic of mast derrick integral structure.

The construction of mast derrick is portable and straightforward, but its bearing capacity is inadequate. It is only suitable for mobile drilling rigs and workover rigs.

Figure 2.4 is a retractable mast derrick of XJ250 workover rig.

– Basic parameters of derrick

The basic parameters of derrick include

– Maximum hook load
  The maximum hook load of derrick refers to the maximum lifting weight of hook without wind load and stands load, the deadline fixed in the specified position, using the specified number of drilling ropes. Max hook load includes the self-weight of traveling block and hook (max hook load of drilling rig does not include the self-weight of traveling block and hook).
– Stand load
  Stand load refers to the horizontal direction force produced by the self-weight of stands and wind load it bears to the fingerboard of the racking platform.
– Height of the derrick
  The height of the derrick is defined based on its type.

a. The height of tower derrick: the vertical height from the bottom of the derrick leg floor to the bottom of the crown block beam;
b. The height of Π-type derrick and A-type derrick: the vertical height from the pin-hole center of derrick lower base angle to the bottom of the crown block beam;
c. The height of the mast: the vertical height from the contact point of the skid seat or wheels and the ground to the bottom of the crown block beam;
d. The effective height of derrick: the vertical height from the drilling floor to the bottom of the crown block beam;
e. The height of the racking platform: the vertical height from the drilling floor to the racking platform;
f. The capacity of the racking platform: the numbers of drill pipes deposited on the racking platform (installed on the minimum height);

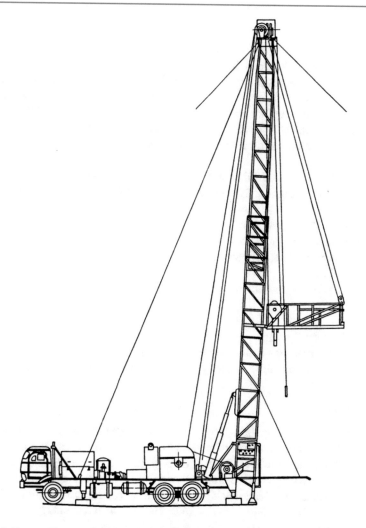

Fig. 2.4 Retractable mast derrick of XJ250 workover rig

g. The size of the upper base and the size of the lower base: the size of the upper base and the size of the lower base of tower derrick, respectively, refer to the horizontal distance between the upper and lower base axes of derrick adjacent legs. For a single angle steel leg, it refers to the distance between the outer edges of angle steel.

h. The height of V-door: The V-door opening height of tower derrick is the vertical height from the bottom of the drilling floor at the top of the V-door opening. The height of the V-door opening generally should be higher than 8 m to pull the joint to the drilling floor.

Drawworks

The drawworks is one of the core components of the hoisting system. The power is inputted to drawworks reduction gearbox through a coupling, transferred to sprocket wheel by meshing of reducing gear of reduction gearbox, and then through sprocket transmitted to the axle of drawworks roller clutch. After clutch ventilation, the power is transferred to drawworks to realize the rotation of roller and the up-and-down of traveling block. ZJ60D drawworks is shown in Fig. 2.5.

The main functions of drawworks are as follows:

- Raising and lowering drilling strings, tools and casing strings;
- Controlling the weight on bit and feeding in the drilling tools during the process of drilling;
- With the help of cat head, make-up or break-down joint connection, lifting goods and other auxiliary work;
- Serving as variable speed gear or intermediate transmission mechanism of the rotary table;
- Rigging up or rigging down the entire derrick.

2.1.2.2   Rotary System

To rotate drilling strings in the wellbore and drive the bit to crush the rock, the rotary system of normal drilling rigs consists of a rotary table, swivel, drill string, or top drive system.

Rotary table

The rotary table virtually is a large power bevel gear reducer, whose primary function is to transmit the power of engines to the kelly, drill pipes, drill collars, and the drill bit through muster bushing, driving the bit to rotate, and deepening wellbore. The rotary table is the crucial device of a rotary drilling rig and also one of the three primary working machines of a drilling rig. Figure 2.6 shows that the ZP-700 rotary table mainly consists of horizontal axis assembly, rotary table assembly, main and auxiliary bearings, seal and shell, etc.

The requirements of drilling technology to the rotary table are as follows:

- Enough torque and a certain speed to rotate the drill string, drive the bit to crush the rock, and can meet the requirements of special operations such as fishing, makeup, back-off, thread making, or milling;
- Seismic loads resistance, impact resistance, and corrosion resistance, especially the main bearing should possess sufficient strength and longevity, and its bearing capacity is not less than the maximum hook load of drilling rig;
- The ability to rotate clockwise and counterclockwise and a reliable brake mechanism;

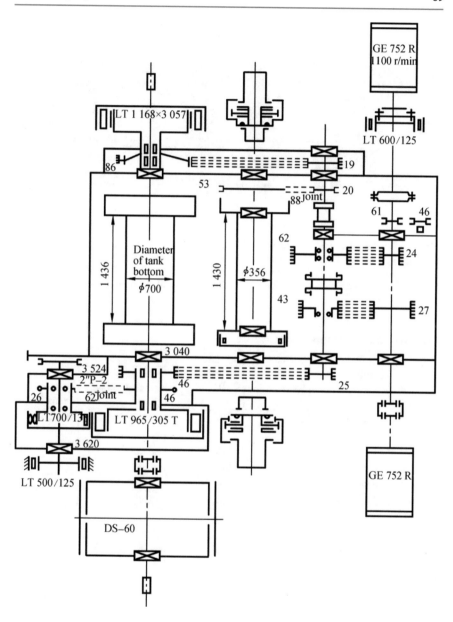

Fig. 2.5 ZJ60D drawworks

– Proper sealing and lubrication performance to prevent the external drilling
  fluid and dirt into the inner of the rotary table to damage the main and
  auxiliary bearings.

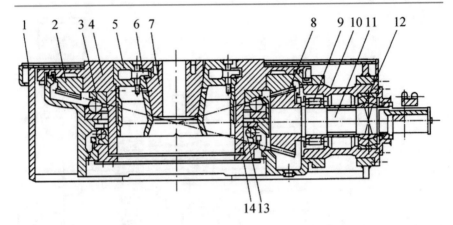

Fig. 2.6 ZP-700 rotary table. 1—Shell; 2—Large bevel gear; 3—Main bearing; 4—Rotary table; 5—Large quadrel; 6—Large quadrel and master bushing locking mechanism; 7—Master bushing; 8—Small bevel gear; 9—Cylindrical roller bearing; 10—Sleeve; 11—Fast axis (horizontal axis); 12—Double row centripetal spherical roller bearing; 13—Auxiliary bearing; 14—Adjusting nut

Table 2.1 Basic parameters of a rotary table

Type	Via diameter [mm(in)]	Center distance (mm)	Maximum static load (kN)	Rotating platform	
				Maximum working torque (N)	Maximum speed (r·min^{-1})
ZP-175	444.5(17^1/$_2$)	1118	1350	13,729	300
		1353			
ZP-205	520.7(20^1/)	1353	3150	22,555	300
ZP-275	698.5(27^1/)	1353	4500	27,459	300
ZP-375	952.5(37^1/)	1353	5850	32,362	300
ZP-495	1257.3(49^1)	1353	7250	36,285	300

The technical parameters of the rotary table are shown in Table 2.1.

Drilling tools

According to the different drilling wells, the components of drilling tools are also various. It generally includes kelly, drill pipe, drill collar, drill bit, stabilizer, shock sub, and crossover sub. Among them, the drill bit is a tool to crush the rock directly; the weight per unit length of the collar is large enough to apply weight on the bit; the drill pipe connects the surface pieces of equipment to the bottomhole types of equipment and transfers the torque, i.e., provide rotational action; the cross section of kelly is usually square and hexagonal, used through the rotary table to drive the entire drill string and bit.

Swivel

The upper part of the swivel is connected with a swivel belt, and its lower part is attached to kelly, through which the high-pressure drilling fluid is guided to the rotating drill string. Meanwhile, the rotary drilling tools bear most and all the weight of the drilling tools and provide the support of the upper bearing for the rotary drilling tools. It is a typical component of a conventional rotary drilling rig.

Top drive system

The top drive system is a top drive drilling device which is mounted on a derrick interior space and is suspended by the traveling block. The conventional swivel and drilling motor are combined, which is equipped with a novel makeup and breakout device of drilling pipe (or string treatment device). It can rotate the drill string from the upper derrick space directly.

As shown in Fig. 2.7, generally speaking, the top drive drilling system is composed of the motor swivel assembly, device of drill pipe guide way assembly, balance system, cooling system, control system, and ancillary equipment. The top drive system includes a liquid motor top drive, AC-SCR-DC top drive, and AC frequency conversional top drive system. The difference among them is that the driving motor is a liquid motor or a direct current motor or an alternating current motor. So, the structure and composition of the three kinds of top drive systems are not fundamentally different.

Compared with the conventional rotary table-kelly rotary drilling system, the top drive system has the following advantages:

- It can reduce the time to connect a single joint. Since the top drive system does not use the kelly and is not constrained by the kelly length, it can avoid the trouble of connecting a single joint of about 9 m in length. This ability to use a stand drilling greatly reduces the time required for connection.
- The back reaming can prevent pipe from sticking. The use of the top drive system can continue to rotate the drill string and circulate drilling fluid during the tripping-out process, which has the back reaming ability with 28 m stand or more. So, the device can be used to circulate and rotate without increasing the drilling time to pull the drilling tools out of the hole, reducing the risk of sticking.
- It can be reaming during a trip-in process. The device can drill through the sand bridge and the shrinkage point without kelly. In tripping, using the top drive system can connect the drill string in a few seconds and then immediately wiper trip, which reduces the risk of sticking.
- It can save directional drilling time. The top drive system can circulate drilling fluid through 28 m stand or more, which reduces the directional time of the downhole motor.

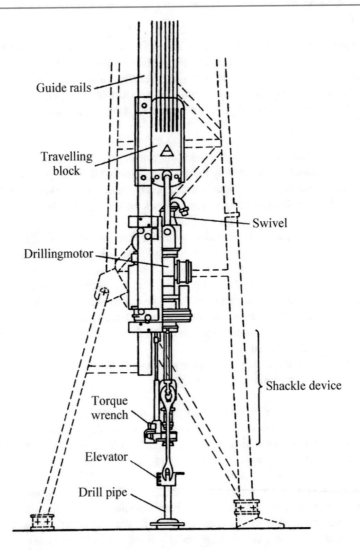

Fig. 2.7 Schematic diagram of the structure of the top drive system

- The personnel safety can be improved. When using the top drive system to connect the single joint, only the backup tong is needed, which can reduce the connecting frequency of a single joint for the drilling workers and significantly reduce the probability of accidents. Also, the inclined device of makeup and breakout assembly can make the elevator down to the mouse hole or up to the racking board, which dramatically reduces the risk of the operator's injury.

- It can control makeup and breakout torque. Due to the use of drilling motor for makeup and breakout of drill pipes, the makeup and breakout torque can be set from the driller's station or be observed from torque indicator which can avoid interference or insufficient of torque. It can make good use of the equipment and prolong the service life of the equipment.
- It can improve the safety of well control. When the top drive system is used for tripping, the makeup and breakout of drill pipes can connect the main shaft to the drill string to break circulation at any position of derrick if it is found abnormal, such as blowout warning. The double internal blowout preventer can safely control the pressure in the drill string.
- The coring length can be increased. Drilling with a top drive system can drill 28 m or more in continuous coring, and makeup of connection is not needed, which increases core recovery.
- Rotary casing running. A crossover sub is added between the casing and the spindle, so that the casing can be rotated and circulated into the well, thus reducing the friction of the shrinkage section and quickly passing through the hole with reduced size.

The top drive system is one of the four new technologies for the development of drilling equipment since the 1980s (top drive, disk brake, hydraulic mud pump, and AC variable frequency drive). The drilling practice shows that the advantages of the top drive system can save 20–25% of drilling time, significantly reduce the sticking accident, control kick, and avoid a blowout. The comprehensive economic benefit is particularly significant when it is used in directional drilling for deep/ultra-deep well, inclined well, and all kinds of difficult wells.

2.1.2.3 Circulating System

To carry the cuttings from the bottom of the hole to the surface, avoid debris settlement, cool drill bit, protect the hole walls, and prevent drilling accidents, a rotary drilling rig should be equipped with a circulating system.

The circulating system includes mud pumps, surface pipe, drilling fluid tanks, and drilling fluid purifying equipment. The surface pipe includes high-pressure pipe, vertical pipe, and flexible hose. The drilling fluid purifying equipment includes shale shaker, desander, desilter, and centrifuge.

Mud pump sucks the drilling fluid from the drilling fluid tank. The drilling fluid pressurized by the pump flows into the faucet after the surface pipe, standpipe, and flexible hose. Drilling fluid is injected into the bottom hole through the hollow drilling string and ejected from the bit nozzles. The cuttings are lifted from the bottom to the surface through annular space between the wellbore and drilling tools. The returned drilling fluid from the hole gets treated by solid particle and gas treatment equipment.

During the drilling process with downhole power drilling tools, the circulating system also provides the power to drive downhole turbine tools or screw drills to drive bits to crush rocks. The mud pump is the core of the circulating system, and it is one of the three primary working machines of the drilling rig.

Classification of mud pumps

According to the structural characteristics, the oilfield mud pump can be roughly classified as follows:

- According to the number of cylinders: It is divided into the single-cylinder pump, double-cylinder pump, triplex-cylinder pump, four-cylinder pump, and so on.
- According to piston-type and plunger-type: where the working mechanism is directly related to the working liquid.
    - The piston pump is a seal which is composed of a piston with a sealed component and a fixed metal cylinder sleeve.
    - The plunger pump is a seal which consists of a metal plunger and a fixed sealing component.

- According to the stroke: There is mainly a single-acting and double-acting pump according to the mode of action.

    - The piston or plunger of a single-acting pump is reciprocating in the hydraulic cylinder, and the cylinder is sucked once and discharged once.
    - The hydraulic cylinder of the double-acting pump is divided into two chambers by the piston or plunger, including a former tank without piston and back tank with the piston. Each tank has a suction valve and a discharge valve. When the piston is reciprocating for one time, the suction and discharge of the hydraulic cylinder are doubled.

- According to the layout of the cylinder and its position: It is divided into horizontal pump, vertical pump, V type, or star pump.
- According to the drive or drive mode, the common one is:

    - Mechanical drive pumps, such as crank one connecting rod drive, cam drive, rocker drive, wire rope-driven reciprocating pump, and diaphragm pump;
    - Steam-driven mud pump;
    - Hydraulic-driven mud pump, etc.

Triplex-cylinders single-action and double-cylinder double-action horizontal piston pumps are widely used in oil fields. In contrast, three-cylinder and five-cylinder single-action horizontal piston pumps and other types of reciprocating pumps are commonly used in fracturing, cementing, and water injection.

Working principle of a mud pump

Figure 2.8 is a schematic diagram of a horizontal single-cylinder single-acting reciprocating piston pump. It is mainly composed of a hydraulic cylinder, piston, suction valve, discharge valve, valve chamber, crank (crankshaft), connecting rod, crosshead, piston rod, gear, pulley, and transmission shaft. When the crank, driven by the belt and gear transmission generated by the power machine, begins to rotate from the left horizontal position as shown in the figure, the piston moves to the right, that is, the power end of the pump. A vacuum level is formed in the hydraulic cylinder. The liquid in the suction tank pushes the suction valve under the pressure of the liquid surface. It enters the hydraulic cylinder until the piston moves to the right dead point, which is the suction process of the hydraulic cylinder. The crank continues to rotate, and the piston begins to move to the left, that is, the hydraulic end. The liquid in the liquid chamber is squeezed, and the pressure rises. The sucking valve is closed, and the discharge valve is pushed away. The liquid is discharged into the exhaust tank through the discharge valve and exhaust pipe until the piston moves to the left dead end. This is the discharge process of the hydraulic cylinder. The crank rotates continuously, and the piston moves reciprocally once in a reciprocation. The cylinder of the single-action pump is used to complete one suction and exhaust process.

The technical specifications of mud pumps

The domestic mud pumps used for oil and gas drilling have been standardized in China. Most of the mud pumps currently used is triplex-cylinder single-acting horizontal piston pumps. And the technical specifications of common mud pumps are shown in Table 2.2.

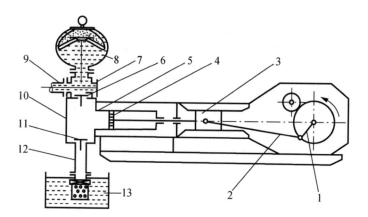

Fig. 2.8 Working diagram of horizontal single-cylinder single-acting reciprocating piston pump. 1—Crank; 2—Connecting rod; 3—Crosshead; 4—Piston; 5—Cylinder liner; 6—Discharge valve; 7—Exhaust spoon; 8—Air chamber; 9—Outlet pipe; 10—Valve box (hydraulic cylinder); 11—Sucking valve; 12—Suction pipe; 13—Sucking tank

Table 2.2 Technical manuals for conventional mud pumps

Titles	Qingzhou pump				Baoshi pump					Lanshi pump		
Types of device	SL3NB-500	SL3NB-1000A	SL3NB-1300A	SL3NB-1600A	F-500	F-800	F-1000	F-1300	F-1600	3NB-800	3NB-1000	3NB-1300
Input power (kW)	368	735	956	1176	368	588	735	956	1176	588	735	956
Stroke length (mm)	180	305	305	305	191	229	254	305	305	216	235	254
Rated punching (min^{-1})	110	120	120	120	165	150	140	120	120	160	150	140
Maximum cylinder diameter (mm)	170	180	180	190	170	170	170	180	180	160	170	170
Maximum working pressure (MPa)	35	35	35	35	26.77	27.26	32.85	30.60	37.65	33	35	35
Maximum displacement (L·s^{-1})	24.5	46.54	46.54	51.85	36.72	41.51	43.22	50.41	50.41	34.5	40.0	40.4
Diameter of function pipe (mm)	203	250	300	300	203	254	305	305	305	250	250	305
Diameter of discharging pipe (mm)	83	123	123	123	101.6	127	127	127	127	101.3	127	127
Tear ratio	4.696	3.657	3.957	3.657	4.286	4.185	4.207	4.206	4.206	2.51	2.658	2.868
Shape size (Length × width × height)/(mm × mm × mm)	3385 × 2280 × 2080	4600 × 2720 × 2470	4300 × 2750 × 2525	4720 × 2822 × 2600	3658 × 2709 × 2227	3963 × 3025 × 2351	4269 × 3162 × 2591	4426 × 3262 × 2688	4426 × 3262 × 2688	3995 × 2360 × 1541	4575 × 2600 × 1700	4900 × 2690 × 1800
Weight (kN)	92.943	189.14	203.84	265.58	95.746	142.10	184.142	240.806	242.952	132.69	166.60	201.439

2.1.2.4 Well Control System

In the process of drilling, to prevent the formation fluids from entering into the well, the drilling fluid hydrostatic pressure within the wellbore is slightly larger than the formation pressure, which is called the primary well control of oil and gas well. However, during drilling operations, the primary well control of oil and gas well is often failed due to various factors, which leads to formation fluids flowing into the wellbore. So it is necessary to rely on well control equipment to shut and or kill well in case of emergency to resume the well pressure balance. The well control system is a set of specialized equipment, instruments, and tools to implement the pressure control technology of oil and gas wells.

The function of well control system

The well control system has the following functions:

- To detect the overflow in time. Monitoring the oil and gas wells to identify the signs of a kick in and taking the control measures as soon as possible before it is developed into a blowout situation;
- Shut-in and carry out well killing operation rapidly to establish the pressure balance when overflow, kick, and blowout happen;
- To deal with complex conditions. Carry out firefighting and rescue operations in the case of a blowout.

The well control system is the basic method of pressure control of oil and gas wells to prevent, monitor, control, and deal with the accidents. It is a reliable guarantee for safety in the drilling site and essential system equipment of drilling operations.

The components of well control system

The components of the well control system include

- Wellhead blowout preventer as the main body, mainly including hydraulic blowout preventer group, casing head, drill spoon, counter flange, and so on;
- Blowout preventer control system mainly includes the driller console, remote console, and auxiliary remote control console;
- Well control manifold mainly includes choke manifold and hydraulic choke control box, kill and choke line, water injection pipeline, fire extinguishing pipeline, and reverse circulation pipeline.
- Blowout preventers inside the drill string. It mainly includes a check valve, top and bottom cocks of kelly, drop-in check valve, and bypass valve.
- Well control instruments mainly for monitoring and prediction of abnormal formation pressure. It includes drilling fluid temperature monitoring and alarm device, the drilling fluid density monitoring and alarming device, drilling fluid flow rate monitoring and alarm instrument, drilling fluid-level monitoring and alarm device, wellbore liquid-level monitoring and alarm device, and monitoring and alarm device of pump strokes.

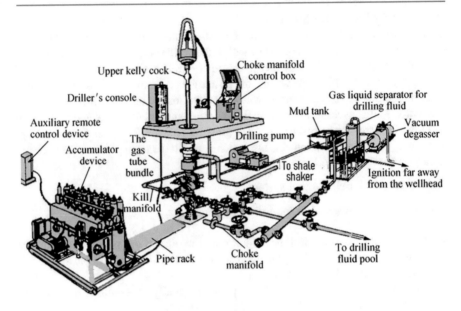

Fig. 2.9 Schematic diagram of well control equipment

- Drilling fluid weighting, degassing, and filling equipment. It mainly includes drilling fluid weighting equipment, drilling fluid/gas separator, conventional or vacuum drilling fluid degasser, and automatic fluid filling device.
- Equipment for uncontrolled blowout and special operations. It mainly includes forced tripping device without killing, self-sealing head, rotary blowout preventer, fire extinguishing device, dismantling, and assembling wellhead equipment and tools.

A typical well control system is shown in Fig. 2.9.

2.1.2.5 Power and Transmission System

The power system provides the energy for the whole set of units (three large working units and other auxiliary units). The engine of a drilling rig is usually a set of diesel engines, sometimes motors. For electric driving rig, it is usually powered by diesel engines, driving the generator to generate electricity and providing power for drilling equipment by cable.

Diesel engines are suitable for drilling in remote areas where power networks are not available. AC motors rely on industrial power networks or require diesel engines to drive AC generators to generate alternating current. DC motors need diesel engines to drive DC generators to generate direct current. At present, the more common situation is that the diesel engine drives the AC generator to produce alternating current, which is then converted into DC by the silicon-controlled rectifier.

According to the different power equipment used in the drilling rig, it can be divided into three kinds: mechanical drive, electric drive, and compound drive.

Mechanical drive

According to the different driving characteristics of the driving unit, the mechanical drive is divided into two types: diesel engine drive-mechanical transmission and diesel engine drive-hydraulic transmission.

- The diesel engine driving-mechanical transmission. With two to four diesel engines as the power, through the combination of V transmission belts, gears, chains, Cardan shafts, and other mechanical transmission components, it realizes the parallel operation, deceleration, torque increasing, and reversing, to drive the drawworks, the rotary table, and the mud pumps. Because of the diesel engine and the mechanical transmission characteristics, working machines can only get effective speed regulation with limited gears.
- Diesel engine drive-hydraulic transmission. With diesel engines as the power, diesel engines are first combined with the hydraulic torque converter (or coupler) to form the diesel-hydraulic drive unit and then through the combination of V belt, gear, chain, universal shaft and other mechanical transmission elements in various forms, it can realize the combination of turning, deceleration, reversal, and backing-up to drive drawworks, rotary table, and mud pumps. Because the hydraulic torque converter can change speed and torque, it belongs to flexible transmission.

Electric drive

Compared with the traditional mechanical drive, electric drive has the advantages of high transmission efficiency, load adaptability, proper installation, and transport capacity, accident handling capacity and protection ability of the equipment, easy to realize the control of torque, speed, acceleration, deceleration and position, easy to realize automatic and intelligent drilling.

The electric drive drilling rig can be divided into four kinds according to its development course:

- Alternating current (AC) driving drilling rig, i.e., alternator (or industrial power grid)—AC motor drive (AC–AC);
- Direct current (DC) driving drilling rig, i.e., DC generator—DC motor drive (DC–DC);
- SCR rectifier DC driving drilling rig, i.e., AC generator—SCR rectifier DC motor drive;
- AC variable frequency electric driving drilling rig, i.e.,—alternator—variable frequency governor—AC motor drive (AC-VFD-AC).

Combination drive

According to the work environment and performance requirements of the three central units of the rotary, drawworks, and mud pump, the compound drive can flexibly choose the suitable driving mode to get the best performance with the most economical power configuration.

There are two main types of compound drives: electromechanical compound drive and AC/DC electric compound drive.

- Electromechanical compound drive

Electromechanical compound drive mainly has two forms: The first one is that a diesel engine plus a coupler is used to drive the mud pump, drawworks, and alternator; the other one is that alternating current generated by alternator controlled by the frequency converter is used to control AC variable frequency motor to drive the winch and turntable. As an independent pump unit, the mud pump is driven by a mechanical drive.

- AC and DC electric compound drive

AC and DC compound drives mean a diesel engine drives the alternator. A part of the alternating current generated by alternator controlled by the frequency converter is used to control AC variable frequency motor to drive drawworks and rotary table. Another part of the alternating current is converted into controllable DC power through a silicon-controlled rectifier, which can control DC motor to drive mud pump.

2.1.2.6 Monitoring and Control System

The rig is equipped with a control system to ensure the coordinated work of the three units of the rig to meet the requirements of the drilling process. Conventional control methods include mechanical control, gas control, electrical control, and hydraulic control. At present, centralized control is commonly used on drilling rigs. The driller can accomplish almost all the rig control through the driller's console, such as clutch release, the merging of the power engines, start and stop of the drawworks, rotary table and mud pumps, and high- and low-speed control of the drawworks.

2.1.3 The Classifications and Characteristics of Drilling Rigs

2.1.3.1 Classifications of Drilling Rigs

According to the features of each drilling machine manufacturer, the classifications of drilling rigs are not the same. Generally speaking, drilling rigs can be classified according to the following methods.

Drilling method

Drilling rigs can be divided into percussion drilling rig and rotary drilling rig according to the different drilling methods. The percussion drilling rig, also known as cable rig, was initially used to make water wells, and in 1859 American Drake introduced it to oil well drilling. The rotary drilling rig is divided into the surface drive rotary drilling rig and the downhole drive rotary drilling rig according to the power source of the bit.

Drilling depth

According to different drilling depth, drilling rigs can be classified into shallow drilling rigs (drilling depth is no more 1500 m), intermediate drilling rigs(drilling depth is 1500–4500 m), deep drilling rigs (drilling depth is 4500–6000 m), and ultra-deep drilling rigs(drilling depth is more than 6000 m).

Environment

According to the area used for drilling rig, it is divided into the land drilling rig, offshore drilling rig, desert drilling rig, swamp drilling rig, low-temperature environment drilling rig, and jungle helicopter hoisting rig.

Mobile mode

Drilling rigs can be divided into block-mounted drilling rig, truck-mounted drilling rig, and trailer drilling rig according to the different mobile modes.

Drive and transmission form

According to the different types of driving equipment, drilling rigs can be classified into diesel-driven drilling rigs (diesel-driven drilling rigs can be divided into diesel-driven mechanical drive drilling rigs and diesel hydraulic drive drilling rigs) and electric drive drilling rigs (electric drive drilling rigs can be divided into DC drive rigs and AC drive rigs).

2.1.3.2 Characteristics of Conventional Drilling Rigs

DC drive rig

Adopting advanced AC-SCR-DC electric drive technology, drawworks and mud pumps are separately driven on their own by DC motors to achieve infinitely variable speeds.

The parallelogram integral lifting base can be divided into two structures: double lift (slingshot) and twist-up lift, and the drill floor has the height of 9 m and 10.5 m.

The drawworks adopts the integral chain drive design mode with mechanical shifting device, equipped with primary hydraulic disk type brake system and electromagnetic eddy current auxiliary brake (or optional pneumatic disk brake), and independent automatic bit feed system; the rotary table can be driven from the drawworks by the chain box and also can be driven by a single DC motor.

AC variable frequency drive rig

The drilling adopts advanced all-digital AC frequency conversion technology, taking the PLC logic control technology as the core, and controls by the integrated design of electricity, gas, and liquid. The drawworks and mud pumps are driven independently by wide frequency and high-power AC frequency motors, which can realize the whole process of continuously variable speeds.

Drawworks is driven by single drum gear with a first gear variable speeds regulation, the main brake is hydraulic disk brake, and the auxiliary brake is motor energy consumption brake and can be controlled by computer quantitative, positioning braking torque; the turntable AC variable frequency motor is driven by a two-stage variable speed transmission box, which can realize the whole journey of continuously variable speed, high transmission efficiency, and meet the requirements of high speed and large torque.

Combined drive rig

The main module adopts the layout of anterior high and posterior low, and the power and the transmission system are installed at a low position; whereas, the base takes the box block type or the front desk twisting up the backstage blocking structure. The power transmission employs "diesel engine + hydraulic coupling + integral chain compounding transmission" driving the drawworks and mud pumps. At the same time, compounding transmission can be equipped with an energy-saving generator and automatic fan. The rotary table adopts AC frequency conversion motor or DC motor driving independently. The drawworks employs the design of whole chain drive structure and long-distance gas-controlled machinery shifting gear transmission, equipped with hydraulic disk-type main brake system (alternatively choosing belt brake), and electromagnetic eddy auxiliary brake.

Mechanical drive rig

The main module adopts the block or box structure and the layout of from high to low design; the power transmission employs "diesel engine + hydraulic coupling + integral chain compounding transmission" or "diesel engine + reduction gearbox + belt parallel locomotive interlocking machine" and "diesel engine + Allision gearbox + gear box" to drive mud pumps and drawworks; the power of drawworks is transferred to the driving box of the rotary table by climbing Cardan shaft or climbing chain box. The drawworks is divided into two types of internal speed transmission and external speed transmission. The main brake adopts hydraulic disk brake or belt brake, and the auxiliary brake employs electromagnetic eddy current brake.

Truck-mounted rig

The truck-mounted rig is a special kind of drilling rig used in field drilling, which can be used for exploration and development of oil, gas, and water injection wells and is suitable for areas with good road conditions. Its main characteristics are the main parts of a drilling rig, such as power, transmission, drawworks, rotary table transmission, derrick, liquid gas control system, and auxiliary devices, are all assembled on one carrier chassis; using the hydraulic mechanism to raise and lay down the derrick, it has the advantages of a lightweight, functional mobility, reliable function, convenient site installation, high efficiency, low transportation cost, and so forth.

Truck-mounted rigs mainly have several types, such as 1000 m, 1500 m, 2000 m, 3000 m, and 4000 m. Truck-mounted rigs are primarily composed of self-propelled chassis and necessary drilling equipment, and these two parts share a set of power systems. Necessary drilling equipment consists of hoisting system [(including drawworks and brake system, auxiliary brake, sheaves system (crown block, wireline, traveling block and hook), derrick and ancillary equipment (rings, elevator, slips, and tongs)], rotary system (rotary table, swivel and drilling string), transmission system (mechanical tran‹smission includes Cardan shaft, angular transmission box, clutch, and chain transmission), control system (for liquid, gas, and electric), and attachment, etc. The operation of the overall machine focuses on the hydraulic console and driller's console.

The truck-mounted rig is the generally hydraulic mechanical drive, which is a high-speed diesel engine + hydraulic mechanical transmission box. Its installed power is selected according to the rated power of the drawworks. Its transmission mode widely adopts Allison hydraulic transmission box to be gearboxes. Allison hydraulic transmission box directly connects the diesel engine through the flexible disk and then the transfer case through the transmission shaft.

2.1.4 Series of Drilling Rigs

2.1.4.1 Basic Parameters of Drilling Rigs

The basic parameters of drilling rigs refer to technical indexes reflecting the essential working performance of drilling rigs, also called characteristic parameters. It is the primary technical basis for designing, manufacturing, selecting, using, maintaining, and reconstructing the drilling rig.

According to the system, the basic parameters of drilling rigs are mainly classified into main parameters, hoisting system parameters, rotary system parameters, circulating system parameters, driving system parameters, and so forth.

Main parameters

In the basic parameters of drilling rigs, the most critical parameter is selected as the main parameter, which should possess the following characteristics—it can directly reflect the drilling capacity and the first performance of drilling rig—it influences and determines other parameters—it can be used to calibrate the type of drilling rig and be used as the primary technical basis for the design and selection of drilling rig.

In China, the standard of drilling rigs adopts the nominal drilling depth D to be the main parameter. This is because the maximum drilling depth of drilling rig affects and determines the size of other parameters.

- Nominal drilling depth. Nominal drilling depth is the maximum well depth drilled by $\Phi 127$ mm (5 in) drill pipe under the standard numbers of drilling ropes.

- Nominal drilling depth range. The nominal drilling depth range is the range between the minimum drilling depth economically available to the rig and the maximum drilling depth. The lower limit of the nominal drilling depth range overlaps that of the previous stage, and the upper limit is the nominal drilling depth of the rig of that stage.

Hoisting system parameters

Hoisting system parameters include

- Maximum hook load. The maximum hook load is the maximum load on the hook that is not allowed to be exceeded under the maximum number of ropes stipulated in the standard, when running casing or carrying out special operations such as drill string stuck freeing.

  Maximum hook load determines the ability of casing pipes and dealing with accidents, which is the main technical basis for checking the static strength of hoisting system components and calculating the static load of main bearing of the rotary table and swivel.
- Maximum drill string weight. The maximum drill string weight is the maximum weight of a drill string in the air that the hook can bear under the standard numbers of drilling ropes in normal drilling operations or tripping operations.
- Number of drilling rope of the hoisting system and the maximum number of ropes. The number of drilling rope of hoisting system refers to the number of effective lifting ropes used in the traveling system during the regular drilling. The maximum number of ropes is the maximum number of effective ropes provided by gear train of the traveling system, used in heavy-load operations, like casing running or pipe stick releasing operations, etc.

In addition, hoisting system parameters also include the lifting speed of each gear of drawworks, the gear number of drawworks, the maximum fast rope pulling force of drawworks, the diameter of wire rope, rated input power of drawworks, the effective height of derrick, drill floor height, and so forth.

Rotary system parameters

Rotary system parameters contain the diameter of the rotary table opening, the rotating speed of each gear of the rotary table, the gear number of the rotary table, the rated input power of the rotary table, and so forth.

Circulating system parameters

Circulating system parameters are defined by the mud pump rated output pressure, rated flowrate, rated input power, and so on.

Driving system parameters

Driving system parameters are the single machine rated power and the assembled engine rated power, etc.

2.1.4.2 Standard Series of Domestic Oil Drilling Rigs

The expression of the oil drilling rig model in China is as follows (Fig. 2.10):

For example, 5000 m AC VFD rig can be expressed as ZJ50/3150DB.

According to GB/T 23505-2009, oil and gas drilling rigs are divided into ten levels based on the nominal maximum drilling depth and the maximum hook load. The main basic parameters of each rig level should comply with the regulations in Table 2.3.

2.2 Drill Bit

2.2.1 Summary

Drill bits are the primary tool to break rocks in well drilling. The performance of a drill bit has a direct influence on the rate of penetration, drilling quality, and drilling cost. The good or bad of the drill bit operating performance is not only directly related to the bit structure design and manufacturing quality, but also depends on whether the formation lithology and other drilling technology conditions are adapted, and whether its use is reasonable, and so on. Therefore, it is critical to design and manufacture high-quality drill bit and to select and use drill bits reasonably according to lithology and drilling process technology requirement.

At present, bits used in oil and gas drilling can be divided into full face drilling bit, core bit, and reaming bit according to the purpose of drilling. According to the

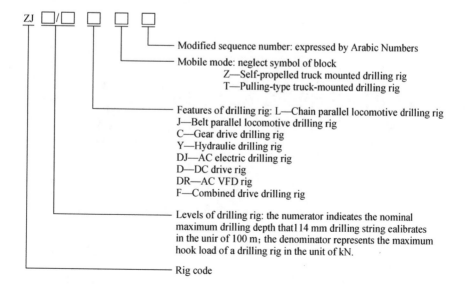

Fig. 2.10 Expression of oil drilling rig model in China

Table 2.3 Basic parameters of oil drilling rigs

Level of driling rig		ZJ10/600	ZJ15/900	ZJ20/1350	ZJ30/1800	ZJ40/2250	ZJ50/3150	ZJ70/4500	ZJ90/6750	ZJ120/9000	ZJ150/11,250
Maximum hook load (kN)		600	900	1350	1800	2250	3150	4500	6750	9000	11,250
Nominal drilling range (m)	ϕ127 mm drilling pipe	500–800	700–1400	1100–1800	1500–2500	2000–3200	2800–4500	4000–6000	5000–8000	7000–10,000	8500–12,500
	ϕ114 mm drilling pipe	500–1000	800–1500	1200–2000	1600–3000	2500–4000	3500–5000	4500–7000	6000–9000	7500–12,000	10,000–15,000
Rated power of drawworks	kW	110–200	257–330	330–500	400–700	735(1100)	1100(1470)	1470(2210)	2210(2940)	2940(4400)	4400(5880)
	hp	150–270	350–450	450–680	550–950	1000(1500)	1500(2000)	2000(3000)	3000(4000)	4000(6000)	6000(8000)
Number of ropes of traveling system	Number of drilling ropes	6	8	8	8	8	10	12	14	14	16
	Maximum number of ropes	6	8	8	10	10	12	14	16	16	18
Nominal diameter of drilling wire rope	mm	19,22	22,26	26,29	29,32		32,35	35,38	42,45	48,52	
	in	3/4,7/8	7/8,1	1,1$^1/_8$	1$^1/_8$,1$^1/_4$		1$^1/_4$,1$^3/_8$	1$^3/_8$,1$^1/_2$	1$^5/_8$,1$^3/_4$	1$^7/_8$,2	
Single unit power of the mud pump is not less than	kW	368	588		735		956	1176		1617	1617,2205
	hp	500	800		1000		1300	1600		2200	2200,3000
Diameter of rotary table opening	mm	381,444.5		444.5,520.7698.5			698.5952.5		952.5,1257.3,1536.7		1257.3,1536.7
	in	15,17$^1/_2$		17$^1/_2$,20$^1/_2$,27$^1/_2$			27$^1/_2$,37$^1/_2$		37$^1/_2$,49$^1/_2$,60$^1/_2$		49$^1/_2$,60$^1/_2$
Light of drilling platform	m	3,4	4,5		5,6,7.5		7.5,9,10.5		10.5,12		12,16

Notes The values in brackets of the rated power of drawworks are not optimal

structure and working principle of bits, it can be divided into three categories: roller cone bit, diamond bit, and PDC bit. Each type of bit has a variety of different models and size to meet the requirement of varying lithology and drilling technology.

The indexes of the operating performance of drill bit are rate of penetration (ROP), bit footage, working life, and unit footage cost. Bit footage refers to the total length of a borehole drilled by a bit. Working life refers to the overall working time when a drill bit is out of use. ROP refers to the ratio of bit footage to working life, which can be expressed as follows:

$$v_{pc} = \frac{H}{t} \tag{2.1}$$

where,

v_{pc} Rate of penetration, m/h;
H Bit footage, m;
t Bit working life, h.

The calculation formula of the unit footage cost is as follows:

$$C_{pm} = \frac{C_b + C_r(t + t_t)}{H} \tag{2.2}$$

where

C_{pm} Unit footage cost, CNY/m;
C_b Single bit cost, CNY;
C_r Drilling rig operation fee, CNY/h;
t_t Tripping time, pipe connection time, h.

In fact, it is not enough to evaluate the operating performance of the drill bit through ROP, working life, or bit footage. Some drill bits have a high ROP, but a short working life and less bit footage, so it needs more bits and multiple tripping time. While some drill bits have a long working life and high bit footage, but low ROP, which makes the drilling speed slow and the circulation long. Therefore, the evaluation of the operating performance of the drill bit is mostly based on the unit footage cost.

This section focuses on the knowledge of structure and type, the working principle, and the reasonable use of commonly used roller cone bit, diamond bit, and PDC bit, to improve the bit structure design, rational selection, and use of the bit.

2.2.2   Roller Cone Bit

Since 1909, when Howard Hughes in the USA obtained the first patent for roller cone bit, the roller cone bit has been widely used in the oil drilling industry as an efficient rock-breaking tool. After more than 100 years of continuous improvement, numerous sizes and models of roller cone bit have been developed that meet the requirements of various drilling technology and adapt to the multiple formations from soft to hard. At present, domestic and foreign bit manufacturers and related research institutions continue to carry out the research on bit materials, structural design, and manufacturing process, and continuously develop new products.

2.2.2.1   The Structure of Roller Cone Bit

The roller cone bit by the tooth type can be divided into two kinds: milling tooth (steel tooth) roller cone bit and tungsten carbide tooth (tungsten carbide insert) roller cone bit; by the number of cones can be divided into single-cone, double-cone, tri-cone, and multi-cone bit. At present, the most widely used and the most common type is the tri-cone bit.

The structure of the roller cone bit is shown in Fig. 2.11. The upper part of the drill bit has a thread for connection with the drill string; bit leg (also called palm) carries the cone axis (axis neck); the cone is mounted on the cone axis, and the cone has teeth to break the rock. There are bearings between each cone and cone axis; the fluid hole (nozzle) is the channel for drilling fluid; the oil storage seal compensation

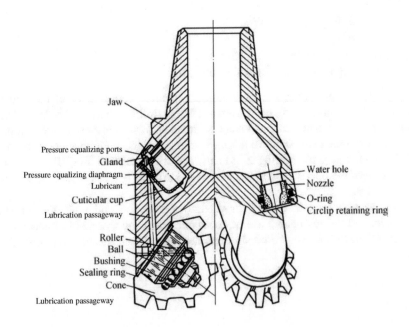

Fig. 2.11 Structure of the roller tri-cone bit

system stores and replenishes the lubrication oil in the bearing chamber, preventing the drilling fluid from entering the bearing chamber and preventing grease leakage.

Cone and tooth

- Cone. The cone is a conical roller with teeth, which is the fundamental element of rock-breaking of the roller cone bit. The cone is made of alloy steel (generally 20CrMo) after forging, the exterior conical surface of cone either milling teeth (milling bit) or mounted tungsten carbide insert (insert bit), and the inner chamber of the cone has bearing race, cog, and seal ring groove.

The exterior conical surface of the cone has two kinds of taper. The single-awl cone is composed of the inner cone and back cone; the multi-awl cone is composed of the inner cone, heel cone, and back cone. However, some have two heel cones, as shown in Fig. 2.12. The primary function of the back cone is to repair and maintain the wellbore wall.

- Milling teeth. The teeth of the milling tooth bit are processed by milling teeth rough, and the tooth shape is limited by the process, which is basically wedge-shaped. To enhance the abrasive resistance of milling teeth, the cemented carbide abrasive-resistant layer is usually welded on the surface of milling teeth (Fig. 2.13).

The main structural parameters of the milling teeth include tooth height, tooth width, and tooth pitch (Fig. 2.14). The determination of these parameters is beneficial to the strength of breaking rock and teeth. In general, the tooth height, tooth width, and tooth pitch for soft formations are considerably large, while those for hard formations are opposite.

According to the different positions and functions of teeth on the cone, it can be divided into inner row teeth, gage teeth, and trim teeth, as shown in Fig. 2.15. The trim teeth are tungsten carbide insert and can enhance the gage protection.

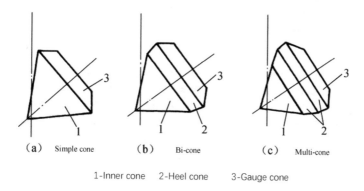

(a) Simple cone   (b) Bi-cone   (c) Multi-cone

1-Inner cone   2-Heel cone   3-Gauge cone

Fig. 2.12 Single-awl cone and multi-awl cone

Fig. 2.13 Welding cemented carbide abrasive milling teeth

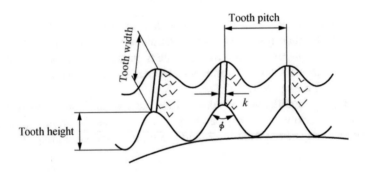

Fig. 2.14 Milling teeth structural parameters

- Tungsten carbide tooth. Milling teeth have poor abrasive resistance due to the limitation of cone material. Although welded cemented carbide layer is applied, its wear resistance still cannot meet the requirements completely, especially in a hard formation, the working life of the tooth is very low. In 1951, the tungsten carbide insert bit was used for the first time in oil and gas drilling, which achieved excellent results in the hard formation. At present, insert cone bits have been widely used in from soft to hard formations.

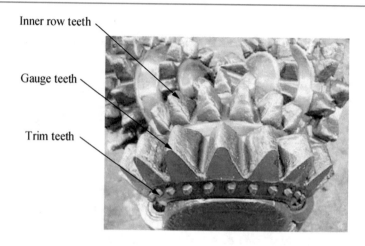

Inner row teeth

Gauge teeth

Trim teeth

Fig. 2.15 Teeth layout on the cone

Fig. 2.16 Tungsten carbide tooth cone

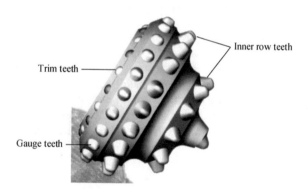

Trim teeth

Inner row teeth

Gauge teeth

The insert cone bit is to insert the teeth made of tungsten carbide into the hole through interference fit after drilling a hole in the cone, as shown in Fig. 2.16.

The hard metal used for the tungsten carbide tooth is tungsten carbide (WC)–cobalt (Co) series hard metal. It takes tungsten carbide powder as the base material, metal cobalt powder as the binder, pressed and sintered by powder metallurgy method. With the increase of cobalt content in the alloy, the hardness gradually decreases, that is, abrasive resistance decreases, but the bending strength and pounding toughness increase progressively. Without changing the content of tungsten carbide and cobalt, increasing the particle size of tungsten carbide can improve the ductility of cemented carbide, and its hardness and abrasive resistance remain unchanged.

The cemented carbide material and properties used in domestic tungsten carbide tooth bit are shown in Table 2.4.

Table 2.4 Domestic cemented carbide materials and properties

Grade	Cemented carbide composition (%)		Hardness HRA	Density (g/cm)	Bending strength (MPa)
	WC	Co			
YG8	92	8	89	14.4–14.8	15
YG8C	92	8	88	14.4–14.8	17.5
YG11C	89	11	87	14.0–14.4	20

The shape of the tungsten carbide insert is usually referred to as the tooth profile, which has a significant influence on the rate of penetration and bit footage of the drill bit. The body of the tooth is a cylinder, which is the part of the perforation inserted into the tooth of the cone shell. The tooth profile refers to the shape and height of the exposed outside part of the cone shell body. The primary basis for determining the tooth profile is the rock property. At the same time, the material properties, intensity, and inserting technology of the teeth must be considered. The tooth profiles of universal tungsten carbide inserts are shown in Fig. 2.17.

– Wedge-shaped tooth: the tooth profile is wedge-shaped, and the tooth angle is 65°–90°; it is suitable for the soft formation and medium-hard formation of high plasticity. It is ideal for soft formation with a smaller tooth angle, while a larger tooth angle is used for hard formation. The tooth-tip part is made into a circular arc, and all edge angles are rounded up to prevent the tooth-tip collapse. In the medium-hard formation, the tooth-tip part has a larger arc (known as blunt chisel insert) or a wider tooth (known as wide chisel insert).

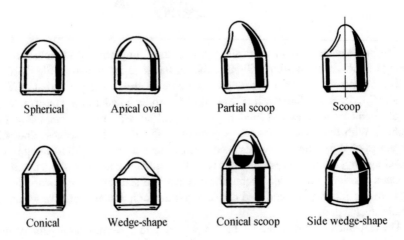

Spherical   Apical oval   Partial scoop   Scoop

Conical   Wedge-shape   Conical scoop   Side wedge-shape

Fig. 2.17 Cemented carbide teeth shape profile

- Conical tooth has long cones, short cones, single cones, double cones, and other shapes, break the rock by crushing, the intensity is higher than wedge-shaped teeth. The middle conical teeth with a taper angle of 60°–70° are used to drill into medium-hard formation, such as limestone, dolomite, and sandstone. The 90° conical and 120° double-cone teeth are used to drill high abrasiveness hard rocks, such as hard sandstone, quartzite, flint, and so on.
- Spherical tooth: the top is a half-sphere, breaking high abrasive resistance hard formation by crushing and pounding, such as flint, quartzite, basalt, and granite, its intensity and abrasive resistance are relatively high.
- Apical oval tooth: a deformation of spherical teeth has a larger tooth depth and certain intensity, also used in a hard formation with high abrasivity.
- Scoop-type tooth: specific tooth profile introduced by US Hughes company in the 1980s, it is a kind of asymmetric wedge-shaped tooth, the formation of cutting surface is concave spoon-shaped, and the back is a slightly convex arc. This structure can improve the stress condition of the tooth, which can not only improve the breaking efficiency, but also enhance the tooth intensity; it can break extremely soft to medium-soft formation rocks with high efficiency. Based on the scoop-type teeth, tip scoop-type teeth and taper scoop-type teeth were further developed. The top of tip scoop-type teeth is offset by a distance relative to its axis with its concave surface facing the formation to be cut, so that it can further improve the stress distribution of tooth stress surface, and improve the breaking efficiency and working life of teeth; taper-shaped teeth are generated on the basis of conical teeth, which is concave on the working face of cutting formation, and on the back is a circular arc of slightly convex arc.

In addition, there is a flat-topped tooth, which is a cylinder with an inverted angle at the top. It is only used in the back-cone of roller cone bit to prevent the back-cone wear and tear and can achieve the purpose of maintaining diameter and improving bit working life.

Bearing

The roller cone bit bearing is composed of a cone chamber, bearing race, tooth neck, and locking element. The bearings include large, medium, small, and thrust bearing. According to the bearing seal or not, the drill bit bearing can be classified into two types: sealing and non-sealing. According to the structure of bearings, the drill bit bearings are divided into two kinds of anti-friction bearing and plain bearing (large bearing). The structural property of different bearing is as shown in Fig. 2.18 and Table 2.5.

For ball bearing, roller bearing, and plain bearing, the contact between the bearing pairs is point contact, line contact, and surface contact, respectively. Therefore, the latter has a larger bearing pressure area, uniform load distribution, and better shock absorption. For larger load-bearing roller cone bit, apparently, the latter is more favorable. Therefore, roller bearing and plain bearing are used for the large bearing and small bearing of roller cone bit. It should be noted that if the drill bit bearing is not well lubricated, the plain bearing will soon fail.

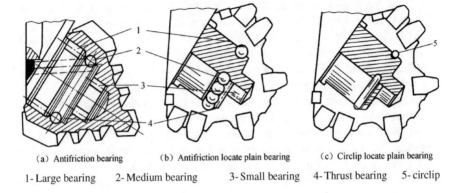

(a) Antifriction bearing   (b) Antifriction locate plain bearing   (c) Circlip locate plain bearing

1- Large bearing   2- Medium bearing   3- Small bearing   4- Thrust bearing   5- circlip

Fig. 2.18 Drill bit-bearing structure

Table 2.5 Different kinds of bearing structure

Structure	Large bearing (bearing radial load)	Medium bearing (latch and locate)	Small bearing (bearing radial load)	Thrust bearing (bearing axial load)
Anti-friction bearing	Roller bearing	Ball bearing	Roller bearing	Sliding bearing
	Roller bearing	Ball bearing	Ball bearing	Sliding bearing
Plain bearing	Sliding bearing	Ball bearing	Sliding bearing	Sliding bearing
	Sliding bearing	Sliding bearing (circlip)	Sliding bearing	Sliding bearing

The center bearer is used to lock the cone, so it is essential. If the center bearer is worn, the cone will be departing from the axle neck and lead to a lost cone accident. Even if the center bearer wear does not reach the degree of separation of the cone from the axle neck, it will also lose its positioning function, and the loosening of the cone and axle neck will aggravate the wear of the bearing. Ball bearing is generally used as the center bearer, some drill bit use circlip to replace ball bearing (Fig. 2.18), thus can further increase the large bearing area, and simplify the structure of bearing and processing technology.

Sealing lubrication system

The sealing lubrication system of a roller cone bit can not only guarantee the lubrication of bearings but also effectively prevent the drilling fluid from entering into the drill bit bearings, thus greatly improve the working life of bearing and drill bit.

The sealing lubrication system of a roller cone bit is shown in Fig. 2.11. The pressure compensation film, also known as the oil storage capsule, is made of oil-resistant rubber, and the protective film cup is installed outside or on it, and the protective film cup is pressed tightly with the gland. The whole oil storage device is installed in the oil storage hole of the tooth claw, connected with the external pressure transmission hole, and combined with the extended oil hole in the bearing cavity. The sealing ring is fixed at the root of the shaft neck to seal the inner cavity of the roller bearing.

When the drill bit is working, the teeth on the roller will be affected by the reaction force of the roller along the axis, resulting in the high-frequency vibration along the axis of the roller, resulting in the pressure difference between the inside and outside of the bearing chamber, and causing the suction and drainage in the bearing chamber. Because of the seal ring, drilling fluid will not be sucked into the bearing cavity. The grease in the bearing chamber will not flow out of the drill bit, while the lubricating oil in the storage chamber will be pumped into the bearing chamber. The sealing lubricant system also makes the pressure of lubricating oil in the bearing cavity consistent with the pressure of drilling fluid outside the drill bit through pressure transmission hole and pressure compensation film, so that the sealing ring works under a small pressure difference to ensure the sealing effect.

Sealing ring is the central part that controls the sealing effect. The sealing ring includes a dished sealing ring, O-ring, and metal sealing ring, etc. The metal sealing ring is a new sealing element that successfully developed by US Hughes Company in recent years. It uses high-grade stainless steel to process. This sealing ring is used in the newer ATM series drill bit, to let it adapt to high ROP drilling conditions, and improve the drill bit working index.

Bit ports

The bit ports are the flow channel of drilling fluid flow out of the bit and toward the bottom hole. The standard bit ports are fixed at the proper position on the bit body and welded with nozzle socket.

In drilling, to make full use of the hydraulic power of the bit, the high-speed liquid can be jetted directly into the bottom hole to entirely remove the cuttings and improve the drilling efficiency. This drilling technology is called jet drilling. The bits suitable for jet drilling are called jet bits. The jet bit is installed with cemented carbide nozzle at the ports. Before using the drill bit, the nozzle with suitable size (inner diameter) should be selected reasonably, and it should be clamped to the port on the drill bit with stainless steel elastic retaining ring, and the nozzle can be detached and reused after the drill bit is worn out.

The layout of cone and tooth

The layout of the roller cone bit is mainly including non-self-clean no-slip design, self-clean non-offset layout, and self-clean offset layout, as shown in Fig. 2.19, could be applied in a different formation.

(a) No self-clean, no-slip layout (b) Self-clean, non-offset layout (c) Self-clean, offset layout

Fig. 2.19 Cone layout program

- Non-self-clean no-slip layout

The main characteristic of this layout is the single-cone structure of the cone. The teeth generally adopt shorter spherical and apical oval tungsten carbide teeth. Therefore, the ring gear of adjacent cone is not embedded with each other; the axes of the three-cone wheels intersect with the bit centerline of the same point, no cone overhang, and no offset. The roller cone bit with this structure is suitable for hard formation.

- Self-clean non-offset layout

The main characteristics of this layout are the multi-cone structure of the cone. The teeth generally adopt medium-length conical and scoop-shaped tungsten carbide teeth. The ring gear of adjacent cone is embedded with each other; the cutting between the ring gear of one cone can be removed from the teeth of another cone gear ring, thus preventing balling up from layup debris. The axial line of the three cones all pass through the centerline of a drill bit, but the apex of the heel cone extends beyond the center of the drill bit by a certain distance. The roller cone bit with this structure is suitable for medium-hard formation.

- Self-clean offset layout

The main characteristics of this layout are the multi-cone structure of the cone. The teeth generally adopt larger size wedge-shaped tooth. The ring gear of adjacent cone is embedded with each other, the axial line of the three cones does not pass through the centerline of a drill bit, but has a certain distance to the direction of rotation of the drill bit (called bit offset). The apex of the heel cone is beyond the center of the bit to a certain distance. Therefore, the structure of the roller cone bit has the characteristics of the ring gear embedment, cone overhang, and offset, which is suitable for soft formation.

The alignment and layout of the tooth on the cone directly affect the drilling efficiency of the drill bit, so it is essential. The layout of teeth usually follows the following principles:

- During each turn in the drilling process, the teeth should break the full bottom of the hole to ensure that there is no residual unbroken circle of the bottom rock.
- When the cone is repeatedly rolling, the tooth should not fall into another old pit where it has been broken by teeth. Therefore, in a bit rotation round, it should make the ratio of the perimeter of each ring gear rolling at the bottom hole to pitch not as an integer, and the spacing of each ring gear should not be greater than the width of the bottomhole broken pit.
- The number of teeth on each cone ring gear should make each tooth bearing the task of bottle hole rock crushing evenly. Therefore, the outer ring should have more teeth, and the inner ring should have less.

2.2.2.2   The Rock-Breaking Principle of the Roller Cone Bit

The movement patterns of the roller cone bit at the bottom of the hole

The movement patterns of the roller cone bit at the bottom of the hole have five different kinds, namely

- Revolution: the teeth together with the cone along with drill bit rotate clockwise around the drill bit axis. The revolution speed n_b is the rotating speed of the drill bit driven by the rotary table or mud motor.
- Rotation: the teeth together with the cone rotate counterclockwise around the cone axis. Rotation is the result of the interaction between the teeth and the bottomhole rock when breaking the rock. The rotary speed n_c is related to the bit rotation speed, cone structure, tooth layout, and the force of tooth on the bottomhole rock. The revolution and rotation of cone are shown in Fig. 2.20.
- Longitudinal vibration: The cone will make the cone and drill bit vibrate longitudinally at the same time while rolling on the bottom hole. This type of vibration is caused by the axial rise of the cone with a single tooth landing and the axial drop with double teeth landing, as shown in Fig. 2.21. The vibration frequency is proportional to the tooth number of the cone (outer ring teeth) and the rotation speed of the cone. The vibration amplitude is proportional to the radius of the cone and inversely proportional to the tooth number of the outer ring. The impact velocity of the vibration is proportional to the radius and rotational speed of the cone and inversely proportional to the teeth of the outer ring. In practice, in addition to the higher-frequency vibration caused by single and double teeth interlacing contact, there are also the lower-frequency and larger amplitude vibration in the longitudinal direction, which are caused by the uneven bottom hole or convex platform.

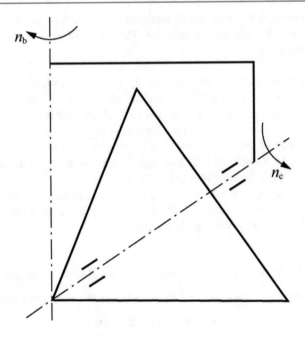

Fig. 2.20 Schematic diagram of the revolution and rotation of the cone

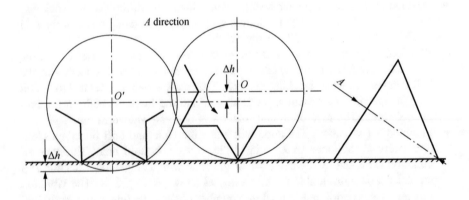

Fig. 2.21 Schematic diagram of cone longitudinal vibration

- Tangential slip: The tooth of the cone overhang layout roller cone bit will produce tangential slip while rolling at the bottom hole, which is a tangential slip of cone.

Figure 2.22 shows the velocity distribution of the cone overtop layout of single cone and the bottomhole contact busbar $\overrightarrow{aa'}$. v_{bx} refers to the linear velocity of cone

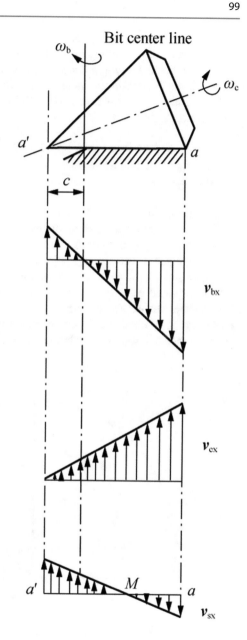

Fig. 2.22 Tangential slip lead by cone overhang

revolute with the drill bit. v_{cx} is the linear velocity at which the cone rotates around the cone axis. From the "synthetic movement of the point," we can obtain that at any point on the contact busbar relative to the bottomhole movement velocity $v_{sx} = v_{bx} + v_{bx} \cdot v_{sx}$ shows the linear distribution and intersects with the contact busbar $\overrightarrow{aa'}$ at M point, the resultant velocity of the point is zero. This point has no sliding, called a real rolling point. Therefore, when the single-cone cone bit with overtop structure is working at the bottom of the well, the teeth on the cone are centered on the real rolling point at the bottom of the well, and the tangential sliding speed will increase with the increase of overtop distance.

Multi-cone with more than two cones generally have cone overhang, so there will have a tangential slip.

- Axial glide: The roller cone bit with offset layout, its tooth will produce a slip along the cone axis while rolling at the bottom hole, which is the axial glide of the cone.

The velocity distribution of the single-cone and bottomhole contact busbar is shown in Fig. 2.23. When the cone revolutes, due to the offset of the cone, the linear velocity v_{sa} at any point a of the cone and rock contact busbar can be decomposed into the tangential velocity v_{ba} perpendicular to the cone axis and the centripetal velocity v_{sa} along the cone axis. When the cone rotates by its own axis, it produces only a tangential velocity perpendicular to the cone axis, which is opposite to v_{ba}. The resultant velocity in tangential direction may or may not be zero, but cannot be zero in the axial direction. It can be seen from the above that when the cone with offset composition works at the bottom hole, the teeth on the cone always generate a

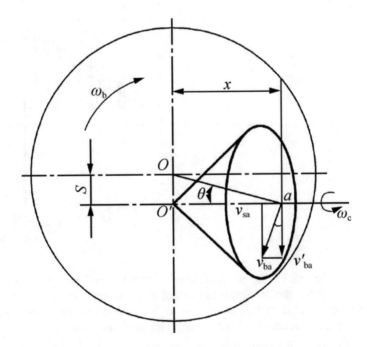

Fig. 2.23 Axial slip lead by the cone offset

centripetal slip along the cone axis, and the axial slip velocity increases as the offset distance increases.

The breaking effect of the roller cone bit

- Percussion and crush action

When the roller cone bit works, the cone rotates around its axis, and alternately contacting the uneven bottom hole with single tooth and double teeth, causing longitudinal vibrations of the cone and drill bit. In each vibration, the drill bit bumps up and compresses the lower drill string to store the potential deformation energy. When the drill bit goes down, the compressed lower drill string resumes its original length and the potential energy can be converted to the dynamic load of the bit to percuss the rock. Dynamic loads and static loads act on the rock through the teeth, forming impact and crushing action on the rock at the bottom of the hole. This action is the main way that the rock is broken by the roller bit.

The impact load produced by bit work is beneficial for rock breaking. Still, it can also prematurely damage the drill bit bearing, causing the teeth breaking down (especially the tungsten carbide insert), even leaving the drill string working under unfavorable conditions.

- Slip shear action

The cone overhung, multi-cone and offset structure of roller cone bit makes the cone rolling at the bottom of hole, at the same time produces the tangential slip and axial slip from tooth to bottom hole, shearing the rock between teeth. The tangential slip of the teeth can shear the rock between the crushing pits of adjacent teeth in same ring gear. The axial slip of the teeth can shear the rock between ring gears.

The slip of the teeth can shear the bottomhole rocks to improve the cursing efficiency, but it also correspondingly increases tooth wear. The axial slip caused by offset leads to the wear at the inner part surface of the tooth, while the tangential slip caused by cone overhang and multi-cone results in the side wear of tooth. Therefore, the consolidation of teeth (especially milling teeth) should be treated differently depending on the situation.

2.2.2.3 The Selection and Reasonable Use of Roller Cone Bit

The classification and model code of the roller cone bit

After a long period of development, there are many types of roller bit for users to choose. To make it easier for users to select the right type of bit, the cone bit has been systematically classified at home and abroad, and the model code has been named.

Table 2.6 Domestic tri-cone bit series

Category	Series full name		Code
	Full name	Abbreviation	
Milling teeth bit	Conventional tri-cone bit	Conventional bit	Y
	Jet tri-cone bit	Jet bit	P
	Roller sealing-bearing jet tri-cone bit	Sealing bit	MP
	Roller sealing-bearing gage protection jet tri-cone bit	Sealing gage protection bit	MPB
	Slip sealing-bearing jet tri-cone bit	Slip-bearing bit	HP
	Slip sealing-bearing gage protection jet tri-cone bit	Slip gage protection bit	HPB
Insert bit	Carbide insert roller sealing-bearing jet tri-cone bit	Insert sealing bit	XMP
	Carbide insert slip sealing-bearing jet tri-cone bit	Insert slip sealing bit	XHP

Table 2.7 Type and applicable formation of domestic tri-cone bit

Formation property		Extra soft	Soft	Medium soft	Medium	Medium hard	Hard	Extra hard
Type	Type code	1	2	3	4	5	6	7
	Previous type code	JR	R	ZR	Z	ZY	Y	JY
Suitable rock examples		Mudstone, gypse, salt rock, soft shale, chalk, soft limestone		Medium-soft shale, hard gypse, medium-soft limestone, medium-soft sandstone	Hard shale, limestone, medium-soft limestone, medium-soft sandstone	Quartz sandstone, ganite, hard limestone, marble		Flint stone, ganite, quartz, basalt, pyrite
Bit body color		Opal	Yellow	Light blue	Gray	Dark green	Red	Brown

- Domestic tri-cone drill bit classification and model representation

The domestic roller cone bit standard regulates that, according to the characteristics of the bit structure, the drill bit is divided into two categories: milling teeth drill bit and tungsten carbide tooth drill bit, a total of eight series, as shown in Table 2.6. The type and applicable formation of the drill bit can be seen in Table 2.7.

The domestic cone bit type can be expressed as follows.

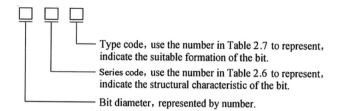

For example, the model of tungsten carbide tooth slip seal-bearing jet-type tri-cone drill bit, which is used in the medium-hard formation with a diameter of 8 ½ in (215.9 mm), is 8 ½XHP5 or 215.9XHP5.

- Classification method and numbering of IADC roller cone bit

Throughout the world, cone bits are manufactured by many manufacturers, with numerous types and structures. To facilitate the selection and use of roller cone bit, the International Association of Drilling Contractors (IADC) established the world's first roller cone bit classification standard in 1972. Since then, although each bit manufactured by different manufacturers has its code, it also marked with corresponding IADC code. In 1987, IADC modified and improved the original classification methods and codes, forming the current classification and numbering method.

IADC stipulates that each type of bit shall be classified and numbered by four codes, with the meaning of each code as follows:

The first code is series code, using digits from 1 to 8 to represent eight series, which indicates the tooth characteristics of the drill bit and its applicable formation.

1—Milling teeth, soft formation with low compressive resistance, high drillability;

2—Milling teeth, medium to medium-hard formation with high compressive resistance;

3—Milling teeth, hard formation with medium abrasiveness or abrasiveness;

4—Tungsten carbide tooth, soft formation with low compressive resistance and high drillability;

5—Tungsten carbide tooth, soft to medium-hard formation with low compressive resistance;

6—Tungsten carbide tooth, medium-hard formation with high compressive resistance;

7—Tungsten carbide tooth, hard formation with medium abrasiveness or abrasiveness;

8—Tungsten carbide tooth, extremely hard formation with high abrasiveness.

The second code is the lithology level code, using digit 1 to 4, respectively, represent the four levels of formation used in the first code from soft to hard.

The third code is the drill bit structural characteristic code, with a total number of nine digits, in which 1–7 refers to the drill bit bearing and gage protection characteristic, 8 and 9 are used for the new structural characteristic drill bit in the future. The meaning of 1–7 are as follows:

1—Non-seal rolling bearing;
2—Air wash, cooling, rolling bearing;
3—Rolling bearing, gage protection;
4—Rolling, sealing bearing;
5—Rolling, sealing bearing, gage protection;
6—Slip, sealing bearing;
7—Slip, sealing bearing, gage protection.

The fourth code is the bit additional structural characteristic code used to represent the characteristics that the first three codes cannot express. This code represents in English letter. At present, IADC defines 11 features, indicated by the following letters:

A—Air cooling;
C—Center nozzle;
D—Directional drilling;
E—Extended nozzle;
G—Additional gage/bit body protection;
J—Nozzle deflection;
R—Enhance weld line (used for cable drilling);
S—Standard milling teeth;
X—Wedge tungsten carbide tooth;
R—Conical tungsten carbide tooth;
Z—Other shape tungsten carbide tooth.

In some drill bit whose structure may have a variety of additional structural characteristics, one should choose a major characteristic symbol to represent.

Selection of roller cone bit

The roller cone bit is one of the most widely used drill bit. Due to the large variety types of roller cone bit, it can be applied to different property formations, but only one type of bit can be used at a time. When the type of bit used is suitable for the drilled formation, a higher ROP and bit footage can be obtained. If the type of drill bit is not suitable for the formation, the ROP will be lower and the bit life will be shorter. Therefore, in the actual production, it is necessary to select the appropriate bit type according to the property of the drilled formation.

If the formation lithology and hardness are different, the breaking mechanism and the requirement of the bit structure will be different.

For soft formation, the rock failure mechanism depends mainly on the slip of the teeth penetrating deeper formation to shear and scrape the rock. Therefore, the soft formation should select milling teeth or tungsten carbide teeth with both the largest offset and cone overtop, multi-cone structure, large and pointed tooth, less tooth number, in order to fully bring the shearing and breaking effect of drill bit on rock.

For hard formation, it is mainly depending on impact effect and cataclasis under the dynamic and static load to break the rock. At this point, the slip shear breaking effect is poor, which may aggravate the wear of the tooth. When the tooth is more pointed or outcrop too high, it will be easier to wear and fracture. Therefore, hard formation should select the spherical tungsten carbide tooth and the non-overhang, non-offset, and single-cone roller cone bit.

On the medium-soft to medium-hard formation, it is better to break the rock by the combined action of the impact, crush, and slip shear of tooth. Therefore, a multi-cone drill bit with both relatively long steel tooth or tungsten carbide tooth and suitable offset and overhang can be used. As the hardness of the rock increases, the offset and overhang value should be reduced, and the tooth should be shorter and denser.

When drilling the abrasive formation such as quartz sandstone, the drill bit wear fast, especially the wear at gage teeth (external teeth), gage compact and palm top are more serious, which reduce the diameter of drill bit. Therefore, the abrasive formation should select insert bit with enforced gage protection.

In addition, the depth of the drilling interval should be considered when selecting the bit type. In shallow well interval, the rocks are generally soft, and the round trip time is short, so ROP should be the main consideration, and drill bit with higher ROP should be used as far as possible. In deep well interval, the rocks are generally hard, and the round trip time is long, so the raise of bit footage should be the main consideration, and a bit with longer bit life should be selected.

Whether the selected bit is suitable for the formation to be drilled should be concluded through the practical inspection. For several types of drill bits used in the same formation, the unit footage cost is usually used as a criterion to evaluate whether the bit selection is reasonable under the premise of ensuring the quality of the wellbore; the calculation formula is as follows (Eq. 2.2).

Optimization of drilling parameters

Drilling parameters refer to the parameters that can be controlled in the process of drilling, including weight on bit (WOB), rotary speed, drilling fluid property, and hydraulic parameters such as pump pressure, flow rate, nozzle diameter, and so on. Whether the drilling parameters are reasonable will directly impact the breaking efficiency and working life of drill bits. Therefore, after selecting the bit type, we should consider the formation characteristics and bit performance, considering the influence of weight on bit, rotary speed, and hydraulic parameters on the breaking efficiency of the drill bit and wearing of teeth and bearing, to optimize the drilling parameters and achieve the best effect of use. The theory and method of optimization of drilling parameters will be discussed in detail in Chap. 4.

The wear grading of roller cone bit

The wear of roller cone bit includes tooth wear, bearing wear, and gage loss of bit. The International Association of Drilling Contractors (IADC) has established a uniform grade scale of the roller cone bit wear.

- Tooth wear grading

The tooth wear is divided into eight grades, and the letter T represents wear degree on the tooth, but the standards for milling teeth and tungsten carbide teeth are different. The wear grading of milling teeth is determined by the relative wear height (the ratio of wear height to the original tooth height). The wear height within 1/8 is grade 1, between 1/8 and 2/8 is grade 2, and so on. Tooth wear is divided into eight categories in total. The wear grading of tungsten carbide tooth is determined by the ratio of the number of broken, fracture, and fall-off teeth to the original number of teeth. The number of broken and fall-off teeth within 1/8 is grade 1, between 1/8 and 2/8 is grade 2, and so on.

Tooth wear code	Milling tooth	Tungsten carbide tooth
T_1	Tooth height wear 1/8	1/8 of tooth collapse or falling
T_2	Tooth height wear 2/8	2/8 of tooth collapse or falling
T_3	Tooth height wear 3/8	3/8 of tooth collapse or falling
T_4	Tooth height wear 4/8	4/8 of tooth collapse or falling
T_5	Tooth height wear 5/8	5/8 of tooth collapse or falling
T_6	Tooth height wear 6/8	6/8 of tooth collapse or falling
T_7	Tooth height wear 7/8	7/8 of tooth collapse or falling
T_8	Tooth height wear 8/8	8/8 of tooth collapse or falling

- Bearing wear grading

The grading of bearing wear is evaluated by the ratio of the use time of the bit to the bearing life (hour), which is divided into eight grades, the bearing wear is indicated by letter B. The use time of the bit reaches 1/8 of the bearing life is grade 1, reaches 2/8 is grade 2, and so on. In total, the bearing wear can be divided into eight grades. The bearing life is obtained from the statistical analysis of the same type of bit used in the same area.

B_1	Bearing life use 1/8
B_2	Bearing life use 2/8 (slight wear)
B_3	Bearing life use 3/8
B_4	Bearing life use 4/8 (medium wear)

(continued)

(continued)

B_5	Bearing life use 5/8
B_6	Bearing life use 6/8 (bearing sway)
B_7	Bearing life use 7/8
B_8	Bearing life use 8/8 (bearing freeze-in or ball fall-off)

The sealing of bearings is divided into three categories: sealing effective (SE), sealing fail (SF), and sealing questionable (SQ).

- Drill bit diameter wear grading

The drill bit remains the original diameter is represented by the letter I; the drill bit diameter decrease is represented by the letter O, and the number of diameter wear is represented by digits (in).

2.2.3   Diamond Bit

A diamond bit is a fixed tooth bit with the diamond as a working edge. In the 1950s, the diamond bit began to be used in oil and gas drilling. Diamond is the hardest and most wear-resistant material known by humans. It is especially suitable for drilling hard abrasive formation. Therefore, the diamond bit was mainly used to drill hard abrasive formation early on. As the oil and gas exploration and development continue to develop into deep formation, the number of a deep well and ultra-deep well keeps increasing. For deep well and ultra-deep well drilling, the life and safety of the drill bit become more and more critical because of the long round trip time, difficulty, and inefficiency in handling downhole accidents. The roller cone bit is limited by tooth material and bearing, the bit life is shorter, the unit bit footage is less, and run the risk of lost cones. Compared with the roller cone bit, the diamond bit has the advantages of long working life in downhole, shorter tripping times, and higher downhole safety. Since the diamond bit has no moving parts such as bearing, high-speed drilling can be used to make up for the shortcutting amount per turn. The ROP has reached the level of the roller cone bit, sometimes even exceeding. The diamond bit with high-speed turbine drill drilling has gradually developed into an important technology to effectively increase the drilling speed of deep well and ultra-deep well.

2.2.3.1   Diamond

Diamond physical and mechanical performance

The diamond is a crystal formed by carbon under high temperature and high pressure, and its crystal structure is a positive tetrahedron, as shown in Fig. 2.24. In a unit cell, the carbon element locates at the top and center of the tetrahedron. Each carbon atom shares four pairs of valence electrons with its four adjacent carbon

Fig. 2.24 Crystalline
structure of diamond

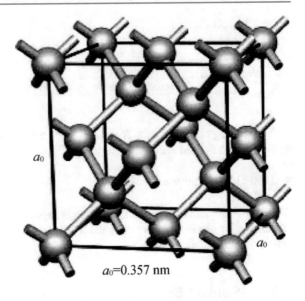

a_0

a_0

a_0=0.357 nm

atoms, forming four covalent bonds. Since the covalent bond has a strong binding
force as well as the stable structure of the tetrahedral crystal, the diamond has a high
hardness and high abrasive resistance. The diamond has a Mohs hardness of 10 and
microhardness of 100 GPa, which is four times of corundum and eight times of
quartz. The abrasive resistance is 100 times of cemented carbide and 5000 times of
steel. The compressive resistance is as high as 88,000 kg/cm^2.

The most common diamond crystal shapes are octahedron, followed by rhombic
dodecahedron and cube, as shown in Fig. 2.25. As a result of the actual formation
of crystals of irregular shape, there are round, oval, and other shapes.

The diamond used in the drill must be of a firm and regular shapes, such as
dodecahedron, octahedron, cube, or other similar shapes to a sphere.

Besides the highest hardness and abrasive resistance, diamond has some disadvantages. Like all hard materials, diamond is more brittle and fragile when
subjected to impact loads. Another problem of the diamond is its thermal

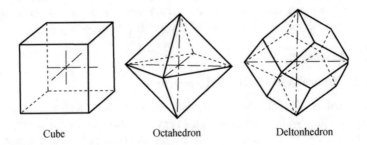

Cube   Octahedron   Deltonhedron

Fig. 2.25 Common diamond crystalline shape

sensitivity, which occurs when the air is graphitization combust at a temperature between 450 and 860 °C. However, when the diamond is sintered with tungsten, tungsten carbide, or in a graphite mold, its oxidation rate can be reduced by 50–90%. Under the inert gas protection or vacuum conditions, the diamond will not burn. But at about 1430 °C, its structure will be transformed into a hexagonal planar structure of graphite; the hardness and strength are significantly reduced.

Drill bit diamond

The diamond used in a drill bit can be divided into natural diamonds and artificial diamonds.

The origin of natural diamonds is mostly concentrated in South Africa, West Africa, Congo, and other regions. According to the varieties, it can be divided into four categories: carbon (also called black diamond), Ballas, Boartz, and Congo (also called Congo Boartz). Carbon diamond has the highest grade, rare, and precious, so it is rarely used in drill bits. The performance and prices of Ballas are second only to that of carbon, and it is also seldom used in drill bits. Boartz diamond is rounded, with high hardness, multi-edge and other features, as well as lower price, so it is the main variety used in drill bits. Congo diamond is mostly granular, and the hardness is less than Boartz, but the price is cheaper, so it is well chosen and usable.

Natural diamond is used as the earliest and always being used, but the price is high, and the bit cost is high. Since the 1950s, synthetic diamond technology has emerged, and synthetic diamond has gradually replaced natural diamond, which has been widely used in the industry.

Synthetic diamonds can be divided into a single crystal and polycrystalline. A synthetic crystal is an essential variety of synthetic diamond from the diamond crystal and some carbonaceous materials such as graphite and metal catalyst (Ni, Mn, Co, Fe, Cr, and cu alloy) with ultra-high pressure (5–10 GPa) and artificial technology of high temperature (1100–3000 °C), which granularity ranges from several microns to several millimeters with the typical morphology of cube (hexahedron), octahedral and six—octahedral and their transitional forms.

Polycrystalline is a kind of polycrystalline diamond with larger particle sintered by fine single-crystal diamond particle (diameter approximately 1–100 μm) and metal bonding agent at high temperature (around 1600 °C) and high pressure (7000 MPa). The shape of polycrystalline can be made into cylindrical, triangular, or other multilateral shapes as needed. General polycrystalline diamond is mostly used metal Co as the metal bonding agent. Due to the significant difference in the thermal expansion coefficient between the Co and diamond, the thermal stability of polycrystalline with Co as a bonding agent is poor. It is generally considered that the thermal resistance should not exceed 700 °C. To improve the thermal stability of polycrystalline diamond, the American G.E company adopts a special acid treatment process to remove the Co binder phase in the Geoset polycrystalline block, to form a thermally stable polycrystalline diamond (TSP) named Bllaset, and its heat resistance can be up to 1200–1300 °C. Recently, the British De Beers Company successfully developed the Syndax 3 thermally stable diamond polycrystalline. It replaces the original binder phase Co with b-SiC whose thermal expansion coefficient is close to the diamond, thus significantly improving the heat resistance property of synthetic diamond. The thermally stable polycrystalline bit diamond has now become the main material used in drill bits manufacturing.

2.2.3.2 The Structure and Design of the Diamond Bit

The diamond bits are composed of diamond, matrix, steel body, nozzle and channel, and gage protection, as shown in Fig. 2.26. The matrix is the sinter of WC-Co, whose main function is to mount the diamond and provide high hardness, setting strength, abrasion resistance, and erosion resistance. The crown part of the drill bit is covered with diamond and has a nozzle and channel. The steel body is ordinary carbon steel or low alloy steel, and the upper part is the thread, which is connected with the drill string, while the lower part is sintered with the matrix.

According to the different diamond materials, diamond bits can be divided into two categories: natural diamond bit and synthetic diamond bit. According to the different setting ways of the diamond, it can be categorized as the diamond set bit and diamond impregnated bit. According to various structures and functions, it can also be classified as a full hole bit and core bit. Diamond bits that use heat-stabilized polycrystalline diamond as the cutting edge are also known as TSP diamond bits.

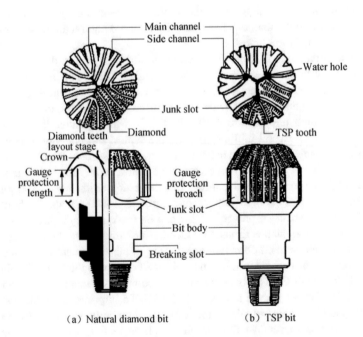

(a) Natural diamond bit   (b) TSP bit

Fig. 2.26 Damond bit structure

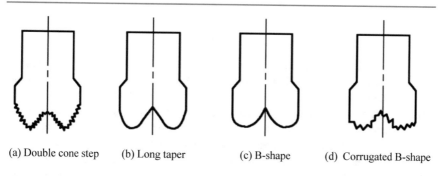

(a) Double cone step   (b) Long taper   (c) B-shape   (d) Corrugated B-shape

Fig. 2.27 Crown shapes of the diamond bit

The structural design of the diamond bit mainly includes profile shape design, hydraulic structural design, diamond tooth layout design, and gage protection design, etc.

Profile shape design

Profile shape refers to the shape of the outline of the crown surface of the bit. It is the most basic and essential work to design the profile shape of the diamond bit reasonably according to the rock characteristics and drilling conditions to improve the use effect of the diamond bit. The typical profile shape of the diamond bit is shown in Fig. 2.27.

(a) **Double-cone step**.

In addition to those two cones, the crown shape also has a staircase or spiral staircase, characterized by the sharp shape at drill bit top, and the stress at the top of the diamond is larger than other parts during the work. After the top of the bit starts drilling into the formation, the diamond on the external conical step also drills into the formation accordingly. Due to the existence of the stepped shape profile, the free surface of the rock is increased, which is favorable for improving the crushing efficiency of the bit. The bits with this shape are suitable to drill soft to medium-hard formations, such as anhydrite, mudstone, sandstone, and limestone.

(b) **Long taper**.

When the long taper bit drills in harder and denser rocks (such as harder sandstone, limestone, dolomite), the diamond on the top and the steps are easily broken and show more weak points. In this type of formation, the bit with long taper profile shape is favorable. The working face of this bit consists of three parts: inner cone, outer cone, and top arc. The inner cone angle is generally between 60° and 70°, and the outer cone angle is between 40° and 60°.

(c) **B-shape**.

Although the crown shape of the above two bits is different, the shape of the top of the bit is sharp, and the load borne by the top diamond is higher than that of other bits. Therefore, in the hard formation, as the hardness of the rock increases, the stress on the diamond and the impact load caused by the drill string vibration also increase correspondingly. In order to make the diamond bearing on each part of the bit as even as possible during drilling and prevent early local damage, the "B" face is adopted. The "B" type working face is composed of an inner cone and an arc surface, with an inner cone angle of no less than 90°. Its structure is characterized by a wide and gentle top, suitable for hard strata such as hard sandstone and dense dolomite.

(d) **Corrugated (or ridged) B-shape**.

The shape of the crown is the same as B-shape, except that the inner cone and the arc surface have spiral corrugated grooves, and the diamond is embedded in the wave crest of ripple. This type of bit is suitable for hard formation, such as quartzite, flint, volcanic rock, and hard sandstone.

• Hydraulic structural design

The diamond bit adopts nozzle—channel-type hydraulic structure, the drilling fluid flows out through the water hole, through the channel to the bit working face. Flush the debris in front of the diamond and cool, lubricate the diamonds, at last, and carry debris to flow into annular space through side-channel and junk slot.

When the drill bit is working, the diamond will press on the formation rock and move at a high-speed relative to the rock surface, thus generating a lot of friction heat and increasing the temperature of the diamond. Due to the poor thermal stability of the diamond, if the drilling fluid does not cool the diamond in time, the diamond will "burn out". If the cuttings cut out before diamonds being cleaned in time, they will lead to the blockage of the working surface of the bit and the local high temperature of the diamond end, thus gradually "burning out" the diamond. Therefore, the hydraulic structure of the diamond bit must provide guarantee conditions for cooling, lubrication, and cleaning of each diamond.

There are four hydraulic structures commonly used in the diamond bit, as shown in Fig. 2.28.

• Forced pressure flow channel

This hydraulic structure of the channel is distributed on the working face of a diamond bit, including a high-pressure channel and a low-pressure channel. The inlet cross-sectional area of the high-pressure channel is smaller than that of the

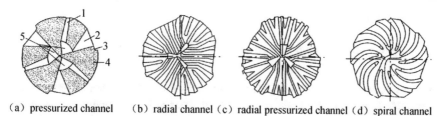

(a) pressurized channel   (b) radial channel (c) radial pressurized channel (d) spiral channel

Fig. 2.28 Hydraulic structure form of the diamond bit. 1—high-pressure flow channel; 2—low-pressure flow channel; 3—chip groove; 4—diamond; 5—jet port

low-pressure channel. As the channel extends outward, the cross-sectional area of the high-pressure channel decreases while the cross-sectional area of the low-pressure channel increases gradually. Therefore, there is a certain pressure difference between the high- and low-pressure channels. Under the effect of this pressure difference, part of the drilling fluid flows from the high-pressure channel to the low-pressure channel through the diamond working face, which can effectively wash, cool, and lubricate each diamond. This type of channel is generally used for soft formation drill bits.

- Radial channel

The channel is radially shaped and uniformly distributed on the working surface of the bit. The working face of the diamond is very narrow (usually only 1–2 rows of diamonds are put in). Therefore, the drilling fluid can thoroughly wash the cuttings and cool the diamond after flowing out of the water hole into the channel. The channel is generally used in soft to medium-hard formations.

- Radial forced pressure channel

This is a combination of the above two channel structures, commonly used in medium-hard to hard formation drill bits and turbo diamond bits.

- Spiral channel

The structure is an anti-spiral flow channel, which forces the drilling fluid to flow through the diamond working face under the condition of a high bit revolutionary speed. This channel is commonly used at high rotating speeds.

Among the above four channel structures, the radial forced pressure channel has the best performance.

Diamond tooth layout design

- The grain size and exposure amount of diamond

The diamond size of the bit depends on the formation. For the soft formation, it is advisable to use a larger grain size. For harder strata, smaller diamond size should be used. See Tables 2.8 and 2.9 for the commonly used particle size ranges of surface-set natural diamond bits and impregnated diamond bits.

Regardless of particle size, the maximum diamond exposure is about one-third of its diameter. If too much exposure, the diamond cladding will not be firm.

- Amount of diamond

The amount of diamond is related to the nature of the rock drilled. The formation is hard, the tooth density is higher, and the amount of diamond is larger. The formation is soft, and the tooth density is low, the amount of diamond is small. In addition, different bit structure and size, the amount of diamond is also different. Table 2.10 lists the amount of diamonds used for diamond-impregnated drill bits from Hughes Christensen, USA. Table 2.11 gives the reference value of the diamond concentration of impregnated bit.

- Arrangement of diamonds

At present, three common arrangements of diamonds on drill bits are radial arrangement, spiral arrangement, and concentric arrangement, as shown in Fig. 2.29.

- Gage design

The gage part of the diamond bit can be used to support the drill bit and keep the well diameter from shrinking. When using the method of setting diamond on the

Table 2.8 Grain size recommendations for surface-set natural diamond bit

Lithology	Medium-hard	Hard	Competent
Grain size (grains.ct^{-1})	3–4	5–8	10–15

Table 2.9 Grain size recommendations for impregnated diamond bit

Lithology		Medium-hard to hard		Hard to competent	
Grain size (mesh)	Natural	20–30	30–40	40–60	60–80
	Sythetic	>46	46–60	60–80	80–100

Table 2.10 Diamond grain size and dosage of a surface-set diamond bit of American Hughes Christensen company

Bit type	MD-34	MD-33	MD-41	MD-24
Suitable formation	Soft	Medium to medium-hard	Hard	Competent
Grain size (grains.ct−1)	1–2	4	8	12
Bit diameter (mm)	Diamond usage (ct)			
152.4	120–143	220–265	248–298	190–220
193.6	180–221	340–409	380–460	294–354
215.9	220–270	410–500	462–582	356–434
244.5	330–340	516–630	580–708	448–544
311.1	436–535	800–991	908–1114	698–880

Table 2.11 Recommended diamond concentration for an impregnated diamond bit

400% normal concentration (%)	44	44	75	100	125
100% normal concentration (%)	11	12.5	18.8	18.8	18.8
Diamond content (ct.cm^{-3})	1.93	2.20	3.30	4.39	5.49
Suitable formation	Hard to competent, weak abrasiveness		Medium-hard to hard, medium abrasiveness		Hard to competent, strong abrasiveness

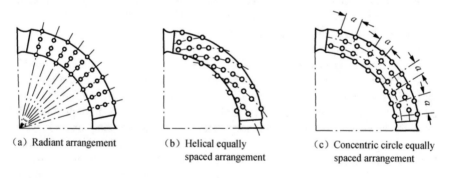

(a) Radiant arrangement

(b) Helical equally spaced arrangement

(c) Concentric circle equally spaced arrangement

Fig. 2.29 Diamond set bit mode of arrangement

side of the drill bit to achieve the gage protection purpose, the density and quality of the diamond can be determined by the abrasiveness and hardness of the drilled rock. For hard and abrasive formations, the quality of diamond at the gage protection part should be higher, and the density should be larger. The structure of the gage protection part includes groove type, mounted type, and combined type, as shown in Fig. 2.30.

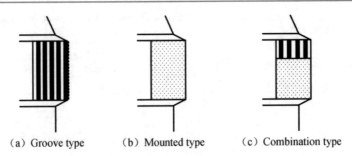

(a) Groove type   (b) Mounted type   (c) Combination type

Fig. 2.30 Gage protection structural shape of diamond bit

2.2.3.3   The Rock-Breaking Principle of the Diamond Bit

Diamond particles accomplish the rock-breaking action of diamond bits. Under the act of weight on bit and rotary torque, the diamond particles continuously impose positive load and tangential force to the rock of the bottom of the hole and break the rock. The main rock-breaking methods are crushing, shearing, cutting, grinding, and so on, and these patterns will demonstrate variously for different types of drill bits and rock properties.

When surface-set diamond bits break hard and brittle rock, crushing and volume shearing are the main ways of breaking rock. The volume of the broken rock is much larger than the amount of diamond particles penetrating and rotating. When crushing hard and medium-hard rocks with considerable plasticity, the primary breaking method is cutting, which is characterized by "plow", and the cutting depth is equal to the penetration depth of diamond particles.

The diamond impregnated bit has fine grain diamonds distributed evenly in the drill bit matrix (impregnated layer). As the matrix wear, the diamond is gradually exposed to the surface, forming small blades. Each corpuscle diamond on the drill bit can be compared to an abrasive grain on the grinding wheel, under the weight on bit and rotary torque, continuously cut and grind the rock. Diamond impregnated bit relies on the small and many hard points (diamond) to cut and grind the rock, which requires high rotary speed work to achieve the high rock-breaking efficiently. Besides, the matrix of the impregnated bit should be adapted to the formation abrasiveness level, so that the matrix can be worn properly, and ensure the diamond can continue exposure to maintain the good self-sharpening ability.

In addition to lithology and external factors that influence lithology (such as pressure, temperature, formation fluid properties), the weight on bit is also an essential factor influencing rock-breaking efficiency. The diamond bit is the same as the roller cone bit, and both have three methods for rock breaking: surface breaking, fatigue breaking, and volumetric breaking. Only when the diamond particles have enough specific weight on bit to penetrate the rock and make the rock break in volume, could the ideal rock-breaking effect be achieved.

2.2.3.4 The Selection and Rational Use of Diamond Bits

Bit selection

The diamond bit takes the hardest diamond particle as the working edge and has advantages of high resistance, high specific weight on bit, easy to penetrate the rock. But the smaller diamond particle size directly affects the possible cutting amount per revolution, and ROP is restricted. Therefore, the diamond bit is mainly used to drill medium-hard to hard and high abrasiveness formations in oil drilling.

The surface-mounted diamond is usually used for drilling medium-hard to the hard integrated homogeneous formation. The harder the formation is, the smaller the diamond particle size is. The higher the abrasiveness of the rock, the higher the diamond grade should be.

Diamond impregnated bits are suitable for hard-competent formation, uneven formation, fractured formation, high abrasiveness formation, and competent formation. The hardness of the drill bit matrix should be selected according to the abrasive, breaking degree, hardness, particle size, and other factors. In general, if the rock formation is high abrasiveness, brittle, softer, coarse particles, the hard matrix should be used. On the contrary, a soft matrix should be chosen for low abrasiveness, homogeneous, high stiffness, and the fine grain size of rocks.

The indexes of evaluating the diamond bit are ROP, bit footage, unit footage diamond consumption, and unit footage cost. Among them, the unit footage cost is the technical index that can best reflect the overall drilling efficiency of the diamond bit.

Drilling recommendations

- Weight on bit

Weight on bit has a significant impact on the ROP and wear of the diamond bit. When the weight on bit is in a low level, it cannot reach the threshold of the rock breaking. The breaking process mainly depends on surface grinding caused by the friction between the diamond and the rock, so the ROP is low. When the weight on bit increases or reaches to the compressive resistance of rock, the ROP increases significantly and increases linearly with the increase of the weight on bit. At this time, the rock breaking presents as volume breaking. When the weight on bit reaches a particular value, the ROP increases slowly due to the limit of diamond outcrop height. If the weight on bit continues to rise, the distance between the diamond matrix and rock surface will decrease, which blocks the flow of drilling fluid, influences the bit cleaning and cooling effect, and boosts the wear effect on bit matrix and diamond. When the weight on bit reaches the compressive resistance of the diamond, the diamond will collapse or crush. Therefore, there is an optimal value for weight on bit when drilling with a diamond bit. The optimal weight on bit depends on the compressive resistance of the rock and the characteristics of the diamond (compressive strength, grain size, quantity, and exposure amount of the diamond).

- Rotary speed

The bit rotary speed is also one of the crucial parameters that affect the drilling efficiency of the diamond bit. Since the diamond particles are generally small, with shallow penetration in the rock, higher efficiency can only be achieved through multiple breaking in unit time. Therefore, high rotary speed is usually suitable for diamond bit drilling. However, both indoor experiments and field practices have proved that, at a certain speed range, the ROP increases linearly with the rotary speed of the bit. When the rotating speed reaches a certain limit, ROP does not go up any more, which is caused by the damage of the diamond due to the severe friction. Therefore, in actual production, reasonable rotary speed should be determined according to the property of drilled formation and structure and quality of the drill bit.

- Hydraulic parameters

In the process of breaking rock, the work of the diamond bit along the direction of the motion is used to break the rock and overcome the friction. The conversion of friction energy into the heat energy can reduce the intensity of the diamond, graphitization and lead to early wear and tear. Hence, the diamond bit must be sufficiently cooled and lubricated when drilling.

In the drilling, the clearance under the working face of the bit is minimal, and the large grain cuttings are broken and ground again in the small clearance to wear out the matrix and diamond. The crushed rock also holds up the drill bit, preventing it from breaking rock, which requires drilling fluid to carry it out of the hole in time.

Therefore, the diamond bit drilling requires high volume circulating drilling fluid to enhance the effect of hydraulic cleaning and cooling.

- Correctly use.
- It is required to remove hard debris such as broken metal from the bottom hole before running the bit down the hole, to ensure there is no hard debris that can directly damage the diamond at the bottom hole, then the diamond bit can be tripped in.

The trip-in speed should not be too fast to avoid shock trip-in. Reaming trip-in is also not allowed to prevent the damage of the diamond by bumping the hard rock at the borehole wall.

When the drill bit is close to the bottom hole, we should flush the bottom hole first with high volume, to prevent the bit from balling.

After the new drill bit trip into the bottom hole, it is used to grind the bottom hole with a small weight on bit and low rotary speed, then smoothly and uniformly increase the weight on bit and rotating speed to average values.

2.2.4 PDC Bit

PDC bit is the abbreviation of the polycrystalline diamond compact bit. It takes sharp, wear-resistant, self-sharpening polycrystalline diamond composite compact as cutting element, under the action of weight on bit and torque, continuously cutting broken rock. When the PDC bit drills in the soft to the medium-hard formation, it has incomparable advantages of fast drilling speed (2–4 times of roller cone bit), great footage (4–6 times of roller cone bit), less downhole drilling accidents, and so on. Now, it has become the main force of the rock-breaking tool in oil drilling with quick developments.

2.2.4.1 Polycrystalline Diamond Compact

The PDC drill bit uses the polycrystalline diamond compact as the cutter. Polycrystalline diamond composite is a kind of super hard composite material developed in the 1970s. It is composed of two parts. One is the polycrystalline diamond layer as the wear-resistant working body, the other is the tungsten carbide matrix as the supporting body, as shown in Fig. 2.31. The general manufacturing method of the polycrystalline diamond compact is mix the synthetic diamond powder and bonding metal (cobalt, etc.) by a certain percentage and then place on the tungsten carbide base, to form a certain thickness of the diamond layer, and finally be sintered under high-temperature (1500 °C) and high-pressure (6 GPa) conditions.

The polycrystalline diamond composite compact is the complex of the diamond polycrystalline layer and hard alloy. The polycrystalline layer of diamond is composed of many small diamond crystals with irregular orientation, and there is no

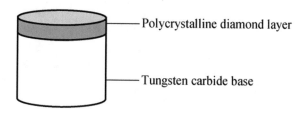

Fig. 2.31 Schematic diagram of a polycrystalline diamond compact

inherent cleavage plane of the single-crystal diamond, so its strength and wear resistance are very high, and it is not easy to fracture. The carbide substrate supports the polycrystalline layer of diamond, making it more impact-resistant and weldable. In the process of drilling and crushing rocks, due to the faster wear of hard alloy than a diamond polycrystalline layer, the polycrystalline diamond composite compact can always maintain a sharp cutting edge. It is this self-sharpening effect that makes the PDC bit have high rock-breaking efficiency and fast drilling speed.

Both diamond and cemented carbide are brittle materials, so the polycrystalline diamond compact is also fragile. It has a weak impact resistance and cannot bear a large impact load. Besides, the thermal stability of the polycrystalline diamond compact is poor. Research shows that when the operating temperature exceeds 350 °C, and its wear rate is significantly accelerated. When the operating temperature exceeds 730 °C, the intensity of the polycrystalline diamond layer may be failed. This is mainly because the thermal expansion coefficient of the bonded metal cobalt (1.22×10^{-4} mm/K) is much larger than that of the diamond (3.3×10^{-5} mm/K) in the polycrystalline layer. Different expansion rates will produce internal stresses in the polycrystalline diamond layer, cause intercrystallite cracks, and reduce the structural strength.

Due to the excellent performance of the polycrystalline diamond compact, it has been developed and produced by competition both at home and abroad. Table 2.12 lists the models and specifications of the currently used polycrystalline diamond compacts.

2.2.4.2 The Structure and Design of PDC Bit

PDC bit is composed of polycrystalline diamond compact, bit body, nozzle and channel, gage protection and joint, etc., as shown in Fig. 2.32. The PDC bit generally uses the blade cutting structure. The polycrystalline diamond compact is brazed into the prefabricated cavities of the blade. The channel is placed between the blades, and water holes arranged in the channel are communicated with the

Table 2.12 Models and specifications of the currently used polycrystalline diamond compacts

Type	Diameter (mm)	Length (mm)	Thickness (mm)	Blade inverted angle	Suitable formation
0808	8.00	8.0	2	0.25 * 45°	Hard
1308	13.44	8.0	2	0.25 * 45°	Homogenous medium-hard
1313	13.44	13.2	2	0.25 * 45°	Heterogenous medium-hard
1608	16.00	8.0	2	0.25 * 45°	Homogenous medium
1613	16.00	13.2	2	0.25 * 45°	Heterogenous medium
1908	19.05	8.0	2	0.25 * 45°	Homogenous soft
1913	19.05	13.2	2	0.25 * 45°	Heterogenous soft
1916	19.05	16.0	2	0.25 * 45°	Heterogenous gravel soft

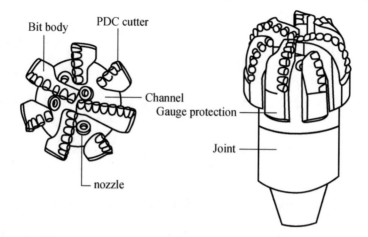

Fig. 2.32 Schematic diagram of the PDC bit structure

inner cavity of the drill bit. The nozzle is installed in the water holes to provide the flow channel for the drilling fluid. The diamond or polycrystalline diamond compacts are welded at the gage protection part to improve its wear resistance and prevent bit diameter reduction. The upper of the joint is welded with the bit body; the lower part has the thread, which is connected with the drill string.

PDC bits can be divided into steel bits and matrix bits according to different bit materials. The bit body of steel bit is generally made of alloy steel and is made by mechanical processing. In order to improve the wear resistance and erosion resistance of drill bit steel, a layer of hard alloy wear-resistant material is usually applied on the surface of the steel. To improve the wear resistance and erosion resistance of a bit body, a layer of hard alloy wear-resistant material is usually used on the surface of the steel. The bit body of PDC bit is a sintered body of WC-CO, which is made by the powder metallurgy process. In order to enhance the bending strength of the blade and facilitate welding with the joint, the drill head tire body is embedded with a steel core, one end is sintered with the tire body, and the other end is welded with the joint.

PDC bit is an integral construction of a fixed bit. There are no moving parts, and the structure and manufacturing process of the bit is more straightforward than the roller cone bit.

The structural design of the PDC bit includes profile shape design, cutter layout design, hydraulic structure design, and gage protection design, etc.

Profile shape design

The profile shape of the PDC bit is the shape of the contour of the bit working face. The profile shape of the PDC bit determines its cutter layout area and has an essential influence on the performance of the bit. Initially, the profile shape design of the PDC bit refers to the profile shape of the diamond bit. After years of

Fig. 2.33 Profile shape of
PDC bit

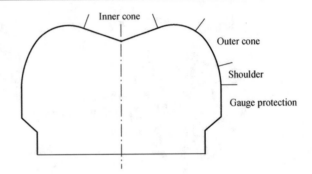

improvement and development, the profile shape of PDC bits has gradually formed, as shown in Fig. 2.33.

The profile shape of the PDC bit is composed of five parts: inner cone, crown, outer cone, shoulder, and gage protection. The inner cone shape is generally a straight line, and its primary function is to resist the lateral force of the drill bit and prevent the bit from moving horizontally, which makes the bit work smoothly. The inner cone angle values range from 130° to 150°, the harder the formation, the larger the inner cone angle.

The crown shape is often designed as a circular arc. For soft formation, the radius of the crown arc is smaller to improve the ability of the bit to penetrate the formation. For hard formation or hard and soft interbedded formation, the radius of the crown arc is larger, so that the cutter is more uniform and avoids the single cutter bearing too much force to damage prematurely. The shape of the outer cone is mostly parabola. The length of the outer cone determines the area of the cutter layout. For soft formation, the cone is shorter to reduce the number of teeth, to improve the ability of the bit to penetrate the formation. For hard formation, the outer cone is longer with an increasing number of the cutters to enhance the abrasive resistance of the drill bit.

The shoulder is the part of the outer cone passes through the gage protection. The turning radius of this part is the largest, the friction distance is the longest, and the impact load is increased, which is the weakest part of the PDC bit, so the cutter layout needs to be strengthened. The role of the gage protection is to prevent the bit diameter reduction due to wear and to make the bit work smoothly, therefore to form a high-quality borehole. The length of the gage protection is generally designed to be 50–100 mm, usually set diamond or polycrystalline diamond compact to enhance its wear resistance, and prevent hole shrinkage.

Over the years, many design theories and methods of the PDC bit profile shape are studied, but it has not yet formed a design method widely recognized. So when making specific profile shape design, there are still lack of theoretical basis, mainly relying on experience or comparison within the section shape of the same kind of bit cone angle, the rotation radius of the crown, the crown radius of a circular arc, outside the cone height, and the length of the gage and related parameters.

The cutter layout design

The cutter layout design is to arrange the PDC cutters (polycrystalline diamond compact) to each blade based on specific regulations.

The design includes the size and type of cutter, the number of cutter, the location coordinates of the cutter, and the determination of the working angle.

Whether the cutter layout design is reasonable or not directly determines the performance of the drill bit. An ideal cutter layout design should meet the following requirements:

- The bottom hole is well covered, and the drill bit does not leave unbroken crock ridge with one rotation.
- The rate of wear of each cutter is equal, to avoid the influence of partly early damage on the bit life.
- The adjacent cutter on the same blade does not interfere with each other, and appropriate spacing is retained.
- The distribution of blade and cutter is helpful to improve the effect of hydraulic cleaning and cooling.
- The drill bit can reach equilibrium under stress and make the bit work smoothly.
- Fast drilling speed, long bit life, and low cost.

The first step of cutter layout design is to select the appropriate PDC cutter based on the properties of formation to be drilled. Generally, for high abrasive formation (such as quartz sandstone), cutting teeth with a high wear ratio should be selected. For the hard and soft interbedded formation or gravel-rich formation, the cutting teeth with strong impact resistance should be chosen. For the harder formation, so the cutting teeth with smaller diameter should be selected to improve the ability of the bit to penetrate the formation. In order to improve drilling speed and bit footage, cutting teeth with larger diameter should be selected.

The second step of the cutter layout design is the cutter layout design in the radial direction, which is according to specific regulations (such as equal cutting volume, same cutting capacity, or equivalent rate of wear) to arrange cutters along with the profile lines, to determine the number of cutters and the radial coordinate, height coordinate and setting angle of each cutter. Then get the radial cutter layout figure which can reflect the cutter layout density and the bottomhole coverage conditions, as shown in Fig. 2.34.

The general principle of radial cutter layout design is

- Cutter can completely cover the bottom hole.
- The rate of wear of each cutter is as equal as possible, and its typical feature is that the density of cutters increases gradually from the center to the outer of the bit.
- Determine the appropriate cutter layout density (number) according to the property of the drilled formation and the drilling condition.

Fig. 2.34 Cutter layout of
the PDC bit in the radial
direction

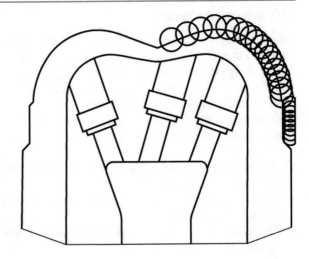

The bit has a longer working life with high tooth density, a large number of cutting teeth, small cutting amounts of each cutter, but a lower rate of penetration. Therefore, for soft formation, medium-deep well, etc., the density of the cutter layout should be lower. For harder, abrasive formation and deep well, the density of cutter layout should be higher.

The third step is the circumferential cutter design. Firstly, determine the number, shape of the blade, and the circumferential position of each blade, and then arrange cutters to each blade according to specific regulations (such as the distance between adjacent cutters in the same blade should be equal or no less than a certain number). Determine the circumferential position coordinate of each cutter, as shown in Fig. 2.35.

The general principle of circumferential cutter layout design is

- Adjacent cutters in the same blade do not cause interference from each other.
- The number of blades can meet the requirement of the number of cutter layout.
- The cutter distributions are balanced so that the bit forces are equilibrium.
- There are plenty of channel spaces between the blades.

The last step is the design of the cutter working angle. The working angle of the cutter refers to the back rake angle α and side rake angle β, as shown in Fig. 2.36. The back rake angle refers to the included angle of the cutter working face and the outer normal of the bottomhole floor. The side rake angle is the included angle of the cutter working face and the bit radial line across the center of the cutter surface.

The critical role of the back rake angle is to make the working blade of the cutter penetrate the rock effectively and has a better performance of impact resistance. The back rake angle is one of the key design parameters of the PDC bit, which has a significant impact on bit performance. Both the laboratory experiments and field practical experience show that the reasonable value range of the back rake angle of

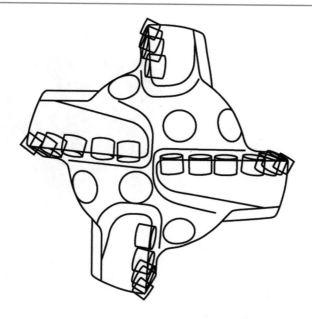

Fig. 2.35 Cutter layout of the PDC bit in the axial direction

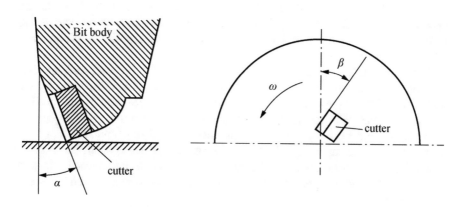

Fig. 2.36 Schematic diagram of back rake angle and side rake angle of PDC cutter

the PDC cutter is 10°–25°. When the back rake angle is smaller, the cutter is more capable of attacking the rock; when the back rake angle is larger, the ability of the cutter to attack the rock is weakened, but the impact resistance increases and not easy to fracture.

The key role of the side rake angle is to produce a lateral thrust force to the front of the cutter while the cutter is cutting the formation to improve the cutting

Fig. 2.37 Hydraulic
structure of PDC bit

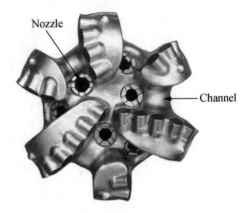

transportability and prevent the bit from balling. The side rake angle generally takes $0°–10°$. In the case of good hydraulic cleaning effect of the drill bit, the bit is not easy to ball up; if the side rake angle of the cutter does not have a significant positive impact on the bit performance, then the side rake angle can be zero.

The hydraulic structure design

The main function of the hydraulic structure of the PDC bit is to effectively remove cuttings from the bottom of the hole, cool the cutters, and prevent the bit balling.

The PDC bit generally adopts a hydraulic structure with combined nozzle and open channels, as shown in Fig. 2.37. The hydraulic structure design parameters include the width and depth of the channel, the number, size, position, azimuth, and angle of the nozzle.

The width of the fluid channel is determined by the bit size, and number, position, and width of the blades; the depth of the channel is determined by the height of the blades. Generally speaking, the number of blades should be as small as possible, and the width should be as narrow as possible under the condition that the cutter layout and intensity meet the requirements (what requirements). For hard formations, the drilling speed is low, and the cuttings amount is small; the channel should be shallow to ensure that the cutter can be completely exposed. For soft formations, the drilling speed is high, the cuttings amount is large, and the channel should be deeper.

The number, position, azimuth, and angle of the nozzle are designed according to the number of blades, location, and bit size. Generally, a nozzle is installed near the center of the drill bit in each channel, and the outlet direction of the nozzle is parallel to the PDC cutter working face and points to the two cutters between the inside of the blades. For larger size bit and soft formations, to enhance the hydraulic effect, reinforced hydraulic structures with two nozzles are installed in the long channel, and one nozzle installed in the short channel is usually being used.

Research and experience show that the influence of pump pressure and displacement on the ROP of roller cone bit and PDC bit is different. For the roller cone bit, the pump pressure has a significant influence on the ROP; for the PDC bit, the

Fig. 2.38 Gage protection structure of PDC bit in directional drilling

displacement has a more substantial influence on the ROP. Therefore, when the hydraulic parameter design of the PDC bit is conducted, large displacement should be a priority, and then the high pump pressure should be considered.

In recent years, with the rapid development of computational fluid dynamics (CFD), major bit manufacturers in the world have gradually begun to use CFD numerical simulation technology to optimize the design of PDC bit hydraulic structure. Through the numerical simulation analysis of the bottomhole flow field, the PDC bit nozzle combination and arrangement scheme are optimized, and reasonable hydraulic structure parameters are determined to make the hydraulic energy distribution of the bottomhole flow field more reasonable, to obtain the best hydraulic cleaning and cooling effect.

Gage design

The role of bit gage protection is to prevent the bit diameter from wearing down, resulting in hole shrinkage and wall trimming, and to form a smooth hole. To enhance the wear resistance of the gage protection part of the PDC bit, it is usually build up welding a cylindrical diamond compact or mount polycrystalline diamond compact. To use the PDC bit in the directional drilling, flatted PDC cutter (Fig. 2.38) at the shoulder of both ends of the gage protections is usually installed, to improve its ability of gage protection and side cutting.

The length of the gage protection of the PDC bit is generally 50–100 mm, which affects the stability and the guiding performance of the bit. The bit with long gage protection has good stability, and the borehole drilled is smooth, but it has poor guiding performance, and the maximum build-up rate obtained is small.

2.2.4.3 The Rock-Breaking Principle of PDC Bit

The PDC bit breaks the rock by cutting. Under the combined action of weight on bit and rotary torque, the bit rotates and advances at the bottom of the hole, continuously invades and shears to break the rock. It strips the rock layer by layer from the bottom of the hole, thereby deepens the wellbore continually.

When the PDC bit works at the bottom of the hole, the PDC cutter simultaneously applies a positive load perpendicular to the rock surface and a cutting force parallel to the rock surface, as shown in Fig. 2.39. Under the combined action of positive load and the shear force, the rock at the blade tip is firstly crushed due to the accumulation of stress. The cutter working blade penetrates the rock to a certain depth and then compresses the rock at cutting edges. With the increase of the contact force between the blade and rock, the rock before the blade is first crushed into powder and compressed into a dense core, and the load is transmitted to the surrounding rock through the dense core. When the maximum shear stress

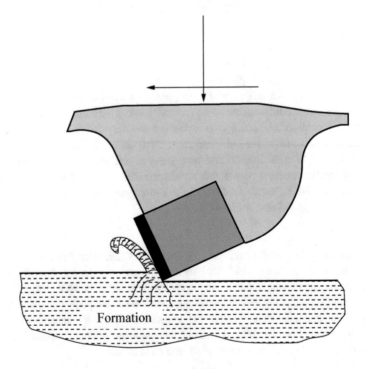

Fig. 2.39 Schematic diagram of rock breaking of PDC cutter

generated in the rock exceeds the shear strength of the rock, a shear fracture is generated and propagates rapidly to the free surface along a certain trajectory (approximate logarithmic spiral), resulting in a bulk shear fracture, and complete a cutting process of rock breaking.

If the rock properties are different, the characteristics of cutting rock-breaking process of PDC bit are different. For plastic rock, the PDC cutter penetrates the rock for a certain depth under the combined action of positive load and shear force to press the cutting edge rock, causing plastic damage, and the cuttings continue to shear and slip out to the free surface. The cutting process is continuous, smooth, less fluctuation of cutting force, similar to the process of cutting metal with a turning tool. For brittle rock, the breaking process is composed of three stages cycle, namely impact crush, forming dense core, and shear collapse, and the cutting process is leaping, very unstable, and large fluctuation of cutting force. The worse the homogeneity of the formation, the faster the rotary speed of the bit is, the larger the fluctuation of cutting force is. When the peak of the cutting force exceeds the impact strength of the PDC cutter, the cutter will collapse.

Compared with the roller cone bit, the rock-breaking process of the PDC bit has the following characteristics:

- The primary rock-breaking method of PDC bit is shearing rather than crushing, taking advantage of the low shear strength of the rock.
- If a PDC cutter is subjected to both positive and tangential loads at the same time, it is easier to penetrate rock than the case of a pure positive load.
- In the drilling process, the PDC cutter is always in contact with the bottomhole rock, the rock-breaking process is continuous, and the energy utilization rate is high.

Therefore, the rock-breaking efficiency of the PDC bit is much higher than the roller cone bit in soft-medium-hard formation. But in hard formation, the rock-breaking efficiency is relatively low since PDC bits cannot impact the rock at the bottom of the well as roller cone bit do. When the PDC cutter cannot penetrate the rock effectively, the "slipping" phenomenon will happen.

2.2.4.4 The Selection and Reasonable Use of the PDC Bit

PDC bit classification

The PDC bit is an efficient one with a fixed tooth. Initially, the PDC bit is mainly used to drill the soft homogeneous formation.

With the improvement of PDC cutter performance and the modification of bit structure, the PDC bit has been able to drill medium-hard and hard interbedded formation.

At present, the PDC bit has replaced the roller cone bit as the main rock-breaking tool in soft to the medium-hard formation.

There are many PDC bit manufacturers from domestic and foreign, which provide users with various structures and types of PDC bits. To help users select the bit type according to the drilled formation correctly, the IADC revised the *Fixed Cutter Bit Classification Standard* developed in 1987, and proposed a new classification system at the IADC/SPE Drilling Conference in 1992.

The classification and numbering method of the new IADC fixed cutter bit (including diamond bit) is used to number each kind of drill bit by four codes, and the meanings of each code are as follows.

- The first code is the drill bit-type code, using capital letters *M*, *S*, and *D* to represent the material of bit body and bit type.

  M—Matrix PDC bit;
  S—Steel PDC bit;
  D—Diamond bit PDC bit.

- The second code is the level code of cutter density and applicable formation.

Numbered 1 to 4, respectively, represents the applicable formation level and the cutter layout density of the PDC bit.

  1—Extra-soft formation, the cutter layout density coefficient <3.7;
  2—Soft formation, the cutter layout density coefficient = 3.7–5.0;
  3—Soft to medium formation, the cutter layout density coefficient = 5.0–6.2;
  4—Medium formation, the cutter layout density coefficient >6.2.

The density of a cutter layout is expressed by the density coefficient of the cutter layout. The larger the density coefficient of cutter layout, the more the number of cutter layout of bit, the harder the formation to be drilled. The density coefficient of the cutter layout with different sizes of cutter and bits is calculated by the following formula:

$$K = \frac{\sum_{i=1}^{n} d_i n_i}{R_b} \tag{2.3}$$

where

K Cutter layout density coefficient;
d_i Diameter of a certain size cutter, mm;
n_i Number of a certain size cutter;
R_b Bit radius, mm.

Number 6 to 8 represents the applicable formation level and density of cutter layout of diamond bit:

Table 2.13 Size and type code of the cutter

Code	PDC cutter size (mm)	Diamond (bit) type
1	24	Natural diamond
2	19	TSP
3	13.3	mixed
4	8	Impregnated bit

6—Medium-hard formation, the diamond grain size <3 grains/ct;
7—Hard formation, the diamond grain size = 3–7 grains/ct;
8—Extra-hard formation, the diamond grain size >8 grains/ct.

The diamond grain size (grains/ct) represents the density of the cutter layout of the diamond bit. The larger the value, the smaller the diamond size is; the more the diamond layouts, the harder the formation to be drilled.

- The third code is the size and type code of the cutter; see Table 2.13. For PDC bit, use the number 1–4 to represent the size of PDC cutter (diameter); for a diamond bit, use the number 1–4 to represent the type of diamond (bit).
- The fourth code is the profile shape code of the crown of the drill bit, using the number 1–4 to represent:

    1—Fishtail shape (flatter shape);
    2—Short cone shape, the height of crown <58 mm;
    3—Medium cone shape, the height of crown = 58–114.3 mm;
    4—Long cone shape, the height of crown >114.3 mm.

The reasonable use of PDC bit

- Trip-in
    - Before the bit trip-in, the bottom of the hole should be clean and free of metal junks.
    - The thread of bit should be coated with standard lubrication grease, and the bit breaker should be used to remove the bit.
    - The bit should be upright and slowly dropped when trip-in, to prevent damage to the PDC cutters; when passing the choke ring position, the speed should be decreased and pay attention to the collision.
    - When the tripping-in is being blocked and reaming, the weight on bit should be controlled below 20 kN (less than 10 kN when the bit diameter is less than 215.9 mm), the rotary speed should be lower than 50 RPM. Extensive segment drilling and non-circulating drilling fluid drilling are strictly prohibited.
    - When the drill bit is close to the bottom hole, the pump displacement will be maximized to clean the bottom of the hole fully; then, start the rotary table, control the speed at 30–60 RPM, and drill slowly until the bit touches the bottom hole.

- Bottomhole surface build-up

After confirming that the bit is in contact with the bottomhole rock surface, the bottom hole is drilled with a small working weight on bit (5–20 kN), low rotary speed (50 RPM) to a bit footage of 0.5–1.0 m, so that the shape of the bottom hole completely matches the shape of the bit surface, which is called the bottomhole surface build-up.

- Drilling parameters
  - Weight on bit (WOB). One of the main characteristics of the PDC bit is that high rotary speed can be achieved with a low WOB, and its WOB is generally about 30% of other roller cone bit with the same size. For soft formation, the WOB is relatively small; for harder formation, the WOB is relatively large. The optimal WOB value should be when the WOB reaches this value, the ROP of the bit no longer increases with the increase of the WOB, or the bit torque has already reached the maximum torque value.
  - Rotary speed. The PDC bit is a fixed tooth bit, with no bearing and other moving parts, so the rotary speed has no limit theoretically. In the soft-medium-hard homogeneous formation, the higher the rotary speed, the higher the drilling speed is. The compound drilling with PDC bit plus screw drill can significantly improve the drilling speed. However, in the harder and more abrasive formation and hard and soft interbedded, gravel-rich hetero- geneous formation, high bit rotation speed can easily lead to the cutter wear out prematurely.
  - Hydraulic parameter. The ROP of the PDC bit is fast, and the amounts of cuttings produced in unit time are large. The hydraulic energy at the bottom of the hole must ensure timely cuttings lifting and sufficient cooling of the PDC cutter. Therefore, under the permitted pump conditions, it is necessary to drill as large flow rate as possible to achieve the best bottomhole cleaning effect and bit cooling effect.

2.3 Drill String

The drill string is the general name of the tools above the drill bit and below the hook. It mainly includes kelly, drill pipe, drill collar, adapter subs and stabilizer, and other downhole tools.

The drill string is an essential tool for drilling, and it is the hub connecting surface and downhole tools and types of equipment. In rotary drilling, it is used to transfer the energy required to break rocks, to apply the drilling weight on bit to the bottom of the well, as well as to circulate drilling fluid. In the process of downhole motor-driven drilling, the downhole motor is delivered to the bottom of the hole by drill string to withstand the reverse torque. Meanwhile, the liquid energy required by the drill bit and drilling tools is also transported to the bottom of the well through the drill string.

With the increase of drilling depth, the requirements for the performance of the drill string are becoming more critical. The working conditions of the drill string with thousands of meters in the downhole are terrible, which is the weak link in the drilling equipment and tools. Drilling tools break and stickings are common drilling accidents and often lead to complicated downhole conditions. Therefore, according to the working conditions and technical requirements of the drill string in the downhole, the reasonable design and use of the drill string are of great significance for preventing drill string accidents, realizing fast and high-quality drilling, and successfully completing various downhole operations.

2.3.1 Function and Composition of Drill Sting

2.3.1.1 Functions of Drill Sting

Main functions of the drill String in the drilling process

- Providing a channel for drilling fluid from the wellhead to the drill bit;
- Applying drilling load required to break rock on the bit;
- Transferring the surface power (torque, etc.) to the bit and driving the bit to break rock;
- Tripping-out and in;
- Determining well depth according to the length of drill string;

Functions of drill sting

- Observing the working condition of the bit, wellbore condition, and the formation situation;
- Special operations such as coring, leakage test of formation, cement squeeze, handling of downhole accidents, and so on;
- In the course of drilling, testing, and evaluating the formation fluid and pressure condition, which is called drill stem test (DST).

2.3.1.2 Composition of Drill Sting

The drill string consists of kelly, drill string, BHA, and fit joints, as shown in Fig. 2.40. BHA is mainly composed of drill collar and, sometimes, contains stabilizer, absorber, bumper jar, hole opener, and other special tools. In directional and horizontal wells, some or all drill collars are replaced by heavyweight drill pipes and bent shell screw drilling tools, MWD, LWD, or rotary steering drilling tools installed above the bit.

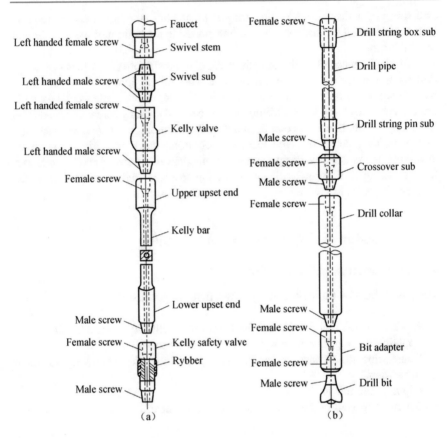

Fig. 2.40 Makeup of string (all connections between the thick end and the bit are right-handed)

Drill pipe

The drill pipe is a seamless steel pipe with a joint at both ends. The length is about 9.45 m (31 ft.), and the thickness of the wall is generally about 10 mm. Drill pipe is an essential component of the drill string. Its primary role is to bear tension load, transfer torque, and drilling fluid and to deepen the hole by the gradual lengthening of drill pipe.

• Structure and specification of drill pipe

The drill pipe is composed of pipe body and joints, which is connected by the butt welding method. The tool joint is divided into a tool joint pin and tool joint box, which are welded to both ends of the drill pipe body, respectively. During drilling, joints should be removed frequently, and the surface of joints is greatly affected by the force of tongs. Therefore, the thickness of drill pipe joints is large, the outer

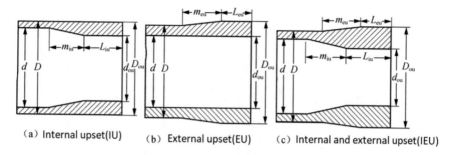

Fig. 2.41 Drill string thickening type

diameter of joints is larger than the outer diameter of pipes, and the alloy steel with higher strength is adopted, such as 35 C'rM alloy steel. In order to enhance the connection strength of the pipe body, both ends of it can be upset. There are three kinds of upset forms including internal upset, external upset, and internal/external upset, as shown in Fig. 2.41.

At present, the API standard drill pipe is used in the oil industry. The type and structural parameters of the new drill pipes are shown in Table 2.14.

The API standard drill pipe joint is a rotary step-type joint. The step is its only seal area, and the thread only serves as the connection, as shown in Fig. 2.42. According to the different connection thread, drill pipe joints can be divided into internal flush tool joint (IF), full hole tool joint (FH), regular tool joint (REG), and NC style tool joint (NC). The four kinds of drill pipe joints all use V-shaped thread, but the thread profile, thread pitch, and angle of taper are different. Due to the fillet transition at the root of NC style connection thread, the stress state of the thread has been significantly improved, and the strength of the joint has been improved too. Therefore, the API standard drill pipes adopt the NC style tool joint.

The NC style tool joint is represented by the letter NC and the first two digits of the intermediate diameter of the threaded base surface (the unit is in). For example, NC26 represents a NC style tool joint with an intermediate diameter of 2.668 of the threaded base surface, and NC38 represents a NC style tool joint with an intermediate diameter of 3.808 of the threaded base surface. The NC style connection thread has a flat top of 0.065 in. and round thread bottom of 0.038 inches. V-0. 038R is used to represent the thread type. It can be connected with the V-0.065 thread (the trimmed width of thread top is 0.065 in.). All API standard internal flush (IF) tool joints and 4-in. full hole (4FH) tool joints are V-0.065 thread, so that they can be used interchangeably with the NC style tool joints.

There still are various fit joints (used to connect different sizes or different types of string), saver sub (protecting the thread at the regular removal location, such as saver sub of kelly) except drill adapter among the drill strings. Also, kelly, drill collar, drill bit, and other downhole tools are connected by the thread. What's more, all kinds of joints and thread types mentioned above are consistent with the standard of drill pipe sub.

Table 2.14 API new drill pipe type and structural parameters (API RP 7G)

| Drill collar | | | | | Drill adapter | | | | | |
Basic size [mm(in)]	Nominal (weight) (N·m⁻¹) (lb·ft⁻¹)	Internal diameter (mm)	Upset type	Steel grade	Joint type	External diameter of joint (mm)	Internal diameter of joint (mm)	Pin/box sub length (mm)	Make-up torque (kN·m)	Ratio of box sub to torsion strength
60.3 (2⁷/₈)	97.12 (6.65)	46.10	EU	E-75	NC26 (2³/₈IF)	85.7	44.5	381	4.39	1.11
				X-95						0.87
				G-105						0.79
73 (2⁷/₈)	151.86 (10.40)	54.6	EU	E-75	NC31 (2⁷/₈IF)	104.8	54.0	406.4	8.05	1.03
				X-95			50.8		10.00	0.90
				G-105						0.82
				S-135		111.1	41.3		10.48	0.82
88.9 (3¹/₂)	194.14 (13.30)	70.20	EU	E-75	NC38 (3¹/₂IF)	120.7	68.3	469.9	12.28	0.98
				X-95		127.0	65.1		13.78	0.87
				G-105			61.9		15.06	0.86
				S-135			54.0		17.64	0.80
	226.22 (15.50)	66.10		E-75	NC38 (3¹/₂IF)	127.0	65.1		13.78	0.97
				X-95			61.9		15.06	0.83
				G-105			54.0		17.64	0.90
				S-135	NC40 (4FH)	139.7	57.2	431.8	22.33	0.87
101.6 (4)	204.34 (14.00)	84.40	IU	E-75	NC40 (4FH)	133.4	71.4	431.8	15.92	1.01
				X-95			68.3		17.40	0.86
				G-105		139.7	61.9		20.42	0.93
				S-135			50.8		24.65	0.87
			EU	E-75	NC46 (4IF)	152.4	82.6		22.80	1.43
				X-95						1.13
				G-105						1.02
				S-135			76.2		25.59	0.94

(continued)

Table 2.14 (continued)

Drill collar			Drill adapter							
Basic size [mm(in)]	Nominal (weight) (N·m⁻¹) (lb·ft⁻¹)	Internal diameter (mm)	Upset type	Steel grade	Joint type	External diameter of joint (mm)	Internal diameter of joint (mm)	Pin/box sub length (mm)	Make-up torque (kN·m)	Ratio of box sub to torsion strength
114.3(4½)	242.30 (16.60)	97.20	IEU	E-75	NC46 (4IF)	158.8	82.6	431.8	23.05	1.09
				X-95			76.2		26.88	1.01
				G-105						0.91
				S-135			69.9		30.42	0.81
	291.95 (20.00)	92.50		E-75			76.2		26.88	1.07
				X-95			69.9		30.42	0.96
				G-105			63.5		33.65	0.96
				S-135			57.2		36.56	0.81
114.3(4½)	242.30 (16.60)	97.20	EU	E-75	NC50 (4½IF)	161.9	95.3	431.8	25.54	1.23
				X-95						0.97
				G-105						0.88
				S-135			88.9		30.28	0.81
	291.95 (20.00)	92.50		E-75			92.1		27.95	1.02
				X-95			88.9		30.28	0.96
				G-105						0.86
				S-135		168.3	76.2		37.77	0.87
127(5)	284.78 (19.50)	108.6	IEU	E-75	NC50 (4½IF)	161.9	95.3	431.8	25.54	0.92
				X-95		165.1	88.9		30.28	0.86
				G-105		168.3	82.6		34.88	0.89
				S-135			69.9		42.29	0.86
	372.40 (25.60)	101.6		E-75		161.9	88.9		30.28	0.86
				X-95		165.1	76.2		37.77	0.86
				G-105		168.3	69.9		42.29	0.87
	284.78 (19.50)	108.6		E-75	5½FH	177.8	95.3	457.2	41.59	1.53
				X-95						1.21

(continued)

Table 2.14 (continued)

Drill collar				Drill adapter					Make-up torque (kN·m)	Ratio of box sub to torsion strength
Basic size [mm(in)]	Nominal (weight) (N·m⁻¹) (lb·ft⁻¹)	Internal diameter (mm)	Upset type	Steel grade	Joint type	External diameter of joint (mm)	Internal diameter of joint (mm)	Pin/box sub length (mm)		
				G-105						1.09
				S-135		184.2	88.9		49.14	0.98
	372.40 (25.60)	101.6		E-75		177.8	88.9		41.59	1.21
				X-95						0.95
				G-105		184.2	88.9		49.14	0.99
				S-135			82.6		52.30	0.83
139.7(5½)	319.71 (21.91)	121.40	IEU	E-75	5½FH	177.8	101.6	457.2	37.92	1.11
				X-95			95.3		41.59	0.98
				G-105		184.2	88.9		49.14	1.02
				S-135		190.5	76.2		59.09	0.96
				E-75		177.8	101.6		37.92	0.99
	360.52 (24.70)	118.60		X-95		184.2	88.9		49.14	1.01
				G-105						0.92
				S-135		190.5	76.2		59.09	0.86

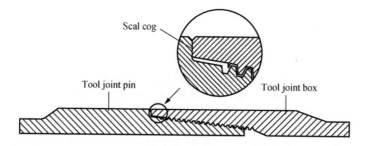

Fig. 2.42 Rotating step-type joint structure and connection form

- Steel grade and strength of drill pipe

The steel grade of the drill pipe refers to the grade of steel making the drill pipe, which is determined by the minimum yield strength of the steel of drill pipe. The steel grades of drill pipe which are commonly used include E-75, X-95, U-105, and 5-135. The higher the steel grade of the drill pipe, the greater the yield strength of the pipe is and the higher the strength of the drill pipe is (including tension strength, torsion strength, external pressure resistance, and internal pressure resistance). The strength characteristics of various drill pipes are specified in the API RP 7U standard. Table 2.15 lists the strength data of the new API drill pipe. In the strength design of the drill string, the method of improving the steel grade is generally recommended to improve the bearing capacity of the drill string, instead of increasing the thickness of the wall.

- Makeup torque of drill pipe sub

The makeup torque is a torque that makes the thread fasten. If the thread is not tight enough, it will continue to fasten during drilling process. Accidents will occur like joint burst, thread off, and so on. If the thread is too tight, it will damage the threads and even lead to breakage. Therefore, torque reasonability directly affects the reliability and safety of the thread connection. The API KP 7U standard specifies the makeup torque for various sizes and levels. The recommended value of the makeup torque for the new drill pipe joint is shown in Table 2.14. The minimum torque shall not be less than 90% of the recommended value, but the maximum should not exceed the recommended value.

Drill collar

The drill collar is located at the bottom of the drill string and is the main component of the lower part of drill string assembly. Compared with drill pipe, the main feature of drill collar is the large wall thickness (usually 38–53 mm, equivalent to 4–6 times of wall thickness of drill pipe). It has large weight and rigidity, and it can bear large compressive stress without bending. The main roles of drill collar in the drilling process are as follows:

- Apply WoB;
- Ensure the necessary strength under the compression condition;
- Reduce the vibration, swing, and bit bounce and make the bit work smoothly;
- Control well deviation.

Drill collar can be divided into slick drill collar, fluted drill collar, kelly drill collar drill, and so on according to its shape. The most commonly used drill collars are slick and fluted drill collar. The fluted drill collar is made of three rotate-right helical channels on the cylindrical surface of a slick drill collar. Its contact area with the borehole wall can be reduced by 40–50%, and its gravity will only decrease by 7–10%. The small contact area can reduce the possibility of differential pressure pipe sticking. The connecting threads (box and pin) of drill collar are directly chased at both ends of the drill pipe without extra joints. There are many specifications for the drill collar. API standard drill collars are shown in Table 2.16. The drill collar-type code in the table is made up of two parts. The first part is the NC-type thread code, and the second part is the number (the first two numbers of the outer diameter within. unit), which indicates the outside diameter of the drill collar. The middle part of the type code is separated by a short line.

Heavyweight drilling pipe (HWDP)

The wall thickness of HWDP is 2–3 times larger than that of the ordinary drill pipe, but it is thinner than that of the drill collar. The joint is longer than the regular drill pipe joint. There is thickening special grinding roller in the middle of the pipe, as shown in Fig. 2.43.

The HWDP stiffness is larger than that of the ordinary drill pipe which allows accepted load. But its stiffness, weight, and contact area with borehole wall are less than that of drilling pipe. It can reduce the friction and the possibility of differential pressure sticking. Therefore, the main functions are in the following aspects in oil and gas drilling:

- Be used to ease the sudden change of cross section and stiffness in the transition zone between drilling pipe and collar, to reduce the fatigue failure of the drill string.
- In the deep well-drilling process, a part of the drill collar is replaced to reduce the torque and lift load.
- In controlled directional well, especially in high angle deviated well and horizontal well, most or all parts of the drill collar are replaced to reduce the friction and the risk of differential pressure pipe sticking.

Table 2.15 New API drill pipe strength data (API RP 7G)

External diameter of drillpipe		Nominal weight		Torque yield strength (kN·m)				Calculated anti-tension according to minimum yield strength (kN)				Minimum collapse strength (MPa)				Calculated internal pressure according to minimum yield strength (kN)			
Mm	in	N/m	lb/ft	E-75	X-95	G-105	S-135	E-75	X-95	G-105	S-135	E-75	X-95	G-105	S-135	E-75	X-95	G-105	S-135
60.3	2³/₈	97.12	6.65	8.46	0.71	11.85	15.25	615.02	779.06	861.02	1107.00	107.58	136.27	150.62	193.65	106.69	135.17	149.38	192.07
73.0	2⁷/₈	151.86	10.40	15.64	19.82	21.90	28.16	953.75	1208.10	1335.27	1716.79	113.86	144.20	159.38	204.96	114.00	144.34	159.58	205.17
88.9	3¹/₂	138.71	9.50	19.15				864.40				69.24				65.66			
		194.14	13.30	25.12	31.84	35.17	45.12	1208.41	1350.66	1691.73	2175.11	97.30	123.31	136.27	175.17	95.17	120.55	133.24	171.13
		226.22	15.50	28.55	36.16	39.97	51.39	1436.28	1819.27	1010.78	2585.29	115.65	146.55	161.93	208.20	116.14	147.10	162.55	290.03
101.6	4	172.95	11.85	26.36				1206.77				58.00				59.31			
		204.31	14.00	31.53	39.49	44.15	56.75	1269.77	1605.45	1777.66	2285.60	78.27	99.17	109.65	139.10	74.69	94.62	104.55	134.41
114.3	4¹/₂	200.71	13.75	15.07				1201.56				49.65				54.48			
		242.30	16.60	41.71	52.83	58.39	75.07	1470.90	1863.09	2058.24	2647.58	71.65	87.93	95.32	115.86	67.78	85.86	94.90	122.00
		291.95	20.00	49.97	63.28	69.94	89.92	1843.89	2324.18	2568.83	3302.76	89.38	113.24	125.17	160.89	86.48	109.58	121.10	155.72
127.0	5	284.78	19.50	55.73	70.59	78.03	100.32	1760.31	2229.17	2464.39	3168.51	68.96	82.83	89.58	108.27	65.52	83.03	91.72	118.00
		372.40	25.60	70.74	89.61	99.05	127.36	2358.97	2988.08	3302.58	4246.19	93.10	117.93	130.34	167.58	90.48	114.62	126.76	162.89
139.7	5¹/₂	319.71	21.91	68.66	86.97	96.12	123.58	1945.06	2463.72	2723.05	3501.08	58.21	68.96	74.04	87.85	59.38	75.24	83.17	106.96
		360.5	24.70	76.59	97.02	107.23	137.87	2212.49	2802.48	3097.49	3982.30	72.14	89.10	96.55	116.70	68.27	86.48	95.58	122.96

Table 2.16 API drilling collar standard (API SPEC 7)

Drill collar type	External diameter		Internal diameter		Length		Nominal weight		Make-up torque (kN·m)	
	mm	in	mm	in	m	ft	lb/ft	N/m	Minimum	Maximum
NC23-31	79.4	$3^1/_8$	31.8	$2^1/_4$	9.1	30	22	321	4.45	4.9
NC26-35($7/_8$IF)	88.9	$3^1/_2$	38.1	$1^1/_3$	9.1	30	27	394	6.25	6.9
NC31-41($2^7/_8$IF)	104.8	$4^1/_8$	50.8	2	9.1	30	35	511	9.00	9.9
NC35-47	120.7	$4^3/_4$	50.8	2	9.1	30	50	730	12.50	13.5
NC38-50($33^1/_2$IF)	127.0	5	57.2	$2^1/_4$	9.1	30	53	774	17.50	19.0
NC44-60	152.4	6	57.2	$2^1/_4$	9.1	30–31	83	1212	31.65	35.0
NC44-62	158.8	$6^1/_4$	57.2	$2^1/_4$	9.1–9.2	30–31	91	1328	31.50	35.0
NC44-62(4IF)	158.8	$6^1/_4$	71.4	$2^{13}/_{16}$	9.1–9.2	30–31	83	1212	30.00	33.0
NC46-65(4IF)	165.1	$6^1/_2$	57.2	2	9.1–9.2	30–31	99	1445	38.00	42.0
NC46-65(4IF)	165.1	$6^1/_2$	71.4	$2^{13}/_{16}$	9.1–9.2	30–31	91	1328	30.00	33.0
NC46-67(4IF)	171.5	$6^3/_4$	57.2	$2^1/_4$	9.1–9.2	30–31	108	1577	38.00	42.0
NC50-70($4^1/_2$IF)	177.8	7	57.2	$2^1/_4$	9.1–9.2	30–31	117	1708	51.50	56.5
NC50-70($4^1/_2$IF)	177.8	7	71.4	$2^{13}/_{16}$	9.1–9.2	30–31	110	1606	43.50	48.6
NC50-72($4^1/_2$IF)	184.2	$7^1/_4$	71.4	$2^{13}/_{16}$	9.1–9.2	30–31	119	1737	43.50	48.0
NC56-77	196.9	$7^3/_4$	71.4	$2^{13}/_{16}$	9.1–9.2	30–31	139	2029	65.00	71.5
NC56-80	203.2	8	71.4	$2^{13}/_{16}$	9.1–9.2	30–31	150	219	65.00	71.5
$6^5/_8$REG	209.6	8	71.4	$2^{13}/_{16}$	9.1–9.2	30–31	160	2336	72.00	79.0
NC61-90	228.6	9	71.4	$2^{13}/_{16}$	9.1–9.2	30–31	195	2847	92.00	101.0
$7^5/_8$REG	241.3	$9^1/_2$	76.2	3	9.1–9.2	30–31	216	3153	119.50	–
NC70-100	254.0	10	76.2	3	9.1–9.2	30–31	243	3548	142.50	156.5
NC70-110	279.40	11	76.2	3	9.1–9.2	30–31	299	4365	194.00	214.5

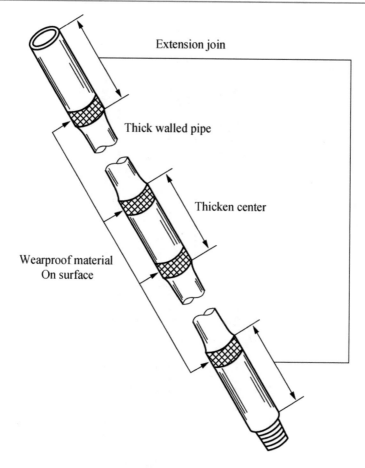

Fig. 2.43 Heavyweight drilling pipe (HWDP)

Kelly

The kelly is located at the top of the drill string with the shape of quadrate and hexagon. During the drilling process, kelly with master bushing and rotary kelly bushing can transmit the surface rotary table torque to the drill pipe. The standard kelly is 12.19 m in length, and the drive part with the rotary table is 11.25 m. To meet the needs of drill string coordination, the kelly has a variety of sizes and types of joints. Standard kelly specification is shown in Table 2.17. The thickness of the kelly is generally three times larger than that of the drill pipe. It is made of high-strength alloy steel. Therefore, it has greater tensile yield strength and torsional yield strength (Table 2.18). It can bear the weight of the whole drill string and the torque needed to rotate the drill string and drill bit.

Table 2.17 Normal kelly specifications

Kelly type		Drive part length		Full length l		Upper female connection					Lower male connection			Internal diameter	Drive part					Mass
		Standard	Choice	Standard	Choice	Specifications and types LH		External diameter D_v		Length L_v	Specifications and types RH	External diameter D_v	Length L_L	d	Across corner D_C	Across corner D_{CC}	Square size D_{FL}	Edge radius R_C	Edge radius R_{CC}	
						Standard	Choice	Standard	Choice											
mm	in	mm	mm	mm	mm	in	in	mm	mm	mm	in	mm	mm	mm	mm	mm	mm	mm	mm	kg
63.5	1	11,250	–	12,190	–	RRG	RRG	196.8	146.0	420	$2\frac{3}{8}$IF	85.7	520	31.8	83.3	82.6	63.5	7.9	41.3	402
76.2	2	11,250	–	12,190	–	RRG	RRG	196.8	146.0	420	$2\frac{7}{8}$IF	104.8	520	44.4	100.0	98.4	76.2	9.5	49.2	500
88.9	3	11,250	–	12,190	–	RRG	RRG	196.8	146.0	420	$3\frac{1}{2}$IF	120.6	520	57.2	115.1	112.7	88.9	12.7	56.4	597
108.0	4	11,250	15,500	12,190	16,460	RRG	RRG	196.8	146.0	420	4IF	152.4	520	71.4	141.3	130.7	108.0	12.7	6.8	829
											$4\frac{1}{2}$IF	155.6		69.8						
133.4	5	11,250	15,500	12,190	16,460	RRG	–	196.8	–	420	$5\frac{1}{2}$FH	177.8	520	82.6	175.4	171.4	133.4	15.9	85.7	1255

Table 2.18 Square kelly strength (API RP 7G)

Kelly size			Lower screw thread		Minimum external diameter recommended		Tensile yield strength (kN)		Torsional yield strength (kN)		Bending strength (kN)	
mm	in		Type	External diameter (mm)	mm	in	Upper male sub buckle	Drive part	Lower male sub buckle	Drive part	Opposite angel of drive part	Opposite site of drive part
63.5	$2^1/_2$		NC26	85.70	114.30	$4^1/_2$	1850	2420	13.10	20.60	20.45	30.00
76.2	3		NC31	104.80	130.70	$5^1/_2$	2380	3170	19.60	32.60	30.10	49.35
88.9	$3^1/_2$		NC38	120.70	168.30	$6^5/_8$	3220	3940	30.80	48.00	48.95	75.00
108.0	$4^1/_4$		NC46	158.80	219.10	$8^5/_8$	4680	5820	53.30	83.50	85.40	131.90
108.0	$4^1/_4$		NC50	161.90	219.10	$8^5/_8$	6320	5700	77.60	85.30	87.30	133.70
133.4	$5^1/_4$		$5^1/_2$FH	177.80	244.50	$9^5/_8$	7150	9250	99.00	167.50	170.40	257.80

When the kelly is rotated, the upper end is always above the rotary table and the lower end is below the rotary table. The connection thread between the upper end of kelly and faucet is the left-hand thread to prevent the rotary kelly screwing off. The connection thread between the lower end of kelly and drill bit is right-hand thread. When kelly drives the rotating drill string, the thread gets tighter and tighter. To reduce the abrasion of the lower part of the kelly joint thread (often removing parts), the protective joint is installed.

Other drilling tools

- Stabilizer

Stabilizer, also known as a centralizer, is a tool for stabilizing drilling tools and controlling wellbore trajectory. It is an important component of the lower part of the drilling assembly. During straight hole drilling, a certain number of stabilizers properly positioned at the bottom of drill pipe can be regarded as packed-hole assembly and dropping assembly. It can prevent and correct well deviation. When drilling a directional well, a stabilizer can be used to change the stress state of the lower part of BHA, to control the well trajectory. In addition, stabilizers can effectively limit the lateral heave of drill bits, improve the stability of bits, and prolong the service life of bits, which is especially crucial for diamond bits.

The stabilizer can be divided into rotating blade type, non-rotating rubber sleeve type, and roller type, as shown in Fig. 2.44.

The rotating blade stabilizer can be divided into the spiral blade type and the straight blade type according to the shape of the blade. It also has a long type and short type in order to adapt to all kinds of formations and technique requirements. It is the most widely used stabilizer.

The main features of the non-rotating rubber sleeve stabilizer are that the rubber sleeve does not rotate when it is in contact with the borehole wall. Meanwhile, it will not cause damage to the borehole wall, and the friction torque is small. But it cannot repair the borehole wall. The rubber sleeve cannot bear high temperature and its service life is low, so the application scope is narrow.

The main advantage of roller stabilizer (also known as gear wheel reaming) is that it has a strong ability to repair the borehole wall, and it can keep the borehole shape, which is mainly used in abrasive formation.

- Shock sub

The shock sub is a tool to dampen the vibration of the downhole drilling equipment. During drilling, downhole drilling equipment will produce longitudinal and lateral vibration, especially when drilling in hard formations, fractured formations, and soft-hard interlaced formations. The vibration of downhole drilling equipment will bring a lot of harm to the drilling work, such as the previous damage of the drill bit and the fatigue failure of the drilling tools. The use of shock sub can significantly reduce the vibration amplitude and impact load on drilling tools, improve the

				Non-rotating rubber sleeve	Roller type
Rotating blade					
Spiral blade		Straight blade			
Short type	Long type	Short type	Long type		

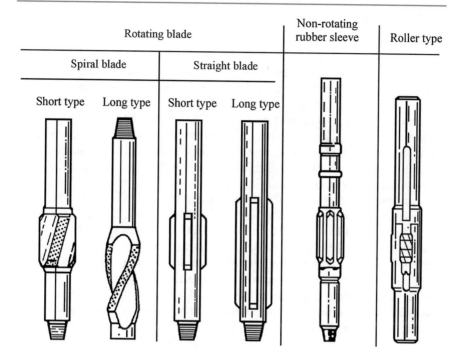

Fig. 2.44 Basic types of stabilizer

working life of downhole equipment (including bits), and decrease the accident of drilling tools.

Shock subs can be divided into one way (can only absorb a direction of vibration, such as longitudinal vibration) and bidirectional (can absorb two directions of vibration at the same time) according to the function; in accordance with shock absorber, they can be divided into spring type, hydraulic type, and rubber type; according to the applicable working temperature, it can be divided into ordinary type (less than or equal to 120 °C) and high-temperature type (120–180 ° C). The spring shock absorber is simple in structure, insensitive to downhole temperature, and able to absorb longitudinal and torsional vibration at the same time.

In principle, the shock sub should be as close as possible to the bit to enhance the absorption effect and protect the bit and drill string. For different makeup of string, the shock sub can be placed in the following principles:

- For the slick assembly or pendulum assembly, the shock sub is installed between drill collar and drill bit.
- For the rigid packed-hole assembly, the shock sub is usually installed on the stabilizer.
- When the mud motor is used, the shock sub is installed on the mud motor.

- Drilling jar

The drilling jar is a downhole connected to the drilling tool employed to release a downhole pipe sticking. In deep or offshore, especially in directional drilling, the drilling jar is often installed in a certain position of the drill string. Once the lower drilling tool is stuck, the jar can be manipulated to produce upward or downward vibration and impact, to achieve the purpose of releasing the stuck pips.

There are two kinds of jars, hydraulic type and mechanical type, and the latter one is more popular. The working principle is as follows: When it is necessary to hit the drill string up quickly, the elongation of the drill string accumulates a lot of energy. When the tension exceeds the safety locking force of the locking mechanism and the lock is released, the elastic force of the drill string makes the shock head impact and produces a strong impact vibration. When it is necessary to lower the drill string, quickly lower the drill string and use the gravity of the upper drill string to unlock the locking mechanism. At this time, the accumulated energy of the upper drill string during the compression process is suddenly released, and the shock head will produce a downward shock effect.

The ideal location to install a drilling jar is at one–two single pipes above possible stuck point, usually at the neutral point of the drill string. During normal drilling, the jar should be in tension state as far as possible, and a short time compress state can also be applied, but the force should be less than the adjusted down-break unlocking force.

- Downhole motor

The downhole motor is a special tool installed at the bottom of the well to drive the drill bit directly to break the rock, also known as mud motor. Compared with the rotary table drilling, the mud motor has a higher rotation speed which can significantly improve the speed of mechanical drilling, control wellbore trajectory easily, and reach the target quickly and accurately. It is especially widely used in directional drilling.

According to the operating principle, the down motors can be divided into positive displacement type, turbine type, and electric type. Electrically driven downhole motors, due to their complex structure, have not yet passed the technical standards and have not been widely used. At present, screw motors and turbine motors are mainly used in petroleum drilling, especially screw ones.

Positive displacement motor (PDM) is a kind of mud motor, which utilizes the drilling fluid as the power and converts the liquid energy into the mechanical energy. The positive displacement motor is composed of four parts: the bypass valve, the screw motor, the universal shaft, and the transmission shaft (Fig. 2.45). Its core is the screw motor. The screw motor consists of two parts of the rotor and the stator, which are divided into two types: single-lobe motor and multi-lobe motor. The rotor of the motor is a solid shaft twisted into a spiral shape. The upper end of the rotor is free, and the lower end is connected with the universal shaft. The cross section of the single-lobe motor shaft is round, and the cross section of the

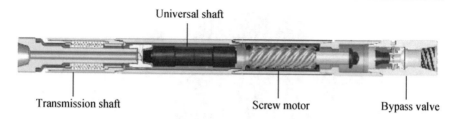

Fig. 2.45 Diagram of positive displacement motor

multi-lobe motor shaft is the shape of a plum blossom. The number of plum blossom determines the number of motor lobes. The stator consists of a molded rubber bushing and a motor housing. The cross section of the rubber bushing for the single-lobe stator is oblong (semicircle at both ends and oblong in the middle). The cross section of the rubber bushing of the multi-lobe motor stator is also quincunx-shaped, but it has one more quincunx lobe than the matching rotor. When the rotor and stator are fitted together, the geometric difference between the two forms a cavity. When circulating drilling fluid, the high-pressure drilling fluid flows through the cavity between the rotor and stator, pushing the rotor to rotate, and passing the speed and torque to the drill bit through the universal shaft and transmission shaft, driving the drill bit to rotate and break the rock.

The rotational speed of the positive displacement motor is directly proportional to the displacement of the drilling fluid, but has nothing to do with the torque. The speed can be adjusted by the pump rate. The output torque is proportional to the pressure drop of the drilling fluid through the screw motor. When the pump rate is fixed, torque increases with the increment of WOB. Meanwhile, the pressure drop of positive displacement motor increases too, but the speed remains unchanged.

A turbo drill is a downhole power machine that converts pressure and kinetic energy of drilling fluid into rotational mechanical energy. The main components of the turbine are hundreds of stages, each stage of which is composed of a stator and a rotor (Fig. 2.46). The stator is fixed on the housing as well as the rotor fixed on the spindle. The stator and rotor blades are basically the same shape, but bend in the opposite direction. When drilling fluid circulates from the stator blade to the rotor blade, the direction and speed of the fluid flow change, which generates thrust on the rotor blade and makes the rotor rotate, thus driving the spindle and the drill bit to turn. Compared with the screw drill, the turbo drill has a higher speed up to 400–800 rpm.

2.3.2 Working State and Force Analysis of the Drill String

In the process of drilling, the drill string is mainly working in normal drilling and tripping conditions. Sometimes it also used to DST, cement squeeze and downhole accident treatment, and other special operations. Under different drilling operation

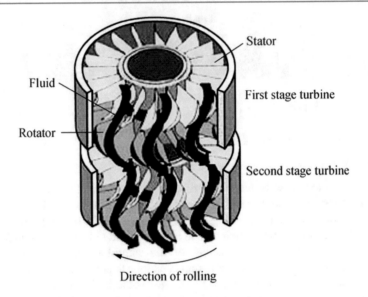

Stator

Fluid

First stage turbine

Rotator

Second stage turbine

Direction of rolling

Fig. 2.46 Diagram of turbine drill

conditions and different drilling methods (rotary table drilling and downhole motor drilling), the drill string has different working conditions and bears various external loadings. To design and use the drill string correctly, it is necessary to understand the working state and stress condition of the drill string during the whole drilling process.

2.3.2.1 Working State and Force Analysis During Tripping Process

The drill string is either suspended from the wellhead or suspended from the hook during the trip and does not contact the bottom of the well. At this point, the drill string is mainly affected by its gravity, in the tensile state.

When there is no drilling fluid in the well, the drill string is only affected by the axial tension caused by its self-weight. Figure 2.47 shows the distribution of axial tension of the drill string in a straight well. The tensile force at the wellhead is the largest, and the tension force at the bottom is zero. The tension at any section of drill string can be calculated by following equation:

$$F_0 = q_p L_p + q_c L_c \tag{2.4}$$

where,

q_p, q_c Nominal weight of a drill pipe and drill collar in the air, kN/m;
L_p, L_c Length of the drill string and collar, m.

Fig. 2.47 Distribution of axial tension of the drill string in a straight well

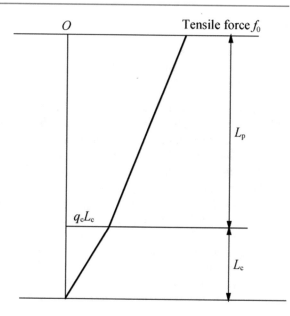

In general, the well is filled up with drilling fluid, and the drill string immersed in the drilling fluid is subject to the buoyancy and the lateral extrusion of the drilling fluid. The combined effect of buoyancy and lateral extrusion is equivalent to the decrease of unit weight of the drill string. The decrease degree can be expressed by the buoyant coefficient K_B.

$$q_m = K_B q \tag{2.5}$$

$$K_B = 1 - \frac{\rho_d}{\rho_s} \tag{2.6}$$

where,

q_m The unit weight of a drill string in a drilling fluid, known as a line floating weight, kN/m;

q The unit weight of a drill string in the air, kN/m;

K_B Buoyant coefficient;

ρ_d Well-drilling fluid density, g/cm³;

ρ_s Steel density of drill string, g/cm³.

Therefore, in the vertical well with drilling fluid, the axial tension on the cross section of the drill hole can be calculated by

$$F_m = K_B \left(q_p L_P + q_c L_c \right) \tag{2.7}$$

where,

F_m The axial tension on the cross section of a drill hole when a well is filled with drilling fluid, which is equal to the gravity of the drill string below the cross section in the drilling fluid, kN.

The gravity of the drill string in the drilling fluid is called the buoyant weight. It is known by Eq. (2.7) that the buoyant weight of the drill string is equal to the gravity of the drill string in the air multiplied by the buoyancy factor. The method of calculating the axial tension of the drill string is called the method of buoyancy factor. In addition, during drill string tripping, the drill string is also under the inertia force action caused by the speed change due to lifting or lowering (called dynamic load) and the frictional force generated by the wall and drilling fluid on the drill string. Especially during the process of quickly lifting and stopping, the axial tension applied to drill string is greatly increased. During the tripping process in the use of the slip, the wellhead part of the drill string by hoop force caused by slip is in biaxial stress state including tension and collapse. The tensile strength of the drill string is reduced, and it should be taken into account especially in deep well drilling. In the crooked hole, the drill string will deduce additional bending stress.

2.3.2.2 Working State and Force Analysis During Drilling Process

During drilling process, most of the drill string's gravity is lifted by the hook. It's called hook load. The gravity of a small part of drill string is mainly applied to the bit, which is called weight on bit (WOB). At this time, the upper section of drill string is subjected to the axial tension caused by gravity, and it is in the tensile state. In contrast, the lower drill string is subjected to WOB in the compression state. When the rotary table is used to drive the drill string and the bit to break the rock, the whole drill string is also subjected to a torque, and it is in a continuous rotating state.

Axial force and neutral point of the drill string

Under normal drilling conditions, the upper drill string is subjected to the axial tension force caused by the force of gravity (buoyant weight), and the lower drill string (mainly drill collar) is subjected to the compressive force caused by WOB. The counteracting force of WOB makes the lower drill string compressed and reduces the axial tension of the upper drill string by a corresponding value. At this time, the axial force at any section of the drill string can be expressed as follows:

$$F_W = K_B \left(q_p L_P + q_c L_c \right) - W \tag{2.8}$$

where

F_w The axial force at the section of drill string, kN;
W Weight on bit(WOB), kN.

Fig. 2.48 Distribution of the axial force of the drill string

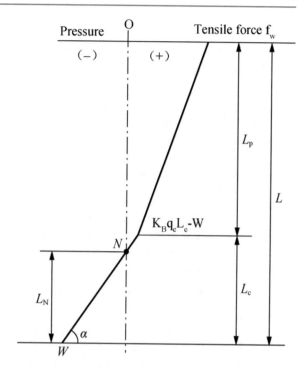

Under the combined action of force of gravity and WOB, the distribution of the axial force of the drill string is shown in Fig. 2.48. The upper drill string is subjected to tensile force, and the maximum force occurs at the wellhead and decreases downward. The lower drill string is subjected to compressive force, and the maximum force occurs at the bottom of the hole and decreases upward gradually. At a certain location, the drill string is neither in tension nor compressed, and its axial force is equal to zero.

The point in which the axial force of the drill string is equal to zero is called the neutral point of the drill string. Obviously, in vertical drilling, the drill string is divided into two segments at a neutral point. The buoyant weight of the upper drill string is equal to the hook load. The buoyant weight of the lower drill string is equal to WOB.

According to the definition of drill string neutral point, assuming that the neutral point is on the drill collar and the frictional influence of shaft wall is not considered, the neutral point height of single drill collar meets the following force equilibrium relations:

$$W = K_B q_c L_N \cos \alpha \tag{2.9}$$

where,

L_N The neutral point height from the bottom of well hole, m;
α Deviation angle, (°).

The neutral point of the drill string is of great significance in the actual work. During drilling process, it is hoped that the neutral point will always locate on the drill collar with strong stiffness and bending resistance, instead of locating on the weaker strength drill pipe, so as to prevent the drill pipe from being subjected to compression, bending, and alternating stress. Therefore, the length of the drill collar cannot be less than the neutral point height, that is to say, the buoyant weight of drill collar cannot be less than weight on bit, which is called the "buoyant weight principle." At present, the drill string design follows this principle to determine the length of drill collars. In addition, in the treatment of stucking accidents, the drill string neutral point is usually adjusted to the fishing position in order to guide the operation of fishing. When fishing with a pin tap, the neutral point should be adjusted to the top of the fish so as to make thread easy.

Torque and shear stress of the drill string

In the rotary drilling process, the rotary table transmits the torque to the drill string through the kelly and drives the bit to crush the rock. At this time, the whole drill string is in a state of continuous rotation. In addition to the axial tension and compression, the force acting on the drill string is torque. The torque of a drill string depends on the power efficiency of the drill string transmitted by the rotary table. The torque on the drill string depends on the power transferred from the rotary table to the drill string. The power transferred from the rotary table to the drill string consists of two parts: the power required for the idle rotation of the drill string and the power necessary for the rotary breaking of rock by the drill bit, i.e.,

$$N = N_s + N_b \tag{2.10}$$

where,

N The power to rotate drill string, kW;
N_s The power required for the idle rotation of the drill string, kW;
N_a The power required for breaking rock by the drill bit, kW.

The torque acting on the drill string is the largest at the wellhead and decreases gradually with energy consumption. The torque at the bottom of the hole is the smallest. The wellhead torque is

$$M = \frac{9549N}{n} \tag{2.11}$$

where

M The drill string torque at the wellhead, N·m;
n The rotational speed of drill string, r/min.

The torque of the rotary table produces shear stress on the cross section of the drill string, and the maximum shear stress on the cross section of the wellhead can be calculated by following formula.

$$\tau = \frac{M}{W_n} \tag{2.12}$$

$$W_n = \frac{\pi\left(d_{po}^4 - d_{pi}^4\right)}{16 d_{po}} \tag{2.13}$$

where

τ The shear stress on the cross section of the drill string at the wellhead, MPa;

W_n The anti-rotation cross-sectional coefficient of the drill string at the wellhead, cm³;

d_{po}, d_{pi} The outer diameter and inner diameter of the drill pipe, cm.

During normal drilling process, the power (N) transferred to the drill string by the rotary table is related to the bit type and diameter, rock property, drill string size, WOB, rotational speed, drilling fluid performance, and wellbore quality.

The power required to rotate the idle drill string:

$$N_s = 4.6 C \rho d_n d_{po}^2 L \times 10^{-3} \tag{2.14}$$

The power required for the bit breaking rock:

$$N_b = 0.0785 \, W n d_n d_{po} L \times 10^{-8} \tag{2.15}$$

where

L The length of the drill string, m;

C Coefficient related to deviation angle.

$C = 18.8 \times 10^{-5}$ when deviation angle equals to 0, $C = 31 \times 10^{-5}$ when deviation angle equals to 6°, $C = 38.5 \times 10^{-5}$ when deviation angle equals to 15°, $C = 48 \times 10^{-5}$ when deviation angle equals to 25°.

In the downhole dynamic drilling process, the torque of the drill string is the reactive torque of the dynamic drilling tool, which is the largest at the bottom of the well and gradually decreases upward. The maximum torque is equal to the torque required for rock breaking.

In the downhole dynamic drilling process, the torque of the drill string is the reactive torque of the dynamic drilling tool, which is the largest at the bottom of the well and gradually decreases upward. The maximum torque is equal to the torque required for rock breaking.

Bending of drill string

In the normal drilling process, the drill string below the neutral point is subjected to axial force. According to the theory of rod stability, when the axial force is less than a critical value, the compression part of the drill string will remain stable state in a straight line. When the load increases to a critical value, the drill string will lose straight stability and bend. It will contact with wellbore wall at a certain point (usually called the tangent point). This is the first bending of the drill string (curve I in Fig. 2.49). If the weight on bit is further increased, the bending shape changes and the tangent point gradually moves down (curve II in Fig. 2.49). When the weight on bit rises to a new critical value, the bending axis of the drill string presents the second half waves which are the second bending of the drill string

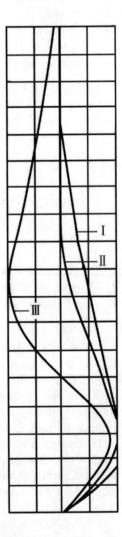

Fig. 2.49 Diagram of the compression bending of a drill string

(curve III in Fig. 2.49). If the weight on bit is further increased, the third or more bending of the drill string will occur.

In vertical wellbore, the critical weight on bit of the first bending for a slick drill collar can be calculated by

$$W_{cr} = 2.04 q_{cm} \sqrt[3]{\frac{EI}{q_{cm}}} \qquad (2.16)$$

where

W_{cr} The critical drilling load of first bending for a slick drill collar, kN;
q_{cm} The buoyant weight of unit length drilling collar, kN/m;
EI Modulus of rigidity of drilling collar, kN.

At present, the drilling load used in rotary drilling generally exceeds the critical weight on bit of drill collar. If no other measures are taken, the lower drill string will bend inevitably.

When the drill string eccentrically rotates around the axis of the well, the centrifugal force will be generated. The action of centrifugal force will aggravate the bending deformation of the lower part of the drill string. The upper part of drill string under tension may also bend under the action of centrifugal force. In addition, the flexural deformation of the drill string is also possible to occur through the controlled directional well and the horizontal well. When the bending drill string rotates in the downhole, it is generally not possible to be in a plane bending state under the action of torque, but in a space helical form with variable pitch.

The bending of the drill string will produce bending stress in the drill string. The bending stress is caused by centrifugal force on the upper part of drill string, while it is caused by the combined action of drill string compression bending and centrifugal force on the lower part of drill string. So, the bending stress of lower part of drill string is the largest. The value of the bending stress is related to the stiffness of the drill string, the length of the bending part, and the maximum bending deflection. Because the bending deformation of the downhole drill string is very complicated, the calculation of bending moment and bending stress is also very complicated, and it is not discussed here.

Movement form of drill string

How does a bent string rotate in the well? This is a very complex problem that has not been thoroughly studied up to now. According to the analysis of the actual wear condition and working condition of the drill string, the movement form of the drill string in the borehole may be as follows:

- Self-rotation. The drill string is like a flexible axis that revolves around its axis. When the drill string rotates, it contacts the wellbore on the whole circle to generate uniform wear along the drill string. The flexural drill string is subjected to the action of alternating bending stress during self-rotation, and it is prone to fatigue damage. In the flexural well section of the soft formation, the key slot is easily formed on the borehole wall by the rotation of the drill string, and the accident of pipe sticking will occur easily.

- Revolution. The drill string is like a rigid body that revolves around the axis of the well and slides along the borehole wall. When the drill string is revolving, it is not subjected to the action of alternating bending stress. Still, it produces ununiformed unidirectional wear (eccentric wear) which speeds up the wear and destruction of the drill string.
- The combination of self-rotation and revolution. The drill string rotates around the axis of the well, and it revolves around its axis at the same time, that is, not to slide along the borehole wall but to roll. In this case, the drill string is worn evenly and is subjected to alternating bending stress, but the cycle index is much lower than that of the self-rotation.
- Vibration. In the process of drilling, the drill string may also produce longitudinal vibration, torsional vibration, and lateral vibration. The longitudinal vibration of the drill string is mainly caused by the longitudinal vibration of the drill bit. The longitudinal vibration of the drill string will cause the drill string near the neutral point to be subjected to alternating axial stress, which is prone to fatigue damage. When the frequency of the longitudinal vibration and the inherent vibration frequency of the drill string itself are the same or multiple, the resonance phenomenon will occur, and the amplitude of the vibration will increase sharply, which is called the bit bounce. Serious bit bounce often causes the rapid fatigue damage of the drill string and the early damage of the drill bit.

The torsional vibration of the drill string is caused by the constant change of the rotating resistance of the downhole rock to the drill bit and the rotating resistance of the borehole wall to the drill string. Especially when the chipping-type bit (such as PDC bit, etc.) is used to drill into soft-hard formation, the torsional vibration of drill string is more serious, and even the so-called bit bouncing phenomenon occurs.

The lateral vibration of a drill string refers to the irregular lateral vibration of the whole or part of the drill string at a critical speed, just like a wave of strings. This movement form is very unstable, which often causes the strong vibration of the drill string. The vibration of the drill string makes the drill string under the action of alternating stress (alternating axial stress, shear stress, and bending stress), which can easily cause fatigue damage. Intense vibration often causes the damage of the bit and the drill string.

In downhole motor drive drilling process, the rotary torque for the bit to break the rock comes from downhole motor. The upper drill string is generally not rotating. There is no centrifugal force and alternating bending stress caused by bending and lateral vibration of drill string, which makes the drill string force distribution simpler.

In addition, whether it is rotary drilling or downhole dynamic drilling, the drilling fluid should be circulated during the normal drilling process. During the circulation of drilling fluid, the pressure drop of the drill string and bit port will produce additional axial tensile stress on the drill string, which is the equivalent of a tensile load acting on the drill string.

2.3.2.3   Working State and Force Analysis During Other Special Drilling Processes

Drill stem test (DST)

Drill stem test, also known as DST, means an operation that if a good indication of oil and gas is found, the drilling process will stop, and some test for the possible oil–gas reservoirs will be operated. During the DST process, the specialized downhole tool is installed at the lower part of drill string (including packer and formation tester). The bit is tripped in the test layer, and the packer is installed at the upper part of the test layer. The drilling fluid of the other layer and annulus space is separated. In the drill string, fluid column pressure is reduced by water cushion the air (or nitrogen) cushions and a certain pressure difference is formed. Then the downhole test valve is opened by surface control, which induces the test formation fluid into the test valve and drill pipe and pressure measurement and sampling are conducted until the fluid reaches the surface.

During the DST process, the drill string is suspended at the wellhead, and the whole drill string is affected by the axial tension caused by self-weight. In addition, the fluid pressure column is low because the water cushion or air cushion exists in the drill string. The pressure difference between inside and outside of the drill string leads to external pressure on the drill string. The external pressure at the bottom of drill string is the largest. The maximum external pressure can be calculated according to the condition of a full hollow drill string, that is

$$p_{cmax} = 0.00981 \rho_d D \tag{2.17}$$

where,

p_{cmax} The maximum external pressure on drill string, MPa;
ρ_d Drill fluid destiny in wellbore, g/cm^3;
D Well depth, m.

Squeezing cement

After cement setting, if empty space is found near the casing shoe or channeling is near the reservoir, the drill pipe is sent down and set at the upper part of the cementing well section, and a packer is used between drill pipe and casing pipe which can isolate the wellbore. Then well cementation is conducted through the drill pipe pump, and the cement is squeezed into annular space from the bottom of the casing or the bullet hole on casing wall.

In cement squeeze operation, drill string is not only subject to axial pull caused by gravity, but also subjected to internal pressure caused by pressure difference between inside and outside pipe (pressure inside the drill string is greater than that in drilling fluid column pressure outside the drill string). When the cement is squeezed by high pressure, the pressure at the bottom of the well is required to exceed the fracture pressure of the formation. At this time, the effective internal pressure of the drill string at the wellhead is maximum, which can be calculated by

$$p_{imax} = 0.00981(\rho_f - \rho_d)D \tag{2.18}$$

where,

$p_{i\ max}$ The maximum internal pressure of drill string, MPa;
ρ_f The equivalent density of formation fracture pressure, g/cm^3.

In addition, during handling pipe sticking accidents, the lifting of the drill string for jam release will make the drill string subjected to an axial tensile load that is much greater than the gravity of the drill string.

From the above analysis, it is known that the force distribution upon the drill string is more complicated during the rotary drilling process. These loads can be divided into unchanging and alternating types according to their properties. The tensile stress, compressive stress, and shear stress generated by axial load and torque belong to the unchanging stress. The shear stress caused by bending stress and torsional vibration belong to alternating stress, so does the tensile stress and compressive stress caused by the longitudinal stress. In the whole length of the drill string, the load characteristic is that the string at the wellhead is mainly affected by the unchanging loads (tension stress), and near the bottom of wellbore, it is mainly affected by the alternating load (tensile and compressive bending stress, etc.), which is the main reason of the drilling string fatigue failure.

From the above analysis, it is easy to see that the several positions of the drill string are as follows:

- Wellhead. The drill string at the wellhead is affected by maximum tensile force and maximum torque. During slip tripping process, the upper part of the drill string at the wellhead is also acted on by slip extrusion force. It is in the biaxial stress state of tensile and extrusion, which decreases the tensile strength of the drill string.
- Lower part of drill string. The lower part of drill string is subjected to axial force, torque, and bending stress at the same time during the process of drilling. The bending and lateral vibration of the drill string will also produce alternating bending stress, which often leads to the fatigue failure of the drill string. The suddenly stuck of drill bit will increase the shear stress of the lower part of drill string.
- Near the neutral point. Due to the existence of factors such as formation lithology change, drill bit impact, and longitudinal vibration, the drill pipe pressure is uneven. Therefore, the drill string near the neutral point is subjected to the action of tension pressure alternating load, and it is easy to cause fatigue damage.

2.3.3 Design of Drill String

Reasonable drill string design is an important requirement to ensuring high-quality, rapid, and safe drilling. Especially for deep well drilling, the downhole working conditions of the drill string are very complex, so the design of the drill string is very important.

The design of drill string includes the selection of drill string size, the length design of the drill string, and the strength design of the drill string.

In the design process, the following three principles are generally followed:

- Meet the technology requirements to ensure the smooth operation of the drilling operation;
- Meet the requirements of strength (tension resistance, collapse resistance, etc.) to ensure the safety of the drill string;
- Minimize the gravity of the drill string in order to drill the deeper well under the existing anti-load capacity condition.

2.3.3.1 Size Selection of Drill String

For a well, the size selection of drill string depends on the size of the drill bit and the lifting capacity of the drilling rig. At the same time, the characteristics of each area should be considered, such as geological conditions, wellbore structure, drilling tools supply, well deviation control, and so on. In the long-term production practice, the relationship between the drill bit size and drill string size is formed as shown in Table 2.19.

It can be seen in the table that a size of the drill string can use two sizes of drilling tools. The specific selection is based on the actual conditions. The basic principles are as follows:

- The torque and pulling force of the kelly are maximum, and the large size kelly should be chosen as far as possible when the supply is possible.
- In the case of the drilling rig's lifting capacity, it is advantageous to select a large size drill pipe. Because of the large size of drill pipe, its strength and nozzle are

Table 2.19 Matching of bit size with drill string size

Drill bit diameter [mm(in)]	Drill collar external diameter [mm(in)]	Drill pipe external diameter [mm(in)]	Kelly width [mm(in)]
>299($11^3/_4$)	203(8)	168($6^3/_8$)	152(6)
248–299($9^3/_4$–$11^3/_4$)	178–203(7–8)	140($5^1/_2$)	133,152 ($5^1/_4$,6)
197–248($7^3/_4$–$9^3/_4$)	152–178(6–7)	114,127($4^1/_2$,5)	108,133 ($4^1/_4$,$5^1/_4$)
146–216($5^3/_4$–$8^1/_2$)	146($5^3/_4$)	89($3^1/_2$)	89,108 ($3^1/_2$,$4^1/_4$)

large. The flow resistance of drilling fluid is small, which is beneficial to improve the bit hydraulic power. The drill string structure in the wellbore should be simple, in order to facilitate the tripping operation. At present, 127 mm (5 in) pipe is used in most of the domestic oil fields in China.

- The drilling collar size is generally equal to or close to the outer diameter of the drill pipe sub and sometimes the diameter of the drill is selected according to the well deviation control technology.

The drill collar size determines the effective diameter of the borehole. In order to ensure that the casing or milling can smoothly go down the drilling hole, the outer diameter of the bottom part of the drill collar should not be less than the allowable minimum diameter. Hodge proposed a formula for calculating the minimum diameter of the drill collar:

$$\text{The allowable minimum diameter} = 2 \\ \times \text{ casing coupling diameter} - \text{bit size} \tag{2.19}$$

When the stabilizer is used between the drill collars, it is possible to select the smaller drill collar.

In the wellbore whose diameter is more than 241.3 mm, the tapered drill collar string is recommended which can control the well deviation. Under normal conditions, the difference between the outer diameters of the two adjacent drilling collars is less than 25.4 mm, so as to avoid excessive stress concentration and fatigue damage at the connection point between two adjacent segments (including the connection between the last section and drill pipe). The maximum outside diameter of drilling collar should be guaranteed to be milled in the fishing operation.

2.3.3.2 Length Design of Drill Collar

The length of drill collar depends on weight on bit and size of drill collar. The design principle is that the buoyant weight of drilling should satisfy the required WOB and ensure that the drill pipe does not bear the compression load under the maximum WOB, that is, neutral point is always within the drill collars section. For a single size of drill collar, the length design formula is as follows:

$$L_c = \frac{S_N W_{\max}}{K_B q_c \cos \alpha} \tag{2.20}$$

where:

L_c The length of drill collar, m;

S_N Safety coefficient, to prevent the neutral point moving to the drill pipe section when the accidental additional force (dynamic load, wall friction resistance, etc.) occurs, is usually 1.15–1.25;

W_{max} Maximum weight on bit designed, MPa;
K_{B} Buoyancy coefficient;
q_{c} The weight of a meter drilling collar in the air, kN/m;
α Deviation angle, $\alpha = 0°$ when the well is straight.

2.3.3.3   Strength Design of Drill String

According to the force analysis of drill string, we can see that the forces acting on drill string include axial tension, pressure, torque, external force, internal pressure, and bending stress during drilling process. Among them, the force which is always on the drill string and has greater effect is the axial tension caused by the gravity of the drill string (buoyant weight). It is the main force acting on the drill string. Moreover, the deeper the well is, the longer the drill collar and the greater the pulling force of the upper drill string. Therefore, the strength design of drill string is generally based on the calculation of tensile strength. The intensity calibration is conducted accordingly for other strength, such as torque, external pressure, and internal pressure.

Strength condition of drill string design

In the strength design of drill string mainly based on tensile strength, the dead load caused by drill string gravity is mainly considered. Some additional loads, such as dynamic load, frictional resistance, slip extrusion force, and lift up force, are considered through a certain design coefficient.

- Considering the impact of dynamic load and friction resistance in the process of tripping

During tripping process, the drill string is not only affected by the axial tension caused by the gravity, but also affected by the dynamic load and the friction force effect. In the design of drill string, an assurance factor S_t is used to consider the influence of dynamic load and frictional resistance. This design method is called assurance factor method, and its strength condition is

$$F_m S_t \leq F_p \tag{2.21}$$

where

F_m An axial tension caused by a buoyant weight on any section of a drill string, kN;
S_t The assurance factor of tensile resistance, the value of which is 1.3;
F_p Maximum allowable tension, kN.

The maximum allowable tension of the drill collar depends on the yield strength of the drill string steel, the size of the drill string, and the actual working conditions of the drill string. First, the tension of the drill string cannot exceed the tensile force

under the minimum yield strength, which is called the yield tension. The formula is
as follows:

$$F_y = 0.1\sigma_y A_p \tag{2.22}$$

where,

F_y Yield tension of drill string, kN;
σ_y Minimum yield strength of drill string steel, MPa;
A_p The cross-sectional area of drill string, cm^2.

When the tensile force on the drill string reaches its yield tensile, the material
will yield and produce a slight permanent extension. In order to avoid this situation,
90% of the yield force is generally taken as the maximum allowable tension of the
drill string. That is

$$F_p = 0.9F_y \tag{2.23}$$

- Considering the impact of the slip banding force

In the tripping process using slip, drill pipe hangs on the slip. Drill pipe is not only
affected by the axial tension caused by the gravity, but also affected by the large
banding force, which is in biaxial stress state of tensile squeeze. When the resulting
stress (greater than the tensile stress) approaches or reaches the minimum yield
strength of the material, it will result in the drill pipe collapse caused by slip. In
order to prevent the drill pipe from being destroyed by slip, the axial tension is
required to stay less than its maximum allowable tension. In the strength design of
drill string, a design safety coefficient is usually used to consider the influence of
slip banding force. This is called design coefficient method, and its strength condition is

$$F_m S_t \leq F_p \tag{2.24}$$

$$S_f \geq \sqrt{1 + \frac{d_{po}K_s}{2L_s} + \left(\frac{d_{po}K_s}{2L_s}\right)^2} \tag{2.25}$$

where,

S_f Design safety coefficient, which can be calculated according to the drill string
         collapse condition;
d_{po} Drill string outside diameter, cm;
L_s Slip length, cm;
K_s Lateral compression coefficient of the slip; $K_s = 1/\tan(\alpha + \phi)$;

α Slip taper angle;
ϕ Frictional angle; $\phi = \arctan f, (^\circ)$;
f Frictional coefficient $f \approx 0.08$.

In order to facilitate the application, the calculation results of minimum design coefficient are listed in Table 2.20.

• Considering the impact of lifting force for pipe sticking

Considering the impact of lifting force for pipe sticking, the drill string design should meet the requirements that axial tension caused by the buoyant weight of drill string should be less than the maximum allowable tensile force of drill string. The value is regarded as margin of pull force (MOP) to ensure that the drill string cannot be broken. This design method is called as MOP method. The condition of the strength is

$$F_m + \text{MOP} \leq F_p \qquad (2.26)$$

where

MOP   Margin of pull force, 200–500 kN.

Emerging Eqs. (2.21, 2.23, 2.24 and 2.26), the strength condition of drill string can be known as follows:

$$F_m = F_n \qquad (2.27)$$

where

F_n The maximum allowable static tension, kN.

Table 2.20 Design assurance factor for preventing slip crushing

Slip length [mm(in)]	Frictional coefficient f	Lateral compression coefficient K_S	Drill pipe size (mm)						
			60.3	73.0	88.9	104.6	108.0	127.0	139.7
			Design safety coefficient S_f						
304.8	0.06	4.35	1.27	1.34	1.43	1.50	1.58	1.66	1.73
	0.08	4.00	1.25	1.31	1.39	1.45	1.52	1.59	1.66
	0.10	3.68	1.22	1.28	1.35	1.41	1.47	1.54	1.60
	0.12	3.42	1.21	1.26	1.32	1.38	1.43	1.49	1.55
	0.14	3.18	1.19	1.24	1.30	1.34	1.40	1.45	1.50
406.4	0.06	4.36	1.20	1.24	1.30	1.36	1.41	1.47	1.52
	0.08	4.00	1.18	1.22	1.28	1.32	1.37	1.42	1.47
	0.10	3.68	1.16	1.20	1.25	1.29	1.34	1.38	1.43
	0.12	3.42	1.15	1.18	1.23	1.27	1.31	1.35	1.39
	0.14	3.18	1.14	1.17	1.21	1.25	1.28	1.32	1.365

For safety factor method:

$$F_a = 0.9F_y/S_t \tag{2.28}$$

For design coefficient method:

$$F_a = 0.9F_y/S_f \tag{2.29}$$

For tension margin method

$$F_a = 0.9F_y - \text{MOP} \tag{2.30}$$

Generally, when drilling pipe string is designed, the force F on drill pipe is calculated by Eqs. (2.28, 2.29, and 2.30). The minimum of the three calculated F is taken as the maximum allowable static tension on the drill pipe. Then the permissible length of the drill pipe is calculated by Eq. (2.27).

Design methods of drill pipe

- Single drill pipe design

A single drill pipe means a drill pipe string with the same size, wall thickness, and steel grade. The steps of the design are as follows:
    The first step is to choose a type of drill pipe.
    The second step is to calculate the maximum allowable static tension of the drill pipe string.
    The third step is to calculate the allowable length of the drill pipe string according to the strength conditions.

$$L_p = \frac{F_a/K_B - q_c L_c}{q_p} \tag{2.31}$$

where

L_p Allowable length of the drill pipe string, m;
q_p Nominal weight in the air, kN/m;
L_c The length of the drill collar, m.

The fourth step is to analyze whether the allowable length of the drill pipe string meets the requirements of the drilling depth designed. If it does not, a higher strength of drill pipe should be chosen until the requirements are met.

- Composite drill pipe design

In deep and ultra-deep well drilling, a composite drill pipe string is often adopted, which means a drill string consists of different sizes (upper part is larger than the lower part), or different wall thickness (upper part is thicker than the lower part), or different steel grades (upper part is higher than the lower part). Compared with single drill pipe, the composite drill string has many advantages. It can meet the strength requirements and also can reduce the total weight of drill string, which allows drilling a deeper well at the certain rig load capacity.

When designing a composite drill pipe string, we should start with the first drill pipe on the drill collar and calculate the allowable length of each drill pipe from above down. Then the actual length of the drill pipe is determined. The strength of the first drill pipe on the top of drill collar is low which increases upward gradually.

Strength Check of Drill Pipe String

- Collapse resistance

When external pressure of the drill pipe is greater than internal pressure, the pipe body will be collapsed. In order to avoid the drill pipe body to be collapsed, the effective maximum external pressure of the drill pipe should be less than the minimum collapse resistance. For safety, it is generally required that the ratio of the minimum collapse resistance to the maximum external pressure of the drill pipe is not less than a certain safety factor. That is

$$\frac{P_{pc}}{p_{cmax}} \geq S_c \qquad (2.32)$$

where

P_{pc} Minimum collapse resistance of drill pipe, MPa;
S_c Safety factor of anti-collapse which is 1.125 general.

The maximum allowable external pressure of the drill string is calculated according to the full hollow of the drilling fluid situation in the pipe (Eq. 2.17). The most dangerous position is located at the bottom position of the drill pipe.

- Internal pressure resistance

Occasionally, the drill pipe string will also be subjected to greater net internal pressure. When the internal pressure of the drill pipe string is greater than the minimum yield strength (the minimum internal pressure resistance), the drill pipe body will be fractured. In order to prevent the drill pipe from being burst, the ratio of the minimum internal pressure and the maximum internal pressure should not be less than a certain safety factor. That is

$$\frac{P_{pi}}{p_{imax}} \geq S_i \qquad (2.33)$$

where

p_{pi} Minimum internal pressure resistance, MPa;
p_{imax} Maximum allowable internal pressure of drill pipe, MPa;
S_i Internal pressure safety factor, which is usual 1.1.

The maximum internal pressure of the drill string can be calculated by Eq. (2.4), and the most dangerous section is at the wellhead.

- Torsional resistance

Torque is one of the main loads on the downhole drill string. Especially in directional well, horizontal well, deep well drilling, borehole enlargement, and drilling accident, the drill string may be subjected to large torque. When the drill string torque exceeds the minimum torsional strength, the drill pipe will be broken. In order to ensure the safety work of the drilling string in downhole, the maximum torque of the drill string should meet the following strength conditions:

$$\frac{M_p}{M_{max}} \geq S_m \tag{2.34}$$

where

M_p Torsional strength of drill string, which is the minimum value of pipe body and joint torsional strength, kN·m;
M_{max} The maximum torque of drill string, kN·m;
S_m Anti-torque safety factor.

The torque of drill string in the normal drilling conditions can be calculated by Eq. (2.11), and the maximum torque occurs at the drill string top which is at rotary drive bushing or the lower part of Kelly saver sub.

2.3.4 Examples of Typical Drill String Design

2.3.4.1 Design Datum
The well depth is 5000 m, the well diameter is 215 mm (8½ in.), the drilling fluid density is 1.2 g/cm³, the maximum weight on bit is 180 kN, the maximum allowable well deviation angle is 3°, the margin of pull force is 200 kN, the slip length is 406.4 mm, the tensile safety factor is 1.3, and the design safety factor is 1.42.

2.3.4.2 Design of Drill Collar String
A drill collar string is chosen with 158.75 mm (6½ in.) outside diameter and 57.15 mm (2¼ in.) inner diameter, nominal weight q_c is 1.35 kN/m, safety factor S_N is 1.18, buoyant coefficient K_B is 0.846.

The drill string can be calculated as follows:

$$L_c = \frac{S_N W_{max}}{K_B q_c \cos \alpha} = \frac{1.18 \times 180}{0.846 \times 1.35 \times \cos 3°} = 186.228(m)$$

According to the length of 9.2 m per drill collar, 21 drill collars are needed, and the total length is 193.2 m.

2.3.4.3 Design of Drill Pipe String
Chose the first section of drill pipe string (connect with the drill collar string)

The new grade E of drill pipes is chosen with outside diameter 127 mm and nominal weight 284.78 N/m, the minimum tensile resistance F_y is 1760 kN, and the maximum allowable static tension can be calculated as follows:

$$F_{a1} = 0.9F_y/S_t = 0.9 \times 1760/1.30 = 1218.46 \, (kN)$$
$$F_{a1} = 0.9F_y/S_f = 0.9 \times 1760/1.42 = 1115.49 \, (kN)$$
$$F_{a1} = 0.9F_y - MOP = 0.9 \times 1760 - 200 = 1384 \, (kN)$$

It can be seen that the maximum allowable static tension under the condition of the slip bending force is the smallest; the allowable length of the first drill string is as follows:

$$L_p = \frac{F_a/K_B - q_c L_c}{q_p} = \frac{1115.49/0.846 - 1.35 \times 193.2}{284.78/1000} = 3714.187(m)$$

According to the length of 9.1 m per drill string, 408 joints of grade E are required; the actual length is 3712.8 m. Obviously, it is necessary to add a high-strength drill string to reach the depth designed.

Chose the second type of drill pipe string

The new grade X-95 drill pipes are chosen with outside diameter 127 mm and nominal weight 284.78 N/m, the minimum tensile resistance F_y is 2229.7 kN, and the maximum allowable static tension can be calculated as follows:

$$F_{a2} = 0.9F_y/S_t = 0.9 \times 2229.7/1.30 = 1543.645 \, (kN)$$
$$F_{a2} = 0.9F_y/S_f = 0.9 \times 2229.71/1.42 = 1413.196 \, (kN)$$
$$F_{a2} = 0.9F_y - MOP = 0.9 \times 2229.71 - 200 = 1806.739 \, (kN)$$
$$L_{p2} = \frac{F_a/K_B - q_{p1}L_{p1} - q_c L_c}{q_p}$$
$$= \frac{1413.196/0.846 - 284.78/1000 \times 3712.8 - 1.35 \times 193.2}{284.78/1000}$$
$$= 1237.072(m)$$

Table 2.21 Final designed results of drilling assembly

Specification	Length (m)	Empty weight (kN)	Buoyant weight (kN)
Drill collar: external diameter-158.75 mm, internal diameter-57.15 mm, nominal weight-1.35 kN/m	193.2	260.82	220.65
First drill pipe: external diameter-127 mm, line weight-284.78 N/m, grade E	3712.8	1057.33	894.50
Second drill pipe: external diameter-127 mm, line weight-284 N/m, grade X-95	1094	311.45	263.49
Total	5000	1629.60	1378.64

According to the length of 9.1 m per drill collar, 136 joints of grade X-95 drill pipes are needed, and the total length is

$$L = 193.2 + 3712.8 + 1237.6 = 5143.6(m)$$

The final design result of the entire drill string is shown in Table 2.21.

Questions and Exercises

1. What are the requirements of the rotary drilling technology for the function of a drilling rig?
2. What are the composition of a drilling rig and the main components of each system?
3. According to the different classification standards, what types of drilling rigs are there?
4. What are the basic parameters and main parameters of the drilling rig? What drilling rigs are there in the domestic rig series?
5. What are the classifications of commonly used derrick types? What are the advantages and disadvantages of various types of derricks?
6. What types of power-driven systems are there on a drilling rig?
7. Describe briefly the functions of the drawworks, the mud pump, the rotary table, and the top drive system in a drilling rig.
8. What are the indicators for evaluating bit performance?
9. What is the difference between a milled-teeth bit and an inserted-teeth bit?
10. What are the types of bearings for a cone bit? What are the characteristics of the plain bearings?
11. How many parts of the oil storage and seal system of the cone bit? What is the function of it?
12. Which slip direction does the cone overhang, offset, and cone of the wheel?
13. What are the series of domestic tri-cone bit? Try to explain the meaning of 8% HP5.
14. What are the protruding advantages of a diamond bit?
15. What are the hydraulic structures used in a diamond bit? What are the advantages of the structures?

16. How does a natural diamond bit break the rock? What formation is suitable for drilling?
17. What is the meaning of PDC? What are the features of it?
18. How are the back rake and side rake angles of the PDC bit cogging defined, and what are the functions?
19. How does a PDC bit break rock? What formation is suitable for drilling?
20. What are the main parts of a drill string? What are the main functions of it?
21. What are the API steel grades of the drill string?
22. Why are drill collars installed at the bottom of the drill string rather than drill pipes?
23. What are the main differences between the three types of internal flush, full hole, and regular tool joints?
24. What are the possible movements of the drill string in the downhole?
25. What are the forces acting on the drill string in the downhole? What is the most important force?
26. What is the neutral point of the drill string? Why should the neutral point locate in the drill collar section?
27. Which parts of the drill string are most severely affected by the load? What do the loads include?
28. What requirements should the drill string design meet?
29. Why should a large size of drill string be chosen as much as possible under the allowed condition?
30. What is the composite drill string? What are the advantages of using a composite drill string?
31. What kind of force is mainly considered in the drill string design? What is the strength condition?
32. The bit is 9½ in, the drill collar is 7 in (q_c = 1632 N/m), the drill pipe is 5 in (q_c = 1632 N/m), the designed weight on bit is 180 kN, the drill fluid density is 1.28 g/cm³, the safety factor is 1.3, the allowable well deviation angle is 3°, so calculate the length of drill collar string needed?
33. The well depth is 5000 m, weight on bit W is 14 t, the drilling fluid density is 1.38 g/cm³, and the drill collars are 8½ in and 6½ in with 200 m length (q_c = 136.24 kg/m). The drill pipe is 5 in with 1300 m length (q_p = 29.04 kg/m). Please calculate the depth of the neutral point?
34. A composite drill pipe string consisting of a 4½ in grade E (IEU), with the wall thickness of 10.92 mm, and a 5½ in grade E (IEU) drill pipe with the wall thickness of 10.54 mm, what is the maximum running depth of the drill string in the drilling fluid with a density of 1.2 g/cm³? If the 5½ in drill pipe is changed into a 4½ in grade X-95 drill pipe (IEU) with a wall thickness of 10.92 mm, what is the running depth again?
35. A well is drilled by the 12¼ in bit to 3500 m. Please design the drill string for the well according to the following conditions:

The drilling fluid density is 1.25 g/cm^3, the weight on bit W is 200 kN, the tension margin *MOP* is 450 kN, the tensile safety factor S_t is 1.3, the slip length L_s is 406.4 mm, regular lubrication, and well deviation angle is 3°. The reserved drill string includes 9 in drill collar with 27 m in length (q_c = 2860 N/m), 7 in drill collar with 81 m in length (q_c = 1632 N/m), 6½ in drill collar is enough (q_c = 2860 N/m), 5 in grade E drill pipe with 900 m length (q_p = 284.78 N/m), and grade X-95 and grade G-105 drill pipe with 5 in diameter are enough supplied (q_p = 284.78 N/m).

Drilling Fluids

3

Abstract

The commonly used drilling fluids and their functions and properties are briefly introduced in this chapter. The term of "drilling fluids" generally refers to all operational fluids that are used in the circulation during drilling. It is also named as drilling muds or simply muds. Drilling fluids can be in forms of either liquid or air. That is why it is termed as "drilling fluids" instead of "drilling liquids." The importance of muds to drilling activities is similar as that of blood to human body. No drilling operation can proceed without using muds. Use of proper drilling fluids has a direct impact on the drilling progress and the associated drilling cost. The performance of drilling fluids is one of the key factors to the success of drilling operation.

3.1 Composition and Category of Drilling Fluids

3.1.1 Main Functions of Drilling Fluids

The circulation process of drilling fluid is presented in Fig. 3.1. It shows that the drilling fluid starts flowing from the mud pumps and goes through the surface pipeline, the standpipe, the rotary hose, then gets into the drill pipe. After that, the drilling fluid ejects to the well bottom through the bit nozzles and carries the drill cuttings up to the surface through the annular space (or annulus) between the drill pipe and the formation (or casing). After returning to the surface, the drilling fluid is directed through shale shakers and other solid control equipments, where the drill cuttings are removed. Finally, the mud flows back into the mud tanks and remains in the circulation.

© China University of Petroleum Press and Springer Nature Singapore Pte Ltd. 2021   173
Z. Guan et al., *Theory and Technology of Drilling Engineering*,
https://doi.org/10.1007/978-981-15-9327-7_3

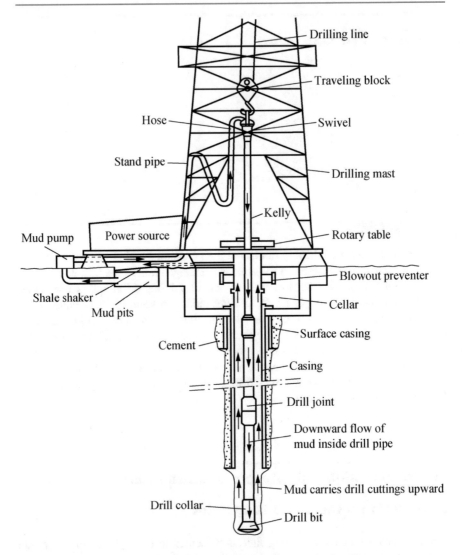

Fig. 3.1 Diagram of the circulation process of drilling fluid

3.1.1.1 Remove Drill Cuttings from Well Bottom

The rock chips crushed by the drill bit are called drill cuttings. As drilling progresses, it is necessary to transport cuttings timely from the well bottom to avoid repetitive drill bit shearing, which enhances the rate of penetration and reduce drilling accidents. Although drill cuttings tend to slip down due to gravity, if the upward velocity of the drilling fluid in the annulus is greater than the slipping velocity of the cuttings, the cuttings can still be moved up the wellbore. Many factors influence the efficiency of cutting removal, such as mud density, viscosity, shear stress, and rheological model.

3.1.1.2 Cool and Lubricate Drill Bit and String

During drilling, heat is generated from the friction at the contact between the drill bit and the drill string and the formation. It is difficult to dissipate heat through the formation. However, the heat can be conducted from the friction part of the drill bit to the drilling fluid and then transported to the surface via circulation.

Drilling fluid contains many types of additives. In particular, lubricants are often used to enhance the lubrication properties of the drilling fluid, which helps to reduce the torque and drag of the drill string, increase the service life of the bit, decrease the pump pressure, and increase the rate of penetration.

3.1.1.3 Enhance Filtration Control and Maintain Wellbore Stability

Drilling fluid with good performance can form a layer of mudcake around the borehole, which has a much lower permeability than the formation. It not only maintains wellbore stability and prevents wellbore from caving, but also keeps the mud filtrate from invading the formation. The filtration properties of the mud can be enhanced by adding bentonite clay and treating with chemicals, improving the particle size distribution of bentonite, increasing the concentration of colloid particles, and using anti-sloughing agents.

3.1.1.4 Control Formation Pressure

Proper control of formation pressure mainly depends on the mud density. At normal conditions, the formation pressure is equal to the hydrostatic pressure of the formation. Generally, the mass of drill cuttings plus the hydrostatic pressure of the mud column is enough to balance the formation pressure. However, abnormal formation pressure happens sometimes, which may cause well kick, well blowout or lost circulation. Therefore, the mud density needs to be adjusted according to the monitored formation pressure. The formula below is used to calculate the mud density for maintaining the formation pressure:

$$\rho_d = \frac{P_p}{0.00981D} \tag{3.1}$$

where

ρ_d Mud density, g/cm^3;
P_p Formation pressure, MPa;
D Well depth, m.

Normally, the hydrostatic pressure of the mud column needs to be larger than the formation pressure; therefore in practice the mud density can be calculated by Eq. 3.1, and then plus an extra value $\Delta \rho$ (g/cm^3). For gas-bearing formations, the extra value is set as 0.07–0.15 g/cm^3; for oil-bearing formations, the extra value is set as 0.05–0.10 g/cm^3.

3.1.1.5   Suspend Cuttings and Weighting Materials

High performance drilling fluids should have gel strength when circulation is stopped, which is the ability to suspend cuttings and weighting materials; otherwise the solid particles will settle down to the well bottom and will cause accidents such as repeated grinding and pack-off pipe sticking. When circulation resumes, the mud returns to a fluid state and carries the coarse particles such as cuttings up to the surface.

3.1.1.6   Acquire Formation Data

Based on the analysis on the cuttings released from the drilling fluid and the oil & gas show, it is possible to assess whether the drilled formation bears economic hydrocarbon resources. Monitoring the mud properties and analyzing the mud filtrate can warn us when drilling into salt, gypsum, or salty water formations.

3.1.1.7   Transmit the Hydraulic Power

Drilling fluid can transmit hydraulic power from the mud pump to the drill bit. The hydraulic power generates a high-pressure jet through the bit nozzles and maximizes the hydraulic impact on the rocks, hence improve the rate of penetration. Furthermore, when turbo drill tools are used, the drilling fluid flows through the turbine blades at a high speed and rotates the driving turbine, which impels the drill bit to crush rocks.

The continuous advancement of drilling technology requires the drilling/completion fluids to have more versatile and better performance. The drilling/completion fluids that possibly encounter the oil & gas formation should have good capability of formation protection. The mud for drilling complex formation and extra-deep wells should have good properties such as high-temperature resistance, salt/gypsum contamination resistance, carbon dioxide or hydrogen sulfide contamination resistance, and corrosion resistance. Furthermore, with the increasing awareness of environment protection, people expect drilling fluid to have better properties in reducing the environmental footprint. The cost of drilling fluids should be as low as possible while still meeting the performance requirements, but the focus should be put on increasing the overall drilling efficiency.

3.1.2   Basic Composition of Drilling Fluids

Broadly defined, the basic composition of drilling fluid includes disperse medium, disperse phase, and mud agents. The disperse medium of drilling fluid can be water, oil, or gas. The disperse phase can be suspended particles, emulsion, and/or foam. The regular suspended particles mainly consist of clays, weighting materials and/or lost circulation materials and drill cuttings, etc. The emulsion is normally comprised of water and oil, while the foam is air/gas.

The mud agents are chemicals used for adjusting the mud properties. Based on their type, mud agents can be classified into inorganic (including inorganic acid, alkali, salt, oxide, etc.) and organic (e.g., surfactant, polymer, and so on) mud agents. They can also be classified based on their functions into 15 categories which are mud pH control agent, mud thinner, mud viscosifier, mud fluid loss reducer, mud flocculent, shale inhibitor (or sloughing preventer), calcium removal agent, mud corrosion inhibitor, mud lubricant, pipe free agent, temperature stabilizing agent, mud foaming agent, mud emulsifier, weighting material, and lost circulation material. The last two types of agent in the 15 categories are called material because their dosage is very high and normally higher than 0.05 g/mL (over 5 g of agents in every 100 mL).

3.1.3 Types of Drilling Fluids

With the development of drilling fluid technology, there are increasing types of drilling fluids. The major components and properties of different mud system have significant difference. The types of mud system can be selected based on the actual requirements of drilling engineering to the mud properties, the operation cost, and the environment protection regulations.

At present, there are different ways to categorize drilling fluids, but the most common method is classifying based on the base fluid and primary ingredients, by which it is generally classified to aqueous drilling fluids, non-aqueous drilling fluid, and gaseous drilling fluids.

Aqueous drilling fluids are the most used muds in field operations. They can be classified into two types according to their capabilities of shale inhibition: inhibitive mud and non-inhibitive (highly dispersive) mud. The inhibitive mud can be subdivided by the agents into calcium-treated drilling fluid, potassium salt drilling fluid, brine drilling fluid, silicate drilling fluid, polymer drilling fluid, mixed-metal hydroxide (MMH) drilling fluid, etc. The polymer drilling fluid can be subdivided into anionic drilling fluid, zwitterionic drilling fluid, cationic drilling fluid, and non-ionic drilling fluid.

Figure 3.2 shows the detailed classification of mud system based on the principles introduced above.

Other than above, the following methods can also be used to classify drilling fluids: based on density, it can be classified into unweighted and weighted mud; based on solid content, it can be classified into high solid mud, low solid mud, and non-solid mud; based on the dispersity of clay particles, it can be classified into fine-disperse mud, coarse-disperse mud and non-disperse mud. Sometimes it can also be classified or named according to the special drilling requirements, such as ultra-high-temperature drilling fluid, extended-reach horizontal well drilling fluid, and so on. Broadly defined, the working liquids in contact with reservoir are all called working fluids, so the fluids used to drill into oil and gas formations are called drilling and completion fluids.

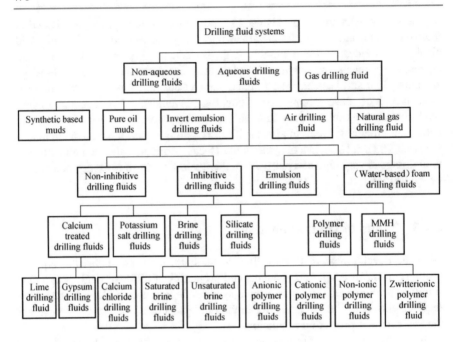

Fig. 3.2 Classification of drilling fluid systems

3.1.3.1 Aqueous Drilling Fluids

Aqueous drilling fluids use water as the dispersing medium. It is normally comprised of water, bentonite, and agents. It also can be subdivided into non-inhibitive drilling fluid and inhibitive drilling fluid, emulsion drilling fluid, and (water-based) foam drilling fluid.

Non-inhibitive drilling fluid

Non-inhibitive drilling fluid is also known as fine-disperse drilling fluid, which is one of the oldest types of mud system and has the lowest cost. It is a type of water-based drilling fluid comprised of water, bentonite, and agent used to disperse clays and drill cuttings (known as dispersant). There are many types of dispersant, including dispersant for thinning (thinner) and dispersant for fluid loss reducing (e.g., CMC, polyanionic cellulose, modified lignite, etc.). One of its basic features is to meet the rheology and filtration requirements by the high dispersion of clay in water.

This drilling fluid has a lot of advantages such as high density (over 2.0 g/cm^3), tight and firm filter cake, low filtrate volume, high-temperature resistance, etc. However, it does not have good capability in inhibition, formation protection, and anti-contamination of calcium and salt. Furthermore, due to the high content of submicron clay particles (diameter lower than 1 μm), this drilling fluid has unfavorable effect on the rate of penetration.

To maintain high performance, the bentonite content of the non-inhibitive drilling fluid should be kept lower than 10% and reduced with the increasing of density and temperature. In the meantime, the salt content should be lower than 1%; pH must be greater than 10, to activate the function of mud thinner.

To increase the high-temperature resistant and salt-resistant capability of fine-disperse drilling fluid, researchers in China have developed a mud system that comprised of sulfonated tannin extract (SMK), sulfonated lignite, and sulfonated phenolic resin. This mud system is a typical deep well fine-disperse drilling fluid, which can be used at over 200 °C.

This drilling fluid is suitable for drilling deep well (over 4500 m of well depth) and high-temperature well (over 200 °C of well temperature), but is not good for drilling into oil, salt, and gypsum formation and large sections of shale. The bottom temperature of the ultra-deep well in China can reach to 240 °C (e.g., Bishen No. 1 Well).

Inhibitive drilling fluid

It is a water-based drilling fluid that uses shale inhibitor as the main agent. This drilling fluid is also called coarse-disperse mud because the shale inhibitor keeps the clay particles in coarse dispersion. It is developed for overcoming the disadvantages (high content of submicron particles and low salt resistance) of non-inhibitive drilling fluid.

This drilling fluid can also be subdivided by the type of shale inhibitors into the following systems:

- Calcium-treated drilling fluid

It is a water-based drilling fluid that uses calcium-treated agents as the major additive. The common calcium-treated agents includes lime, gypsum, calcium chloride, etc. The concentration of calcium ions that these agents can provide increases successively. Correspondingly, they are called lime-treated drilling fluid (lime drilling fluid), gypsum-treated drilling fluid (gypsum drilling fluid), and calcium chloride-treated drilling fluid (calcium chloride drilling fluid). Calcium drilling fluid needs to add dispersant to control the strong flocculating capability of calcium ions, so that it can keep the clay particles in a controlled flocculation and maintain a good stability of drilling fluid. The dispersants include thinning dispersant (thinner) and fluid loss reducing dispersant (e.g., CMC, modified lignite, etc.)

The calcium-treated drilling fluid has advantages of resisting calcium contamination, stabilizing shale, and maintaining the clays of drilling fluid in coarse dispersion. It is especially good for using in drilling gypsum formation.

- Potassium salt drilling fluid

It is a water-based drilling fluid that using potassium salt polymer or ammonium salt polymer and potassium chloride as the major agents.

The potassium salt drilling fluid has the advantages of inhibiting shale from hydration and swelling, preventing formation from collapsing, and reducing the content of sub-micron clay particles in the drilling fluid. It also helps formation protection.

To ensure the drilling fluid working properly, the pH of drilling fluid should be within the range of 8–9, and the concentration of K^+ in mud filtrate should be greater than 18 g/L. Potassium salt drilling fluid is mainly used for drilling collapsible shale formation.

- Brine drilling fluid

It is a water-based drilling fluid that using salt (sodium chloride) as the major agents. The salt content is from 10 g/L (Cl^- concentration is about 6 g/L) to saturation (Cl^- concentration is about 18 g/L). The drilling fluid with saturated salt content is called saturated-brine drilling fluid.

The brine drilling fluid has advantages of salt contamination resistance, calcium contamination resistance, and magnesium contamination resistance. It has strong anti-collapsing capability in shale formations. The filtrate also has low damage to the oil and gas formations.

To ensure that the drilling fluid is working properly, it is better to use salt-resistant clays (e.g., sepiolite) and agents that have salt resistant, calcium and magnesium ion-resistant capability. To protect the drilling pipe from brine corrosion, the corrosion inhibitor should be added. The salt crystallization inhibitor should be added in the saturated-brine drilling fluid to prevent the salt from crystallization.

Brine-based drilling fluid is suitable for offshore drilling or drilling inshore and other areas lacking fresh water. Saturated-brine drilling fluid is mainly used to drill salt formation, unstable shale formation, and mixed formation of salt, gypsum, and shale.

- Silicate drilling fluid

It is a water-based drilling fluid using silicate as the major agents. The silicate ions can react with the calcium and magnesium ions in the wellbore surface and formation water to produce calcium silicate and magnesium silicate precipitate, and sediment on the borehole surface to form a protection layer. Therefore, the silicate drilling fluid has the capability of resisting calcium contamination and preventing shale from hydration and swelling. When using silicate drilling fluid, it requires the pH value to be within 11–12, because the silicate ion will change into silicic acid if pH is lower than 11, which will deactivate the agent. This type of drilling fluid has problems of controlling rheology and filtration, and the filter cake is soft and thick.

Silicate drilling fluid is especially suitable to be used in gypsum and mixed formation of gypsum and shale.

- Polymer drilling fluid

It is a water-based drilling fluid that using polymer as the major agents. Due to the selective flocculating ability of polymer, the clay particles in drilling fluid can maintain their large size. In the meantime, the surface of drill cuttings is protected by the adsorptive layer (encapsulation), so the cuttings cannot disperse to fine particles, which is because of the adsorption property of polymer. Therefore, a higher rate of penetration can be obtained by using polymer drilling fluid. Polymer drilling fluid does not belong to the category of non-disperse low solid drilling fluids.

According to the type of polymer used, polymer drilling fluid can be subdivided into four categories:

- Anionic polymer drilling fluid. It is a water-based drilling fluid that using anionic polymer as the major agent. This type of drilling fluid has high cutting-carrying capability, low content of submicron clay particles, low nozzle viscosity (the high-speed shearing viscosity in bit nozzle), anti-collapse and O&G formation protection capability. This drilling fluid provides better anti-collapse capability if working with potassium salts.

To ensure the high performance of anionic polymer drilling fluid, the mud solid content should be lower than 10% (best to be lower than 4%); the mass ratio of cuttings and bentonite should be within the range of 2:1–3:1; the ratio of yield point and plastic viscosity should be around 0.48 Pa/ (mPa · s). This type of drilling fluid is used to drill into the formations that have the well depth lower than 3500 m and temperature lower than 150 °C.

- Cationic polymer drilling fluid: It is a water-based drilling fluid that using cationic polymer as the major agent. Because cationic polymers have the ability to neutralize the negative electricity on the surface of clay and encapsulate it, they have a strong ability to stabilize shale. In addition, cationic surfactants can also be added, which can diffuse to clay intergranular layers where cationic polymers cannot diffuse and stabilize shale. Cationic polymer drilling fluids are especially suitable for drilling in collapse-prone shale formations.
- Zwitterionic polymer drilling fluid: It is a water-based drilling fluid that using zwitterionic polymer as the major agents. The cationic group of the zwitterionic polymer is capable of stabilizing shale, while the anionic group can enhance the stability of drilling fluid by hydration. This type of polymer is compatible with other agents; therefore it is a high-performance drilling fluid.
- Non-ionic polymer drilling fluid: It is a water-based drilling fluid that using ether-type polymer as the major agents. Other than that, there is a new type of high performance amino water-based drilling fluid. It not only has strong

anti-collapsing capability, it has also outstanding lubrication, anti-bit balling and pipe-freeing capability, which is good for increasing the rate of penetration.

Non-ionic polymer drilling fluid is non-toxic, non-pollutive, *and lubricative*. It also has strong capability of *anti-bit balling, anti-pipe sticking, borehole stabilizing*, and reservoir protecting. It is especially suitable for offshore drilling and drilling in collapsible shale formation.

- MMH drilling fluid

It is a water-based drilling fluid that using MMH as the major agents. MMH is a type of mixed-metal hydroxide crystal particle with positive charge, which can form a gel structure with bentonite to perform high shear thinning ability. This type of drilling fluid has the advantages of high cutting carrying and high wellbore stabilizing capability, good formation protection ability. It is especially suitable for drilling horizontal well and drilling into oil and gas formations.

Emulsion drilling fluid

Adding oil and oil-in-water emulsifier in water-based drilling fluid can make emulsion drilling fluid. The oil using for making emulsion drilling fluid can be mineral oil or synthetic oil. It has to be added with the water-soluble surfactant to make emulsion drilling fluid. Emulsion drilling has high lubricity, low fluid loss, and good capability of formation protection, so it is suitable for drilling the problematic formations that easy to get bit balling and pipe sticking.

Foam drilling fluid

Adding foam agent or/and foam stabilizer in the water-based drilling fluid and mixed with gas can make foam drilling fluid. Because it uses water as the disperse medium, it belongs to water-based mud. The gas that is used for making foam drilling fluid can be air, nitrogen, and carbon dioxide. The uniform and stable microfoam drilling fluid is recyclable.

Foam drilling fluid has low friction, high cutting-carrying capability, and good protection for low-pressure formations. It is mainly used for drilling low-pressure and collapse-prone formations.

3.1.3.2 Non-aqueous Drilling Fluid

This type of drilling fluid use oil (or synthetic organic chemicals) as disperse medium. It is comprised of oil (or synthetic organic chemicals), organic clays and agents. Oil-based mud also contains water, and it can be subdivided by the content of water.

Pure oil drilling fluid

The oil-based mud that has water content lower than 10% is called pure oil drilling fluid. The oil used for making pure oil mud can be mineral oil (e.g., crude oil, white oil, diesel, etc.) or synthetic organic chemicals. The organic clay used for making pure oil mud is bentonite treated with quaternary fatty-acid amine surfactant. The major agents in pure oil mud include fluid loss reducing agent (e.g., oxidized asphalt) and emulsifier (e.g., calcium stearate).

Pure oil mud has the advantages of temperature resistant, anti-collapsing, anti-pipe sticking, corrosion resistant, good lubrication, anti-contamination, and reservoir protection. However, the disadvantage is high cost, environmental pollution, and unsafe for using. The pure oil mud is suitable for drilling shale formation, salt formation, and gypsum formation, especially good for high-temperature formation and hydrocarbon-bearing formation.

Invert emulsion drilling fluid

Adding water (water content greater than 10%) in oil based mud and invert emulsifier can make invert emulsion drilling fluid. The emulsifier used for making invert emulsion is organophilic surfactant. Invert emulsion drilling fluid has the properties of pure oil mud, but the cost is lower than pure oil mud. It has the same range of application as the pure oil mud.

Synthetic oil based drilling fluid

The drilling fluid that uses synthetic organic chemicals (known as synthetic oil) as continuous phase is called synthetic-based drilling fluid, which is the substitute of conventional oil-based mud. The synthetic organic chemicals include straight-chain paraffin, straight-chain olefin, poly α-olefin, esters, and ethers. These are all environmental-friendly. This type of drilling fluid maintains the excellent properties of oil-based mud, also significantly reduces environment pollution, or prevents the interference of cutting fluorescence logging, especially suitable for drilling in offshore and other high-risk area, to obtain a desirable overall efficiency.

3.1.3.3 Gaseous Drilling Fluid

The drilling fluid that uses gas (e.g., air, nitrogen, carbon dioxide, natural gas, etc.) as disperse medium is called gaseous drilling fluid. To ensure efficient cutting carrying, the annulus return velocity of gaseous drilling fluid should be higher than 15 m/s.

Gaseous drilling fluid has low density and high drilling rate, so it can protect the oil and gas formation and effectively prevent the lost circulation. However, gaseous drilling fluid also has the shortcomings of dry friction, inflammable, or explosive. When the formation has water, it is easy to cause serious accident like pipe sticking, wellbore collapsing, etc. Under the premise of stable wellbore, gaseous drilling fluid is suitable for drilling low-pressure oil and gas formation, easy-leakage formation, and heavy oil formation.

3.2   Basic Properties of Drilling Fluids

3.2.1   Density of Drilling Fluid

The density of drilling fluid is the mass of drilling fluid in a unit volume, expressed as kg/m^3 or g/cm^3. Mud density is an important property of drilling fluid. A proper mud density can be used to balance the pressure of oil, gas, and water formation, and prevent oil, gas, and water from entering into the well which may cause well kick or blowout. In the meantime, the hydrostatic pressure of drilling fluid can offset the lateral pressure of formation rocks, maintain wellbore stability, and prevent the collapse of wellbore. The density of drilling fluid should not be too high; otherwise it could cause the problem of lost circulation; meanwhile mud density also has significant impact on the rate of penetration (ROP). In order to increase the ROP, mud density should be as low as possible if the formation condition is secure.

If the sign of lost circulation is observed, the mud density should be lowered immediately. Measures such as mechanical desanding should be taken as well as adding fresh water, air, oil, and flocculant to settle down the solid particles in the drilling fluid. If increasing mud density is required to prevent well blowout, then different weighting materials can be added to meet the requirement, such as calcium carbonate ($CaCO_3$, density is equal or larger than $2.7 \ g/cm^3$), barite ($BaSO_4$, density is equal or larger than $4.2 \ g/cm^3$), ilmenite powder ($TiO_2 \cdot FeO$, density is equal or larger than $4.7 \ g/cm^3$),etc.

3.2.2   The Rheological Properties of Drilling Fluid

The actual fluids can be divided into Newtonian fluids and non-Newtonian fluids, while non-Newtonian fluids can be subdivided into plastic fluids, pseudo-plastic fluids, and dilatant fluids. The rheological curves of the three flow types are shown in Fig. 3.3.

Fig. 3.3 Four basic types of fluid flow

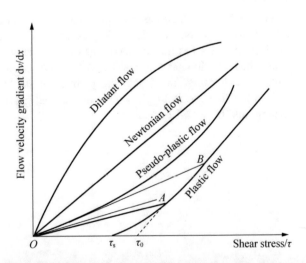

3.2.2.1 Plastic Flow

Most drilling fluids belong to plastic flow and some belong to pseudo-plastic flow. The drilling fluids made with starch agent sometimes belong to dilatant flow. The following sessions introduce the flow characteristics of plastic fluid that closely related to drilling fluid.

Properties of plastic flow

The properties of plastic flow are:

- No movement of the drilling fluid occurs until the applied force exceeds a minimum yield stress τ_s. This minimum yield stress τ_s is called static shear stress, also known as gel strength. When shear stress increases over τ_s, the plastic fluid dose not deform completely. The part that is close to the side wall deforms first, while the central part dose not deform or start to deform gradually. This fluid state is like a plug, so it is also called plug flow.
- As the shear stress keep increasing, the flow curve goes to the straight section. The reverse elongated line dose not cross the origin point; instead, it crosses with the axis of shear stress at τ_0. τ_0 is called dynamic shear stress or yield point.

Rheological equation of plastic flow

Once brought in yield point, the rheological flow state of plastic fluid can be described by the following equation:

$$\tau - \tau_0 = \mu_{pv} \frac{dv}{dx} \tag{3.2}$$

where

μ_{pv} plastic viscosity.

This equation is called Bingham equation, which is the basic equation of plastic fluid.

The rheological parameters of plastic flow and their physical meaning

- τ_s (static shear stress, gel strength): The minimum shear stress that makes drilling fluid start to flow, which is a measurement of the strength of the continuous space grid structure formed on unit area when mud is static. The continuous space grid structure is also known as gel structure, so static shear stress is also called gel strength. The value of static shear stress is related to the mud capability of cutting and weighting material suspension when mud circulation is stopped. However, the gel strength cannot be too high, otherwise it is difficult to start the pump, sometimes it could lead to formation leakage.
- τ_0 (yield point): τ_0 is obtained by elongating the straight section of rheological curve to cross with the shear stress axis, which is a hypothetical value. It is a

measurement of the strength of the continuous space grid structure when mud is in the state of laminar flow. The factors affect τ_0 and the method to adjust it is the same as τ_s. The yield point and gel strength of mud can be adjusted by changing the clay content and its dispersity, also by using inorganic or organic flocculant and mud thinner.

- Plastic viscosity μ_{pv} is obtained from Eq. 3.2.

$$\mu_{pv} = \frac{\tau - \tau_0}{\frac{dv}{dx}}$$

It is the reciprocal of the slope of plastic fluid rheological curve. It does not change with shear stress, which means it is the viscosity where the rate of breaking the structure is equal to the rate of recovering in the system. Plastic viscosity is formed in the friction that generated between the solid particles, between the solid particles and ambient liquid, and between the liquid molecules when the fluid system is in the state of laminar flow. Plastic viscosity is determined by the solid content, the shape and dispersity of solid particles, surface lubricity, and the viscosity of the liquid phase. It is directly related to the cutting carrying capacity of mud, so the value should be balanced. If the plastic viscosity of drilling fluid is too high and need to be reduced, it is common to use the solid control equipment to reduce the solid content. Increasing the system inhibitive ability and decreasing the dispersity of active solid, such as bentonite, is also useful. If it is urgent to decrease the plastic viscosity, water can be used to dilute (don't try to use this method). If plastic viscosity needs to be increased, add high-molecule organic polymer viscosifier.

- Apparent viscosity μ_{av}: Also known as effective viscosity is the ratio of the shear stress at a certain flow velocity gradient and the corresponding flow velocity, as shown in Fig. 3.3. The apparent viscosity of point A $\mu_{avA} = \tau_A / \left(\frac{dv}{dx}\right) A$, and the apparent viscosity of point B $\mu_{avB} = \tau_B / \left(\frac{dv}{dx}\right) B$. This means that the apparent viscosity of drilling fluid is not identical under different flow velocity gradient. For ease of comparison, it is required to measure the apparent viscosity at 600 rpm (see the section of measurement of rheological parameters), with mPa $\cdot$ s as the measuring unit.

As previously mentioned, the plastic flow is represented by $\tau = \tau_0 + \mu_{pv} \frac{dv}{dx}$. According to the definition of apparent viscosity, $\mu_{av} = \tau / \frac{dv}{dx}$, so $\mu_{av} = \tau_0 / \frac{dv}{dx} + \mu_{pv}$. Yield point τ_0 is related to the density and strength of the system grid structure in laminar flow, therefore, $\tau_0 / \frac{dv}{dx}$ can be considered as the structure viscosity of drilling fluid, so the apparent viscosity of plastic flow is

$\mu_{av} = \mu_{pv} + \mu_{structure} \cdot \tau_0/\mu_{pv}$ is the ratio of yield point and plastic viscosity, which indicates the proportional relation of mud structure strength and plastic viscosity. It also determines the flow state of drilling fluid in the annulus, and it is closely related to the cuttings carrying efficiency of drilling fluid. Normally, τ_0/μ_{pv} should be between 0.36 and 0.48. If τ_0/μ_{pv} is too low, it could lead to the peak shape laminar flow and cause the rotation of cuttings; if τ_0/μ_{pv} is too high, the mud structure is highly strengthened, which could increase the pump pressure. The adjustment of apparent viscosity can use the same method of adjusting the yield point and plastic viscosity.

3.2.2.2 Pseudo-Plastic Flow and Dilatant Flow

The rheological curves in Fig. 3.3 show that the rheological curves of pseudo-plastic flow and dilatant flow are both passing through the origin point, which means applying a very low shear stress will lead to flowing. The difference is the pseudo-plastic fluid will become thinner with the increase of shear rate, while the dilatant fluid will become thicker with the increase of shear rate. These two flow models can be described by the following exponential equation:

$$\tau = K / \left(\frac{dv}{dx}\right)^n \tag{3.3}$$

In the equation, n indicates the non-Newtonian level of the fluid in a certain range, so it is called flow index. When $n < 1$, it is pseudo-plastic fluid; when $n > 1$, it is dilatant fluid. K is related to the viscosity of fluid at flow velocity gradient of $1\ s^{-1}$. The higher of K, the higher of viscosity; therefore, it is called consistency coefficient. To correspond to the value required by the ratio of yield point and plastic viscosity, the value of n should be between 0.4 and 0.7.

3.2.2.3 The Measurement of Rheological Parameter

It normally uses the six-speed rotation viscometer to measure the rheological parameters of drilling fluid; the measure method is as follows.

Static shear stress

To determine the thixotropy (drilling fluid flows when sheared and becomes thick under static condition) of drilling fluid, it usually tests the initial gel (10 s gel) and final gel (10 min gel).

Determining initial gel: Stir the drilling fluid at 600 rpm for 10 s, then shut motor off and wait for 10 s. After 10 s, turn on and record the dial reading at 3 rpm. Multiply the reading by 0.511 to get the initial gel (Pa).

Determining final gel: Stir the drilling fluid at 600 rpm for 10 s, then shut motor off and wait for 10 min. After 10 min, turn on and record the dial reading at 3 rpm. Multiply the reading by 0.511 to get the final gel (Pa).

The difference on the value of initial gel and final gel reflects the thixotropic properties of drilling fluid; normally the value should be rational.

Other rheological parameters

Use six-speed rotational viscometer to measure the dial reading at 600 and 300 rpm (ϕ_{600}, ϕ_{300}), then calculate the following rheological parameters:

Apparent viscosity (mPa $\cdot$ s) $\mu_{av} = \frac{1}{2}\phi_{600}$

Plastic viscosity (mPa $\cdot$ s) $\mu_{pv} = \frac{1}{2}\phi_{600} - \phi_{300}$

Dynamic shear stress (yield point) (Pa) $\tau_0 = 0.511\left(\phi_{300} - \mu_{pv}\right)$

Flow index $n = 3.321\text{g}\frac{\phi_{600}}{\phi_{300}}$

Consistency coefficient (mPa $\cdot$ s^n) $K = \frac{0.511\phi_{600}}{1022^n}$.

3.2.2.4   The Relation of Mud Rheology with Drilling Engineering

The mud rheology is related closely to the drilling engineering. Neither laminar flow nor turbulent flow is good for carrying cuttings, but modified laminar flow is beneficial for carrying cuttings and cleaning the wellbore. It also causes very light erosion effects to the wellbore. It requires the ratio of yield point to plastic viscosity $\frac{\tau_0}{\mu_{pv}}$ to be between 0.36 and 0.48 (or flow index $n = 0.4$–0.7). Drilling fluid has the property of shear thinning (the apparent viscosity decreases with the increase of shear rate), and low turbulent friction in bit nozzle is good for increasing the rate of penetration, and good for the suspension of cuttings and weighting materials. However, the gel strength should not be too high; otherwise it could affect the pump and generate the pressure surge. For different area, different drilling type requires different value of rheological parameters, which can be used according to the actual drilling conditions.

3.2.3   The Filtration Properties of Drilling Fluid

3.2.3.1   Filtration Property

When the bit drills through the porous permeable formation, the hydrostatic pressure of drilling fluid is normally greater than the formation pressure. Due to the pressure difference, the liquid phase in the mud starts to infiltrate the formation, and this process is called mud filtration. In this process, a layer of particle sediment is formed on the surface of borehole—filter cake. The creation of filter cake starts from plugging the large pores with large particles, then the medium sized particles plug the pores between the large particles. The plugging process goes forward until the smallest pores are plugged. Generally speaking, the permeability of filter cake is orders of magnitude lower than the permeability of formation rock. Therefore, the filter cake can prevent mud filtrate from invading into the formation and protect the wellbore as well. The property of formation of filter cake is called wall-building ability.

3.2.3.2   Concepts of Mud Filtration

Different drilling conditions cause different types of filtration loss. The brief introduction is as follows.

Spurt loss

At the moment when drill bit crushes the bottom rock and forms the borehole, drilling fluid infiltrates into the pores of formation rapidly. The fluid loss that occurs in the period before the formation of filter cake is called spurt loss. In static filtration test, when the pressure gas valve is open, the graduated cylinder starts to collect the mud or cloudy fluid, which is the spurt loss.

Spurt loss can be beneficial or not. It is beneficial to increase the ROP, but when the hydrocarbon-bearing formation is drilled, the spurt loss can be detrimental to the reservoir and decrease the permeability of hydrocarbon-bearing formation. At this time, the spurt loss needs to be controlled.

The factors affecting the spurt loss include the pore size of formation, mud solid content and particle size distribution, mud viscosity and filtrate, etc.

Dynamic filtration

As the drilling progresses, after spurt loss happened, another layer of filter cake is formed on the wellbore. The filter cake is getting thick and strengthened; meanwhile the filter cake encounters the damage from the erosion of drilling fluid, scratch, and collision from drill pipe. When the speed of filter cake formation equals to the speed of erosion, the thickness of filter cake does not change, and the fluid loss rate remains constant. The mud filtration process in the circulation is called dynamic filtration. Dynamic filtration is characterized by its thin filter cake and large filtrate volume. The rheological parameters of mud are not only affected by the formation condition, pressure difference, mud viscosity, solid content, and particle size distribution, but also related to dynamic filtration. The thickness of filter cake is related to the velocity and flow state of mud. The higher the velocity is, the worse the filter cake erosion is. The filter cake will be thinner, and the filtrate volume will be greater. The erosion generated by turbulent flow is worse than laminar flow, so the filtrate volume is greater than the one from laminar flow.

Static filtration

In the action of run in hole (RIH) and pull out of hole (POOH) or dealing with the accidents, the mud circulation has to stop; the filter cake on the borehole cannot be swept away by mud. As the filtration process goes on, the resistance from filter cake becomes higher, the filtration rate becomes lower, and hence the filtrate volume becomes lower. The filtration process happened after the mud circulation is stopped is called static filtration. Comparing with dynamic filtration, static filtration is simple. The normally mentioned filtration process is static filtration. Static filtration equation is:

$$V_f = A\sqrt{\frac{K\left(\frac{C_c}{C_m} - 1\right)\Delta p t}{\mu}}$$ (3.4)

In the equation

V_f Mud filtrate volume, mL;
A Filtrate area, cm^2;
K Permeability of filter cake, μm^2;
C_c Solid content in filter cake (volume fraction), %;
C_m Solid content in mud (volume fraction), %;
Δp Pressure difference, 10^5 Pa;
t Filtration time, s;
μ Mud filtrate viscosity, mPa · s.

The most effective way to control fluid loss is using fluid loss agents to control the permeability of filter cake.

3.2.3.3   Fluid Loss Agents and Their Function Mechanism

The addition of fluid loss agents effectively reduces the loss of drilling fluid. The mechanisms of fluid loss protection are listed as following:

Colloid protection effect

To form a low permeable filter cake, the solid particles should distribute widely, and the mud system should have a lot of colloid particles. The function of fluid loss agent is to be adsorbed on the surface of clay particles and form adsorption layer to prevent the clay particles to flocculate. Another function is to adsorb the fine particles, which are broken up by the mud stirring, onto the molecular chains, where the fine particles can not bond together to grow. Thus, the number of fine particles has increased significantly, which enables the mud to form a thin and tight filter cake, hence to reduce the fluid loss. This function is called colloid protection effect.

Increasing the hydration film thickness of clay particles helps to reduce the fluid loss

According to the static filtration equation, if other conditions remain unchanged, increasing the deformability of clay particles and reducing the solid volume fraction of filter cake (reduce C_c/C_m), reduces the fluid loss. Because the fluid loss agent is adsorbed on the clay particles in the mud, which decreases the hydration of clay particles, therefore, the hydration film around the clay particles becomes thicker, to make the filter cake much easier to deform under the pressure difference, hence reduces the filter cake permeability.

Increase filtrate viscosity to reduce fluid loss

According to the static filtration equation, filtrate volume is reversely proportional to mud filtrate viscosity to the power of ½. Most of the fluid loss agents are long-chain chemical compounds, which increase the filtrate viscosity if added in mud, hence reduce fluid loss.

The blocking effect of the molecules of fluid loss agent

The molecular size of most polymer fluid loss agents is in the range of colloid scale. Adding these agents increases the content of colloid particles in the mud. They have blocking effect on the filter cake. The macromolecules block the pore space in two ways, the first way is to insert the long molecular chain into the pores of filter cake, and the second is to curl the long molecular chain into the coil, so that the micropores of filter cake are blocked, which make the filter cake thin and tight, hence reduces fluid loss.

Other than that, blocking agent such as asphalt can enhance the quality of filter cake, and forms a thin and tight filter cake with high toughness, that will efficiently reduce the high-temperature high-pressure fluid loss.

The commonly used fluid loss agents include Na-CMC (sodium carboxymethyl cellulose), SMP (sulfonated phenolic resin), NH4HPAN (ammonium hydrolyzed polyacrylonitrile), Ca-HPAN (calcium hydrolyzed polyacrylonitrile), SPNH (sulfonated lignite resin) and sulfonated asphalt powder, etc.

3.2.3.4 The Measurement of Fluid Loss and Technical Requirement

There are two types of fluid loss measured in laboratory and onsite:

Static fluid loss

Static fluid loss is also known as API fluid loss, which is measured by API filter press. The filtrate volume (mL) measured for 30 min at room temperature with 0.7 MPa pressure difference is the static fluid loss. To save time, normally use the filtrate volume that measured in 7.5 min and multiplied by 2 to get API fluid loss. Because $V_{30} = f(\sqrt{30}) = f(2\sqrt{7.5}) = V_{7.5} \times 2$.

In the drilling process, different time and different formation has different requirements for the API fluid loss. For the upper formation and hard formation, the API fluid loss can be higher; for the formations prone to collapse, API fluid loss should be low; when drilling oil and gas formations, API fluid loss should not be higher than 5 mL.

High-temperature high-pressure fluid loss

To simulate the formation temperature and pressure condition, it must use high-temperature high-pressure (HTHP) filter press to measure the HTHP fluid loss of drilling fluid. It requires that the filtrate volume measured for 30 min at 150 °C (or simulate formation temperature), 3.5 MPa of pressure difference and multiplied by 2 (the HTHP filtration area is half of API), to get the HTHP fluid loss. The requirement to HTHP fluid loss is the same as API fluid loss. Normally the HTHP fluid loss during drilling reservoir should not be higher than 15 mL.

3.2.4   Solid Content and Control in Drilling Fluid

The sources of solids include clay, drill cuttings, weighting material, solid of agents, etc. Solid particles in drilling fluid have significant effect on the mud density, viscosity, and shear stress, while these properties are directly related to the hydraulic parameters, ROP, drilling cost, and complex downhole situation. The solid control is one of the most important technical areas of drilling fluid technology.

High solid content of drilling fluid can cause many problems: generating thick filter cake that lead to differential pressure sticking; causing high filter cake permeability and high fluid loss, which lead to reservoir damage and wellbore instability; causing serious abrasion of drill bit and drill pipe; especially causing the decreasing of ROP.

3.2.4.1   The Effect of Solid on ROP

The effect of solid content on ROP

ROP decreases with the increase of mud solid content. If the ROP of drilling with fresh water is 100%, when the solid content increases to 7%, the ROP will be reduced by 50%. Many research results indicate that every 1% reduction of solid content will increase the ROP at least by 10%. The relation of solid content in drilling fluid and ROP is shown as Fig. 3.4.

Fig. 3.4 The correlation between the solid content in drilling fluid and ROP

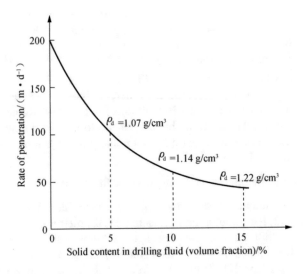

The effect of solid type on ROP

It has been found that different types of solids have different effects on the ROP. Inert solids such as sand or barite have low impact on ROP; drill cuttings and low-quality clay have medium impact while the high-quality clay has the highest impact. When drilling fluids have the same solid content, different particle size will lead to different ROP. The colloid particles with diameter less than 1 μm have the greatest effect on ROP. Experiment results indicate that the particles with diameter smaller than 1 μm have 13 times bigger effect on ROP than the particles with diameter bigger than 1 μm. This can be verified by the big difference of ROP between the disperse drilling fluid and non-disperse drilling fluid that have the same solid content (Fig. 3.5).

Figure 3.4 also indicates that, in different solid content range, the increase amplitude of ROP with the decrease of solid content is different. When solid content is under 7% (equivalent to mud density of 1.08 g/cm³), the ROP increases very quickly with the decrease of solid content; but if solid content is over 7%, to increase ROP by decreasing the solid content will not be effective.

3.2.4.2 Methods of Solid Control

Sedimentation

This is a common method on the rig-site. Utilizing the density difference of solid phase and liquid phase, drill cuttings settle down by gravity in the mud pit.

Removal by mechanical solid control equipment

The commonly used mud solid separation equipment is shale shaker, cyclone separator, and centrifuge. According to the size of removed solid particle, cyclone separator can be subdivided into desander, desilter, and ultra-cyclone separator. The size distribution of particles removed by solid control equipment is shown in Fig. 3.6.

Fig. 3.5 The effect of mud type with the same solid content

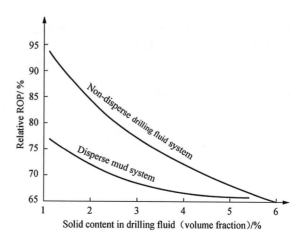

Solid content in drilling fluid (volume fraction)/%

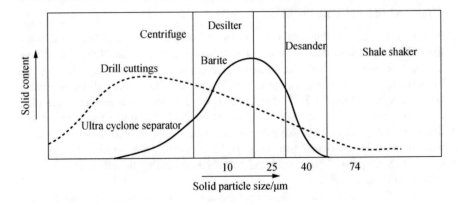

Fig. 3.6 Size distribution of particles removed by solid separation equipment

Mechanical equipment removes solid without increase the mud volume and do not need to supplement a lot of agents; therefore it is beneficial to reduce the drilling cost. In the meantime, it has low effect on the mud properties, which is good for the regular downhole drilling. This method is considered as the best among all the solid control methods.

Chemical control method

Chemical control is the method that flocculating the solid particles by adding flocculant in the mud so that it is easy to be removed by sedimentation or solid control equipment. This method is able to remove the solid particles with diameter of less than 5 μm, while the simple methods like sedimentation and mechanical separation can only remove the particles larger than 5 μm. Combining the chemical method and mechanical solid control equipment can give higher efficiency of solid removal.

Dilution method

Dilution is a simple method that adding disperses medium (water or oil) or low solid mud into the drilling fluid to temporarily decrease the solid content of drilling fluid. Adding disperse medium can affect other properties of drilling fluid and increase the treatment cost. The extra drilling fluid also needs to be discharged; therefore, it is not recommended to use dilution method.

3.2.4.3 Solid Control Equipment

The solid control equipment can be classified into three types: shale shaker, cyclone separator, and centrifuge.

Shale shaker

Shale shaker utilizes the high-frequency vibration to remove the solid from drilling fluid that flows on the screen cloth. The particles with diameter bigger than the sieve size of screen stay on the shaker and move forward with the screen cloth, but the solid particles that are smaller than the sieve diameter will pass through the screen and flow into the mud tank.

The screen of shale shaker has many sieve sizes. Choosing shaker screen should be based on the mesh size that is related to the size of drill cuttings and mud solids, and the percentage of solids with different sizes. Normally shale shakers use the 40-mesh screen or bigger. Shale shaker is mainly used to remove the drill cuttings and sand with particle diameter larger than 0.5 mm. The tightest shaker screen is 200-mesh, which can remove coarse barite particles.

Cyclone separator

Cyclone separator removes the coarse particles under the effect of centrifugal force. According to the removed particles size, cyclone separator can be subdivided into the following three types:

- Cyclone desander: It removes solid particles with diameter higher than 74 μm and also a small proportion of solid particles of 40–74 μm.
- Cyclone desilter: It removes solid particles with diameter between 10 and 74 μm and mostly between 20 and 40 μm.
- Ultra-cyclone separator: It removes solid particles with diameter between 5 and 10 μm. It can separate bentonite from the mud.

Centrifuge

Centrifuges use the centrifugal force to throw the solid particles with bigger mass to the side wall of separation chamber (shell) and settle down, then the spiral conveyor pushes the stack of solid particles to the top of centrifuge, and discharge from the hole on the top, while the mud is discharged from the hole on the other side of centrifuge.

Centrifuge has two functions:

- Remove solid from unweighted mud: In this situation, centrifuge is used to remove coarse particles in the 2 and 5 μm diameter ranges, mostly silt. The mud discharged from the other end of centrifuge will return to the circulation system.
- Recycle the weighting material from weighted mud: After drilling fluid pass through centrifuge, barites and other coarse particles are separated and recycled, while drilling fluid discharged from the other end will be discarded.

To achieve an effective solid control, the above solid control equipment should be combined together, and the process of solid removal must follow the sequence of shale shaker, desander, desilter, centrifuge, ultra-cyclone separator to arrange. If only use some of the equipment, it cannot achieve a good solid control performance. If only use shale shaker and desander, it can only separate coarse particles of a diameter larger than 74 μm; if only use desilter, centrifuge, and ultra-cyclone separator, the equipment will be plugged by coarse particles and will not work properly.

To increase the working efficiency of solid control equipment and remove the colloid particles (smaller than 1 μm) that have the greatest effect on ROP, it is effective to combine with polymer flocculants.

3.2.4.4   Polymer Flocculants

In the end of 1960s, polymer flocculants are introduced into the drilling fluids. The high-performance non-disperse low solid polymer drilling fluid start to be used, which significantly increased the ROP. China started to research and developed this type of drilling fluid and related agents since the year of 1974, and then promoted this mud system in all oil fields over the country. Now it is an indispensable additive in the jet drilling technology. The brief introduction to organic polymer flocculants is as follows.

Types of organic polymer flocculant

According to the interaction of organic polymer flocculant and the mud solid, it can be divided into two types: Complete and selective flocculant.

- Complete flocculant: The polymer that makes all solids (includes bentonite, drill cuttings, etc.) flocculated and settled, e.g., polyacrylamide (PAM).
- Selective flocculant: The polymer that only flocculates drill cuttings and low-quality clays, but does not flocculate bentonite, e.g., partially hydrolyzed polyacrylamide (HPAM) with proper degree of hydrolysis.

The commonly used organic polymer flocculants are PAM, HPAM, VAMA, etc.

Working mechanism of organic polymer flocculant

The flocculation of organic polymer to the mud solid particles has three stages as following:

- Adsorption: Use the adsorption group of molecular chain (e.g., $-OH$, $-CONH_2$) to form hydrogen bond with the oxygen atom or hydroxyl group on the clay surface, so to take place the preferential adsorption, in the meantime, use the iodized group (like $-COONa$) on the molecular chain to take place the electrostatic adsorption with the broken bond from clay particles.
- Bridging: Because the molecular chain of flocculant is relatively long, there are many adsorption groups on the molecular chains; therefore one long chain can

adsorb many clay particles at the same time. This process is the bridging effect of long-chain molecule in the clay particles.

- Flocculation and settle down under gravity: When the bridging effect is completed, the molecular chain of polymer and its chain segment start to rotate and move, to get the small clay particles together, and form flocculated blocks. The flocculated blocks settle down under the effect of gravity and are removed from the mud.

The dosage of polymer flocculant

To achieve good flocculation performance, the dosage of flocculant must be proper. Theories and experiments demonstrate that only when the dosage of flocculant is the half of the saturation adsorption capacity of solid in the mud, the flocculation can achieve the optimal performance.

3.2.5 The Lubrication Property of Drilling Fluid

In drilling process, drilling fluid changes the dry friction between drill pipe and borehole into the wet friction, hence reduce the frictional resistance. The property of drilling fluid that reduces the frictional resistance is called mud lubrication, which includes the lubrication of mud cake and the lubrication of the drilling fluid itself.

The mud lubricity can be measured by coefficient of friction (COF). COF is under certain conditions, the ratio of the force of friction generated by two moving objects (solid) and the normal force. The so-called certain condition is the moving speed of objects in relative motion and if liquid exists between them. Using the filter cake COF tester can measure the COF of drilling fluid. In the simulated wet friction environment between drill pipe and borehole, use the filter cake generated in the filtration test to measure the COF. The extreme pressure lubricity tester can measure the COF of mud. The lower of COF, the higher is the mud lubricity. The mud lubricity has significant effect on the abrasion of drill pipe, prevention of stuck pipe, and increasing of ROP. To most water-based drilling fluid, the COF of mud should be maintained around 0.2, while the COF of horizontal drilling fluid should be kept in the range of 0.08 to 0.10, or even lower. The COF of most oil-based mud is between 0.08 and 0.09.

Adding lubricant in the drilling fluid can improve the lubricity of drilling fluid. The material used to improve the lubricity of mud is called mud lubricant. The common mud lubricant can be classified into liquid lubricant and solid lubricant. Liquid lubricants mostly are oil products, include vegetable oil (e.g., soy bean oil, cotton seed oil, castor oil), animal oil (e.g., lard) and mineral oil (e.g., kerosene, diesel, mechanic lubricant oil, white oil or crude oil, etc.). To form a layer of film on the friction surface, surfactant can be added into the drilling fluid. Solid lubricant includes solid spheres, graphite powder. The commonly used solid spheres are plastic spheres and glass spheres. The plastic spheres are polyamide (nylon) spheres and the copolymer of polystyrene and divinyl benzene spheres; they are all have advantageous temperature resistance, pressure resistance, and chemical inertness; therefore they can be used as the lubricant for all kinds of drilling fluids. Glass spheres can be made of glass with different composition (sodium glass, calcium glass), which also have temperature resistance and chemical inertness. The cost of glass spheres is lower than plastic spheres, but the compressive strength is lower than plastic spheres, and easy to settle down. Graphite powder can be strongly adsorbed on the surface of borehole. It has high-temperature resistance and no fluorescence, no impact on the major mud properties, low dosage, and good lubrication. It also has good filtration control and anti-collapsing and reservoir protection performance. Solid lubricant either forms a lubrication film between drill pipe and borehole or changes sliding friction into rolling friction, hence to reduce the frictional resistance.

3.2.6   Well Collapsing and Anti-collapsing Methods

Well collapsing is the phenomenon that ambient rocks and chips fall into the wellbore, generally called wellbore instability. The wellbore stability is the ability of wellbore to maintain its original state. If wellbore cannot maintain its original state, it is called wellbore instability, mainly indicated as wellbore enlargement, wellbore shrinkage, or formation fracturing. Wellbore instability is a complex technical problem which is related to the safety, quality, ROP, and cost of drilling operation. Wellbore collapsing is an important topic of drilling fluid technology and drilling engineering. Even though a lot of research has been conducted and gained lots of results, there are still many problems need to be studied and solved.

When wellbore collapsing happens, the following phenomenon will occur:

- The mud viscosity, shear stress, density, and sand content are all increased. Pump pressure rise and fall, sometimes pump pressure build up suddenly;
- The amount of drill cuttings returning to surface increases and mix with different types of rocks. Sometimes it returns formation rocks from the upper section, sometimes it returns with large rock cuttings;
- POOH is resisted, RIH cannot reach the bottom. The annulus is blocked sometimes and causes pump pressure build-up and stuck pipe.

Wellbore collapsing causes the drilling to not proceed normally, low ROP, poor cementing quality, and wellbore abandoned when the problem is severe. These problems can lead to huge economic loss.

3.2.6.1   Reasons for Wellbore Collapsing

Wellbore instability is a complex technical problem that involves with mud chemistry, rock mechanics, porous media flow mechanics, geology and drilling technology. The reason that causes wellbore instability has the following three aspects.

- Geological and mechanical factors: Pressure release from abnormal pressure formation, drilling into geological crushed zone, fault, formation containing abundant microfractures and coal seams, etc.
- Engineering factors: High mud flow rate erodes the borehole; POOH and RIH are too fast that cause pressure surge to fracture the formation; the mud hydrostatic pressure is lower than formation pressure; the time of wellbore being soaked in mud is too long, etc.
- Physicochemical factors: The hydration effect generated by the contact of shale and water can cause well instability. The clay mineral content is very high in shale formation. If the shale formation mainly contains swelling clay mineral, it will cause hydration and dispersion of shale formation when contact with water, which normally lead to borehole shrinkage and/or downhole viscosity build-up; if the shale formation only contains non-swelling clay mineral (e.g., illite, kaolinite), when contact with water-base mud, it will partially hydrate and cause sloughing (hole enlargement) or stuck pipe, which cause wellbore instability. Statistically, most of the wellbore instable problems (about 90%) are related to shale formation.

The wellbore instability of complex formation is normally caused by the combination of the above factors.

3.2.6.2 Anti-collapsing Methods

Because the major factors causing wellbore instability are different, the methods (coupling anti-collapsing measures) to control wellbore stability are also different.

For the wellbore instability caused by geological factors, it is practical to increase mud density (on the premise that it can effectively plug borehole and prevent the pressure propagation and the invasion of mud filtrate into the formation), so the hydrostatic pressure of drilling fluid can provide effective pressure to support the wellbore (mechanically stabilize wellbore), or use chemical well stabilizing methods.

For the wellbore instability caused by physicochemical factors (hydration), it mainly uses shale inhibitor to improve the inhibition ability of drilling fluid, so to enhance the blocking ability and properly adjust the mud density.

For the wellbore instability caused by engineering factors, it should take effective measures such as improving drilling technology as a precaution.

The chemical agents that can effectively inhibit shale from hydration and/or dispersion (peeling off) are called shale inhibitor. Shale inhibitor can be divided into the following types:

Inorganic salt and organic salt

Inorganic salt shale inhibitors include sodium chloride, ammonium chloride, potassium chloride and calcium chloride, etc. Organic salt shale inhibitors include formate (sodium formate, potassium formate, cesium formate) and acetate. When over a certain concentration, any water-soluble salts will have the ability to inhibit shale from hydration and swelling. Inorganic salts normally stabilize the shale formation by suppressing the electrostatic double layer on the surface of shale (clay) and decreasing the zeta potential. Although any water-soluble salts have the inhibitive ability, their performance is different. In the water-soluble salts, potassium salt and ammonium salt have the best shale inhibitive ability because their cation diameter is close to the diameter (0.288 nm) of the hexagonal ring made of oxygen atom in the bottom of silica oxygen tetrahedron. When the cation enters the clay lattice space, it will have lattice fixation effect, which bond the clay lattice together; therefore it provides strong inhibition on the hydration and dispersion of shale. Normally salt inhibitor combined with polymer gives better inhibitive performance.

Cationic surfactants

The activating cationic group of cationic surfactants adsorbed on the surface of shale, to neutralize the electronegativity of the surface of shale, and alternate the shale surface wettability into oil wet (hydrophobic). It also inhibits the hydration and dispersion of shale, so to stabilize the wellbore.

Other than that, part of organic cationic chemicals with low molecular weight can also be used as shale inhibitor.

Polymer shale inhibitors

Polymer shale inhibitors normally inhibit the shale by multi-point adsorption (encapsulation). The polymer shale inhibitors include cationic, non-ionic, zwitterionic, and anionic polymer. Especially the amino polymer used in the newly developed high-performance amino water-based drilling fluid has strong ability to inhibit the hydration between the interlayer surfaces of clay, which shows strong inhibitive ability.

Modified asphalt

Asphalt is comprised of small amount of hydrocarbon compound (molecule only has carbon and hydrogen) and large amount of non-hydrocarbon compound (besides for carbon, hydrogen, also has oxygen, sulfur and nitrogen elements). Asphalt includes natural asphalt, petroleum asphalt, and tar asphalt. Natural asphalt is formed by petroleum under the natural condition; petroleum asphalt is obtained from the petroleum refining and processing; tar asphalt is obtained from the dry distillation of wood and coal.

There are two important modified asphalt products: one type is oxidized asphalt, which is the residual oil from atmospheric distillation and reduced pressure distillation, or cracked residue oxidized by air at high temperature (200–220 °C). It is mainly used in oil-based mud; the other type is sulfonated asphalt, which is the asphalt sulfonated by sulfonating agent (e.g., concentrated sulfuric acid, oleum, or sulfur trioxide). The major component is sulfonate asphalt, which can be dispersed in water-based mud.

Modified asphalt product mainly works by adsorbing on the shale surface to block the pore in the shale and forms a hydrophobic oil film. It also can combine with other inert blocking material (e.g., ultra-fine calcium carbonate powder) when necessary to improve the filter cake quality, effectively prevent the invasion of pressure and mud filtrate, reduce the contact of water and shale, hence to stabilize the shale formation.

To achieve desirable anti-collapsing performance, it usually combines inorganic salt inhibitor, polymer, and blocking anti-collapsing agent in the drilling fluid.

3.2.7 Lost Circulation of Drilling Fluid and Plugging

3.2.7.1 Lost Circulation of Drilling Fluid

During drilling, the phenomenon that large amount of drilling fluids flow into the formation under the effect of pressure difference is called lost circulation. Severe lost circulation not only causes the loss of huge amount of drilling fluid, but also leads to wellbore collapsing, even blowout and other major accidents. There are several essential conditions that can lead to lost circulation: mud pressure in the wellbore is higher than the formation pore pressure, which is the pressure difference; formation contains porous flow channel (natural or induced) and space that is enough for containing fluid, where the opening size of the flow channel should be bigger than the size of solid particles in the drilling fluid. Mud pressure in the wellbore includes the hydrostatic pressure of mud, circulation pressure drop, and borehole surge pressure. To solve the problem of lost circulation, we should focus more on prevention. The major prevention method for lost circulation includes the proper design of well structure, decreasing the wellbore mud pressure as low as possible, and blocking the formation to increase the pressure-bearing capacity.

Based on the type of formations, the lost circulation can be divided into three types.

Permeable loss

The loss of circulation happened in highly permeable sandstone formation or gravel formation is called permeable loss. It is characterized by low rate of loss (0.5–10 m^3/h) and normally indicated by the slow drop of mud pit fluid surface.

Fractured loss

The loss of circulation happened in fractured formation is called fractured loss. The fractures that cause lost circulation include the naturally existed fracture (natural fracture) in limestone and sandstone, and the induced fracture formed by the mud pressure fracturing limestone and sandstone. It is characterized by relatively high rate of loss (10–100 m^3/h) and normally indicated by the quick drop of mud pit fluid surface.

Karst cave loss

The loss of circulation happened in the karst cave style formation is called karst cave loss. This type of loss only happens in limestone formation. It is characterized by the sudden total loss (higher than 100 m^3/h) without returning mud.

3.2.7.2 Lost Circulation Plugging

The material used for plugging formations of lost circulation is called lost circulation material (LCM). Different formations need different LCMs. The methods for plugging formation loss can be summarized as: adding the inert bridging LCMs into the mud to plug while drilling, or pump chemical plugging fluid to conduct stage plugging. For the mild lost circulation happened in permeable formation and microfractured formation, it should use the method of plugging while drilling, which is mixing the inert bridging blocking material with different shapes (particles, flakes, fibers) and different sizes (coarse, medium, fine) in the drilling fluid, and pumping into the formation while drilling, hence to block formation and enhance the pressure-bearing capability. For the problematic lost circulation, it can use the chemical stage blocking method.

3.2.8 Pipe Sticking and Stuck Freeing

Pipe sticking is the phenomenon that drilling tools are stuck in the wellbore and can not move. It is a complex situation frequently encountered in the drilling process. The pipe sticking that closely related to the mud properties is referred to as differential sticking. Differential sticking is a complex issue that the drilling pipe adhere to the mud filter cake and get stuck due to the difference between the mud pressure and the formation pressure. The mitigating action that sets free of the sticking pipe is called stuck freeing.

To lower the pulling force needed to free the differential sticking, it can only be achieved by decreasing pressure differential (mud density) and COF of filter cake and drill pipe. The latter method is commonly used, which is adding pipe free agent if the problem is serious. Pipe free agent is the chemical used for reducing the COF of filter cake and drill pipe. It can be liquid lubricants mentioned in the previous sections, which is the diesel viscosified by asphalt. To emulsify the viscosified diesel, it should normally use emulsifiers including water-soluble surfactants (polyoxyethylene alkyl alcohol ethers) and oil-soluble surfactants (calcium oleate).

To avoid the fluorescence matters in diesel and asphalt that can affect the interpretation of mud logging and petrophysical logging, only surfactant-based pipe free agents should be used, for example, alkyl sodium sulfonate or sodium alkyl benzene sulfonate with low molecular alcohol (methanol, propanol) and salt (NaCl). After adjusting the Hydrophile–Lipophile Balance (HLB), it can be made into pipe free agent, but their performance is not as good as the viscosified diesel pipe free agent.

3.2.9 Drilling and Completion Fluids and Formation Protection

Drilling & completion fluid is the type of drilling fluid used in drilling reservoir formations. When drilling into the target formation, the drilling & completion fluid should not only meet the requirements of drilling engineering and geology, but also meet the demands of formation protection to mitigate or prevent formation damage. Generally speaking, any condition that hinders the reservoir fluid flow from the surrounding formation into the wellbore is called formation damage. Severe formation damages can cause very negative impacts on the productivity of an oil and gas well, or even delay the discovery of hydrocarbon reservoirs. Any types of working fluid (completion fluid) or operations that have contact with the hydrocarbon reservoir can possibly cause formation damage.

During drilling, the invasion of mud solid particles and/or the migration of formation particles cause solid plugging. On the other hand, the incompatibility of drilling fluid component with the formation rocks and fluids can cause complex physical and/or chemical reactions (e.g., hydration and swelling of formation clay, precipitation, emulsion blocking, etc.), which can also increase the flow resistance to oil and gas and cause the reduction of formation permeability, which subsequently causes formation damage. Therefore, we must adopt the formation-protecting drilling fluid technology. For instance, before drilling into a reservoir, we should adjust the mud density and rheological parameters into a proper range. Try to conduct near-balanced drilling as much as possible, or use underbalanced drilling if drilling conditions allow, in order to reduce or prevent the invasion of mud solid or liquid into formations; the effective temporary-blocking agents should be selected to improve the mud filtration properties, which can reduce the filtrate loss volume as much as possible; for the water-sensitive formations, we should use highly inhibitive drilling and completion fluid system or use low solid or non-clay systems; the treating agent should be compatible with the reservoir formations.

There are various formulations of the commonly used drilling & completion fluid systems, e.g., non-solid brine completion fluid, low-bentonite polymer completion fluid, modified drilling fluid, etc.

Exercises and Questions

1. What is the relation between drilling fluids and drilling engineering? What are the functions of drilling fluids?
2. Explain the basic composition of drilling fluids and the classification of drilling fluids.
3. What is the flow model for most drilling fluids? Write its common rheology equation.
4. Explain the physical meaning of gel strength, yield point, apparent viscosity, and plastic viscosity. How to adjust these parameters?
5. If there is a laminar flow in the tube, tube length $L = 15$ cm, the pressure difference of two ends $\Delta p = 3$ kPa, calculate the shear stress on the liquid layer where radius $R = 3$ cm.
6. How does the mud flow state affect the drilling operation?
7. If the drilling fluid is measured by rotational viscometer, and get $\phi_{600} = 76$, $\phi_{300} = 54$, calculate μ_{av}, μ_{pv} and τ_0.
8. Explain the meaning of spurt loss, dynamic filtration, and static filtration.

9. How does the fluid loss agent work?
10. Summarize the commonly used mud lubricants and their working mechanisms.
11. Summarize the types of lost circulation and the commonly used blocking agents.
12. What are the factors causing wellbore instability?
13. From the perspective of drilling fluid, what measures should be taken to prevent wellbore from collapsing?
14. Explain the effects of mud solid on ROP.
15. How many types of solid control equipment? What is the working range for each type?
16. What are the major reasons to cause formation damage? How to prevent them?
17. Which completion fluids are suitable for drilling hydrocarbon reservoir?

Drilling Parameters Optimization

4

Abstract

Relationships between drilling parameters of weight on bit, rotary speed, tooth and bearing wear, hydraulic power, and rate of penetration (ROP) as well as drilling bit wear are first analyzed. After that, the method of optimizing mechanical drilling parameters is presented. Finally, the method of optimizing hydraulic parameters is introduced in terms of maximum hydraulic power and jet impact force based on drill bit flow analysis and frictional pressure loss evaluation.

The definition of drilling is to form a well hole with designed trajectory shape from the surface to the underground by means of certain equipment, tools and techniques. In drilling process, a great deal of work is to break rock and deepen the wellbore. During drilling process, the ROP, cost, and quality of drilling will be influenced and restricted by many factors, which can be divided into controllable and uncontrollable factors. Uncontrollable factors are objective factors, such as formation lithology, reservoir depth, and formation pressure. Controllable factors refer to factors that can be artificially adjusted through certain equipment, tools and technical methods, such as surface pump equipment, bit type, drilling fluid performance, weight on bit (WOB), rotation speed, pump pressure, and flow rate. The so-called drilling parameters are important properties of the equipment, the drilling fluid, and the operating conditions which represent the controllable factors in the drilling process. Optimization of drilling parameters refers to using optimization methods to select reasonable drilling parameters to make the drilling process achieve the best technical and economic indicators according to the influence of various factors on ROP under certain objective conditions.

© China University of Petroleum Press and Springer Nature Singapore Pte Ltd. 2021
Z. Guan et al., *Theory and Technology of Drilling Engineering*,
https://doi.org/10.1007/978-981-15-9327-7_4

4.1   Basic Relationship Between Various Parameters in Drilling Process

The premise of parameter optimization is to analyze the main factors affecting drilling efficiency and the basic rules in drilling process, and to establish corresponding mathematical models.

4.1.1   Main Factors Affecting ROP

The rock characteristics and bit types introduced above have a significant impact on ROP. Additionally, WOB, rotation speed, bit tooth wear, hydraulic factors, and drilling fluid performance are also main factors affecting ROP.

4.1.1.1   Effect of WOB on ROP

In drilling process, bit teeth drill into formation and break rocks under the action of WOB. The value of WOB determines the depth of drilling into the rock and the size of rock fragmentation. Therefore, WOB is one of the most direct and significant factors affecting ROP. The effect of WOB on the ROP has been studied for a long time. The drilling practice in oil field shows that the typical relationship between WOB and ROP is shown in Fig. 4.1 [1]as the other drilling conditions remain unchanged.

As can be seen from Fig. 4.1, the WOB is approximately linear with the ROP in a large range of variation. At present, the WOB values in the actual drilling process are generally changed within the linear range of AB in the diagram. This is mainly because, before A point, the WOB is too low and the ROP is very slow; after B point, the WOB is too large, even the bit teeth could completely enter into the

Fig. 4.1 Relationship curve of WOB and ROP

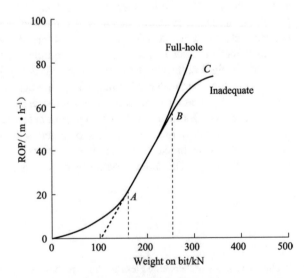

formation. However, the amount of rock cuttings becomes excessive; the cleaning condition of bottom hole is difficult to improve. Also, the drill bit wear intensifies. Hence, as the WOB increases beyond the linear stage, the improvement effect of ROP is not obvious, and even becomes worse. Therefore, in practical application, the quantitative relationship between WOB W and ROP v_{pc} is established on the basis of the linear section in Fig. 4.1.

$$v_{pc} \propto (W - M) \tag{4.1}$$

where

v_{pc} ROP, m/h;
W WOB, kN;
M Threshold WOB, kN.

Threshold WOB is the intercept of extrapolated line of AB, which is equivalent to WOB when tooth begins to press into formation. The value of it is mainly determined by rock property. Threshold WOB in different regions cannot be referenced by each other.

4.1.1.2 Effect of Rotational Speed on ROP

The effect of rotational speed on ROP is a problem that has already been well realized and solved. With the increase of rotational speed, ROP is changed exponentially, but the index is generally less than 1. The main reason is that contact time between bit and rock is shortened, and rock fragmentation depth decreases at each contact after speed is improved. This reflects the effect of contact time on rock fragmentation. Under unchanged WOB and other drilling parameters, relationship between rotational speed and drill speed is shown in Fig. 4.2, and the relation is as follows:

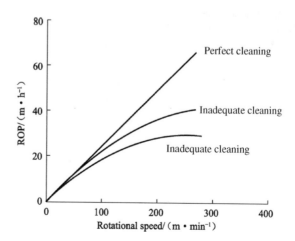

Fig. 4.2 Relationship curve of rotational speed and ROP

$$v_{pc} \propto n^{\lambda} \tag{4.2}$$

where

λ Drill speed ratio, generally less than 1, the value is related with rock property;
n Rotational speed, r/min.

4.1.1.3   Effect of Tooth Wear on ROP

During drilling process, the bit is broken and tooth is worn by formation. With the wear of bit teeth, the bit work efficiency will decrease obviously and ROP will decrease. If WOB and rotational speed remain unchanged, the relationship between drilling rate and amount of tooth wear is shown in Fig. 4.3, and the relation is:

$$v_{pc} \propto \frac{1}{1 + C_2 h} \tag{4.3}$$

where

C_2 the coefficient of tooth wear which is related to bit shape structure and rock property, and needs to be obtained by field statistical data;
h tooth wear amount which is relative wear height of teeth.

The ratio of worn height to original height, $h = 0$ for new drill bit, and $h = 1$ when teeth are all worn out.

4.1.1.4   Effect of Hydraulic Factors on ROP

In drilling process, it is an important means to improve ROP which remove rock fragments from bottom hole timely and effectively to avoid repeated fragmentation.

Fig. 4.3 Relationship curve of drill speed and tooth wear amount

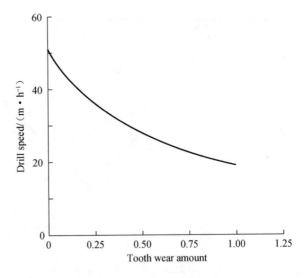

The fragment cleaning of bottom hole is finished through the flushing of drilling fluid jet produced by bit nozzle. The parameters referring to drilling bit and bit flow characteristics are called hydraulic factors. The overall index of hydraulic factors is usually expressed by average hydraulic power (called hydraulic power ratio) on unit area of bottom hole. In 1975, AMOCO Research Center published a relationship curve between ROP and bottomhole ratio of specific hydraulic power (Fig. 4.4).

The figure indicates that total amount of rock cuttings drilled in unit time is constant when a certain ROP is implemented. The rock cuttings need to be completely cleaned out under a certain ratio of hydraulic power. When the hydraulic power is lower than this value, bottomhole cleaning is not perfect. If actual ratio of hydraulic power located on imperfect clean area in Fig. 4.4, then actual ROP is lower than that of perfect area. If hydraulic power increases at this time, bottomhole cleaning conditions can be improved, ROP will increase under other unchanged conditions. Therefore, the effect of hydraulic factors on ROP is mainly manifested by bottomhole hydraulic cleaning ability. Hydraulic cleaning capacity is usually represented by hydraulic cleaning coefficient C_H, which is the ratio of actual ROP and perfect area ROP:

$$C_H = v_{pc}/v_{pcs} = P/P_s \qquad (4.4)$$

where

v_{pcs} ROP when bottomhole cleaning is perfect, m/h;

P Specific hydraulic power, kW/cm^2;

P_s Specific hydraulic power when bottom hole cleaning is perfect, kW/cm^2, the value of P_s could be obtained by regression expression from Fig. 4.4:

Fig. 4.4 Relationship curve of ROP and specific hydraulic power

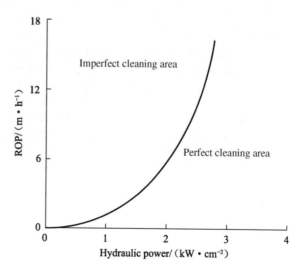

$$Ps = 9.72 \times 10^{-2} v_{pcs}^{0.31} \qquad (4.5)$$

It should be noted that C_H in Eq. 4.4 shall be less than or equal to 1; that is, when actual hydraulic power is greater than hydraulic power required for perfect hole cleaning, $C_H = 1$. The reason is that, with hydraulic power ratio increasing above the required value, ROP will not be further improved due to bottomhole cleaning.

The effect of hydraulic factors on ROP takes another form, which is rock-breaking effect of hydraulic energy. When specific hydraulic power exceeds the value required to clean bottom hole, mechanical ROP is still likely to increase. The effect of hydraulic rock breaking on ROP is mainly to reduce the threshold WOB in the relationship between WOB and ROP.

4.1.1.5   Effect of Drilling Fluid Performance on ROP

The effect of drilling fluid performance on ROP is more complicated, which involves characterizing varying degrees of impact of drilling fluid performance parameters on ROP. It is almost impossible to change any drilling fluid properties without affecting other parameters. Therefore, it is very difficult to evaluate the effect of a drilling fluid properties on ROP. A large number of experimental studies show that density, viscosity, solids content, and dispersion of drilling fluid have different degrees of effect on ROP.

- The effect of drilling fluid density on ROP
  The basic function of drilling fluid density is to maintain a certain liquid column pressure to control formation fluid flow into the well. The effect of drilling fluid density on ROP is mainly the effect of pressure difference between fluid column pressure and formation pore pressure on ROP. Laboratory experiments and drilling practice show that as pressure difference increases ROP decreases obviously. The main reason is that bottomhole pressure difference has chip hold down effect on broken cuttings which hinders timely cleaning of bottomhole fragments and affects drilling bit rock-breaking efficiency. When drilling in low permeability formation, the effect of pressure difference on ROP is greater than that in highly permeable formation, because drilling fluid is difficult to penetrate into low permeability formation pores, which cannot balance up-down pressure difference on fragment in time. Figure 4.5 shows the effect of bottomhole pressure difference on ROP into shale formation in field.
  Boggs (A. T. Bourgoyne) et al. after analyzing and processing large amount of experimental data pointed out that the relation between pressure difference and ROP can be expressed in a straight line in semi-logarithmic coordinates:

$$v_{pc} = v_{pc0} e^{-\beta \Delta p} \qquad (4.6)$$

Fig. 4.5 Relationship curve
of drill speed and pressure
difference in well bottom

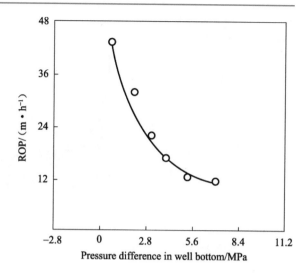

where

v_{pc0}—ROP when press difference is 0, m/h;

β—The index related to rock properties;

Δp—The difference between fluid column pressure and formation pore pressure, MPa.

The ratio of actual ROP and ROP under zero pressure difference is called influence coefficient of pressure difference, C_P, which is:

$$C_p = \frac{v_{pc}}{v_{pc0}} = e^{-\beta \Delta p} \qquad (4.7)$$

- The effect of drilling fluid viscosity on ROP

  Drilling fluid viscosity does not directly affect ROP, but affects ROP indirectly through the effect of bottomhole pressure difference and purifying effect. Under the condition of certain ground power, the increase of drilling fluid viscosity will result in the increase of pressure drop in drill string and annulus, the increase of bottomhole pressure difference, the decrease of hydraulic power of bottom hole, and the reduction of ROP. The relation curve between decreased ROP and increased drilling fluid viscosity is shown in Fig. 4.6.

- The effect of solids content and dispersion of drilling fluid on ROP

  The solids content of drilling fluid, solid-phase type and particle size have great effects on ROP. Figure 4.7 is the curve of solid-phase content effect on drilling parameters obtained from statistic data of more than 100 test wells. It can be seen from the figure that solids content of drilling fluid has a serious effect on ROP and bit wear. Therefore, solid-phase content should be strictly controlled, and drilling fluid with solid content below 4% should be used generally.

Fig. 4.6 Relationship curve of dynamic viscosity in drill fluid and ROP

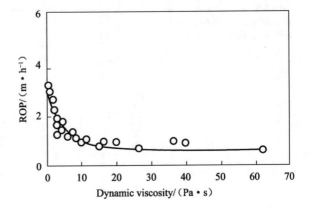

Fig. 4.7 Drilling fluid solids content effect on drilling parameters

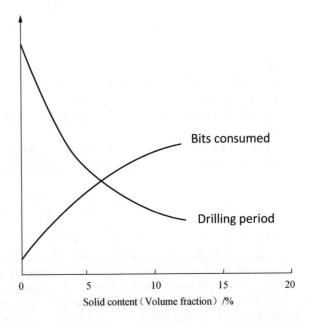

Further studies show that the size and dispersion of solid particles also have an effect on ROP. The results show that the more the colloidal particles with the size of less than 1 μm in drilling fluid, the greater the effect on ROP. Figure 4.8 is a contrast curve of solid particles dispersion effect on ROP. It can be seen that dispersive drilling fluid is lower than that of non-dispersive drilling fluid while solid-phase content is the same. The smaller the solid content is, the greater the difference between both. In order to improve ROP, low solids non-dispersive drilling fluid should be used as much as possible.

Drilling practice has proved that drilling fluid performance is an extremely important factor affecting ROP. However, because of its complex effect on ROP, and the effect of bottomhole working condition on drilling fluid performance, it is

Fig. 4.8 Solids content and dispersion effect on ROP

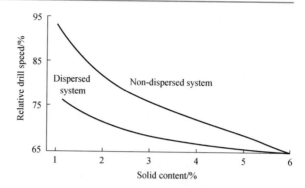

difficult to control strictly. So there is no mathematical model that can accurately describe drilling fluid performance effect on drilling speed as an objective basis for optimizing drilling fluid performance.

4.1.2　Drilling Rate Model

Based on the above analysis of various factors affecting ROP, we can sum up the influential factors and establish the comprehensive relationship between ROP and WOB, rotational speed, tooth worn, pressure difference, and hydraulic factors, namely

$$v_{\text{pc}} \propto (W - M)n^{\lambda} \frac{1}{1 + C_2 h} C_{\text{P}} C_{\text{H}} \tag{4.8}$$

Introducing a proportional coefficient K_{R}, Eq. 4.8 can be written as a drilling rate model:

$$v_{\text{pc}} = KR(W - M)n^{\lambda} \frac{1}{1 + C_2 h} C_{\text{P}} C_{\text{H}} \tag{4.9}$$

Equation 4.9 is the revised F. S. Young model. Considering WOB, rotational speed, and tooth wear effect on ROP, Young presented a drilling rate model that was of the same form in 1969, but did not take into account hydraulic cleaning coefficient C_{H} and pressure differential effect factor C_{P}.

The proportional coefficient K_{R} in Eq. 4.9 is usually called formation drillability coefficient. In fact, K_{R} contains other effect factors on ROP except WOB, rotational speed, tooth wear, pressure difference and hydraulic factors, which are related to rock mechanical properties, drill bit type, and drilling fluid performance. When rock characteristics, bit type, drilling fluid performance, and hydraulic parameters are constant, K_{R}, M, C_2 in Eq. 4.9 are fixed constants, which can be determined by field drilling tests and bit data.

4.1.3   Bit Wear Equation

In drilling process, drilling bit is gradually worn and invalidated while it breaks rock. The analysis and studies of factors affecting drilling bit wear and bit wear rules are of great significance to drilling parameters optimization, drilling indexes prediction, and bit working condition. The wear forms include tooth wear, bearing wear, and diameter wear. The following is mainly about influential factors and rules of cone bit tooth wear and bearing wear.

4.1.3.1   Tooth Wear Rate Model

Bit tooth wear is mainly related to WOB, rotational speed, formation, and tooth condition itself. The wear rate of bit tooth can be measured by differential dh/dt of tooth wear capacity to time.

WOB effects on tooth wear rate

The relationship between wear rate of different diameter bit teeth and the WOB is shown in Fig. 4.9, and the relation expression is:

$$\frac{\mathrm{d}h}{\mathrm{d}t} \propto \frac{1}{Z2 - Z1W} \tag{4.10}$$

where

Z_1, Z_2 WOB influence coefficient, the value is related to teeth drill bit size.

When WOB equals Z_2/Z_1, tooth wear speed is infinitely large, indicating that the value of Z_2/Z_1 is the ultimate WOB of this size bit. The values of Z_1 and Z_2 based on experimental data from American Hughes Company are shown in Table 4.1.

Fig. 4.9 Relationship between WOB and tooth wear speed

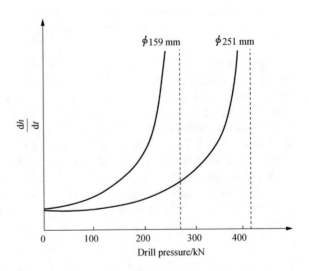

Table 4.1 Effect of bit teeth wear on ROP

Bit Diameter (mm)	Z_1	Z_2
159	0.0198	5.5
171	0.0187	5.6
200	0.0167	5.94
220	0.0160	6.11
244	0.0148	6.38
251	0.0146	6.44
270	0.0139	6.68
311	0.0131	7.15
350	0.0124	7.56

The effect of rotational speed on tooth wear speed

When the rotational speed increases, the tooth wear speed will also accelerate under the condition of constant WOB. The relationship curve between rotational speed and tooth wear speed is shown in Fig. 4.10, which has a relationship expression:

$$\frac{dh}{dt} \propto a_1 n + a_2 n^3 \tag{4.11}$$

where

a_1, a_2—Rotational speed influence factors which are determined by bit types, and the values of a_1 and a_2 for different bit types are shown in Table 4.2.

Effect of tooth wear condition on teeth wear rate

Bit teeth are generally trapezoidal, conical, or spherical with a small top area and a large bottom area. The working area of tooth will increase with the increase of tooth wear. Tooth wear speed will decrease with the increase of tooth wear when all kinds of drilling parameters are constant. The relationship between tooth wear amount and tooth wear speed is shown in Fig. 4.11, and the relationship expression is:

Fig. 4.10 Relationship curve of rotational speed and tooth wear speed

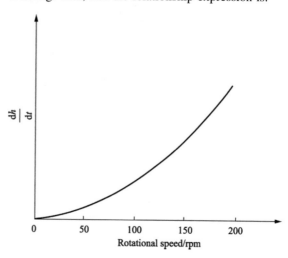

Table 4.2 Rotational speed influence factors

Teeth type	Application formation	No	Type	a_1	a_2	C_1
Milling tooth bit	Soft	1	1	2.5	1.008×10^{-4}	7
			2			
			3	2.0	0.870×10^{-4}	6
			4			
	Middle	2	1	1.5	0.653×10^{-4}	5
			2	1.2	0.522×10^{-4}	4
			3			
			4	0.9	0.392×10^{-4}	3
	Hard	3	1	0.65	0.283×10^{-4}	2
			2	0.5	0.218×10^{-4}	2
			3	0.5	0.218×10^{-4}	2
			4			
Button bit	Very soft	4	1	0.5	0.218×10^{-4}	2
			2			
			3			
			4			
	Soft	5	1			
			2			
			3			
			4			
	Middle	6	1			
			2			
			3			
			4			
	Hard	7	1			
			2			
			3			
			4			
	Very hard	8	1			
			2			
			3			
			4			

$$\frac{\mathrm{d}h}{\mathrm{d}t} \propto \frac{1}{1 + C1h} \tag{4.12}$$

where

C_1 Tooth wear influence coefficient which is related to bit type, and the values are shown in Table 4.2.

Fig. 4.11 Relationship curve of tooth wear amount and tooth wear speed

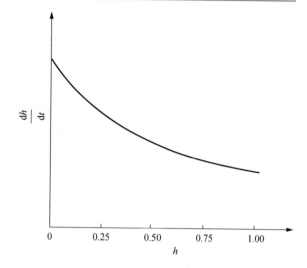

According to the above relations, we can establish a comprehensive relationship between tooth wear rate and influencing factors:

$$\frac{dh}{dt} \propto \frac{a_1 n + a_2 n^3}{(Z2 - Z1W)(1 + C1h)} \tag{4.13}$$

Introducing a proportional coefficient A_f, Eq. 4.13 can be written in equation form of tooth wear velocity.

$$\frac{dh}{dt} = \frac{A_f(a_1 n + a_2 n^3)}{(Z2 - Z1W)(1 + C1h)} \tag{4.14}$$

where

A_f Rock abrasive property coefficient, which needs to be determined according to statistic calculation of field bit data.

Example 4.1.1 Formation abrasive coefficient $A_f = 3.22 \times 10^{-3}$ in a well section of an oil field with depth of 2,800 m, 21-drill bit with $\phi 251$mm is used to drill medium-hard formation, WOB $W = 196$ kN, rotate speed $n = 110$ r/min, try to calculate the amount of tooth wear after 10 h.

Solution: 21-drill bit with $\phi 251$ mm is suitable for medium-hard formation according to Table 4.1 and Table 4.2, correlation coefficients are as follows: $Z_1 = 0.0146$, $Z_2 = 6.44$, $a_1 = 1.5$, $a_2 = 0.653 \times 10^{-4}$, $C_1 = 5$. Based on Eq. 4.14 tooth wear amount can be obtained by transposition and integration:

Integrate Eq. 4.14 and a quadratic trinomial of tooth wear can be obtained.

$$\frac{C_1}{2}h_f^2 + h_f - \frac{A_f(a_1 n + a_2 n^3)}{Z_2 - Z_1 W}t_f = 0$$

Then the root of the equation is,

$$h_f = \frac{-1 + \sqrt{1 + 2C_1 \frac{A_f(a_1 n + a_2 n^3)}{Z2 - Z1W}}}{C1}$$

The related values are substituted into the equation and we can get:

$$h_f = 0.77$$

4.1.3.2   Bearing Wear Rate Model

Bit bearing wear is indicated by letter B. when it is a new drill bit, $B = 0$; when it is a completely worn bit, $B = 1$. Bearing wear rate can be expressed by the differential of bearing wear amount to time dB/dt.

Bit bearing wear rate is mainly influenced by factors such as WOB, rotational speed, and so on. According to J. W. Graham's research, bit bearing wear rate is proportional to the rotational speed and the 1.5 times power of WOB, which is expressed as:

$$\frac{dB}{dt} = \frac{1}{b}W^{1.5}n \tag{4.15}$$

where

B Bearing working coefficient, which is related to drill bit type and drilling fluid performance, and it should be determined by actual field data.

4.1.4   Determination of Related Coefficient in Drilling Equation

Drilling rate model and bit wear equation describing the basic rules of drilling process are obtained by experiment and mathematical analysis under certain conditions. Formation drillability coefficient K_R, threshold WOB M, speed index λ, tooth wear coefficient C_2, rock abrasive coefficient A_f, and bearing working coefficient b in the equation are closely related to actual drilling conditions and environment, which need to be determined according to actual drilling data analysis. The basic steps for determining each parameter are: Firstly, according to ROP test data of drill bit, threshold WOB, rotational speed index, and formation drillability coefficient are obtained, and then rock abrasive coefficient, tooth wear coefficient, and bearing working coefficient are determined according to bit working record.

4.1.4.1 Threshold WOB and Rotational Speed Index

The basic method of calculating threshold WOB and rotational speed index is the five-point method of ROP test. Test conditions are:

Drilling fluid performance in test is constant. Hydraulic parameters are constant and maintained upon the general level of this region, so as to ensure that C_P and C_H are invariable in test. The influence of changing hydraulic rock-breaking condition on M value should be avoided.

Try to make test well section shorter or test time shorter while not affecting test accuracy, so as to ensure that the difference between the tooth wear at the beginning and that in the end is very small.

The steps of five-point method for ROP test are as follows:

According to the WOB and rotational speed ranges which can be used for the well section in local area, the highest WOB W_{max} and the lowest WOB W_{min} in test, the highest rotational speed W_{max} and the lowest speed W_{min} can be obtained. At the same time, select a WOB W_0 and rotate speed n_0 that are close to the average WOB and average rotational speed.

According to the combination of WOB and speed, starting from the first point Point (W_0, n_0), drilling test are conducted according to the order of points with directions shown in Fig. 4.12. Record drilling time at each point per 1 m or 0.5 m drilled. The test is completed until the sixth point is finished.

Fill test data into Table 4.3 and convert drill time to drill speed.

In the experiment, the same WOB and rotational speed of the first and sixth point are set up to obtain relative test error. The relative error of experiments $\frac{|v_{pc1} - v_{pc6}|}{v_{pc1}}$ should be less than 15%, test is then successful.

According to test data of the second and fifth under constant speed, the threshold bit weight under rotating speed is set to M, rotation speed equation can be obtained:

Fig. 4.12 Five-point method ROP experiment

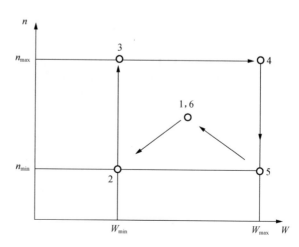

Table 4.3 Five-point ROP test record

Actual point number	WOB	Rotation speed (r/min)		Drilling time (s/m)	ROP (m/h)
1	W_0	n_0		Δt_1	v_{pc1}
2	W_{min}	n_{min}		Δt_2	v_{pc2}
3	W_{min}	n_{min}		Δt_3	v_{pc3}
4	W_{max}	n_{max}		Δt_4	v_{pc4}
5	W_{max}	n_{max}		Δt_5	v_{pc5}
6	W_0	n_0		Δt_6	v_{pc6}

$$v_{pc2} = K_R C_P C_H (W_{min} - M1) n_{min}^{\lambda} \frac{1}{1 + C_2 h} \tag{4.16}$$

$$v_{pc5} = K_R C_P C_H (W_{max} - M1) n_{min}^{\lambda} \frac{1}{1 + C_2 h} \tag{4.17}$$

The invariants in equation can be eliminated through Eq. 4.16 divided by Eq. 4.17 and rearrangement gives:

$$M_1 = W_{min} - \frac{W_{max} - W_{min}}{v_{pc5} - v_{pc2}} v_{pc2} \tag{4.18}$$

Similarly, on basis of test data from the third and fourth points, the threshold bit weight under test speed can be obtained as M_2:

$$M_2 = W_{min} - \frac{W_{max} - W_{min}}{v_{pc4} - v_{pc3}} v_{pc3} \tag{4.19}$$

Threshold bit weight is the average of M_1 and M_2:

$$M = \frac{1}{2}(M_1 + M_2) \tag{4.20}$$

Similarly, according to two pairs of test points recorded in constant WOB condition, rotational speed index λ_1 and λ_2 of two WOB are obtained by substituting test data of second, third, fourth, and fifth points into ROP equation and eliminating invariants in equation:

$$\lambda_1 = \frac{\lg(v_{pc2}/v_{pc3})}{\lg(n_{min}/n_{max})} \tag{4.21}$$

$$\lambda_2 = \frac{\lg(v_{pc5}/v_{pc4})}{\lg(n_{min}/n_{max})} \tag{4.22}$$

The rotational speed index λ is the average of λ_1 and λ_2:

$$\lambda = \frac{1}{2}(\lambda_1 + \lambda_2) \tag{4.23}$$

In the five-point drilling test, threshold bit weight, and rotational speed index are more suitable for formations with faster drillability. For formations with very slow drillability, threshold bit weight and rotational speed index can be obtained by releasing drilling pressing method according to related data.

4.1.4.2 Determination of Formation Drillability Coefficient

According to drilling test data of new drill bit, the mount of tooth wear $h = 0$, the drillability coefficient can be got by drilling rate model Eq. 4.9:

$$K_R = \frac{v_{pc}}{\text{CHCP}(W - M)n^\lambda} \tag{4.24}$$

4.1.4.3 Determination of Tooth Wear Coefficient C_2

It is assumed that rock properties are basically constant during drilling, and drilling parameters are basically consistent, tooth wear amount is h_f when bit is pulled out, and the beginning ROP and pull-out ROP are v_{pc0} and v_{pcf}, respectively. It can be obtained from drilling rate Eq. 4.9:

$$\frac{v_{pc0}}{v_{pcf}} = \frac{1 + C_2 h f}{1 + C_2 h 0} \tag{4.25}$$

Due to tooth wear amount $h_0 = 0$ when drilling process starts, we can get:

$$v_{pc0} = v_{pcf}(1 + C_2 h_f)$$
$$C2 = \frac{v_{pc0} - v_{pcf}}{v_{pcf} h_f} \tag{4.26}$$

4.1.4.4 Determination of Rock Abrasive Coefficient A_f

The integral of tooth wear velocity Eq. 4.14 gives:

$$t_1 = \frac{(Z_2 - Z_1 W)}{(a_1 n + a_2 n^3)}\left(h_f + \frac{C_1}{2}h_f^2\right)$$
$$A_f = \frac{(Z_2 - Z_1 W)}{(a_1 n + a_2 n^3)t_f}\left(h_f + \frac{C_1}{2}h_f^2\right) \tag{4.27}$$

According to drill bit type, its influence coefficient, average WOB, rotational speed, bit working time during drilling process, and tooth abrasion after pulling out, we can get rock abrasive coefficient A_f during drilling process.

4.2   Optimization of Drilling Parameters for Mechanical Rock Breaking

Mechanical rock-breaking parameters in drilling process include WOB and rotational speed. The aim of mechanical rock-breaking parameters optimization is to find a certain combination of WOB and rotational speed parameters, so that drilling process achieves the best technical and economic effect. In order to achieve this goal, it is necessary to establish a standard to measure economic effect of drilling technology, and to combine the standard with the basic rules of parameters' effects on drilling process. Drilling objective function is established. Then, using optimization mathematics theory, we seek for extreme points of objective function under various constraint conditions. The parameters satisfying extreme point conditions are the optimal mechanical rock-breaking parameters in drilling process.

4.2.1   Objective Function Establishment

There are several types of standards for measuring overall drill technical and economic effects. At present, unit footage cost is generally used as a standard, its expression is:

$$C_{pm} = \frac{C_b + C_r(t_1 + t)}{H} \tag{4.28}$$

where

C_{pm} Unit footage cost, RMB/m;
C_b One bit cost, RMB;
C_r Drill rig operating cost, RMB/h;
t_t Tripping and pipe connection time, h;
t Bit working time, h;
H Drill footage, m.

Bit footage and bit working time are related to parameters used in Eq. 4.28 in drilling process. The relationship between parameters H and t are established, and it is substituted in unit footage cost equation. The objective function represented by drilling cost per meter is obtained.

The drilling rate Eq. 4.9 can be written as:

$$v_{pc} = \frac{dH}{dt} = C_H C_p K_R (W - M) n^\lambda \frac{1}{1 + C_2 h}$$

Then we can get:

$$dH = C_H C_p K_R (W - M) n^\lambda \frac{1}{1 + C_2 h} dt \tag{4.29}$$

Bit tooth wear speed Eq. 4.14 can be rearranged:

$$dt = \frac{Z_2 - Z_1 W}{A_f (a_1 n + a_2 n^3)} (1 + C_1 h) dh \tag{4.30}$$

Equation 4.30 is substituted into Eq. 4.29, and we can get:

$$dH = \frac{C_H C_p K_R (W - M) n^\lambda (Z_2 - Z_1 W)}{A_f (a_1 n + a_2 n^3)} \frac{1 + C_1 h}{1 + C_2 h} dh \tag{4.31}$$

Under the condition of constant WOB and rotational speed, Eq. 4.31 is integrated:

$$\int_0^{H_f} dH = \frac{C_H C_p K_R (W - M) n^\lambda (Z_2 - Z_1 W)}{A_f (a_1 n + a_2 n^3)} \int_0^{h_f} (\frac{1 + C_1 h}{1 + C_2 h}) dh$$

The final tooth wear amount is h_f when drill footage is H_f:

$$H_f = \frac{CHCpKR(W - M) n^\lambda (Z_2 - Z_1 W)}{A_f (a_1 n + a_2 n^3)} [\frac{C_1}{C_2} h_f + \frac{C_2 - C_1}{C_2^2} \ln(1 + C_2 hf)] \tag{4.32}$$

In Eq. 4.32,

$$J = CHCpKR(W - M) n^\lambda \tag{4.33}$$

$$S = \frac{A_f (a_1 n + a_2 n^3)}{(Z_2 - Z_1 W)} \tag{4.34}$$

$$E = \frac{C_1}{C_2} h_f + \frac{C_2 - C_1}{C_2^2} \ln(1 + C_2 hf) \tag{4.35}$$

The drill footage expression Eq. 4.32 can be written as:

$$H_f = \frac{J}{S} E \tag{4.36}$$

where J is initial drilling speed when tooth wear amount $h = 0$. S is initial wear rate of bit teeth when tooth wear amount $h = 0$ and its reciprocal value is the theoretical bit life without considering the effect of tooth wear. The meaning of J/S is theoretical drill bit footage without considering the effect of tooth abrasion. E is footage coefficient considering the impact of tooth wear on the ROP and tooth wear rate, which is a function of final tooth wear.

An integral of final tooth wear amount h_f of Eq. 4.30 yields the total tooth bit working time:

$$\int_0^{t_f} dt = \frac{Z_2 - Z_1 W}{A_f(a_1 n + a_2 n^3)} \int_0^{h_f} (1 + C_1 h) dh$$

$$t_f = \frac{Z_2 - Z_1 W}{A_f(a_1 n + a_2 n^3)} \int_0^{h_f} (h_f + \frac{C_1}{2} h_f^2)$$

(4.37)

Equation 4.34 is substituted into Eq. 4.37, and we can get:

$$F = h_f + \frac{C_1}{2} h_f^2$$

(4.38)

That is:

$$t_f = \frac{F}{S}$$

(4.39)

It can be seen from Eq. 4.38 that F and footage coefficient E are similar. Its physical significance is the bit life factor considering the impact of teeth wear on tooth wear rate, it is a function of the ultimate teeth wear.

Substitute Eq. 4.36 and bit working time Eq. 4.39 into cost Eq. 4.28. The target function containing various drilling parameters can be obtained as:

$$C_{pm} = \frac{C_b S + C_r (S t_t + F)}{JE}$$

Assuming that:

$$t_E = \frac{C_b}{C_r} + t1$$

That is:

$$C_{pm} = \frac{C_r}{JE} (t_E S + F)$$

(4.40)

where

t_E The reduced time of bit and tripping cost. When drill bit cost and drilling rig operating cost are fixed, t_E is only related to drilling time, but unrelated to drilling parameters.

If parameters J, E, S and F are substituted into Eq. 4.40, an objective function containing five variables (W, n, h_f, C_H, C_p) can be obtained.

$$C_{pm} = \frac{C_r \frac{t_E A_f (a_1 n + a_2 n^3)}{Z_2 - Z_1 W} + h_f + \frac{C_1}{2} h_f^2}{C_H C_p K_R (W - M) n^\lambda [\frac{C_1}{C_2} h_f + \frac{C_2 - C_1}{C_2^2} \ln(1 + C_2 hf)]}$$

(4.41)

4.2.2   The Extreme Condition and Constraint Condition of Objective Function

The aim of drilling parameters optimization is to determine parameters that achieve the lowest cost of unit footage, that is, to seek optimal parameters when objective function produces (4.41) the minimum value. According to classical optimization theory, the necessary condition for a function to obtain the extreme value is that partial derivative of function to each variable equals zero in its definition domain. A large number of mathematical operations have proven that the point which conforms to the extreme condition of drilling objective function is the minimum point for drilling cost function.

The drilling target function contains five variables, namely W, n, h_f, C_p, and C_H. First, through analyzing where C_p and C_H are located in a function expression, it can be found that, C_p and C_H should be as large as possible for minimizing drilling cost. According to the definitions of these two coefficients, their maximum value can only be 1. So in drilling practice, C_p and C_H values should be 1 for the lowest cost. After determining the optimal value of C_p and C_H, the minimum extreme condition of objective function is:

$$\frac{\partial C_{pm}}{\partial W} = 0, \frac{\partial C_{pm}}{\partial n} = 0, \frac{\partial C_{pm}}{\partial h_f} = 0$$

(4.42)

In the equation, W, n, and h_f are the constraints of objective function, values range of which is determined by actual operating conditions using four groups of inequalities:

Tooth wear amount h

$$0 \leq h \leq 1$$

(4.43a)

Bearing wear capacity B

$$0 \leq B \leq 1$$

(4.43b)

Wob W

$$M < W < Z_2 / Z_1 \ M > 0$$

(4.43c)

$$0 < W < Z_2 / Z_1 \ M < 0$$

Rotational speed n

$$n > 0 \qquad (4.43\text{d})$$

The combinations of drilling parameters which cannot satisfy the above constraints are not feasible. In addition, the inequality of bearing wear is not directly related to objective function in four groups of inequalities mentioned above. But for the same bit, working life of drill bit is also the function of bearing wear and tooth wear amount. The constraint condition of bearing wear can be indicated by corresponding tooth wear amount. Set final wear and tear of bearing as B_f, because

$$t_f = \frac{Z_2 - Z_1 W}{A_f(a_1 n + a_2 n^3)}\left(h_f + \frac{C_1}{2} h_f^2\right)$$

$$t_f = \frac{bB_f}{nW^{1.5}}$$

For the same drill bit, tooth and bearing have the same working hours; thus, from the above two equations we could obtain:

$$B_f = \frac{(Z2 - Z1W)nW^{1.5}}{A_f(a1n + a2n^3)b}\left(h_f + \frac{C_1}{2} h_f^2\right) \qquad (4.44)$$

4.2.3 The Optimal Wear, ROP and WOB

After determining objective function, extreme condition, and constraint condition, it is possible to solve a set of optimal WOB, optimal rotational speed, and optimal wear out of the lowest drilling cost by optimization method. Due to its complicated mathematical derivation and calculation process, it is omitted here and can be consulted in other references.

This paper mainly introduces the method of determining optimal wear, optimal rotational speed, and optimal WOB under the condition of certain parameter combination.

4.2.3.1 Optimal Drill Bit Wear

For a working drill bit under certain WOB and rotational speed, the optimal bit wear amount problem is when bit is worn out which can lead to the lowest drilling cost. According to cost function Expression (4.41) and necessary condition of optimal wear amount $\frac{\partial C_{pm}}{\partial h_f} = 0$, we can derive the expression of optimal wear amount, namely

$$\frac{C_1}{2}h_f^2 + \left(\frac{C_1}{C_2} - 1\right)h_f - \frac{C1 - C2}{C_2^2}(1 + C_2h_f)\ln(1 + C_2h_f)$$
$$-\frac{A_f t_E(a_1 n + a_2 n^3)}{Z_2 - Z_1 W} = 0 \tag{4.45}$$

Equation 4.45 is a three-dimensional nonlinear equation, which consists of a surface in three-dimensional space called optimal wear surface. Theoretically, each group of W and n values correspond to a point on the optimal wear surface; that is, each group of W and n numerical value that is substituted into (4.45) can be used to solve for an optimal wear h_f. However, because drilling cost function is limited by constraint condition of Eqs. 4.46 and 4.47, it is undesirable to get optimal worn amount beyond restricted range. At this time, the limit wear amount of bit tooth or bearing can be used as the optimal bit wear amount.

4.2.3.2 Optimal Rotational Speed

In three-dimensional space of the constraint conditions of W–n–h_1, rotational speed n with the lowest drilling cost is the optimum rotational speed for the given WOB and wear amount. From cost function Expression (4.41), making $\frac{\partial C_{pm}}{\partial n} = 0$, one can derive optimal rotational speed surface equation:

$$n^3 + \frac{(1 - \lambda)a_1}{(3 - \lambda)a_2}n - \frac{1}{3 - \lambda}\frac{F(Z_2 - Z_1 W)}{t_E A_f a_2} = 0 \tag{4.46}$$

Equation 4.49 has three solutions, only the following real solution is meaningful to drill parameters:

$$n_{opt} = \sqrt[3]{\frac{V}{2} + \sqrt{\left(\frac{V}{2}\right)^2 + \left(\frac{U}{3}\right)^3}} + \sqrt[3]{\frac{V}{2} - \sqrt{\frac{V}{2} + \left(\frac{U}{3}\right)^3}} \tag{4.47}$$

where:

$$V = \frac{F(Z_2 - Z_1 W)\lambda}{t_E A_f a_2(3 - \lambda)}$$

$$U = \frac{(1 - \lambda)a_1}{(3 - \lambda)a_2}$$

Equation 4.47 is the formula of the optimum rotational speed according to given WOB W and bit wear amount h_f.

4.2.3.3 Optimal WOB

Similar to optimal rotational speed, in W–n–h_1 three-dimensional space, an optimal WOB W with the lowest drilling cost is derived by the given pair of n and h_f under a constraint condition. Through cost function expression (4.41) and the extreme conditions of pressure for determining optimal WOB $\frac{\partial C_{pm}}{\partial W} = 0$, we can derive the optimal WOB equation:

$$W^2 - 2\left[\frac{Z_1}{Z_2} + \frac{t_E A_f(a_1 n + a_2 n^3)}{Z_1 F}\right]W$$
$$+ t_E A_f \frac{(a_1 n + a_2 n^3)}{Z_1 F}\left(\frac{Z_1}{Z_2} + M\right) + \left(\frac{Z_2}{Z_1}\right)^2 = 0 \qquad (4.48)$$

Eq. 4.48 has two solutions for WOB. One is greater than Z_2/Z_1, and the other one is less than Z_2/Z_1. The solution of WOB which is less than Z_2/Z_1 is:

$$W_{opt} = \frac{Z_2}{Z_1} + \frac{R}{F} - \sqrt{\frac{R}{F}\left(\frac{R}{F} + \frac{Z_2}{Z_1} - M\right)} \qquad (4.49)$$

where:

$$R = \frac{t_E A_f(a1n + a2n^3)}{Z_1}$$

Equation 4.49 is the optimal WOB for given rotational speed n and bit wear amount h_f.

In practice, it is generally based on the same type of drill data in adjacent wells or in the same well to first determine reasonable wear amount for teeth or bearings. And then one can determine allowable range of rotational speed according to rig equipment conditions. Finally the drilling cost of different WOB and rotational speed can be calculated, and the lowest optimal WOB and rotational speed can be found.

Example 4.2.1 The formation drillability coefficient of a well section is $K_R = 0.0023$, abrasive coefficient $A_f = 2.28 \times 10^{-3}$, threshold bit weight $M = 10kN$, rotational speed index $\lambda = 0.68$. By using 21-drill bit with $\phi 251mm$ suitable for medium-hard formation, $C_2 = 3.68$, $C_H = 1$, $C_p = 1$, single bit cost $C_b = 900$ Yuan, drilling rig operating fee $C_r = 250$ RMB/h, tripping time $t_1 = 5.75$ h; Rig rotary speed has only three options, $n_1 = 60r/min$, $n_2 = 120r/min$, $n_3 = 180$ min, respectively. According to an adjacent well data, the selected bit tooth wear in this section is generally T_6 grade ($h_f = 0.75$). Calculate the optimal WOB and rotational speed combination and its working index.

Solution: Suitable parameters of 21-type drill bit with $\phi 251$ mm in medium-hard formation are:

$$Z_1 = 0.0146, Z_2 = 6.44, a_1 = 1.5, a_2 = 6.53 \times 10^{-5}, C_1 = 5$$

$$t_E = \frac{C_b}{C_r} + t_1 = \frac{900}{250} + 5.75 = 9.35h$$

$$E = \frac{C_1}{C_2}h_f + \frac{C_2 - C_1}{C_2^2}\ln(1 + C_2 h_f)$$

$$= \frac{5}{3.68} \times 0.75 + \frac{3.68 - 5}{3.68^2}\ln(1 + 3.68 \times 0.75) = 0.89$$

$$F = h_f + \frac{C_1}{2}h_f^2 = 0.75 + \frac{5}{2} \times 0.75^2 = 2.156$$

Optimal WOB at different rotational speeds can be obtained by Eq. 4.49, i.e.

$$W_{opt} = \frac{Z_2}{Z_1} + \frac{R}{F} - \sqrt{\frac{R}{F}\left(\frac{R}{F} + \frac{Z_2}{Z_1} - M\right)}$$

$$R = \frac{t_E A_f (a1n + a2n^3)}{Z_1}$$

The calculation results can be seen in Table 4.4.

It is shown in Table 4.4 that optimal WOB values are only the local optimal values when rotational speed is 120 r/min and 180 r/min, and the optimal WOB at $n = 60$ r/min achieves the lowest drilling cost for this drilling system. In addition, determined WOB should conform to the requirements of wellbore trajectory control.

Table 4.4 Optimal WOB and its working index at different rotational speeds

n (r/min)	60	120	180
$a_1n + a_2n^3$	104.105	292.838	650.830
R	152.008	427.584	950.301
R(F)	70.505	198.323	440.770
W_{opt}(kN)	323.544	286.109	261.950
S	0.1383	0.2951	0.5673
t1(h)	15.59	7.31	3.80
n(r/min)	60	120	180
J	11.67	16.47	19.80
Hf(m)	75.10	49.67	31.06
C_{pm}(Yuan/m)	83.02	83.83	105.84

4.3   The Optimal Design of Hydraulic Parameters

One important condition during the drilling operation is that the drilling cuttings should be removed at a timely manner for safe and fast drilling. There are two steps in this process. The first step is to remove cuttings from the bottom and cuttings enter the annulus; the second step is the drilling fluid carry cuttings to move forward. People used to believe the first step is easy to achieve and the second step is harder. Thus, people put more effort in improving well flushing to accelerate the moving speed of cuttings. It does improve the moving forward speed. But the well flushing technique is limited by the well wall flushing issue and the limitation of surface pumps. Besides, in practice, people notice that when the bit nozzle is damaged, the flow rate does not reduce but the drilling rate drops dramatically. This phenomenon motivates people to study these two steps from different aspects. After years of study, it is realized that the second step is not difficult to achieve. The difficult step is to remove the cuttings from the bottom hole. Cuttings at the bottom which are not removed timely are the main factor that affects ROP. To address this issue, bits with high-speed jet nozzles are invented. The drilling fluid is ejected at a very high speed on the cuttings to remove them quickly. By using this kind of bit, the bottom hole can be cleaned and the pressure of high-speed fluid can help break rocks at the same time.

The optimal design of hydraulic parameters comes along with the utilization of jet bit. Hydraulic parameters are used to describe the hydraulic characteristics of jet bits and the jet flow, the specifics of the surface hydraulic equipment including the power of drilling pump, the flow rate, pump pressure, bit hydraulic horsepower, hydraulic pressure drop, diameter of jet, jet impact force, jet speed, annulus velocity, and so on. The aim of the design is to seek for the best combination of hydraulic parameters and the optimal hydraulic energy distribution to reach the best cleaning effect and improve the ROP. However, the energy distribution is restricted by the chosen jet bit, the energy loss of circulatory hydraulic system, and the surface pumps. The optimal design is based on the understanding of characteristics of bits, the energy loss of the systems, and hydraulic parameters of surface pumps.

4.3.1   The Hydraulic Characteristics of Bits

The main characteristics of the jet bits are that the jet flow nozzle with specific structures is installed on the bit. After flowing through the nozzle, the drilling fluid forms a high-speed jet. The jet creates strong impact force on the bottom to remove cuttings and break rocks.

4.3.1.1   Jet Flow and Its Impact on Hole Bottom

The characteristics of jet flow

The jet flow is the fluid which flows through nozzle without contacting with borehole walls. The jet flow can be classified by relationship between the flow and the surrounding medium including submerged jet (the density of jet flow is equal to or smaller than that of surrounding medium), non-submerged jet (the density of jet flow is larger than that of surrounding medium). According to the limitation of walls, jet flow can be classified as free jet flow (without the wall restriction) and non-free jet flow (with wall restriction). According to the jet flow pressure, they can be classified as continuous flow (the pressure of a point in the flow is stable) and pulse flow (the pressure of jet flow is unstable). Under the bottomhole condition, the drilling fluid is ejected from the nozzle and submerged in the drilling fluid already in the bottom hole. The movement of jet flow is restricted by the borehole walls and bottom; thus, it is submerged flow.

When the jet flow is ejected from the nozzle, due to the friction, the jet flow is interchanging the momentum with surrounding fluid, driving other fluid moving, and making the perimeter diameter of the jet flow increase. The angle on the perimeter of the jet flow section profile is called angle of flare (α on the Fig. 4.13). This angle shows the density of the jet flow. Obviously, the smaller α is, the larger the flow density is and the more intensive the energy is.

The jet velocity at the section profile is almost the same, and they are all at original velocity v_1. With the development of the jet flow, due to momentum exchange, the velocity distribution is influenced and the influence region moves forward to the jet center. The diameter of the flow region that at the original velocity decreases until the velocity at the jet center is smaller than the original velocity. These streams at the original velocity are called jet flow potential core (Fig. 4.14). The length L_0 of the potential core is mainly influenced by the diameter and the internal flow passage of nozzles. As the surrounding medium affects the jet flow

Fig. 4.13 Characteristics of jet flow at bottom hole

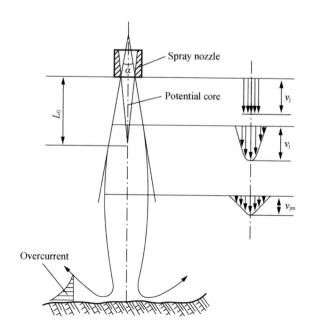

from outside to inside, the highest velocity is at the center of core. The velocity decrease dramatically from the inside to outside, and it becomes 0 at the flowing boundary. Within the potential core, the velocity at the jet flow axis is the same with outlet velocity; the velocity out of the potential core drops significantly. Figure 4.14 shows the law of diminishing velocity on the axis. d_n is diameter of the nozzle, L is the distance of a point form the outlet, v_j is the jet flow outlet velocity, v_{jm} is the maximum velocity with a distance of L from the outlet.

After the flow hitting the bottom hole, the kinetic energy is transformed into pressure energy forming the surge pressure waves. The jet flow is finally converted to high-speed sheet flowing on the bottom.

The basic characteristics of submerged free continuous jet flow include: The jet flow has potential core and flare angle; the maximum speed is at the center of the section profile; out of the potential core, the velocity at the axis drops significantly; after the flow hitting the bottom, surge pressure waves and sheet flow are formed.

The cleaning effects of jet flow on bottom hole

Jet flow generates surge pressure waves and sheet flow which helps clean the bottom hole.

Surge pressure. Surge pressure is not impacted on the hole bottom while it is impacted on the small circle as shown on Fig. 4.15. For the bottom hole, the pressure within the jet impact region is relatively high while the pressure out of the region is low. Even within the jet impact region the pressure is not even, the pressure is the highest at the center and then decreases around the center. It should be noticed that as the bit rotates the small circle moves around. Thus, the pressure distribution is changing all the time. As a consequence of that, the pressure impacted on the bottom is very uneven, which gives a turning torque to cuttings to take them away from the bottom hole (see Fig. 4.16). That is the impact and reversal effects that jet flow exerts onto cuttings.

The horizontal pushing of sheet flow. Sheet flow refers to a very thin flow layer with high speed that is formed after the jet flow hitting the bottom. It has the

Fig. 4.14 Law of jet flow attennuation along nozzle centerline

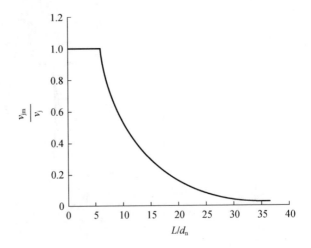

Fig. 4.15 Jet impact area

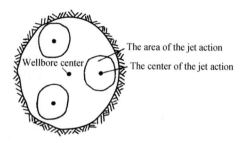

Fig. 4.16 Reversal of cuttings

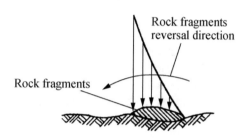

characteristic of boundary jet flow. Study shows that the highest sheet flow speed appears at the height which is smaller than 0.5 mm under the smooth bottom condition. The speed can be 50%–80% of the nozzle outlet speed. The closer the nozzle is to the bottom, the higher the sheet flow speed is. Because of the high-speed sheet flow, cuttings are pushed away from the bottom and carried by the drilling fluid. Consequently, the sheet flow plays an important role in bottomhole cleaning.

The rock-breaking effect of jet flow

It is proved through many years' practice that the jet flow can not only clean the wellbore but also directly breaks or helps break the rocks, when the horsepower of jet flow is strong enough. In the low strength rock layers, when the pressure of jet flow is higher than the fracture pressure of rocks, the jet flow will break the rocks. This kind of rock breaking is very common when drilling the surface layer. In some formations, bit rotation is not necessary and jet flow alone is sufficient. In some high strength formation, bits can only exert mechanical force for creating micro-fractures. The jet flow can enter into these micro-fractures and act as water wedges which can reduce the rock strength and increase the bit efficiency.

4.3.1.2   Hydraulic Parameters of Jet Flow

Hydraulic parameters include jet flow speed, impact force, and power of the jet flow. In terms of the cleaning effects, the parameters of the jet flow arriving at the bottom should be calculated. However, the distributions of jet flow speed and pressure on section profile are different, making them difficult to calculate. Thus, the outlet section of the jet flow is chosen as the calculation point.

Jet velocity

The flow velocity at the nozzle outlet is called the jet velocity.

$$v_j = \frac{10Q}{A_0} \tag{4.50}$$

$$A_0 = \frac{\pi}{4} \sum_{i=1}^{z} d_{ni}^2$$

where

v_j The jet velocity, m/s;
Q The quantity of drilling fluid flowing through the bit;
A_0 Cross-sectional area of jet outlet;
d_{ni} Diameter of the nth nozzle, cm;
z Number of nozzles.

Impact force of the jet flow

The impact force is the total force that exerted on the area. The equation of the impact force can be derived from the theorem of momentum.

$$F_j = \frac{\rho_d Q^2}{100 A_0} \tag{4.51}$$

in which

F_j Impact force of the jet flow, kN;
ρ_d Density of drilling fluid, g/cm^3.

The power of the jet flow

The cleaning, cutting removal, and rock breaking is actually a process that drilling fluid applying work to bottomhole rock and cuttings. In unit time, the more work the jet flow applies, the stronger the cleaning and rock-breaking effects. The work done in unit time is called the hydraulic power of the jet flow, as shown below:

$$P_j = \frac{0.05 \rho_d Q^3}{A_0^2}$$

where

P_j Hydraulic power of the jet flow, kW.

4.3.1.3   The Hydraulic Parameters of the Bit

The hydraulic parameters are meaningful for cleaning effects. The jet flow is generated by the drilling fluid flowing through bit nozzles. Due to the friction in nozzles, a part of the energy is lost. In the design, not only the jet flow energy should be considered but also the lost energy. The parameters, including the bit pressure drop and bit hydraulic power, should reflect these two parts.

The pressure drop across the bit

The bit pressure drop is the pressure difference of drilling fluid passing through the nozzle. When the quantity of drilling fluid and the size of the nozzle are fixed, according to the energy equation, the equation of pressure drop is:

$$\Delta p_b = \frac{0.05 \rho_d Q^2}{C^2 A_0^2} \tag{4.52}$$

where

Δp_b Pressure drop across bit, MPa;
C Discharge coefficient of the nozzle, it is related to the friction of the nozzle, the value is always smaller than 1.

If the nozzle diameter is represented by the nozzle equivalent diameter, the equation is:

$$\Delta p_b = \frac{0.081 \rho_d Q^2}{C^2 d_{ne}^4} \tag{4.53}$$

$$d_{ne} = \sqrt{\sum_{i=1}^{z} d_{ni}^2}$$

where

d_{ne} Nozzle equivalent diameter, cm.

The hydraulic horsepower of the bit

The hydraulic horsepower is the power consumption of drilling fluid flowing through the bit. Most of the hydraulic horsepower is transformed into jet flow hydraulic horsepower, and part of it is used to overcome the friction of nozzles. According to the hydraulics, hydraulic horsepower of the bit is shown as below:

$$P_b = \frac{0.05 \rho_d Q^3}{C^2 A_0^2} \tag{4.54}$$

or

$$P_b = \frac{0.081 \rho_d Q^3}{C^2 d_{ne}^4} \tag{4.55}$$

where

P_b Hydraulic horsepower of the bit, kW.

By comparing Eq. 4.57 and 4.55, we can obtain:

$$P_j = C^2 P_b \tag{4.56}$$

From Eq. 4.56, there is only slightly difference between the hydraulic horsepower of bit and the hydraulic horsepower of the jet flow. C^2 is the energy conversion rate. The hydraulic horsepower of the jet flow is a part of that of the bit. To increase the efficiency of the jet flow, the nozzle with high flux coefficient should be used.

The other hydraulic parameters of the bit are shown as below:

$$v_j = 10C\sqrt{\frac{20}{\rho_d}}\sqrt{\Delta p_b} \tag{4.57}$$

$$F_j = 0.2A_0 C^2 \Delta p_b \tag{4.58}$$

From these two equations, to improve the jet flow speed and the impact force, it is important to choose the nozzle with high flux coefficient.

4.3.2 The Basic Relation of Hydraulic Horsepower Transmissions

The hydraulic horsepower is provided by drilling pump. The drilling fluid flowing out of the pump has hydraulic horsepower called drilling pump output power. The hydraulic horsepower is delivered to the bit through the circulatory system which includes surface pipelines, the drill string, the annulus, and the bit nozzle. When the drilling fluid is flowing, these four parts are all consuming energy thus the pressure is decreasing. When the drilling fluid is circulated to the outlet, the pressure becomes 0. The basic relationship in the transmission is shown as below.

$$p_s = \Delta p_g + \Delta p_{st} + \Delta p_a + \Delta p_b \tag{4.59}$$

where

p_s Pump pressure, MPa;
Δp_g Pressure loss in surface pipelines;
Δp_{st} Pressure loss in drill strings, MPa;
Δp_a Pressure loss in the annulus, MPa;
Δp_b Pressure loss in the bit, MPa.

According to the hydraulics, the hydraulic horsepower is product of multiplying pressure by flow rate.

$$P_s = p_s Q \tag{4.60}$$

where

P_s Output horsepower, kW;
Q Drill pump flow rate, L/s.

As the circulatory system is a single pipeline, the flow rate in the system is the same. From Eqs. 4.60 and 4.3.13, under the same pump pressure or pump horsepower, the bit pressure drop and bit hydraulic horsepower can only be improved by reducing the pressure loss in surface pipelines, drilling string, and annulus. The pressure loss in these three parts is called the circulatory system pressure loss.

4.3.3   The Pressure Loss Calculation

The drilling fluid mainly flows in the drill string and the annulus. The fluid, according to the rheological properties, can be classified as Bingham fluid, power law fluid, and Carson fluid. Based on the flow regime, it can be classified as laminar flow and turbulent flow. Fluids in different flow models flow differently in different zones; thus, the calculation is different. In the system, to calculate the pressure loss, the fluid model and properties have to be identified first. Next, the fluid rheological properties need to be identified as well. The pressure loss equations are chosen based on fluid models and properties.

From the analysis above, it can be noticed that pressure loss calculation is a very complicated problem. The reasons are that firstly drilling fluids are non-Newtonian fluids which means their rheological properties vary a lot. On the other hand, as the geometrical shapes of different parts are various, rheological properties are influenced even with the same flow rate. The drill string is rotating all the time; thus, in the annulus the drilling fluid is not flowing axially. These issues have not been addressed theoretically. In practice, for convenience, the flowing calculation is simplified. In fact, the flow of drilling fluid in pipelines is always laminar flow while in annulus it could be turbulence flow or laminar flow. But the pressure loss mainly happens in pipelines and the pressure loss in the annulus is very small. In conclusion, the flow regime of the whole system can be treated as turbulence flow. Besides, the shearing rate of turbulence flow is high, at which the rheological properties of different drilling fluids are similar. The calculation accuracy is high enough when the drilling fluid is treated as Bingham flow. Therefore, some assumptions are made in the calculation.

– the drilling fluid is Bingham fluid;
– the drilling fluid in each part of the system is isothermal turbulent flow;
– the drill string is located in the center part of the hole;
– the drill string rotation is ignored;
– the well hole diameter has been identified;
– the drilling fluid is incompressible.

4.3.3.1   The Basic Pressure Loss Equation

Based on these assumptions, Fig. 4.17 shows the flowing of drilling fluids in pipelines. The head loss Δh can be described as below:

$$\Delta h = \frac{p_1 - p_2}{\gamma} = \xi \frac{v^2}{2g}$$

where

γ Specific gravity of drilling fluids, N/m^3;
ξ Head loss coefficient;
v Average flowing velocity of drilling fluids in pipelines, m/s;
g Acceleration of gravity, m/s^2.

In the equation, $P_1 - P_2$ is the pressure loss of drilling fluids in pipelines. Thus, the pressure loss Δp_L is

$$\Delta p_L = p_1 - p_2 = \xi \frac{\gamma v^2}{2g} = \xi \frac{\rho_d v^2}{2}$$

It has been proved that, ξ is in direct proportion to pipeline length L and friction coefficient f, and in inverse proportion to the hydraulic radius r_w, namely $\xi = f \frac{L}{r_w}$. The pressure loss equation is shown as below.

$$\Delta p_L = f \frac{\rho_d L v^2}{2 r_w} \tag{4.61}$$

According to the definition of the hydraulic radius, which equals the cross-sectional area divided by the wetted perimeter. Thus, for the internal flow, $r_w = \frac{d_i}{4}$; for the annulus flow, $r_w = \frac{d_h - d_{po}}{4}$. In this equation, d_i is the inner pipe diameter, d_h is the well diameter, d_{po} is the outer diameter of the drill string. The pressure loss equation for interior flow and annulus flow can be concluded by subtitling r_w into Eq. 4.62.

For the internal flow:

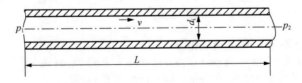

Fig. 4.17 Drilling fluid flow in a pipeline

$$\Delta p_l = \frac{0.2 f \rho_d L v^2}{d_i} \tag{4.62}$$

For the annulus flow:

$$\Delta p_l = \frac{0.2 f \rho_d L v^2}{d_h - d_{po}} \tag{4.62}$$

where

Δp_l the pressure loss, MPa;
f the friction coefficient of pipelines;
ρ_d density of drilling fluids, g/cm^3;
L the length of pipelines, m;
v the average velocity of drilling fluids, m/s;
d_i the inner pipeline diameter, m;
d_h the well diameter, m;
d_{po} the outer diameter, m.

4.3.3.2 The Identification of Friction Coefficient

It can be noticed that from Eqs. 4.63 and 4.64 that the density and average velocity of drilling fluids and the geometric size of pipelines are easy to be identified. The critical step is to identify fraction coefficient f. It is very complicated to identify f; thus, there are a lot of studies about the friction coefficient which shows that the friction coefficient is related to the flowing models, flowing regimes, levels of roughness, and Reynolds number. However, there is no accurate solution of the friction coefficient until now. The friction coefficient f is measured by experiments or calculated using empirical formulas.

To identify the hydraulic coefficient of the circulatory system under drilling condition, someone used the Newtonian fluid to find the relationship between the hydraulic coefficient and Reynolds number (see Fig. 4.18). A study has proven that these results of Newtonian fluid can be applied to the Bingham fluids in laminar flow, and when calculating the Reynolds number, the plastic viscosity has to be changed to laminar viscosity accordingly. The relationship between the equivalent laminar flow viscosity μ_e and the plastic viscosity of Bingham fluid is:

$$\mu_e = \frac{\mu_{pv}}{3.2} \tag{4.64}$$

By substituting Eq. 4.65 into Reynolds number calculation equation, the Reynolds number for Bingham fluids can be obtained.

For the internal flow,

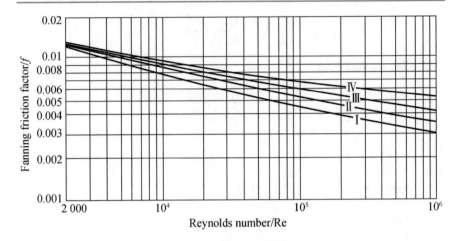

Fig. 4.18 Relationship of f and Re in turbulent flow regime

$$Re = \frac{32\rho_d d_i v}{\mu_{pv}} \tag{4.65}$$

For the annulus flow,

$$Re = \frac{32\rho_d (d_h - d_{po})v}{\mu_{pv}} \tag{4.66}$$

I   Minimum value of cold rolled brass tube;
II   New tubes with fixed joint section area;
III   Annulus zone of the cased well;
IV   Annulus zone of uncased well.

For convenience, the four curves in the figure can be described by one equation below:

$$f = \frac{k}{R_e^{0.2}} \tag{4.67}$$

In the equation, k is the calculation coefficient, referring to Table 4.5.

According to the k value in Eqs. 4.66, 4.70, and Table 4.5, the friction coefficient can be calculated.

- The friction coefficient in drill string and drill pipe:

$$f = 0.0265 \left(\frac{\mu_{pv}}{\rho_d d_i v}\right)^{0.2} \tag{4.68}$$

Table 4.5 k value of different tubes

Number of curves	k
I	0.046
II	0.053
III	0.059
IV	0.062

- The friction coefficient of inner drill pipe joint

$$f = 0.0295 \left(\frac{\mu_{pv}}{\rho_d d_i v} \right)^{0.2} \tag{4.69}$$

- The friction coefficient of the annulus zone:

$$f = 0.0295 \left(\frac{\mu_{pv}}{\rho_d (d_h - d_{po}) v} \right)^{0.2} \tag{4.70}$$

4.3.3.3 The Pressure Loss Equation of the Circulatory System

Substitute Eqs. 4.69 and 4.71 into pressure loss Eqs. 4.63 and 4.64 to obtain the pressure loss calculation equation of each part.

The pressure loss of surface pipelines, including pressure loss in high-pressure pipelines, stand pipes, water hoses, and Kelly bars. The pressure loss calculation equation is:

$$\Delta p_g = 0.51655 \rho_d^{0.8} \mu_{pv}^{0.2} \left(\frac{L_1}{d_1^{4.8}} + \frac{L_2}{d_2^{4.8}} + \frac{L_3}{d_3^{4.8}} + \frac{L_4}{d_4^{4.8}} \right) Q^{1.8} \tag{4.71}$$

where

L_1, L_2, L_3, L_4 Lengths of the high-pressure pipelines, stand pipes, water hoses, and Kelly bars, m

d_1, d_2, d_3, d_4 Radius of the high-pressure pipelines, stand pipes, water hoses, and Kelly bars, cm.

- ·The pressure loss in drill pipes is:

$$\Delta p_{pi} = \frac{B \rho_d^{0.8} \mu_{pv}^{0.2} L_P Q^{1.8}}{d_{pi}^{4.8}} \tag{4.72}$$

where

Δp_{pi} Pressure loss in drill pipes;
d_{pi} Inner radius of drill pipes, cm;

B Constant, for the internal flat drill pipes $B = 0.51655$, for full hole drill pipes $B = 0.57503$;

L Total length of drill pipe, m.

- ·The pressure loss in the annulus:

$$\Delta p_{pa} = \frac{0.57503 \rho_d^{0.8} \mu_{pv}^{0.2} L_p Q^{1.8}}{(d_h - d_{po})^3 (d_h + d_{po})^{1.8}} \qquad (4.73)$$

where

Δp_{pa} Pressure loss in the annulus, MPa;
d_h Well diameter, m;
d_{po} Outer drill pipe diameter, m.

- ·The pressure loss in drill collar

$$\Delta p_{pa} = \frac{0.51655 \rho_d^{0.8} \mu_{pv}^{0.2} L_p Q^{1.8}}{d_{ci}^{4.8}} \qquad (4.74)$$

where

Δp_{ci} the pressure loss in the drill collar, MPa;
L_c the length of the drill collar, m;
d_{ci} the inner diameter of the drill collar, cm.

- ·The pressure loss in the annulus of drill collars

$$\Delta p_{ca} = \frac{0.57503 \rho_d^{0.8} \mu_{pv}^{0.2} L_p Q^{1.8}}{(d_h - d_{co})^3 (d_h + d_{co})^{1.8}} \qquad (4.75)$$

where

Δp_{ca} Pressure loss in the annulus of the drill collar, MPa;
d_{co} Outer diameter of the drill collar, cm.

For convenience, combining Eqs. 4.73 and 4.76 yields the equation of the pressure loss in drill pipes:

$$\Delta p_p = \left[\frac{B}{d_{pi}^{4.8}} + \frac{0.57503}{(d_h - d_{po})^3 (d_h + d_{po})^{1.8}} \right] \rho_d^{0.8} \mu_{pv}^{0.2} L_p Q^{1.8} \qquad (4.76)$$

The pressure loss in drill collars:

$$\Delta p_c = \left[\frac{0.51655}{d_{ci}^{4.8}} + \frac{0.57503}{(d_h - d_{co})^3 (d_h + d_{co})^{1.8}} \right] \rho_d^{0.8} \mu_{pv}^{0.2} L_p Q^{1.8} \tag{4.77}$$

In Eqs. 4.72, 4.77, and 4.78, let the values of K_g, K_p, and K_c equal to the pressure loss coefficient of surface pipelines, the drill pipe, and the drill collar.

$$K_g = 0.51655 \rho_d^{0.8} \mu_{pv}^{0.2} \left(\frac{L_1}{d_1^{4.8}} + \frac{L_2}{d_2^{4.8}} + \frac{L_3}{d_3^{4.8}} + \frac{L_4}{d_4^{4.8}} \right) \tag{4.78}$$

$$K_p = \left[\frac{B}{d_{pi}^{4.8}} + \frac{0.57503}{(d_h - d_{po})^3 (d_h + d_{po})^{1.8}} \right] \rho_d^{0.8} \mu_{pv}^{0.2} L_p \tag{4.79}$$

$$K_c = \left[\frac{0.51655}{d_{ci}^{4.8}} + \frac{0.57503}{(d_h - d_{co})^3 (d_h + d_{co})^{1.8}} \right] \rho_d^{0.8} \mu_{pv}^{0.2} L_p \tag{4.80}$$

For the whole system, the pressure loss equation is:

$$\Delta p_L = \Delta p_g + \Delta p_p + \Delta p_c = (K_g + K_p + K_c) Q^{1.8} = K_L Q^{1.8}$$

where $K_L = K_g + K_p + K_c$ is called the pressure coefficient of the system. Simplify the coefficient K_L:

$$m = \rho_d^{0.8} \mu_{pv}^{0.2} \left[\frac{B}{d_{pi}^{4.8}} + \frac{0.57503}{(d_h - d_{po})^3 (d_h + d_{po})^{1.8}} \right]$$

$$K_L = mL_P$$

$$K_L = K_g + K_p + K_c = K_g + mL_p + K_c$$
$$= K_g + m(D - L_c) + K_c = mD + K_g + K_c - mL_c$$

$$a = K_g + K_c - mL_c$$

Let $a = K_g + K_c - mL_c$ then we get $K_L = mD + a$

By analyzing m and a, it can be concluded that when the surface pipelines, the structures of drill tools and the well, and the drilling fluid are identified, m and a are considered as constants. As a result, the pressure loss coefficient K_L increases with the depth of the well.

Finally, the pressure loss equation of the pressure is:

$$\Delta p_L = K_L Q^{1.8} = (K_g + K_p + K_c) Q^{1.8} = (mD + a) Q^{1.8} \tag{4.81}$$

4.3.3.4   Methods to Improve the Hydraulic Parameters of Bits

According to the transmission relationship discussed before, the drill fluid pressure and hydraulic horsepower provided by the pump is consumed in the bit and circulatory part. Thus, improving the hydraulic parameters means delivering more energy to bits from pumps and reducing pressure loss in circulatory system.

Equation 4.53 can be converted into

$$\Delta p_b = \frac{0.05 \rho_d Q^2}{C^2 A_0^2} = K_b Q^2 \tag{4.82}$$

$$K_b = \frac{0.05 \rho_d}{C^2 A_0^2}$$

where

K_b the pressure loss coefficient of the bit.

Based on the transmission relationship of the pump pressure and pump horse-power, we can get that:

$$p_s = \Delta p_b + \Delta p_L = K_b Q^2 + K_L Q^{1.8}$$

$$\Delta p_b = K_b Q^2 = p_s - K_L Q^{1.8} \tag{4.83}$$

$$P_s = P_b + P_L = \Delta p_b + \Delta p_L = K_b Q^2 + K_L Q^{1.8}$$

$$P_b = K_b Q^3 = P_s - K_L Q^{2.8} \tag{4.84}$$

Analyzing Eqs. 4.84 and 4.85, there are four major methods to improve the hydraulic parameters (Δp_b and P_b).

Improve pump pressure p_s and pump power P_s.

Increasing the pump pressure and pump power can increase the total energy level of the system, but it is limited by the surface condition. The development history of the bitcan be divided into three phases. The first phase is when p_s = 13–15 MPa, the second phase is when p_s = 17–18 MPa, and the third phase is when p_s = 20–22 MPa. With the increasing pressure, the rated pump pressure and rated power of the drill pump increase as well. This provides a good foundation for improving and bit power.

Reduce the pressure loss coefficient K_L of the circulatory system.

As $K_L = K_g + K_p + K_c$, from Eqs. 4.79 and 4.81, the pressure loss coefficient is related to the density of drill fluid, plastic viscosity of drilling fluid, and the pipeline diameter. Consequently, the method to reduce K_L includes: using low-density drill fluids; reduce the plastic viscosity of the drill fluid; increase the pipeline inner radius. Among these methods, the inner radius effects the K_L most significantly. It is recommended to use the inner flat drill pipe with larger radius if possible. The ϕ144 mm drill pipe and the ϕ127 mm drill pipe have very different loss coefficients. Under the same condition, the ϕ144 mm drill pipe has a coefficient that is 66% higher than that of the ϕ127 mm drill pipe.

Increase the pressure loss coefficient of the bit K_b

From Eq. 4.83, to increase K_b the possible method is to increase ρ_d or reduce C and A. But in fact, increasing ρ_d or reducing C is not applicable. When density is increasing, K_L is increasing as well which means the pressure on the bottom is increasing. Reducing C actually increases the pressure loss in nozzles. The only way to increase K_b value is to reduce the diameter of nozzle which have a remarkable influence on improving K_b. For example, when the diameter of the nozzle reduces from 12 to 11 mm, the K_b value increases by 42%.

Optimize the flow rate Q

From Eqs. 4.84 and 4.85, increasing flow rate Q can improve the bit pressure loss and bit power and also contribute to the system pressure loss and system power loss. Thus, the optimal flow rate Q is critical for the energy distribution of the bit and the circulatory system.

4.3.4 The Working Performance of the Drill Pump

Increasing the power and pressure of the surface pump is a common method to improve hydraulic parameters. However, it is not practical to change the surface pump for the pump power improvement. The optimal hydraulic parameter design is based on the existing surface pump, and it is about making best of the surface pump. This requires full understanding of the drill pump working performance.

For each drill pump, there is a maximum output power which is called rated power. In addition, there are several cylinder liner with different diameters. The rated power of the pump is the maximum allowable output power. The flow rate at the rated pump power and pressure is called rated flow rate. The performance characteristics of the most widely used pump 3NB-1000 and 3NB-1300 are listed in the following Tables (Tables 4.6 and 4.7).

Table 4.6 Performance parameters of 3NB-1000

Cylinder liner diameter (mm)	The rated pump stroke	The rated flow rate (L/s)	The rated pump pressure (MPa)
120	150	19.9	33.1
130	150	23.4	28.2
140	150	27.1	24.3
150	150	31.1	21.2
160	150	35.4	18.6
170	150	40.0	16.5

Table 4.7 Performance parameters of 3NB-1300

Cylinder liner diameter (mm)	The rated pump stroke	The rated flow rate (L/s)	The rated pump pressure (MPa)
130	140	23.6	34.3
140	140	27.4	31.4
150	140	31.4	27.3
160	140	35.7	24.0
170	140	40.4	21.3

The relationship between the rated power, rated pressure, and rated flow rate is:

$$P_r = p_r Q_r$$

where

P_r Pump rated power, kW;
p_r Rated pump pressure, MPa;
Q_r Rated flow rate, L/s.

There are two types of working status of a drilling pump, based on the change of flow rate, as shown in Fig. 4.19.

When pump rate Q is lower than rated pump rate, the maximum pump pressure can only reach the rated pressure p_r, since the pump pressure is limited by the allowable pressure within cylinder liner. When the pump flow rate reduces, the pump power drops. This type of pump working status is called the rated pump pressure status. When $Q > Q_r$, as the pump power is limited by rated pump power, the maximum pump power is the rated pump power P_r. As a result, the pump pressure should be lower than the rated pump pressure. With the increase of

Fig. 4.19 Pump working status

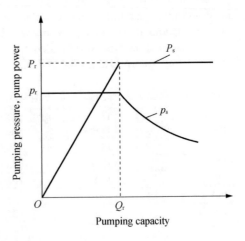

Pumping capacity

flow rate, this working status is called the rated power status. From these two types of working status, only when the pump flow rate equals the rated Q_r, the pump power can reach the rated output power and the maximum allowed pressure simultaneously. Therefore, when choosing the cylinder liner, it is recommended to choose the flow rate that is close to the rated pump flow rate in order to achieve the full capacity of the pump.

4.3.5 The Criteria of the Optimal Design of Hydraulic Parameters

The hydraulic parameters of the jet flow and bits, the jet flow velocity v_j, the impact force F_j, the bit hydraulic power P_b, power of the jet flow P_j, the pressure drop in the bit Δp_b have been discussed. The difference between P_b and P_j is the coefficient C^2, so only P_b needs to be calculated. The other four parameters are related to the surface pump parameters and the system energy loss. The equations can be converted as below:

$$\Delta p_b = p_s - K_L Q^{1.8} \tag{4.85}$$

$$v_j = K_v \sqrt{p_s - K_L Q^{1.8}} \tag{4.86}$$

$$F_j = K_F Q \sqrt{p_s - K_L Q^{1.8}} \tag{4.87}$$

$$P_b = Q(p_s - K_L Q^{1.8}) \tag{4.88}$$

In these equations:

$$K_v = 10C \sqrt{\frac{20}{\rho_d}}$$

And

$$K_F = \frac{C\sqrt{20\rho_d}}{100}$$

These four equations describe the relationships between hydraulic parameters and flow rate Q. Figure 4.20 shows these relationships.

These four parameters are expected to be as large as possible due to the bottomhole cleaning requirements. From Fig. 4.20, it is not possible to choose a flow rate that makes these parameters reach maximum values at the same time. Hence the problem arises in the selection of operational parameters, the nozzle diameter and the flow rate, that which hydraulic parameters out of four should be maximized. Typically, due to the difference in the understanding of wellbore cleaning

Fig. 4.20 Relationships between hydraulic parameters and the flow rate. Q_p—the flow rate for the maximum bit hydraulic power. Q_F—the flow rate for the maximum jet impact force

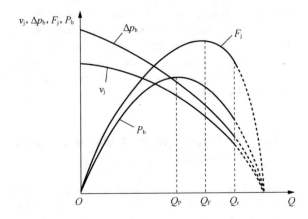

mechanism, there are three most common design criteria, including the maximum bit power, the maximum jet impact force and the maximum jet flow velocity. The former two are most widely used in field drilling operation.

4.3.6 The Maximum Bit Hydraulic Power

Based on the maximum bit hydraulic power criterion, the process of wellbore cleaning and rock breaking are actually the results of the jet flow applying work to the bottom. The hydraulic horsepower should be as high as possible if the surface pump permits.

4.3.6.1 The Condition to Reach the Maximum Bit Hydraulic Power

When the drill pump is working under the rated power, the pump power$P_s = P_r$. The equation of bit hydraulic power is:

$$P_b = P_s - P_L = P_r - K_L Q^{2.8} \tag{4.89}$$

As indicated in Eq. 4.90, when Q increases, P_b reduces; when Q reduces, P_b increases. Under the rated working condition, the way to get the maximum bit hydraulic power is to make Q as small as possible. However, the flow rate cannot be smaller than Q_r; thus, the condition to reach the maximum bit hydraulic power is $Q_{opt} = Q_r$.

When the pump is working under rated pressure condition, $p_s = p_r$, and the bit hydraulic power is:

$$P_b = P_s - P_L = P_r Q - K_L Q^{2.8}$$

Make $\frac{dP_d}{dQ} = 0$

$$\frac{dP_d}{dQ} = P_r Q - K_L Q^{2.8} = 0$$

Thus, the optimal flow rate is obtained:

$$Q_{opt} = \left(\frac{p_r}{2.8K_L}\right)^{\frac{1}{1.8}} = \left[\frac{p_r}{2.8(a+mD)}\right]^{\frac{1}{1.8}} \qquad (4.90)$$

When the optimal flow rate is achieved, $\frac{d^2 P_d}{dQ^2} = -5.04K_L\left(\frac{p_r}{2.8K_L}\right) < 0$. Hence, the corresponding bit hydraulic power with the optimal flow rate is the maximum bit hydraulic power.

Equation 4.91 can be further transformed as below,

$$p_r = 2.8K_L Q^{1.8} = 2.8\Delta p_L$$

$$\Delta p_L = \frac{p_r}{2.8} = 0.357 p_r \qquad (4.91)$$

Equations 4.91 and 4.92 are the conditions to reach the maximum bit hydraulic power.

4.3.6.2 The Variation of Bit Hydraulic Power with Flow rate and the Well Depth

According to the hydraulic power distribution of the entire system:

$$P_b = P_s - P_L = P_s - K_L Q^{2.8} = P_s - (a+mD)Q^{2.8}$$

When $Q > Q_r$,

$$P_b = P_r - (a+mD)Q^{2.8} \qquad (4.92)$$

When $Q \leq Q_r$,

$$P_b = p_r Q - (a+mD)Q^{2.8} \qquad (4.93)$$

For wells with different depth D, the relationships between the bit hydraulic power Pr and the flow rate Q based on Eq. 4.93 and 4.94 have been plotted, as shown in Fig. 4.21, $D_0 < D_1 < D_2 < D_3 < D_{Pc} < D_5 < D_{Pa} < D_7$. When the well depth $D \leq D_{Pc}$, the maximum bit hydraulic power is reached and it is also the optimal bit hydraulic power, namely $Q_{opt} = Q_r$. At this time, the pump is working under rated power condition. When $D > D_{Pc}$, the optimal flow rate is reached when the bit hydraulic power is maximized as computed in Eq. 4.91. At this time, the pump is working under rated pump pressure condition. When the well depth $D > D_{Pa}$, the flow rate required to achieve the maximum bit hydraulic power is smaller than minimum flow rate Q_a required for cuttings removal. At this time, the minimum flow rate Q_a is applied in the drilling operation. Therefore, it can be observed that D_{Pc} and D_{Pa} have special meaning in the selection of optimal flow rate. Typically, D_{Pc} and D_{Pa} are called the first critical well depth and the second critical well depth, respectively.

When $D = D_{Pc}$, the optimal flow rate $Q_{opt} = Q_r$, from Eq. 4.91, the first critical well depth is:

$$D_{Pc} = \frac{p_r}{2.8mQ_r^{1.8}} - \frac{a}{m} \tag{4.94}$$

When $D = D_{Pa}$, the optimal flow rate $Q_{opt} = Q_a$, from Eq. 4.91, the second critical well depth is:

$$D_{Pc} = \frac{p_r}{2.8mQ_a^{1.8}} - \frac{a}{m} \tag{4.95}$$

4.3.6.3 Determination of the Optimal Nozzle Diameter

The maximum bit hydraulic power discussed above refers to the possible hydraulic power when the optimal flow rate is reached. However, whether the bit can reach the maximum power or not also depends on the selection of the nozzle diameter.

When the optimal flow rate is selected, the optimal nozzle diameter depends on the pressure drop at the maximum bit hydraulic power. From Eq. 4.54:

$$d_{ne} = 4\sqrt{\frac{0.081\rho_d Q^2}{C^2 \Delta p_b}}$$

When $D \leq D_{Pc}$, $\Delta p_b = p_s - \Delta p_L = p_r - (a + mD)Q_r^{1.8}$

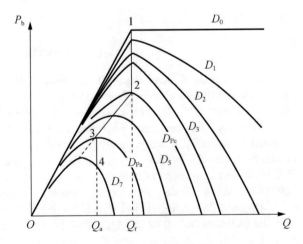

Fig. 4.21 Variation of bit hydraulic power with the changes of well depth and flow rate

$$d_{ne} = 4\sqrt{\frac{0.081\rho_d Q_r^2}{C^2[p_r - (a + mD)Q_r^{1.8}]}} \tag{4.96}$$

When $D_{Pc} < D \le D_{Pa}$, $\Delta p_b = p_s - \Delta p_L = p_r - 0.357 p_r = 0.643 p_r$,

$$d_{ne} = 4\sqrt{\frac{0.126\rho_d Q_{opt}^2}{p_r C^2}} \tag{4.97}$$

When $D > D_{Pa}$, $\Delta p_b = p_s - \Delta p_L = p_r - (a + mD)Q_a^{1.8}$,

$$d_{ne} = 4\sqrt{\frac{0.081\rho_d Q_a^2}{C^2[p_r - (a + mD)Q_a^{1.8}]}} \tag{4.98}$$

From these equations, we can see that when $D \le D_{Pc}$, the nozzle equivalent diameter increases with the increment of the well depth; when $D_{Pc} < D \le D_{Pa}$, the nozzle equivalent diameter increases with the decrease of the well depth; when $D > D_{Pa}$, the nozzle equivalent diameter increases with the increment of the well depth again.

4.3.7 The Maximum Jet Impact Force

According to the maximum jet impact force criterion, the impact force is the key factor in determining the wellbore cleaning performance. The higher the impact force is, the better the cleaning performance is.

4.3.7.1 The Condition to Reach the Maximum Jet Impact Force

Under the rated power working condition, Eq. 4.88 is converted into:

$$F_j = K_F\sqrt{P_r Q - K_r Q^{3.8}} \tag{4.99}$$

From Eq. 4.100, the conditions for F_j to reach the maximum value are $\frac{dF_j}{dQ} = 0$ and $P_L = \frac{P_L}{3.8}$. These are the conditions to reach the maximum jet impact force under the rated power working condition. However, in real practice, it is not appropriate to require $Q > Q_r$ since the pump stroke rate is not supposed to be higher than the rated pump stroke rate. Thus, $Q = Q_r$ is the optimal condition under the rated power working status.

Under the rated pump pressure condition, Eq. 4.88 is transformed into:

$$F_j = K_F\sqrt{P_r Q^2 - K_L Q^{3.8}} \tag{4.100}$$

From $\frac{dF_j}{dQ} = 0$, the maximum jet impact force condition is:

$$\Delta p_b = \frac{p_r}{1.9} = 0.526 p_r \tag{4.101}$$

$$Q_{opt} = \left(\frac{p_r}{1.9K_r}\right)^{\frac{1}{1.8}} = \left[\frac{p_r}{1.9(a+mD)}\right]^{\frac{1}{1.8}} \tag{4.102}$$

4.3.7.2 The Relationship Between the Maximum Jet Impact Force and Flow rate and Well Depth

Substituting equation $K_L = a + mD$ into Eqs. 4.100 and 4.101, we obtain:
When

$$Q > Q_r, F_j = K_F\sqrt{P_rQ - (a+mD)Q^{3.8}} \tag{4.103}$$

When

$$Q \le Q_r, F_j = K_F\sqrt{p_rQ^2 - (a+mD)Q^{3.8}} \tag{4.104}$$

Under varying well depths, we plot the correlation curves between the jet impact force and flow rate Q, by computing F_j using Eqs. 4.104 and 4.105.

Figure 4.22 presents the relationship of the jet impact force with well depth and flow rate.

As indicated in Fig. 4.22, the theoretical working path to reach the maximum jet impact force is $1' \to 2 \to 3 \to 4 \to 5$. As $1' \to 2$, $Q > Q_r$, this is not in favor of the pump working; therefore, the true working path is $1 \to 2 \to 3 \to 4 \to 5$. It should be noticed that the well depth D_{Fc} and D_{Fa} are called the first critical well

Fig. 4.22 Relationship between the jet impact force and the well depth

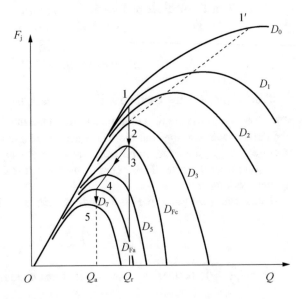

depth and the second critical well depth under the maximum jet impact force, respectively.

$$D_{Fc} = \frac{p_r}{1.9mQ_r^{1.8}} - \frac{a}{m} \tag{4.105}$$

$$D_{Fa} = \frac{p_r}{1.9mQ_a^{1.8}} - \frac{a}{m} \tag{4.106}$$

In this equation, Q_a is the minimum flow rate to carry cuttings.

4.3.7.3 The Determination of Optimal Nozzle Diameter

Similar to the technique for determining the optimal nozzle diameter using the maximum bit hydraulic horsepower criterion, the determination of nozzle diameter can be achieved using the optimal flow rate and the corresponding bit pressure drop under the maximum jet impact force condition.

When $D \le D_{Fc}$, $\Delta p_b = p_s - \Delta p_L = p_r - (a + mD)Q_r^{1.8}$:, then

$$d_{nc} = 4\sqrt{\frac{0.081\rho_d Q_r^2}{C^2[p_r - (a + mD)Q_r^{1.8}]}} \tag{4.107}$$

When $D_{Fc} < D \le D_{Fa}$, $\Delta p_b = p_s - \Delta p_L = p_r - 0.526p_r = 0.474p_r$, then

$$d_{nc} = 4\sqrt{\frac{0.171\rho_d Q_{opt}^2}{p_r C^2}} \tag{4.112}$$

When $D > D_{Fa}$, $\Delta p_b = p_s - \Delta p_L = p_r - (a + mD)Q_r^{1.8}$, then

$$d_{nc} = 4\sqrt{\frac{0.081\rho_d Q_a^2}{C^2[p_r - (a + mD)Q_a^{1.8}]}} \tag{4.113}$$

As shown in the above equations, similar rule follows between the nozzle equivalent diameter and the well depth in this case, *i.e.*, under the maximum jet impact force, as compared to the case of maximum bit hydraulic power.

4.3.8 The Optimal Design of Hydraulic Parameters

The design of hydraulic parameters is selection of the pump working factors (the flow rate, the pressure, and the power), the bit working parameters, the jet flow parameters (velocity, impact force, power) before drilling the well operation. By analyzing all these hydraulic parameters, it can be concluded that when the surface pump, the drilling tools, the well structure, and the drill fluid are identified the only controllable variables are the drill fluid flow rate and the nozzle diameters. Thus the main task is to select the drill fluid flow rate and the nozzle diameter.

4.3.8.1 The Minimum Flow rate Q_a

The minimum flow rate refers to the lowest flow rate to carry cuttings. Once the annulus velocity is identified, the minimum flow rate is then known. There are many different methods to calculate the minimum annulus velocity. One type is based on practical experience; the other type uses empirical equation:

$$v_a = \frac{18.24}{\rho_d d_h} \tag{4.114}$$

where

v_a the minimum velocity of the drilling fluid, m/s;
d_h the wellbore diameter, cm.

In fact, the minimum velocity is related to the carrying capacity of the drill fluid, which is quantified by cutting lift efficiency. The cutting lift efficiency is defined as the ratio of the actual cuttings lift velocity to the average lift velocity.

$$K_s = \frac{v_s}{v_a} \tag{4.115}$$

where

K_s the cutting lift efficiency, a dimensionless number;
v_s the actual cutting lift velocity, m/s.

In field operation, to maintain the equilibrium between the quantity of cutting production and the quantity of the cuttings lifted to surface, the condition of $K_s \geq 0.5$ is typically required. Thus, after the lowest annulus velocity is determined, the lift efficiency should be calculated to make sure that $K_s \geq 0.5$ is achieved.

To compute K_s, the lift velocity v_s should be calculated. Assume that the drop velocity of cuttings in the drill fluids is v_{s1}, then $v_s = v_a - v_{s1}$. The drop velocity is related to the drilling fluid property, and the calculation equation is:

$$v_{s1} = \frac{0.0707 d_s (\rho_s - \rho_d)^{\frac{2}{3}}}{\rho_d^{\frac{1}{3}} \mu_e^{\frac{1}{3}}} \tag{4.116}$$

where

v_{s1} the drop velocity of cuttings in the drilling fluid, m/s;
d_s the cutting diameter, cm;
μ_s the density of cuttings, g/cm^3;
ρ_e the viscosity of the drill fluid, Pa·s.

The equation of μ_e is:

$$\mu_e = K\left(\frac{d_h - d_{po}}{1200v_a}\right)^{1-n}\left(\frac{2n+1}{3n}\right)^n \tag{4.117}$$

where

K the viscosity coefficient, Pa·s^n;
n the liquidity index of the drilling fluid.

If $K_s \geq 0.5$, the calculated minimum velocity can be used; if $K_s < 0.5$, the minimum velocity should be adjusted to make sure $K_s \geq 0.5$.

When the minimum annulus velocity is identified, the minimum flow rate is computed by:

$$Q_a = \frac{\pi}{40}(d_h{}^2 - d_{po}{}^2)v_a \tag{4.114}$$

where

Q_a Minimum flow rate, L/s.

4.3.8.2 The Pressure Loss Coefficient at Different Well Depth

The well is divided into several sections, and the lowest point of each section is selected as the calculation point of well depth. According to the equations discussed above, K_g, K_p, K_c, m, and a are calculated and then the pressure loss coefficient is calculated ($K_L = a + mD$).

4.3.8.3 Determination of the Size of Cylinder Liner

There are two key specifications for each cylinder liner: the rated flow rate and the rated pump pressure. If the allowable pressure of the recycling system is higher than the rated pump pressure of each cylinder liner, and also assume the prerequisite condition that the selected rated flow rate Q_r is larger than the minimum flow rate Q_a, we should always select the cylinder liner with higher rated pressure. The rated pump pressure is considered as the maximum allowable pressure when optimizing the hydraulic parameters. If the allowable pressure of recycling system is lower than the rated pump pressure of each sleeve barrel, we should first compute the maximum bit hydraulic power and the maximum jet impact force under this condition, which ensures that the rated flow rate is higher than the optimal flow rate. In the optimization of hydraulic parameters, the minimum endurable pressure of the system (including surface manifolds, hoses and valves) should be used as the maximum allowable pressure.

4.3.8.4 Computation and Determination of Optimal Flow rate, Optimal Nozzle Diameter and Hydraulic Parameters

Before optimizing the flow rate, we should determine the selection criteria of the optimal hydraulic parameters. The first critical well depth and the second well depth are determined based on the predefined criteria. Based on the optimization criteria, the critical well depths and the conditions to achieve the maximum hydraulic parameters, the optimal flow rate, and the optimal nozzle diameter can be calculated for each well section. And parameters, such as v_j, F_j, Δp_b, and P_b, can be computed for different well sections as well.

Example 4.3.1 A ϕ215.9 mm bit (a nozzle discharge coefficient $C = 0.98$), and the ϕ177.8 mm drill collar (inner diameter 71.4 mm) that is 120 m in length are used to drill a well. The drill pipe has a diameter of 127 mm (the inner diameter is 108.6 mm). There are two 3NB-1600 drill pumps in the field. According to the experience the surface pump pressure is supposed to be not more than 20 MPa; K_g is 1.07×10^{-3} MPa·s$^{1.8}$·L$^{-1.8}$. It is expected that the drill fluid when drilling to 4000 m is about 1.64 g/cm^3 with the plastic viscosity of 0.047 Pa·s. Try to use the maximum bit hydraulic power to design the hydraulic parameters at this well depth. The annulus velocity is required to be higher than 0.7 m/s.

   Solution:

- Determination of the minimum flow rate Q_a.

According to empirical Eq. 4.111, the minimum annulus velocity v_a is:

$$v_a = \frac{18.24}{\rho_d d_h} = \frac{18.24}{1.64 \times 21.59} = 0.515 \text{m/s}$$

Since the minimum annulus velocity is required to be higher than 0.7 m/s, then $v_a = 0.7$ m/s.

- Calculation of the pressure loss coefficient at different well depths.

$$m = \rho_d^{0.8} \mu_{pv}^{0.2} \left[\frac{B}{d_{pi}^{4.8}} + \frac{0.57503}{(d_h - d_{po})^3 (d_h + d_{po})^{1.8}} \right]$$

$$= 1.64^{0.8} \times 0.047^{0.2} \times \left[\frac{0.51655}{10.86^{4.8}} + \frac{0.57503}{(21.59 - 12.7)^3 \times (12.59 + 12.7)^{1.8}} \right]$$

$$= 5.58 \times 10^{-6}$$

$$K_c = \rho_d^{0.8} \mu_{pv}^{0.2} L_c \left[\frac{B}{d_{pi}^{4.8}} + \frac{0.57503}{(d_h - d_{co})^3 (d_h + d_{co})^{1.8}} \right]$$

$$= 1.64^{0.8} \times 0.047^{0.2} \times 120 \times \left[\frac{0.51655}{7.14^{4.8}} + \frac{0.57503}{(21.59 - 17.78)^3 \times (12.59 + 17.78)^{1.8}} \right]$$

$$= 5.34 \times 10^{-3}$$

$$a = K_g + K_c - mL_c$$
$$= 1.07 \times 10^{-3} + 5.34 \times 10^{-3} - 5.58 \times 10^{-6} \times 120$$
$$= 5.34 \times 10^{-3}$$

The pressure loss coefficient at well depth of 4000 m is:

$$K_L = mD + a$$
$$= 5.58 \times 10^{-6} \times 4000 + 5.7404 \times 10^{-3}$$
$$= 0.028$$

- Selection of optimal flow rate.

When the drilling pump is under the rated pressure working condition, p_s is fixed at about 20 MPa. The bit hydraulic power is:

$$P_b = P_s - P_L = p_s Q - K_L Q^{2.8}$$

Let $\frac{dP_b}{dQ} = 0$ and we can get that:

$$\frac{dP_b}{dQ} = p_s - 2.8 K_L Q = 0$$

The optimal flow rate is:

$$Q_{opt} = \left(\frac{p_s}{2.8 K_L}\right)^{\frac{1}{1.8}} = \left[\frac{p_s}{2.8(a + mD)}\right]^{\frac{1}{1.8}}$$
$$= \left(\frac{20}{2.8 \times 0.028}\right)^{\frac{1}{1.8}} = 21.73 (L/s)$$

The pressure loss is:

$$\Delta p_L = (mD + a)Q_{opt}^{1.8} = 0.028 \times 21.73^{1.8} = 7.14 (\text{MPa})$$

- Selection of the cylinder liner.

The working performance specifications of 3NB-1600 drill pump are presented in Table 4.8. According to the pump working specifications, the working rated flow rate is typically 90% of the rated flow rate of the cylinder liner, and the same rule applies to the working pressure. Thus, the rated flow rate of the selected cylinder liner should be:

$$Q_r \geq \frac{21.73}{0.9} = 24.12 \text{L/s}$$

The rated pressure is:

$$p_r \geq \frac{20}{0.9} = 22.22 \text{MPa}$$

Based on Table 4.8, the diameter of cylinder liners that satisfy the above requirements is from 130 to 180 mm.

- Determination of nozzle diameter.

$$d_{ne} = 4\sqrt{\frac{0.081 \rho_d Q_{opt}^2}{C^2 [p_s - (a + mD) Q_{opt}^{1.8}]}}$$

$$= 4\sqrt{\frac{0.081 \times 1.64 \times 21.73^2}{0.98^2 \times [20 - (5.7404 \times 10^{-3} + 5.58 \times 10^{-6} \times 4000) \times 21.73^{1.8}]}}$$

$$= 1.50 \, (\text{cm})$$

- Computation of hydraulic parameters

$$v_j = \frac{10 Q_{opt}}{A_0} = \frac{10 \times 21.73}{3.14 \times 1.50^2 / 4} = 123.03 \text{m/s}$$

$$F_j = \frac{\rho_d Q_{opt}^2}{100 A_0} = \frac{1.64 \times 21.73^2}{100 \times 3.14 \times 1.50^2 / 4} = 4.38 \text{kN}$$

Table 4.8 Parameter table for the specifications of 3NB-1600 drilling pump

Cylinder liner diameter (mm)	The rated pump stroke Stroke/min	The rated flow rate (L/s)	The rated pump pressure (MPa)
130	140	23.6	34.3
140	140	27.4	31.4
150	140	31.4	27.3
160	140	35.7	24.0
170	140	40.4	21.3

$$P_j = \frac{0.05\rho_d Q_{opt}{}^3}{A_0{}^2} = \frac{0.05 \times 1.64 \times 21.73^2}{\left(\frac{3.14}{4} \times 1.50^2\right)^2} = 269.70\text{kN}$$

$$\Delta p_b = p_s - \Delta p_L = 20 - 7.14 = 12.86\text{MPa}$$

$$P_b = \Delta p_b Q_{opt} = 12.86 \times 21.73 = 279.45\text{kW}$$

The results are listed as below (Table 4.9):

Exercises and Questions

1. Type 241 bit with a diameter of 200 mm is selected for drilling a well. The weight on bit W is 196 kW and rotational speed n is 70 r/min. The cleaning condition at the bottomhole is good. After drilling for 14 h, the pipe is pulled out, also $A_f = 2.33 \times 10^{-3}$. What is the tooth abrasion loss h_f?
2. A type 211 bit with the diameter of ϕ215.9 mm is employed for drilling. The weight on bit is $W = 196$ kN. The rotational speed is 80 r/min. After drilling for 14 h, the drill pipe is pulled out. The bearing wear is at level B_6. What is the working coefficient b of the bearing?
3. The experimental results of a five-point well drilling test are listed as below:

Test point	1	2	3	4	5	6
WOB	225	254	254	196	196	225
Rotation speed (r/min)	70	60	120	120	60	70
ROP (m/h)	31	32.5	46	34	24	30

Calculate the threshold weight on bit M and the rotation index λ.
4. A ϕ215.9 mm bit is used for drilling. The nozzle discharge coefficient $C = 0.96$. The density of drilling fluid is 1.43 g/com³, and the flow rate is 16 L/s. If the bottomhole hydraulic power is required to be 0.418 kW/m². Besides, one of the three nozzles has a diameter of 9 mm, and the other two have the same diameter. Try to calculate the diameter of these two nozzles and calculate the hydraulic parameters of the jet flow and the bit.
5. The wellbore diameter is supposed to be 215.9 mm, and a drill pipe with the diameter of 127 mm (inner diameter 108.6 mm) is used for drilling. The drill collar has a diameter of 177.8 mm (inner diameter 71.4 mm) and a length of

Table 4.9 Design values of hydraulic parameters

Well depth (m)	Q_{opt} (L/s)	d_{nc} (cm)	Δp_b (MPa)	P_b (kW)	v_j (m/s)	F_j (kN)	P_j (kW)
4000	21.73	1.50	12.86	279.45	123.03	4.384	269.70

100 m. The drilling fluid density ρ_d is 1.2 g/cm^3 with the plastic viscosity μ_{pv} of 0.022 Pa·s. $\Delta p_g = 0.34$ MPa, and $Q = 22$ L/s. Try to calculate the values of m, a, and the pressure loss of the system at a well depth of 2000 m.

6. The conditions are the same as question 5. Calculate the bit pressure drop, nozzle equivalent diameter (nozzle discharge coefficient $C = 0.96$) and the jet flow hydraulic parameter at the allowable pump pressure of 17.3 MPa and 14.2 MPa, relatively.

7. It is known that a well is drilled by the $\phi 215.9$ mm bit. The diameter of the expansion point of the borehole is 310 mm. The outer diameter of the drill pipe is 127 mm. The flow rate is 21 L/s and drilling fluid density is 1.16 g/cm^3. The value of viscometer at the rotation speed of 600 and 300 are 65 and 39, respectively. The cutting density is 2.52 g/cm^3, and the average cutting particle diameter is 6 mm. Calculate the cutting lift efficiency.

8. A well is drilled by the $\phi 215.9$ mm bit. The nozzle discharge coefficient C is 0.98. A drill pipe with the diameter of 139.7 mm (inner diameter 118.6 mm) is used for drilling. The drill collar has a diameter of 177.8 mm (inner diameter 71.4 mm) and the length of 120 m. There are two 3NB-1300 pumps. According to the experience the pump pressure should be less than 18 MPa. K_g is 1.05 MPa$\times 10^{-3} \cdot$s$^{1.8} \cdot$L$^{-1.8}$. When it is drilled to 4000 m the density of the drilling fluid is increased to 1.45 g/cm^3 and the plastic viscosity is increased to 0.038 Pa·s. Design the hydraulic parameters based on the maximum bit hydraulic power. The annulus velocity should be higher than 0.7 m/s.

9. A well is drilled by roller bits from 2000 to 2500 m. The rated pump flow rate Q_r is 36 L/s and the rated pump pressure p_r is 20 MPa. It is known from the test that m is 5.5×10^{-3} and a is 5.7×10^{-3}. The density of drilling fluid is 1.15 g/cm^3, and the minimum flow rate is 20 L/s. All nozzles have the same diameter. Calculate the equivalent nozzle diameter for the maximum bit hydraulic power (the nozzle discharge coefficient C is 0.96).

10. A well is drilled using the jet flow drilling technology. At the well depth of 2000 m, the flow rate is 28 L/s. The system pressure loss is 8 MPa; when the well depth of 2500 m is reached, the flow rate is 28 L/s, and the pressure loss is 9 MPa; when 2700 m is reached, the flow rate remains the same and the pump pressure is 20 MPa. At this time, the density of drilling fluid is 1.3 g/cm^3. It is known that the three nozzles on the bit have the same diameter. Calculate the equivalent nozzle diameter (the nozzle discharge coefficient C is 0.96).

Well Trajectory Design and Wellpath Control

5

Abstract

Directional drilling technology is one of the most important drilling technologies in the field of oil exploration and development in the world today. It is a drilling technology that makes the drill bit reach the predetermined target underground by effectively controlling the wellpath with special downhole tools, measuring instruments and technologies. In this chapter we will discuss: the applications of directional drilling, the basic parameters of wellpath, wellpath measurement and calculation, the cause of well deviation and the deviation control technology during drilling a vertical well, the design method of well trajectory, calculations for azimuth correction run, and orientation methods of deflection tools.

Well trajectory refers to a predetermined course of the borehole axis before drilling, while wellpath reflects the actual borehole axis after drilling.

As rotary drilling technology emerged at the end of the nineteenth century, the vertical well was primarily planned by drillers based on their knowledge that rotary drilling produced an identical wellpath to percussion drilling. They did not realize that the actual wellpath could be inclined, until about the late 1920s when they found that a new well penetrated the casing of an adjacent old well; two wells with different total measured depths had been drilled in the same oil-bearing layer. Since then, the technology of deviation control in a vertical well was developed, and a series of effective approaches were put forward to reduce or even eliminate well deflection. In the early 1930s, as directional drilling was successfully tested in offshore oil fields, the application of directional drilling was expanded into three major domains as follows.

- **Ground environmental constraints**

Directional drilling intends to reach reservoirs under natural obstructions such as mountains, lakes, swamps, rivers, gullies, oceans, farmlands, or critical buildings,

© China University of Petroleum Press and Springer Nature Singapore Pte Ltd. 2021
Z. Guan et al., *Theory and Technology of Drilling Engineering*,
https://doi.org/10.1007/978-981-15-9327-7_5

which frequently prohibit setting up a rig and drilling a vertical well. For better exploitation and development of these reservoirs to avoid costly setting-up and drilling operations of vertical wells, it is better to drill directional wells.

- **Underground geological constraints**

A larger number of oil-bearing layers can be targeted and penetrated by directional drilling in some fault-screened oil reservoirs. Directional wells and horizontal wells have a larger reservoir contact area, especially in thin oil zone. Some special directional wells, such as sidetracking wells, multilateral wells, extended-reached wells (ERWs), and radial horizontal wells, have considerably enlarged the potential of oil exploitation and development, increased oil production, and eventually enhanced oil recovery.

- **Treatment for downhole drilling accidents**

When a downhole junk or a section of broken drill string (called "fish") cannot be fished out, a directional well can be drilled from the upper section. Especially when a wild well blowout with fire is difficult to deal with, a directional well (called relief well) can be used in the vicinity of the accident well, and the accident well can be connected, discharged, or killed.

In the petroleum industry, directional drilling plays a significant role in the exploitation and development of oil reservoirs. Well trajectory design and wellpath control are two major components in directional drilling. A vertical well can be considered as a special directional well. Deviation control for vertical wells and wellpath control for directional wells embrace the identical technical theory but different applications. The wellpath control has evolved from experience to science and from qualitative technology to quantitative technology. Currently, it is heading to an automatic control phase.

5.1 Fundamentals of Wellpath

Basic concepts of wellpath parameters and the correlation between each parameter are essential for well trajectory design, wellpath measurements and calculations as well as wellpath control.

5.1.1 Basic Parameters of Wellpath

Wellpath refers to an actual shape of the borehole axis when the drilling process is completed. An actual borehole axis is a spatial curve. Wellpath measurement (or wellpath survey) is conducted to know the borehole axis shape and thereby to control it. Rather than continuous measurement, the widely used method involves obtaining a measurement point every certain borehole interval (or survey interval). Based on the wellpath measurement, three significant parameters, well depth, inclination, and azimuth, can be obtained.

- **Well depth**: also called the inclined depth, the total measured depth (TMD), symbolized by $L(m)$ with the increment of ΔL, refers to the wellbore length from the rotary table to the survey station, measured by the length of associated drill string, or cables. The well depth is one of the basic parameters of the survey station and also the only label of the survey station. Two adjacent survey stations consist of a survey interval, where the deeper survey station is called the lower survey station while the other one is called the upper survey station. The well depth increment is the well depth of the lower survey station subtracted from that of the upper one.
- **Inclination**: At a certain survey station, the borehole direction line is a tangent line with the same tendency of the borehole axis. The inclination, symbolized by α with the unit of degree (°), refers to the angle between the borehole direction line and the vertical gravity line. Both the borehole direction line and the gravity line are vectors on a certain plane, and the inclination indicates the severity of the wellbore deflection at the survey station. Basically, the inclination lies between 0° and 90°. The inclination increment, symbolized by $\Delta \alpha$, is the inclination of the lower survey station subtracted from that of the upper one. Note that the inclination may be over 90° in some scenarios for horizontal drilling. As illustrated in Fig. 5.1, there are two survey stations (A and B) located at the two ends of a certain survey interval with inclinations of α_A and α_B, respectively. The inclination increment of these two survey stations can be expressed as $\Delta \alpha = \alpha_B - \alpha_A$.

Fig. 5.1 Diagram of inclination

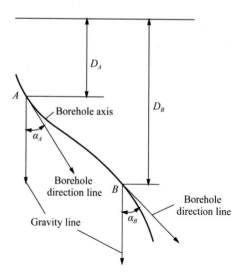

- **Azimuth**: Borehole azimuth line or borehole bearing line refers to the projection of the borehole direction line onto a horizontal plane. Therefore, the azimuth refers to the angle between the true north bearing line and the borehole bearing line measured in the clockwise direction on the horizontal plane. Geographically, the bearing line of the true north represents an elongation of the meridians pointing to the North Pole. That is to say, both the true north bearing line and the borehole bearing line are vectors on a horizontal plane. Note that a bearing line denotes a vector on a horizontal 2D plane while a direction line represents a vector in a 3D space. The bearing line, azimuth line or azimuth, is on a certain horizontal plane. Also, direction or direction line is in a 3D space (sometimes also on a horizontal 2D plane because a plane is a special scenario of a 3D space). The borehole direction line is a tangent line at a certain survey station in the same tendency of the borehole axis, while the borehole azimuth line indicates the projection of the borehole direction line at the same survey station onto a horizontal plane. This issue would also be critical for the calculation of azimuth correction run in Sect. 5.5.

Azimuth, symbolized by ϕ with the unit of degree (°), lies between 0° and 360°, and the azimuth increment, symbolized by $\Delta\phi$, is the azimuth of the lower survey station subtracted from that of the upper one. As shown in Fig. 5.2, azimuths of two survey stations (A and B) are ϕ_A and ϕ_B, respectively, and the azimuth increment can be expressed as $\Delta\phi = \phi_B - \phi_A$.

The widely used magnetic inclinometer utilizes the magnetic north (MN) as a benchmark. It should be noted that the magnetic north does not overlap with the geographical north (true north, TN). The bias angle between these two vectors on the horizontal plane is called the magnetic declination. The magnetic declination comprises the east magnetic declination and the west magnetic declination. If the magnetic north lies to the east of the true north, the angle is the east magnetic declination with a positive value; otherwise, it is the west magnetic declination with a negative value. A magnetic azimuth can be obtained from the magnetic inclinometer, and the true azimuth can be calculated by the following conversion.

$$\text{True Azimuth} = \text{Magnetic Azimuth} + \text{Magnetic declination}$$

Fig. 5.2 Diagram of borehole azimuth

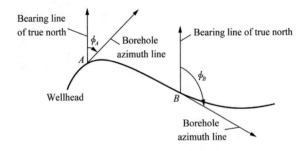

Fig. 5.3 Azimuth in the 90° quadrant scheme

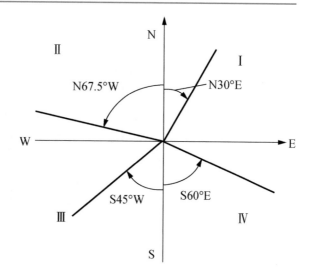

As illustrated in Fig. 5.3, the azimuth can also be expressed in terms of a 90° quadrant scheme, and the degrees are always read from north to east or west and from south to east or west, such as N67.5°W.

5.1.2 Calculated Parameters of Wellpath

Calculated parameters of wellpath obtained based on the basic wellpath parameters can be used to identify the acrual borehole shape and location of wellpath and also be applied for wellpath plotting.

True vertical depth (TVD), symbolized by D, refers to the vertical distance from a certain survey station to the horizontal plane through the wellhead. The increment of TVD is generally symbolized by ΔD. As illustrated in Fig. 5.1, TVDs of two survey stations (A and B) are D_A and D_B, respectively, and the increment of TVD can be expressed as $\Delta D = D_B - D_A$.

N-coordinate and **E-coordinate** represent the two coordinates at a certain survey station on a horizontal coordinate system with the wellhead as its origin. The horizontal coordinate system comprises two coordinate axes with one oriented toward the geographical north (N) and the other pointed to the east (E). As shown in Fig. 5.4, coordinates of two survey stations (A and B) are (N_A, E_A) and (N_B, E_B), respectively, and the corresponding increments of two coordinates are ΔN and ΔE.

Horizontal projection length indicates the length of the horizontal projection of wellbore axis from the wellhead to a certain survey station on a horizontal plane. Horizontal projection length is usually symbolized by P with its increment symbolized by ΔP. As shown in Fig. 5.4, both the horizontal projection length and its increment represent the length of corresponding curves on the horizontal plane.

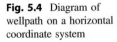

Fig. 5.4 Diagram of wellpath on a horizontal coordinate system

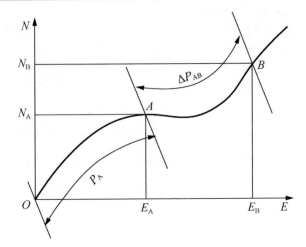

Closure distance, symbolized by C, refers to the distance from a certain survey station to the plumb line through the wellhead. As illustrated in Fig. 5.5, the closure distances of two measurement points (A and B) are C_A and C_B, respectively. As practiced in China, closure distance used to be called as horizontal displacement, and this terminology is still in use today.

It should be noted that although the closure distance is similar to the horizontal projection length for a designed 2D well trajectory, these two concepts are completely different for the actual wellpath.

Closure azimuth, symbolized by θ, refers to the angle on the horizontal plane from the true north azimuth line to the closure azimuth line measured in the clockwise sense. Here, the closure azimuth line indicates the azimuth line from the wellhead to current survey station on the horizontal plane. In Fig. 5.5, closure azimuths of two survey stations (A and B) are θ_A and θ_B, respectively.

Fig. 5.5 Diagram of closure distance, closure azimuth, and vertical section

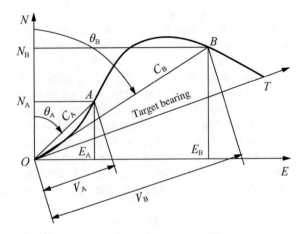

Vertical section, symbolized by V, refers to the projection length of the closure distance on target bearing, and it is crucial for wellpath plotting. As illustrated in Fig. 5.5, vertical sections of two survey stations (A and B) are V_A and V_B, respectively. It must be noted that the vertical section might be negative when striking differences exist between the actual wellpath and the designed well trajectory.

Borehole curvature represents the curvature of the borehole axis. Since the actual wellpath has an arbitrary shape in 3D space, the borehole curvature continually changes as the drilling progresses. For convenience, the average borehole curvature is usually adopted for engineering calculations.

Both inclination and azimuth vary as the drilling progresses, and the overall angle change reflects the changes in both inclination and azimuth. Furthermore, the overall angle change rate or the overall angle change per survey interval length indicates the average borehole curvature, which can be expressed as follows.

$$\gamma = \sqrt{\Delta\alpha^2 + \Delta\phi^2 \sin^2 \alpha_c} \tag{5.1}$$

$$\alpha_c = \frac{\alpha_1 + \alpha_2}{2} \tag{5.2}$$

where

γ Overall angle change within a certain survey interval, (°);

α_c Average inclination in a certain survey interval, (°);

α_1, α_2 The inclination of the upper and lower survey stations, respectively, (°).

Dogleg angle, or dogleg, expressed in the following formula, refers to the angle between borehole direction lines of two adjacent survey stations in the 3D space when the borehole axes at the two survey stations can be considered as a two-dimensional circular curve. The dogleg severity, or overall angle change rate, indicates the average curvature of the borehole.

$$\cos \gamma = \cos \alpha_1 \cos \alpha_2 + \sin \alpha_1 \sin \alpha_2 \cos(\phi_2 - \phi_1) \tag{5.3}$$

Note that both the overall angle change and dogleg are symbolized by γ and subscripts 1 and 2 indicate the upper and lower survey stations, respectively.

The average curvature of borehole within a certain survey interval, symbolized by K with the widely used unit of (°)/30 m, can be easily obtained using the following formula after the dogleg or the overall angle change is calculated:

$$K = \frac{\gamma}{\Delta L} \times 30 \tag{5.4}$$

K refers to the dogleg severity calculated by using Eq. 5.3 or represents the overall angle change rate calculated by using Eqs. 5.1 and 5.2. In conclusion, both expressions represent the average borehole curvature within a certain survey interval.

As the concept of the dogleg is put forward under the assumption of plane arc curve, the calculated dogleg severity technically indicates the minimum value of borehole curvature. Therefore, within the same survey interval, the difference between the overall angle change rate and dogleg severity should be understood, and it should be determined which one can be selected as a benchmark for the engineering application. Western countries, such as the USA, prefer to use dogleg severity which was adopted by China since 2003, according to the Industry Standard.

5.1.3 Graphical Methods of Wellpath

Two ways of illustrating a wellpath are shown in Fig. 5.6. One is a combination of vertical projection and horizontal projection, and the other is a combination of vertical profile and horizontal projection. The horizontal projection is critical to both methods. In China, operators used to select the latter one, but the former one is the most commonly used now.

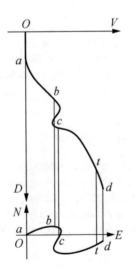

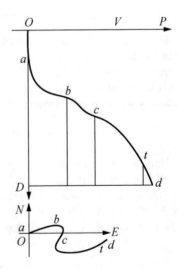

（a）the combination of vertical projection and horizontal projection

（b）the combination of vertical profile and horizontal projection

Fig. 5.6 Graphical methods of actual directional wellpath

Horizontal projection, similar to the top view in the mechanical drawing, refers to the projection of wellpath on the horizontal plane where the wellhead is located at the origin of the horizontal coordinate system with N and E coordinates. Accordingly, if the N and E coordinates of all survey stations on a certain wellpath are known, it is easy to obtain the horizontal projection.

Vertical projection, similar to the side view in the mechanical drawing, indicates the projection of wellpath on a certain vertical plane where the wellhead is located at the origin of the coordinate system with TVD and the vertical section as its coordinates. It is well known that the vertical planes passing through the wellhead are countless and one of them should be selected as the vertical projection plane. According to the Industry Standard in China, the vertical plane passing through the target bearing is an optimal selection. In this way, by comparison between the true vertical projection and the predetermined one, operators can figure out differences between the wellpath and well trajectory and implement wellpath control operations easily. The vertical projection can be easily obtained with the given TVDs and vertical sections at every survey station.

Vertical profile, illustrated in Fig. 5.7, refers to a mathematical cylindrical surface constituted by several plumb lines passing through every survey station on the wellpath. The cylindrical surface can be unfolded to a plane which is called vertical profile. The vertical profile can be easily obtained with given TVD (D) and horizontal projection length (P) of every survey station on the wellpath.

Fig. 5.7 Principle of vertical profile

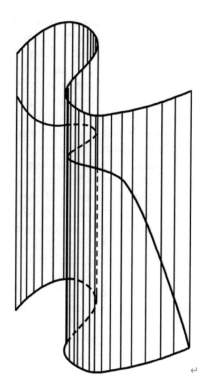

5.2   Wellpath Measurements and Calculations

After drilling, it is necessary to know the actual shape and location of a wellpath and to ascertain whether it has hit the predetermined target. While drilling, it is important to know the shape of a wellpath to judge the tendance of wellpath and then take effective measures in a timely manner to control the wellpath. Therefore, it is crucial to measure the wellpath and calculate unknown parameters from the measured data. In engineering terminology, the professional instrument for wellpath measurement is called an inclinometer or a surveyor.

5.2.1   Introduction to Wellpath Survey Methods and Inclinometer's Principles

According to wellpath survey methods, inclinometers are devided into the single-shot instrument, the multi-shot instrument, and the measurement while drilling (MWD) instrument. The single-shot instrument can be delivered to the downhole either on a steel wire or a cable through the drill string and can make only one record of wellpath parameters at a certain well depth (generally close to the drill bit) in one run. The single-shot instrument is less time-consuming and is used to orient deflection tools in the process of wellpath control. The multi-shot instrument drops down to a certain position approaching the drill bit by a cable in an open hole or through the drill string before a tripping-out operation. The instrument is capable of taking numerous survey records in one run during the static period of screwing off the drill string and pulling a stand aside. Measured data from multi-shot instruments are widely used for the wellpath calculation. The MWD instrument is run into the downhole with the drill string concurrently. The tool is used for continuous records while drilling and real-time measured data are transmitted to the surface receiver for accurate wellpath control. Due to the high expenses of the MWD tool, it used to be merely applied to the wells where the wellpath survey was difficult to conduct. At present, this tool is widely used for directional drilling.

The basic wellpath parameters of a survey station consist of well depth, inclination, and azimuth. The well depth is measured by recording the length of the cable or the drill string, while the inclination and the azimuth are measured by a specialized inclinometer. The development of the inclinometer started from the syphon inclinometer and the hydrofluoric acid bottle inclinometer based on the horizontal principle of static liquid and then progressed to the punching inclinometer based on the principle of a plumb line. However, these tools are only used for measuring the inclination. Around the 1940s, a compass inclinometer emerged which is used for measurement of both the inclination and the azimuth. It was followed by a gyro-clinometer which is used under the condition of strong magnetic interference. In the 1980s, the electronic inclinometer emerged, which combined two kind of vibration-proof solid-state sensors, namely the gravity accelerometer and the fluxgate (or magnetometer). The new generation electronic inclinometers, used for single-shot measurement, multi-shot measurement as well as measurement while drilling, have been widely used for wellpath measurements and calculations for all kinds of directional wells. The principle of the electronic inclinometer is introduced as follows.

5.2.1.1 Measurement Principle of the Inclination by the Gravity Accelerator

The gravity accelerometer is a sensor that measures the inclination. A typical capacitive quartz gravity accelerator is described as follows. As illustrated in Fig. 5.8, a gravity element, made from a quartz slice where two iron-core coils are connected in series on its two sides, is fixed on the shell. A small gap exists between two plates of the capacitive transducer and the quartz slice. Affected by its gravity, the quartz slice revolves around the hinge and produces an angular displacement. After the change is captured by the capacitive transducer, an output signal is produced and transformed into the feedback current through the servo loop, which consists of a capacitive bridge, an amplifier, a demodulator, and a power source. Affected by the magnetic force produced by the iron-coil where the feedback current flows, the quartz slice reverts to its original position. The feedback current indicates the gravity component in the direction of the axis of gravity element.

As shown in Fig. 5.8b, the receptive gravity of the gravity element varies according to its different attitude in 3D space. The receptive gravity reaches the maximum when the angle between the axial direction of the gravity element and the gravity direction is 0°, while the receptive gravity comes to zero when the angle is 90°. The output reading from the gravity accelerator has a certain relationship with the inclination. To measure the inclination, three mutually perpendicular gravity

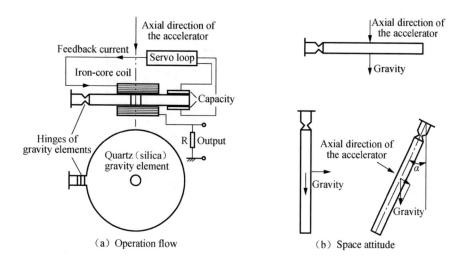

(a) Operation flow (b) Space attitude

Fig. 5.8 Schematic diagram of the gravity accelerator

accelerators need to be installed in an inclinometer. By certain calculations, the inclination can be obtained where the inclinometer is located.

5.2.1.2 Measurement Principle for the Azimuth by the Fluxgate

The fluxgate is a sensor for measuring the azimuth. As illustrated in Fig. 5.9, two parallel iron cores of the same type (identical structures and performance parameters) are wound at one side of the fluxgate by two main coils with the same structure parameters in the opposite winding directions and by one secondary coil at the other side. These two main coils are connected in series and powered by an AC power supply.

Once the AC power is on, both the main coils are energized simultaneously. Given that the Earth's magnetic field does not exist, two separate induced magnetic fields are produced in the two main coils with the same magnitude but in the opposite directions. As a result, both the induced current and the output voltage in the secondary coil would be zero.

In fact, due to the existence of the Earth's magnetic field, the magnetic flux of one main coil increases while that of the other one decreases. The main coil where the magnetic flux is enlarged reaches saturation in advance, a half period faster than the other where the magnetic flux is reduced. Therefore, variation of the magnetic flux on both main coils produces a rate difference, which is illustrated in Fig. 5.10. The asymmetric waveforms generate secondary harmonics with the same phases and amplitudes in the two iron cores which can be superposed, while the produced fundamental waves and third harmonics are symmetrical and offset each other. As a result, the induced current is generated in the secondary coil with the output voltage

Fig. 5.9 Schematic diagram of the fluxgate

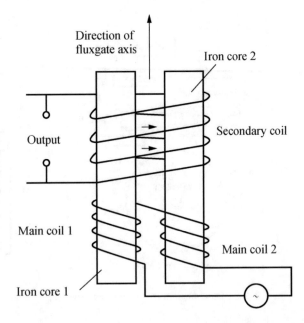

signal due to the superposed secondary harmonics. Significantly, the output voltage from the secondary coil reflects the magnitude and direction of the magnetic flux component of the Earth's magnetic field in the axial direction of the fluxgate. This indicates that the amplitude of the second harmonic is proportional to the axial component of the Earth's magnetic field and the phase is directly opposite to the direction of the field. Therefore, the voltage value of the output of the dependent secondary coil just reflects the magnitude and direction of the magnetic flux on the magnetic fluxgate axis.

Similar to the principle of the gravity accelerator, the magnitude and direction of the magnetic flux of the Earth's magnetic field through the axis of the fluxgate depend on its attitude in the 3D space. Therefore, the magnetic flux reaches the maximum when the direction of the fluxgate axis overlaps that of the Earth's magnetic field (the angle between these two directions is 0°), while the magnetic flux approaches zero when the angle is 90°.

To measure the azimuth, three mutually perpendicular magnetic fluxgates and three mutually perpendicular gravitational accelerometers need to be installed in an inclinometer. The azimuth can then be obtained by a certain calculation.

5.2.1.3   Calculations for Wellpath Parameters and Orientation Parameters

Both the gravity accelerator and fluxgate are installed in a probe tube, a cylindrical shell of the inclinometer. To meet the requirement of wellpath measurements and calculations, a comoving coordinate system is established on the probe tube, as illustrated in Fig. 5.11. The Z-axis is along the direction of the inclinometer axis and points toward the downhole; Y-axis and X-axis are dependent on the right-handed coordinate system.

Note that the formulae for calculating required parameters vary according to the different layout schemes of the gravity accelerator and the fluxgate in the probe tube. Figure 5.11 illustrates one of the widely used layout schemes adopted in China.

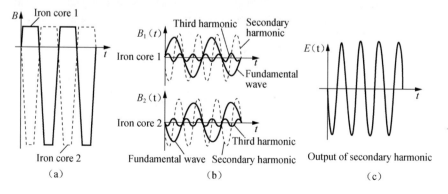

Fig. 5.10 Output of the fluxgate affected by the Earth's magnetic field

Fig. 5.11 Comoving
coordinate system of the
probe tube and the layout of
the gravity accelerator and the
fluxgate in the probe tube

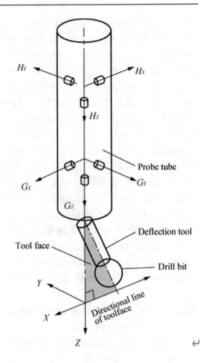

- The positive directions of the gravity accelerators G_X, G_Z are consistent with the positive directions of the coordinate system, respectively, but the gravity accelerator G_Y is opposite to that of the Y-axis.
- The positive direction of the fluxgate H_Y and H_Z is opposite to that of the Y-axis and Z-axis in the coordinate system, while the positive directions of fluxgates H_X are consistent with those of the X-axis, respectively.
- The directional line of the tool face is opposite to the direction of the X-axis.
- According to the layout scheme shown in Fig. 5.11, the inclination, the high-side tool face (or gravity tool face, GTF) angle, the magnetic-north tool face (or magnetic tool face, MTF) angle, and the azimuth at every survey station can be obtained based on measurement values read from three gravity accelerators and three fluxgates.
  - Calculations for the inclination α:

$$\alpha = \begin{cases} 180° + \arctan\dfrac{\sqrt{G_X^2 + G_Y^2}}{G_Z} \,(G_Z < 0) \\ \arctan\dfrac{\sqrt{G_X^2 + G_Y^2}}{G_Z} \,(G_Z > 0) \\ 90°\,(G_Z = 0) \end{cases} \tag{5.5}$$

- Calculations for the high-side tool face angle ω:

$$\omega = \begin{cases} \arctan\dfrac{G_Y}{G_X}(G_X<0) \\[2mm] 180° + \arctan\dfrac{G_Y}{G_X}(G_X>0) \\[2mm] 90°\,(G_X=0 \text{ and } G_Y<0) \\[2mm] 270°\,(G_X=0 \text{ and } G_Y>0) \end{cases} \tag{5.6}$$

- Calculations for the magnetic-north tool face angle ω_m:

$$\omega_m = \begin{cases} \arctan\dfrac{H_Y}{H_X}(H_X>0) \\[2mm] 180° + \arctan\dfrac{H_Y}{H_X}(H_X<0) \\[2mm] 90°\,(H_X=0 \text{ and } H_Y>0) \\[2mm] 270°\,(H_X=0 \text{ and } H_Y<0) \end{cases} \tag{5.7}$$

- Calculations for the azimuth ϕ:

$$\phi = \begin{cases} \arctan\dfrac{U}{-V}(V<0) \\[2mm] 180° + \arctan\dfrac{U}{-V}(V>0) \\[2mm] 90°\,(V=0 \text{ and } U<0) \\[2mm] 270°\,(V=0 \text{ and } U>0) \end{cases} \tag{5.8}$$

$$V = H_X\cos\omega\cos\alpha + H_Y\sin\omega\cos\alpha + H_Z\sin\alpha \tag{5.9}$$

$$U = H_X\sin\omega - H_Y\cos\omega \tag{5.10}$$

It must be noted that the electronic inclinometer is a type of magnetic inclinometer. A major disadvantage of the magnetic electronic inclinometer is that it is subject to magnetic interference from drilling tools made of ferromagnetic material, which is commonly known as magnetic interference. Instead of using a steel drill collar where the inclinometer is located, a non-magnetic drill collar with sufficient length will be a better selection for operations. For the determination of the length of the non-magnetic drill collar and associated problems, please refer to the relevant references.

5.2.2   Regulations on Calculation Data of Wellpath Survey

According to China's Drilling Industry Standard for calculation data of wellpath survey, the following provisions and regulations must be strictly observed.

- The data used for wellpath calculation must be measured by a multi-shot inclinometer.
- The azimuth measured by the magnetic inclinometer must be corrected both by the magnetic declination (see Sect. 5.1.1) and by the meridian convergence (or grid convergence) before implementing the wellpath calculations.
- Numbering survey stations. Although the wellpath survey process is conducted from the bottom to the top, the order of the survey stations is specified upside down. The first survey station whose inclination is not equal to $0°$ is regarded as the survey station number 1, followed by a series of sequential numbers for the rest of survey stations. All wellpath parameters at each survey station are subscripted by its corresponding number.
- Numbering survey intervals. Similar to the numbering sequence of the survey stations, the ith survey interval is located between the $(i-1)$th and the ith survey stations. Accordingly, the total number of survey stations is equal to that of survey intervals, and all wellpath parameters at each survey interval are subscripted by its corresponding number.
- Rules on the pseudo zeroth survey station. According to the numbering sequence of the survey intervals, the first interval is located between the zeroth and first survey stations. Rather than being measured, the pseudo zeroth survey station is set as per requirement. Operators also have to comply with the following regulations. If the well depth at the first survey station is greater than 25 m (approximately 82 ft), the well depth at the zeroth survey station is set as 25 m less than that of the first survey station, and the inclination at the zeroth survey station is set as $0°$. If the well depth at the first survey station is less than or equal to 25 m, the well depth at the zeroth survey station is set as 0 m, and the corresponding inclination is also $0°$.
- The azimuth at a certain survey station does not exist if its inclination is $0°$. Therefore, operators must follow the regulations that ϕ_i is equal to ϕ_{i-1} at the ith survey interval and ϕ_{i+1} is equal to ϕ_i at the $(i+1)$th survey interval.
- The absolute azimuth change in the same survey interval cannot exceed $180°$. Accordingly, the calculations for the average azimuth ϕ_c within the survey interval should be performed as expressed below.

If $\phi_i - \phi_{i-1}$ is larger than $180°$, the following formulae are used for calculating the azimuth increment $\Delta\phi_i$ and the average azimuth ϕ_c.

$$\Delta\phi_i = \phi_i - \phi_{i-1} - 360° \tag{5.11}$$

$$\phi_c = \frac{\phi_i + \phi_{i-1}}{2} - 180° \tag{5.12}$$

If $\phi_i - \phi_{i-1}$ is less than $-180°$, the formulae for calculating the azimuth increment $\Delta\phi_i$ and the average azimuth ϕ_c are modified as follows.

$$\Delta\phi_i = \phi_i - \phi_{i-1} + 360° \tag{5.13}$$

$$\phi_c = \frac{\phi_i + \phi_{i-1}}{2} + 180° \tag{5.14}$$

Further explanation is necessary for the second item above. The azimuth correction refers to a combination of the magnetic declination correction and the meridian convergence correction. As the concept and the correction method of the magnetic inclination have been introduced in the previous sections, only those of the meridian convergence correction are briefly discussed below.

The Earth projection refers to a planar projection of the Earth which contains the position of each point and the distance between every two relevant points on its spherical surface. Among the Earth projection methods, the widely used one is the Gauss projection method, which involves setting a projection zone within every 6 degrees of longitude along the equator. Except for a straight central meridian, the Gauss projection depicts multiple non-parallel meridians where the tangential direction of every survey station points to the true north. The azimuth corrected by the magnetic declination is based on the direction of the true north. The true north at survey stations vary with their different positions. Accordingly, the Gauss planar rectangular coordinate system, known as the grid coordinate system, is widely used in the process of directional well trajectory design, wellpath calculation and plotting. This coordinate system points to the Grid North on its ordinate and the East on its abscissa. In this system, the north at survey stations are identical and parallel to one another (Fig. 5.12).

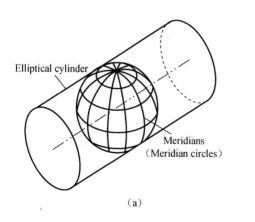

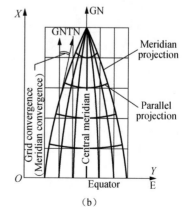

(a)   (b)

Fig. 5.12 Gaussian projection of the Earth

As can be seen, except survey stations on the central meridian and the equator, the azimuth lines of true north at other survey stations are distinct from those of grid north at the same position. This produces a series of bias angles known as the meridian convergence or grid convergence. The meridian convergence can be positive or negative in value. If the azimuth line of the grid north lies to the east of that of the true north, the angle is called the east meridian convergence with a positive value; otherwise, it is called the west meridian convergence with a negative value.

According to the formula derived by the principle of the Gauss projection, the meridian convergence at a certain survey station can be calculated using its longitude and latitude on the spherical surface of the Earth.

Therefore, in the practical engineering calculation, the measured azimuth ϕ_m from the magnetic inclinometer must be corrected both by the magnetic declination δ and by the meridian convergence β, since the azimuth based on the true north cannot be directly used in the Gauss planar rectangular coordinate system. Its expression is shown as follows:

$$\phi = \phi_m + \delta - \beta \qquad (5.15)$$

where

δ Magnetic declination, (°);
β Meridian convergence, (°).

The measured azimuth from a non-magnetic inclinometer, such as gyro-inclinometer, merely requires correction by the meridian convergence ($\delta = 0°$) (Fig. 5.13).

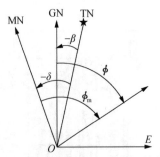

(a) The measured azimuth corrected by the west magnetic declination and the west meridian convergence

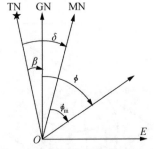

(b) The measure azimuth corrected by the east magnetic declination and the west meridian convergence

Fig. 5.13 Azimuth correction

5.2.3 Wellpath Calculation Methods

5.2.3.1 Process of Wellpath Calculation

Basically, five calculated parameters are required for a certain survey interval, namely TVD increment (ΔD), the increment of the horizontal projection length (ΔP), borehole curvature (K), and both increments of N-coordinate (ΔN) and E-coordinate (ΔE), respectively. Correspondingly, a certain survey station requires seven calculated parameters, namely TVD (D), the horizontal projection length (P), N-coordinate (N) and E-coordinate (E), vertical section (V) in the rectangular coordinate system, as well as closure distance (C) and closure azimuth (θ) in polar coordinates system, respectively.

Coordinates for all the survey stations are essential for the wellpath calculation. Coordinate increments for each survey interval should be calculated first and then added up together to obtain the corresponding coordinates (except C and θ) at each survey station. The specific steps are described as follows.

Starting from the zeroth survey station at the first survey interval with the given wellpath parameters, namely $D_0 = L_0$, $P_0 = 0$, $N_0 = 0$, $E_0 = 0$, the wellpath parameters for the first survey station can be easily obtained by adding the corresponding calculated values for the increments within the first survey interval. The wellpath calculations at the remaining survey stations and survey intervals can be achieved following the same rule that coordinates of the lower survey station at the ith survey interval can be considered as those of the upper survey station at the $(i + 1)$th survey interval. Therefore, the above discussed rule can be expressed mathematically as follows:

$$D_i = D_{i-1} + \Delta D_i \tag{5.16}$$

$$P_i = P_{i-1} + \Delta P_i \tag{5.17}$$

$$N_i = N_{i-1} + \Delta N_i \tag{5.18}$$

$$E_i = E_{i-1} + \Delta E_i \tag{5.19}$$

The closure distance, which is the closure azimuth at the ith survey station, can be calculated by using the above four parameters in the rectangular coordinate system, as expressed below:

$$C_i = \sqrt{N_i^2 + E_i^2} \tag{5.20}$$

$$\theta_i = \begin{cases} \arctan\dfrac{E_i}{N_i}\,(N_i > 0) \\[3mm] \arctan\dfrac{E_i}{N_i} + 180°\,(N_i < 0) \\[3mm] 90°\,(N_i = 0 \text{ and } E_i > 0) \\[1mm] 270°\,(N_i = 0 \text{ and } E_i < 0) \end{cases} \tag{5.21}$$

The vertical section can be calculated by the closure distance, the closure azimuth, and target bearing.

$$V_i = C_i \cos(\theta_0 - \theta_i) \tag{5.22}$$

Or

$$V_i = N_i \cos\theta_0 + E_i \sin\theta_0 \tag{5.23}$$

5.2.3.2   Diversity of Wellpath Calculation Methods

The calculation method of borehole curvature has been introduced in Sect. 5.1, which can be calculated by Eq. 5.4. It is challenging to calculate all coordinate increments within a certain survey interval. Although 20 or more wellpath calculation methods have been proposed, none of them can be regarded as an absolutely accurate technique. As a matter of fact, operators only measure the inclination and azimuth at both ends of a certain survey interval, and they are ignorant of the non-measured points along the same survey interval. In other words, operators do not know the true shape of the survey interval they have measured. The relevant wellpath parameters can be computed based on the assumption of the survey interval shape which definitely has some distinctions from the true one, and different assumptions will lead to different formulae which cannot be absolutely accurate.

Engineers and experts have made a number of comparative studies for a more accurate method to calculate these relevant wellpath parameters. Among the over 20 methods, some are different nominally but identical in essence, some are just a combination of other methods, and others are rejected on account of distinctive errors. Actually, only six or seven methods are practical and economical, and different companies may have various selections. By comparison, China used the angle averaging method and the corrected angle averaging method before 2003, while the cylindrical spiral method (corrected curvature radius method) and the minimum curvature method have been widely used since 2003. These four methods are introduced as follows.

- **Angle averaging method (average angle method)**

In the angle averaging method, the survey interval is considered as a straight line, and the average inclination and the average azimuth within a certain survey interval refer to the arithmetic mean of inclinations and azimuths at the upper and lower survey stations, respectively. Therefore, the four coordinate increments $(\Delta D, \Delta P, \Delta N, \Delta E)$ can be calculated by the following formulae:

$$\Delta D = \Delta L \cos \alpha_c \tag{5.24}$$

$$\Delta P = \Delta L \sin \alpha_c \tag{5.25}$$

$$\Delta N = \Delta L \sin \alpha_c \cos \phi_c \tag{5.26}$$

$$\Delta E = \Delta L \sin \alpha_c \sin \phi_c \tag{5.27}$$

where

α_c Arithmetic average of inclination, $\alpha_c = (\alpha_1 + \alpha_2)/2$, (°);
ϕ_c Arithmetic average of azimuths, $\phi_c = (\phi_1 + \phi_2)/2$, (°).

- **Cylindrical spiral method (corrected curvature radius method)**

In 1975, Prof. Zheng Jiying, a renowned scholar in China, put forward the cylindrical spiral method, which is based on the curvature radius method presented by an American scholar G. J. Wilson in 1968. In this method, the survey interval is assumed as a cylindrical spiral with the lead angle proportional to the course length, namely constant $d\alpha/dL$. The tangential directions at both ends of the cylindrical spiral are identical to the borehole directions at the two ends of the survey interval, respectively. It can be deduced that the borehole axis of the survey interval on the horizontal projection is a circular arc, as well as another circular arc on the vertical profile. Accordingly, the formulae to calculate the four coordinate increments are expressed below.

$$\Delta D = \frac{2\Delta L \sin \frac{\Delta \alpha}{2} \cos \alpha_c}{\Delta \alpha} \tag{5.28}$$

$$\Delta P = \frac{2\Delta L \sin \frac{\Delta \alpha}{2} \sin \alpha_c}{\Delta \alpha} \tag{5.29}$$

$$\Delta N = \frac{4\Delta L \sin \frac{\Delta \alpha}{2} \sin \frac{\Delta \phi}{2} \sin \alpha_c \cos \phi_c}{\Delta \alpha \cdot \Delta \phi} \tag{5.30}$$

$$\Delta E = \frac{4\Delta L \sin \frac{\Delta \alpha}{2} \sin \frac{\Delta \phi}{2} \sin \alpha_c \sin \phi_c}{\Delta \alpha \cdot \Delta \phi} \tag{5.31}$$

Note that the inclination increment $\Delta \alpha$ and the azimuth increment $\Delta \phi$ in the denominator must be converted into the unit of radian. If the denominator in the formula is zero, the following treatments would be conducted.

Scenario 1: $\Delta\alpha = 0, \Delta\phi \neq 0$.

$$\Delta D = \Delta L \cos \alpha_2 \tag{5.32}$$

$$\Delta P = \Delta L \sin \alpha_2 \tag{5.33}$$

$$\Delta N = \frac{2\Delta L \sin \alpha_2 \sin \frac{\Delta\phi}{2} \cos \phi_c}{\Delta\phi} \tag{5.34}$$

$$\Delta E = \frac{2\Delta L \sin \alpha_2 \sin \frac{\Delta\phi}{2} \sin \phi_c}{\Delta\phi} \tag{5.35}$$

Scenario 2: $\Delta\alpha \neq 0, \Delta\phi = 0$

$$\Delta D = \frac{2\Delta L \sin \frac{\Delta\alpha}{2} \cos \alpha_c}{\Delta\alpha} \tag{5.36}$$

$$\Delta P = \frac{2\Delta L \sin \frac{\Delta\alpha}{2} \cos \alpha_c}{\Delta\alpha} \tag{5.37}$$

$$\Delta N = \frac{2\Delta L \sin \frac{\Delta\alpha}{2} \sin \alpha_c \cos \phi_2}{\Delta\alpha} \tag{5.38}$$

$$\Delta E = \frac{2\Delta L \sin \frac{\Delta\alpha}{2} \sin \alpha_c \sin \phi_2}{\Delta\alpha} \tag{5.39}$$

Scenario 3: $\Delta\alpha = 0, \Delta\phi = 0$

$$\Delta D = \Delta L \cos \alpha_2 \tag{5.40}$$

$$\Delta P = \Delta L \sin \alpha_2 \tag{5.41}$$

$$\Delta N = \Delta L \sin \alpha_2 \cos \phi_2 \tag{5.42}$$

$$\Delta E = \Delta L \sin \alpha_2 \sin \phi_2 \tag{5.43}$$

- **Corrected angle averaging method**

This method is an approximate treatment based on the cylindrical spiral method. The formulae are rewritten in a new mathematical way as follows.

$$\Delta D = \left(1 - \frac{\Delta\alpha^2}{24}\right)\Delta L \cos\alpha_c \qquad (5.44)$$

$$\Delta P = \left(1 - \frac{\Delta\alpha^2}{24}\right)\Delta L \sin\alpha_c \qquad (5.45)$$

$$\Delta N = \left(1 - \frac{\Delta\alpha^2 + \Delta\phi^2}{24}\right)\Delta L \sin\alpha_c \cos\phi_c \qquad (5.46)$$

$$\Delta E = \left(1 - \frac{\Delta\alpha^2 + \Delta\phi^2}{24}\right)\Delta L \sin\alpha_c \sin\phi_c \qquad (5.47)$$

The simpler formulae are effective even though either $\Delta\alpha$ or $\Delta\phi$, or both are equal to zero and have produced smaller errors through practical verifications. Note that both $\Delta\alpha$ and $\Delta\phi$ are expressed in *rad* (the unit of radian).

- **Minimum curvature method**

This method assumes that the survey interval is a circular arc on a certain plane in the 3D space, which is identical to the assumption of dogleg severity reflecting the minimum curvature of the survey interval. Thus, the minimum curvature method can be expressed as follows.

$$\Delta D = \lambda_M (\cos\alpha_1 + \cos\alpha_2) \qquad (5.48)$$

$$\Delta N = \lambda_M (\sin\alpha_1 \cos\phi_1 + \sin\alpha_2 \cos\phi_2) \qquad (5.49)$$

$$\Delta E = \lambda_M (\sin\alpha_1 \sin\phi_1 + \sin\alpha_2 \sin\phi_2) \qquad (5.50)$$

$$\lambda_M = \frac{180}{\pi} \frac{\Delta L}{\gamma} \tan\frac{\gamma}{2} \qquad (5.51)$$

$$\gamma = \arccos(\cos\alpha_1 \cos\alpha_2 + \sin\alpha_1 \sin\alpha_2 \cos\Delta\phi) \qquad (5.52)$$

This method does not include the calculation for the horizontal projection length due to the fact that a circular arc on an oblique plane in the 3D space basically produces an elliptical horizontal projection, the length of which is too complicated to be calculated by a simple method.

For the other methods, please refer to relevant references.

Example 5.2.1 Determine the wellpath by the corrected angle averaging method with the given surveying data for a certain horizontal well. Please refer to Table 5.1 listing the raw data and results from 134 groups, some of which are omitted.

Table 5.1 Surveying data and calculated results for a certain horizontal well

No.	L (m)	α (°)	ϕ (°)	N (m)	E (m)	D (m)	K (°/30 m)
1	0	0	42.8	0	0	0	0
2	6.7	0.1	42.8	0	0	6.7	0.45
3	33.4	0.3	76.6	0.06	0.08	33.4	0.26
4	60.1	0.4	72.8	0.12	0.23	60.1	0.12
5	86.8	0.7	102	0.15	0.48	86.8	0.46
6	113.5	0.3	122.8	0.08	0.7	113.5	0.49
Data from No. 7 to No. 62 are omitted							
63	1642.82	0.7	328.1	3.88	10.57	1642.7	0.01
64	1670.96	1.3	151.8	3.7	10.32	1670.83	3.34
65	1698.63	2.5	163.1	2.9	10.76	1698.49	1.36
66	1726.69	3.4	168.1	1.54	11.25	1726.51	1
67	1735	3.5	168	1.06	11.4	1734.8	0.36
68	1735.3	3.54	167.28	1.04	11.4	1735.1	5.78
69	1745.56	4.8	138	0.42	11.8	1745.33	7.24
70	1754.18	6.7	114	−0.01	12.54	1753.91	10.67
71	1773.05	11.1	92	0.39	15.42	1772.55	8.84
72	1782.65	14.5	83	−0.09	17.52	1781.91	12.32
73	1793.37	15.1	78	0.62	20.16	1792.27	3.95
Data from No. 74 to No. 96 are omitted							
97	1974.9	41.85	72.07	35.14	109.58	1944.4	8.03
98	1983	44.62	72.51	37.33	114.68	1950.3	10.32
99	1991	47.69	72.25	39.59	119.99	1955.84	11.53
100	2001.48	52.18	72.42	42.75	127.36	1962.58	12.86
101	2006	54.2	72.42	44.17	130.69	1965.29	13.41
102	2011.26	56.62	72.25	45.87	134.67	1968.28	13.83
103	2021	60.9	69.5	49.34	142.24	1973.33	15.04
104	2030	65	66.6	53.03	149.35	1977.42	16.15
Data from No. 105 to No. 120 are omitted							
121	2144	86.77	68.75	103.4	246.75	2005.53	6.93
122	2150.29	88.18	68.38	106.24	252.35	2005.8	6.94
123	2169	87.94	66.53	115.04	268.85	2006.44	2.99
124	2188.19	88.66	69.11	123.95	285.84	2007.01	4.19
125	2206.67	88.2	68.64	132.22	302.35	2007.51	1.07
126	2217	88.18	68.91	136.87	311.57	2007.84	0.79
127	2226.28	87.91	69.17	141	319.87	2008.16	1.21
128	2235.89	87.78	69.17	145.26	328.48	2008.52	0.41
129	2245	87.34	71.19	149.15	336.71	2008.91	6.8
130	2254.57	87.34	70.22	153.16	345.39	2009.35	3.04
131	2294.2	88.66	68.3	170.67	380.91	2010.73	1.76
132	2303.42	89.67	67.15	174.97	389.06	2010.87	4.98

(continued)

Table 5.1 (continued)

No.	L (m)	α (°)	φ (°)	N (m)	E (m)	D (m)	K (°/30 m)
133	2312.8	90.68	66.9	179.44	397.31	2010.84	3.33
134	2344	91	66.5	194.47	424.65	2010.38	0.49

5.3 Deviation Control for Vertical Wells

In terms of the well trajectory, there are two major types of wells, specifically the vertical well and the directional well. Although a vertical well with a plumb well trajectory does not require special designs in general, the drilling practice has verified that the true wellpath often deviates from the predetermined plumb course, which is known as *Well Deflection*. It is a major challenge to control the vertical wellpath, referred to as *Deviation Control for Vertical Wells* in the drilling terminology, which is even more difficult than that for directional wells.

Generally, no wells can be drilled strictly following the pre-designed trajectory. The operators aim to control the inclination, the closure distance at the downhole or the borehole curvature within the tolerance limit.

5.3.1 Analyzing the Reasons for Well Deflection

Determining the reasons for well deflection will contribute to specific measures for the deviation control. Specifically, the reasons include geological factors, drilling tools as well as borehole enlargement.

5.3.1.1 Geological Factors

Among the reasons for the well deflection, the main ones are inhomogeneous formation drillability and inclined structure of the formation. Combined with the formation dip, the heterogeneity of the formation drillability leads to the well deflection, which is manifested in many aspects as explained below.

- **The heterogeneity of the formation drillability**

The heterogeneity of the formation drillability refers to the inhomogeneity of the formation drillability in different directions. As illustrated in Fig. 5.14, sedimentary rocks have such properties. The drillability in the direction perpendicular to the formation plane is the highest, while the drillability parallel to the formation plane is the lowest. The bit always has a tendency to drill toward the direction with minimum resistance. Generally, if the dip angle is less than 45°, the bit tends to drill toward the direction perpendicular to the formation plane; the opposite tendency to drill downdip occurs if the dip angle more than 60°. Accordingly, the drilling direction is unstable if the dip angle is between 45° and 60°.

Fig. 5.14 Well deflection
caused by the heterogeneity of
the formation drillability

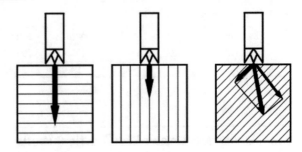

- **Longitudinal variation of the formation drillability**

In the process of sedimentation, changes in the sedimentary environment lead to the
soft-hard interbedding formation with a longitudinal variation of the formation
drillability perpendicular to the formation plane. This refers to the fact that the drill
bit encounters the soft-hard interbedding formation in the direction of its axis. In the
inclined formation, the drilling is faster at one side of the bit where the soft stratum
is encountered than at the other side where the hard stratum is met. Therefore, the
borehole axis deviates from the predetermined direction and well deflection occurs
(Fig. 5.15).

- **Transverse variation of the formation drillability**

As a matter of fact, changes in the formation drillability occur in the directions both
perpendicular and parallel to the formation plane. The transverse variation of the
formation drillability mentioned here indicates the changes in the direction perpendicular to the axis of the drill bit. As illustrated in Fig. 5.16, the drill bit runs
into a vug or a softer stratum at one side, while it encounters a hard or tight layer at
the other side. Therefore, the bit trend will deviate from the pre-designed direction.

Accordingly, both the heterogeneity of the formation drillability and the formation dip result in asymmetrical cutting of the drill bit at the downhole, which
deviates the axis of the drill bit from that of the original wellbore.

Fig. 5.15 Well deflection
caused by longitudinal
variation of the formation
drillability

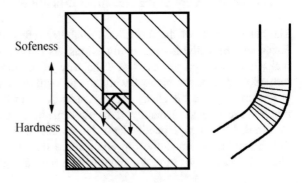

Fig. 5.16 Well deflection
caused by transverse variation
of the formation drillability

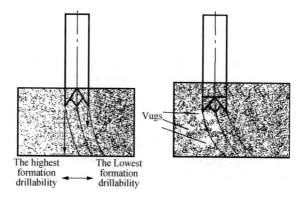

The highest formation drillability ←→ The Lowest formation drillability

Vugs

5.3.1.2   Drilling Tools

Inclining and bending drill tools also contribute to the well deflection. The major influence stems from a near-bit portion of the drill string, known as the bottom hole assembly (BHA). The inclining and bending drill tools may produce two negative consequences. One effect is depicted in Fig. 5.17a, where the tilting drill bit results in asymmetrical cutting at the downhole, with the wellpath trend deviating from the original wellbore. The other is illustrated in Fig. 5.17b, where the drill bit is affected by the side force and compelled to perform side cutting, straying from the initial course of the wellbore.

The specific reasons for inclining and bending drill tools can be listed as follows.

First of all, compared with the borehole diameter, the drill tool with a smaller diameter produces a larger space (gap between the wellbore and the drill tool) for a more inclining and bending drill tool.

Fig. 5.17 Asymmetrical
cutting and side cutting of the
drill bit

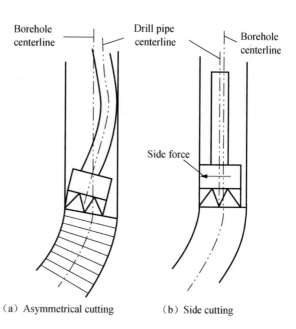

Borehole centerline   Drill pipe centerline   Borehole centerline

Side force

(a) Asymmetrical cutting   (b) Side cutting

Second, affected by the weight on bit (WOB), the BHA leans toward the borehole side. If the axial WOB exceeds the limits, the drill string will bend, causing the near-bit drill tools to tilt more.

Other reasons may be that the drill tools are originally bent before drilling process; three center points of the crane, the traveling block, and the rotary table are not on the same plumb line; and sloping rotary table guides the drill tools to deviate from the vertical direction at the beginning.

5.3.1.3 Borehole Enlargement

Besides geological factors and inclining and bending drill tools, borehole enlargement is a third major cause of the well deflection. In a larger-diameter borehole, the drill bit tends to move to the left or the right or lean toward to just one side. Therefore, the axis of the drill bit cannot overlap the borehole centerline, resulting in the well deflection.

Among the three causes mentioned above, the existing geological factors cannot be changed, while the well deflection caused by the inclining and bending drill tools can be prevented or even eliminated, on which a number of relevant studies have been conducted. Some deviation-prevented and deviation-corrected assemblies have also been put forward. Among them, two widely used ones are the packed-hole BHA and the pendulum BHA. Meanwhile, borehole enlargement takes time, and operators can take advantage of this process to prevent the well deflection.

5.3.2 Deviation Control by Packed-Hole BHA

As mentioned above, well deflection results from the asymmetrical cutting at the downhole, deviation of the bit axis from the borehole axis and side cutting caused by the side force on the bit. Based on these three aspects, the packed-hole assembly is designed.

Given that the diameter of the drill string is identical to that of the drill bit, the deviation control problems mentioned above can be overcome. However, it is not feasible in the drilling engineering because the drilling fluid cannot be circulated in this pattern, causing a series of associated problems. Actually, multi-stabilizers are the best solutions for this key issue.

Several stabilizers are installed at proper locations of the packed-hole BHA within 20 m (65.62 ft) from the drill bit. Proper installment indicates the number, position, and diameter of the stabilizers. Various packed-hole assemblies have been designed with distinctive design ideas which produce similar results. The below section introduces a *YXY* packer-hole BHA put forward by Yang Xunxiao, a renowned Chinese drilling expert.

5.3.2.1 Structure of *YXY* Packed-Hole BHA

As illustrated in Fig. 5.18, this BHA generally comprises four stabilizers, specifically from bottom to top, the near-bit stabilizer, the middle stabilizer, the upper stabilizer, and the fourth stabilizer, respectively.

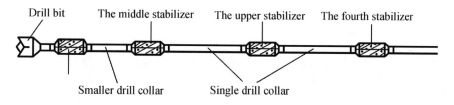

Fig. 5.18 Structure diagram of the YXY packer-hole BHA

- **Near-Bit Stabilizer**

It has a larger diameter that is merely 1–2 mm (0.04–0.08 in) less than the bit diameter. In some cases where the drill bit tends to incline, operators can lengthen the near-bit stabilizer. In other cases where the drill bit tends to severely incline, it is necessary to join two stabilizers in series as a new near-bit stabilizer. Against the side force on the drill bit, the near-bit stabilizer aims to maintain the borehole which has yet to expand, effectively reducing the negative effect resulting from the side cutting. Due to its large diameter, long length, and strong stiffness, the near-bit stabilizer can effectively prevent the tilting bit, eliminating the negative influence caused by the asymmetrical cutting.

- **Middle Stabilizer (Second Stabilizer)**

Rigorous calculations are required to determine the location of the middle stabilizer with identical diameter to the near-bit stabilizer. The main function of the middle stabilizer is to maintain a straight drill string sitting between it and the drill bit, preventing the bit from tilting and minimizing the asymmetrical cutting at the downhole.

- **Upper Stabilizer (Third Stabilizer)**

It is installed at one drill collar above the middle stabilizer. It has an identical diameter to that of the near-bit stabilizer or the middle stabilizer, but it is a slightly less rigorous requirement.

- **Fourth Stabilizer**

The fourth stabilizer with identical diameter to the upper stabilizer is necessary only for the formation where the bit tends to severely tilt. It is installed at one drill collar above the one where the upper stabilizer is located.

Both the upper and the fourth stabilizers increase the stiffness of the bottom drill string and assist the middle stabilizer in preventing the bottom drill string from inclining.

5.3.2.2 Location Calculations for Middle Stabilizer on YXY Packed-Hole BHA

It is essential to determine the location of the middle stabilizer on the packed-hole BHA. Figure 5.19 illustrates a mechanical model for the packed-hole BHA presented by Yang Xunyao. As depicted in the figure, L_p represents the optimum length between the middle stabilizer and the drill bit. The angle between borehole centerline and the bit axis is called the bit tilt angle, symbolized by θ $(\theta = \theta_c + \theta_q)$. Increase in length between the middle stabilizer and the drill bit will contribute to a decreasing angle θ_c and an increasing angle θ_q, and vice versa. Accordingly, there must be an optimum length L_p with a corresponding minimum bit tilt angle θ. On the basis of the mechanical model, the mathematical model for the optimum length L_p can be simplified as follows:

$$L_p = 4\sqrt{\frac{16CEJ}{q_m \sin \alpha}} \qquad (5.53)$$

where

L_p Optimum length between the middle stabilizer and the drill bit, m;
C Half gap between the borehole and the middle stabilizer, m;
d_h Borehole diameter, m;
d_s Outside diameter of the stabilizer, m;
E Elastic modulus of the drill collar, kPa;
J Axial moment of inertia on the cross section of the drill collar, m^4;
q_m Line weight of the drill collar in drilling fluids, kN/m;
α Tolerant maximum inclination, (°).

Example 5.3.1 Determine the optimum length L_p between the middle stabilizer and the drill bit with the given data including the bit diameter $d_h = 216$ mm, the stabilizer diameter $d_s = 215$ mm, the elastic modulus of the drill collar $E = 205.94$

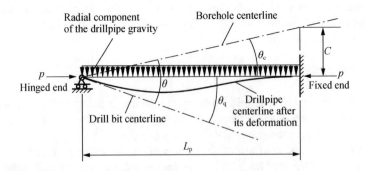

Fig. 5.19 Schematic of the mechanical model of the *YXY* packer-hole assembly

GPa, outside diameter of the drill collar d_{co} = 178 mm, inside diameter of the drill collar d_{ci} = 71.40 mm, density of the drilling fluid ρ_d = 1.25 g/cm^3, line weight of the drill collar q_c = 1.606 kN/m, tolerant maximum inclination angle α = 3°, and density of the drill collar ρ_s = 7.85 g/cm^3.

Solution

$$J = \frac{\pi}{64}\left(d_{co}^4 - d_{ci}^4\right) = \frac{3.14}{64}\left(0.178^4 - 0.0714^4\right) = 0.48 \times 10^{-4}\left(m^4\right)$$

$$q_m = q_c\left(1 - \frac{\rho_d}{\rho_s}\right) = 1.606 \times \left(1 - \frac{1.25}{7.85}\right) = 1.350 \; (kN/m)$$

$$C = \frac{d_h - d_s}{2} = \frac{0.216 - 0.215}{2} = 0.0005 \; (m)$$

The L_p can be calculated by Eq. 5.53 as follows.

$$L_p = \sqrt[4]{\frac{16CEJ}{q_m \sin \alpha}} = \sqrt[4]{\frac{16 \times 0.0005 \times 205.94 \times 10^6 \times 0.48 \times 10^{-4}}{1.350 \times \sin 3°}} = 5.784 \; (m)$$

Example 5.3.2 Determine the optimum length L_p between the middle stabilizer and the drill bit with the given data including the bit diameter d_h = 311 mm, the stabilizer diameter d_s = 309.5 mm, the elastic modulus of the drill collar E = 205.94 GPa, outside diameter of the drill collar d_{co} = 203.2 mm, inside diameter of the drill collar d_{ci} = 71.40 mm, density of the drilling fluid ρ_d = 1.25 g/cm^3, line weight of the drill collar q_c = 2.190 kN/m, tolerant maximum inclination angle α = 3°, and density of the drill collar ρ_s = 7.85 g/cm^3.

Solution

$$J = \frac{\pi}{64}\left(d_{co}^4 - d_{ci}^4\right) = \frac{3.14}{64}\left(0.2032^4 - 0.0714^4\right) = 0.8237 \times 10^{-4}\left(m^4\right)$$

$$q_m = q_c\left(1 - \frac{\rho_d}{\rho_s}\right) = 2.190 \times \left(1 - \frac{1.25}{7.85}\right) = 1.8413 \; (kN/m)$$

$$C = \frac{d_h - d_s}{2} = \frac{0.311 - 0.3095}{2} = 0.00075 \; (m)$$

The L_p can be calculated by Eq. 5.53 as follows.

$$L_{\mathrm{p}} = \sqrt[4]{\frac{16CEJ}{q_{\mathrm{m}} \sin \alpha}} = \sqrt[4]{\frac{16 \times 0.00075 \times 205.94 \times 10^6 \times 0.8237 \times 10^{-4}}{1.8413 \times \sin 3^\circ}}$$
$$= 6.779 \ (\mathrm{m})$$

5.3.2.3   Notes for Packed-Hole Assembly

Operators must pay more attention to the following problems.

– The packed-hole BHA aims to reduce or eliminate the changes (increase or decrease) in inclination rather than to decrease the inclination of the well. Accordingly, instead of controlling the inclination, the packed-hole BHA mainly intends to control the borehole curvature.
– "Packed-hole" indicates that the assembly is significantly affected by the gap between the borehole and the stabilizer, which must be as small as possible. The tolerant gap ($\Delta d = d_{\mathrm{h}} - d_{\mathrm{s}}$) is generally between 0.8 and 1.6 mm. Due to the wear of stabilizer, the gap increases during the drilling process. According to the relevant regulations, the stabilizer needs to be replaced or renovated if the gap reaches or exceeds double tolerances.
– Another key point to achieve the goal of "*Packed hole*" is to prevent borehole enlargement. Borehole protection technique with high performance drilling fluids is required. Even if the performance of drilling fluid does not meet the expectations, the borehole can still stay "*Packed*" if the new borehole is drilled before the borehole enlargement which takes time. It means that the operators need to accelerate drilling. The operating rules are summarized by field operators in China as: Faster drilling keeps packed borehole and packed hole maintains a straight borehole.
– In the soft-hard interbedding formation and the severely inclined formation, less WOB and more frequent reaming are both required to eliminate the dogleg. Here, the term dogleg particularly refers to a wellbore interval with larger borehole curvature on the actual wellpath of the vertical well.

5.3.3   Deviation Control by Pendulum BHA

5.3.3.1   Principle of Pendulum BHA

As illustrated in Fig. 5.20, this BHA has only one stabilizer. If the stabilizer is properly positioned, the drill string below the stabilizer cannot achieve tangency with the borehole and looks like a pendulum. The pendulum force is generated to cut the lower side of the borehole, and thus, the new borehole is less deviated. It is observed that a larger pendulum force G_{c} ($G_{\mathrm{c}} = G \cdot \sin \alpha$) is generated as the inclination α increases.

Fig. 5.20 Schematic of pendulum BHA

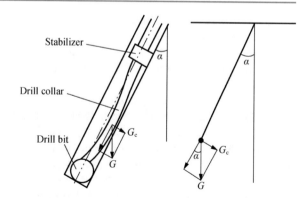

Stabilizer

Drill collar

Drill bit

5.3.3.2 Design for Pendulum BHA

A. Lubinski, an American scholar, first put forward the pendulum BHA through a graphic design method which was too complicated for application. Since then, starting from the basic mechanical models, many scholars have proposed the calculation methods for pendulum BHA through rigorous mathematical deductions. A simpler and more practical method presented by Yang Xunyao is introduced below.

The length L_z between the stabilizer and the drill bit is crucial for pendulum BHA. A smaller L_z contributes to less pendulum force, while a larger L_z causes new tangency point between the stabilizer and the drill bit. Accordingly, the optimum L_z can be computed by the following formula proposed by Yang.

$$L_z = \sqrt{\frac{\sqrt{B^2 + 4AC} - B}{2A}} \tag{5.54}$$

In this formula,

$$A = \pi^2 q_m \sin \alpha$$
$$B = 82.04 Wr$$
$$C = 184.6\pi^2 EJr$$
$$r = (d_h - d_{co})/2$$

where

W	WOB, kN;
d_h	Borehole diameter, m;
d_{co}	Outside diameter of the drill collar, m;
E, J, q_m	are same as in Eq. 5.53.

Considering the stabilizer wear and the borehole enlargement, the practical value of L_z can be 5–10% less than the calculated one.

5.3.3.3 Notes for Pendulum BHA

The pendulum force varies with the inclination. Specifically, a larger inclination leads to a larger pendulum force, and a zero inclination indicates a zero pendulum force. Therefore, the pendulum BHA mainly aims to correct the deviation for severe well deflection.

The performance of the pendulum BHA is extremely sensitive to the WOB. Specifically, a larger WOB results in a larger bending force and a smaller pendulum force. As the WOB continues to increase, the drill string below the stabilizer tends to bend, and even a new tangency point emerges. Therefore, the pendulum BHA fails in the deviation correction. Importantly, operators must strictly control the WOB when using the pendulum BHA for the deviation correction.

Less WOB is employed to maintain a vertical borehole as designed when the inclination is small enough or even zero. Therefore, the packed-hole BHA is mostly used due to a low penetration rate of the pendulum BHA. The pendulum BHA is employed only in the vertical well (or the vertical section) where the wellpath is required to strictly follow the predetermined course.

The gap between the stabilizer and the borehole also exerts an evident effect on the performance of the pendulum BHA. Therefore, the stabilizer should be replaced or renovated if its diameter decreases due to abrasion.

The pendulum BHA with multiple stabilizers requires more sophisticated designs and more complicated calculations.

5.4 Well Trajectory Design for Directional Wells

Well trajectory design is one of the major components of the directional well design.

5.4.1 Types of Directional Well Trajectory

The types of directional well trajectory are determined by the designed trajectory before drilling rather than by the actual wellpath because it is more systematic to classify a regular predetermined well trajectory than to define an arbitrary curve of the wellpath in the 3D space. Specifically, a vertical well refers to the wellbore with a plumb trajectory, even though it may deviate while drilling.

Based on the pre-designed well trajectory, directional wells are divided into 2D directional wells and 3D directional wells. For 2D directional wells, the well trajectory lies in a plumb plane, with changeable inclination and an invariant azimuth. By contrast, 3D directional wells incorporate both variant inclinations and azimuths.

2D directional wells are subdivided into conventional and unconventional 2D directional wells. The conventional ones consist of a series of straight lines and circular arcs, while the unconventional ones require some particular curves, such as catenary curves and quadratic parabolas.

In the drilling process, conventional 2D directional wells are widely used. Therefore, the Drilling Industry Standardization Committee of China has developed standards for well trajectory design of conventional 2D directional wells.

5.4.2 Well Trajectory Design for Conventional 2D Directional Well

5.4.2.1 Design Principles

Several regulations must be followed for the directional well trajectory design.

- **Meet the goal of drilling the directional well**

Directional drilling aims to achieve diverse goals including hitting multiple oil layers to improve the recovery, elongating the target interval for more reservoir contact, reviving old wells and dead wells, sidetracking due to the downhole drilling troubles, repositioning due to surface constraints, or drilling new wells avoiding subsurface obstacles, cluster wells for the economization of land, relief wells for firefighting at adjacent wells, etc. Operators must consider practical drilling purposes before drilling directional wells.

- **Facilitate safe, high-quality and fast drilling**

The kick-off point (KOP) is essential for directional drilling. Directional wells must "kick off" at the formation with moderate hardness and no complex subsurface conditions, such as collapse, shrinkage, high pressure, and leakage. Based on the maximum inclination of a well trajectory, directional wells can be divided into less-deviated directional wells with the maximum inclination between 15° and 30°, moderately deviated directional wells with the maximum inclination between 30° and 60°, and highly deviated directional wells with the maximum inclination over 60°. Wherever possible, the maximum inclination should not be too large to decrease the difficulty of wellpath control. However, the maximum inclination must be larger than 15° so that it would be easier to maintain the azimuth. To select the borehole curvature, operators need to select better deflection tools, making it easier for tripping and running casing and also for shortening the length of the curved interval(s).

- **Meet the requirements of oil production**

The borehole curvature needs to be reduced if possible to maintain a relatively straight wellbore where the tubing and the sucker rod work. In some practical scenarios, the inclination of the target section should be as small as possible (0° is the best), which is beneficial for the electric submersible pump and the packer installations and other downhole operations.

5.4.2.2 Types of Well Trajectory

According to the Drilling Industry Standard of China, the well trajectory types of conventional 2D directional well can be subdivided into (a) Build–hold well trajectory, (b) Multi-target build–hold well trajectory, (c) Build–hold–drop well trajectory, and (d) Build–hold–build well trajectory. As illustrated in Fig. 5.21, point k represents the KOP; point b indicates the end point of the first build section; point t refers to the target point; point c represents the starting point of the drop section in build–hold–drop well trajectory or the second KOP in build–hold–build well

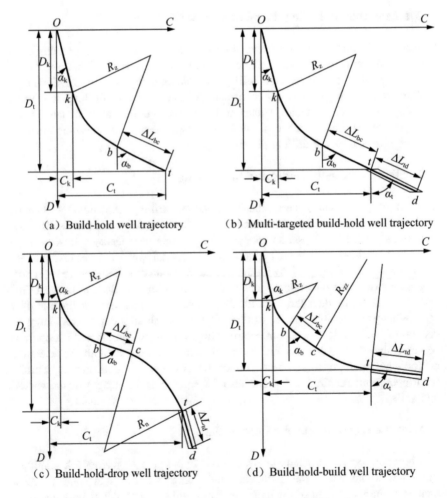

(a) Build-hold well trajectory

(b) Multi-targeted build-hold well trajectory

(c) Build-hold-drop well trajectory

(d) Build-hold-build well trajectory

Fig. 5.21 Four types of well trajectory of conventional 2D directional wells

trajectory; and point d denotes the end point in multi-target build–hold well trajectory. All these mentioned points, known as nodes, are symbolized as subscripts of associated parameters.

Considering the possibility of using the rig with inclined derrick, operators will drill an inclined straight interval starting from the wellhead and then build at the KOP. Therefore, as illustrated in Fig. 5.21, a closure distance C_k and an inclination α_k occur at the KOP. Definitely, both C_k and α_k will be zero while using the rig with vertical derrick.

Design conditions and calculation formulae vary for different types of well trajectory.

5.4.2.3 Given Conditions for Well Trajectory Design

Basically, there are two kinds of given conditions for well trajectory design. The first one includes the prediction for the stratified geological conditions and relevant data from target points or target sections, which are provided by geology and production department. The other one considers the KOP position and build-up rate selected by the drilling engineering department on the basis of relevant design principles and drilling conditions.

Necessary conditions and parameters for well trajectory designs are listed in Table 5.2 where the interpretations for these parameters are also given.

D_t TVD of the target point, m;

C_t Closure distance of the target point, m;

D_k TVD of the KOP, m;

C_k Closure distance of the KOP, m;

α_k Inclination of the KOP, (°);

K_z Build-up rate of the "Build" interval, (°)/30 m;

K_n Build-up rate of the "Drop" interval, (°)/30 m;

K_{zz} Build-up rate of the second "Build" interval in build–hold–build well trajectory, (°)/30 m;

θ_0 Target bearing, (°);

Table 5.2 Necessary conditions and parameters for well trajectory designs

Types of well trajectory	Necessary conditions	Key parameters
Build–hold well trajectory	$D_t \ C_t \ D_k \ C_k \ \alpha_k \ K_z \ \theta_0$	$\alpha_b \ \Delta L_{bc}$
Multi-target build–hold well trajectory	$D_t \ D_k \ C_k \ \alpha_k \ K_z \ \theta_0 \ \alpha_t \ \Delta L_{td}$	$C_t \ \Delta L_{bc}$
Build–hold–drop well trajectory	$D_t \ C_t \ D_k \ C_k \ \alpha_k \ K_z \ \theta_0 \ \alpha_t \ \Delta L_{td} \ K_n$	$\alpha_b \ \Delta L_{bc}$
Build–hold–build well trajectory	$D_t \ C_t \ D_k \ C_k \ \alpha_k \ K_z \ \theta_0 \ \alpha_t \ \Delta L_{td} \ K_{zz}$	$\alpha_b \ \Delta L_{bc}$

α_t Inclination of the target point or the target interval, (°);

ΔL_{td} Length of the target interval, m;

α_{bc} Inclination of the "Hold" interval, (°);

ΔL_{bc} Length of the "Hold" interval, m.

Radiuses R_z, R_n, R_{zz} can be computed based on build-up rates K_z, K_n, K_{zz}, respectively. The conversion formula between the radius of a curved interval and its build-up rate is unified and expressed as follows:

$$R = \frac{5400}{K \cdot \pi} \approx \frac{1719}{K} \qquad (5.55)$$

where

R Radius of a build or drop interval, m;

K Build-up rate of a build or drop interval, (°)/30 m

C_k can be calculated with given D_k and α_k.

$$C_k = D_k \tan \alpha_k \qquad (5.56)$$

"Key parameters" in Table 5.2 must be calculated first before preforming calculations for other associated parameters.

5.4.2.4   Calculations for Key Parameters

The formulae for key parameters vary according to different types of well trajectory.

- **Build–hold well trajectory**
    - The given parameters are the same as those given in Table 5.2. The key parameters α_b and ΔL_{bc} are calculated by the following formulae.

$$D_e = D_t - D_k + R_z \sin \alpha_k \qquad (5.57)$$

$$C_e = C_t - C_k - R_z \cos \alpha_k \qquad (5.58)$$

$$R_e = R_z \qquad (5.59)$$

$$\Delta L_{bc} = \sqrt{D_e^2 + C_e^2 - R_e^2} \qquad (5.60)$$

$$a_b = 2 \arctan \frac{D_e - \Delta L_{bc}}{R_e - C_e} \qquad (5.61)$$

- The given parameters include D_t C_t D_k C_k α_k α_b. The key parameters ΔL_{bc} and K_z can be computed by the following formulae.

$$R_z = \frac{(D_z - D_k)\sin\alpha_b - (C_t - C_k)\cos\alpha_b}{1 - \cos(\alpha_b - \alpha_k)} \tag{5.62}$$

$$\Delta L_{bc} = \frac{D_t - D_k - R_z(\sin\alpha_b - \sin\alpha_k)}{\cos\alpha_b} \tag{5.63}$$

$$K_z = 1719/R_z$$

- The given parameters include D_t C_t K_z α_k α_b. The parameters D_k and ΔL_{bc} can also be computed by the following formulae

$$D_k = \frac{C_t\cos\alpha_b - D_t\sin\alpha_b + R_z[1 - \cos(\alpha_b - \alpha_k)]}{\tan\alpha_k\cos\alpha_b - \sin\alpha_b} \tag{5.64}$$

$$\Delta L_{bc} = \frac{(D_t - D_k) - R_z(\sin\alpha_b - \sin\alpha_k)}{\cos\alpha_b} \tag{5.65}$$

- **Multi-target build–hold well trajectory**

The greatest disparity between multi-target build–hold well trajectory and other trajectories is that the closure distance C_t of the target point is unknown. That is to say, the well location on the surface is unknown. The method, known as *backward design method*, indicates that C_t should be calculated before the well location on the surface is determined.

$$C_t = C_k + (D_t - D_k)\tan\alpha_t - R_z\frac{1 - \cos(\alpha_t - \alpha_k)}{\cos\alpha_t} \tag{5.66}$$

The formula for ΔL_{bc} is identical to Eq. 5.65.

- **Build–hold–drop well trajectory**

The following three formulae are necessary for calculating D_e, C_e, R_e, and then ΔL_{bc} and α_b are computed by Eqs. 5.60 and 5.61.

$$D_e = D_t - D_k + R_z\sin\alpha_k + R_n\sin\alpha_t \tag{5.67}$$

$$C_e = C_t - C_k - R_z\cos\alpha_k - R_n\cos\alpha_t \tag{5.68}$$

$$R_e = R_z + R_n \tag{5.69}$$

- **Build–hold–build well trajectory**

The following three formulae are necessary for calculating D_e, C_e, R_e, and then ΔL_{bc} and α_b are computed by Eqs. 5.60 and 5.61.

$$D_e = D_t - D_k + R_z \sin \alpha_k - R_{zz} \sin \alpha_t \tag{5.70}$$

$$C_e = C_t - C_k - R_z \cos \alpha_k + R_{zz} \cos \alpha_t \tag{5.71}$$

$$R_e = R_z - R_{zz} \tag{5.72}$$

5.4.2.5 Calculations for Nodal Parameters and Interpretations for Design Results

Calculations for nodal parameters involve computing the well depth, the TVD, and the closure distance at each node on the basis of the given conditions and the calculated key parameters. The formulae for the well depth, the TVD, and the closure distance at each node are listed as follows.

- Well depth can be calculated with the given D_k and C_k at the KOP (Point k)

$$L_k = D_k / \cos \alpha_k \tag{5.73}$$

- Parameters at the end of "Build" (Point b) can be calculated by using the following formulae.

$$L_b = L_k + \frac{R_z \pi (a_b - a_k)}{180} \tag{5.74}$$

$$D_b = D_k + R_z (\sin \alpha_b - \sin \alpha_k) \tag{5.75}$$

$$C_b = C_k + R_z (\cos \alpha_k - \cos \alpha_b) \tag{5.76}$$

- Parameters at the end of "Hold" (Point c in the build–hold well trajectory, or Point t in the multi-targeted build–hold well trajectory) can be calculated using the following formulae.

$$L_c = L_b + \Delta L_{bc} \tag{5.77}$$

$$D_c = D_b + \Delta L_{bc} \cos \alpha_b \tag{5.78}$$

$$C_c = C_b + \Delta L_{bc} \sin \alpha_b \tag{5.79}$$

– Parameters at the end of "Drop" (Point t in the build–hold–drop well trajectory) can be calculated by using the following formulae.

$$L_t = L_c + \frac{R_n \pi (a_b - a_t)}{180} \tag{5.80}$$

$$D_t = D_c + R_n(\sin \alpha_b - \sin \alpha_t) \tag{5.81}$$

$$C_t = C_c + R_n(\cos \alpha_t - \cos \alpha_b) \tag{5.82}$$

– Parameters at the end of the second "Build" (Point t) can be calculated using the following formulae.

$$L_t = L_c + \frac{R_{zz} \pi (a_t - a_b)}{180} \tag{5.83}$$

$$D_t = D_c + R_{zz}(\sin \alpha_t - \sin \alpha_b) \tag{5.84}$$

$$C_t = C_c + R_{zz}(\cos \alpha_b - \cos \alpha_t) \tag{5.85}$$

– Parameter at the target (Point d in the multi-targeted build–hold well trajectory, build–hold–drop well trajectory, and build–hold–build well trajectory) can be calculated using the following formulae.

$$L_d = L_t + \Delta L_{td} \tag{5.86}$$

$$D_d = D_t + \Delta L_{td} \cos \alpha_t \tag{5.87}$$

$$C_d = C_t + \Delta L_{td} \sin \alpha_t \tag{5.88}$$

Based on the calculations above, the design results can be shown in a list or a plot.

Example 5.4.1 Design for the build–hold–drop well trajectory with the given data including $D_t = 2530$ m, $C_t = 910$ m, $D_k = 300$ m, $\alpha_k = 0°$, $\alpha_t = 15°$, $\Delta L_{td} = 120$ m, $K_z = 2.7°/30$ m, $K_n = 1°/30$ m.

SolutionFirst, based on Eq. 5.55, we can obtain

$$R_z = 636.67 \text{ m}, R_n = 1719.00 \text{ m};$$

Second, according to Eqs. 5.60, 5.61, 5.67, 5.68, and 5.69, we have

$$D_e = 2674.91 \text{ m}, C_e = -1387.10 \text{ m}, R_e = 2355.67 \text{ m},$$
$$\Delta L_{bc} = 1878.83 \text{ m and } \alpha_b = 24.02°$$

Table 5.3 Design results of a build–hold–drop well trajectory

Well interval	Vertical	Build	Hold	Drop	Target
α (°)	0	0–24.02	24.02	24.02–15	15
ΔD (m)	300	259.11	1716.19	254.70	115.91
TVD (m)	300	559.11	2275.30	2530.00	2645.91
C (m)	0	55.11	819.76	910.00	941.59
Length of the interval (m)	300	266.68	1878.83	270.49	120
L (m)	300	566.68	2445.51	2716.00	2836.00

Finally, all nodal parameters can be calculated by Eqs. 5.73–5.82 and Eqs. 5.86–5.88. All design results are shown in Table 5.3.

5.5 Deflection Tools and Wellpath Control for Directional Well

In Sect. 5.3, deviation control technology of a vertical well has been discussed, which includes the associated reasons for well deflection, the principles, structures, and notes for the deviation-prevented BHAs and deviation-corrected BHAs. That is helpful to understand and use deflection tools and also control the wellpath of the directional well.

5.5.1 Downhole Motor Deflection Tools

The downhole motor includes the turbine motor, the positive displacement motor (PDM) and the electric motor. Among these, the first two kinds are widely used in China. The downhole motor is typically installed between the drill collar and the drill bit. For the turbine motor and the PDM, the circulating drilling fluid flows through the motor and drives the motor to rotate. Then, the motor drives the bit to rotate and break the rock. In deflection drilling, the drill strings above the downhole motor do not rotate. This feature is advantageous for the deflection operation.

5.5.1.1 Types of Mud Motor Deflection Tool
As illustrated in Fig. 5.22, there are three types of mud motor deflection tools.

– **Bent bub (tilting joint)**

The bent sub, which produces a bent angle, is installed between the mud motor and the drill collar. This structure forces the drill bit to tilt, which results in asymmetric cutting at the downhole, thus changing the direction of the wellbore. Moreover, the bending part of the deflection tool is forced to straighten, and the drill bit is affected

Fig. 5.22 Mud motor
deflection tools

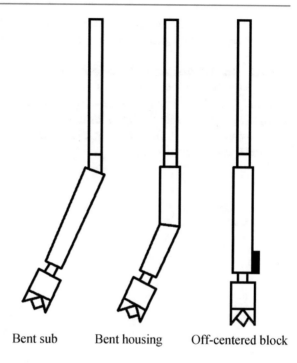

Bent sub   Bent housing   Off-centered block

by the elasticity of the drill string, leading to the side cutting and changes in the actual wellpath.

The build-up rate of the bent sub is related to the following factors. A larger bending angle causes a larger build-up rate. A greater stiffness of the drill string above the bending point leads to a larger build-up rate. A smaller distance between the bending point and the drill bit, or a lower ROP results in larger build-up rate. In addition, the build-up rate is also related to the clearance between the deflection tool and wellbore, formation factors and bit types.

– Bent housing

With a similar deflection principle as that of the bent sub, the mud motor with an elbow-shaped housing is also known as a bent housing motor, which has greater build capability while drilling.

– Eccentric block

A block is welded on one side of the lower end of the mud motor. In an inclined well, the block is located on the lower side of the borehole by orientation and acts as a fulcrum. The bit is subjected to a lever force caused by the gravity of the upper drill string, which produces the side cutting and changes the wellpath. Obviously, a larger eccentric magnitude for the block will also give rise to a greater built-up rate.

It should be noted that a greater built-up rate of the deflection tool also increases the difficulty in running the deflection tool into the wellbore.

5.5.1.2 Structures and Characteristics of Turbine Motor

Figure 5.23 shows a typical design of the turbine motor, including an upper housing with different sizes of heads, a stator mounted inside and tightly pressed by the compacted nipple, a spindle with a lower nipple and connected with the bit, a rotor set on a spindle, and a thrust bearing and a radial bearing. Both the stator and the rotor comprise special blades.

Fig. 5.23 Schematic of
turbine motor

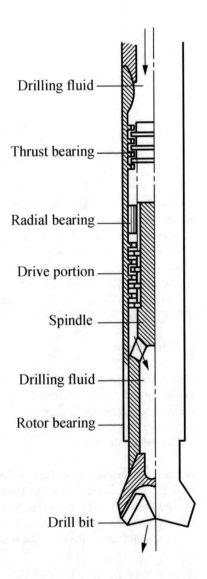

Drilling fluid

Thrust bearing

Radial bearing

Drive portion

Spindle

Drilling fluid

Rotor bearing

Drill bit

Fig. 5.24 Characteristic curve of turbine motor

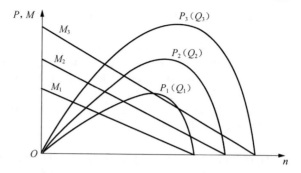

Figure 5.24 shows the output characteristic curve of the turbine motor. The abscissa refers to revolutions per minute (RPM) of the turbine (n), and the ordinate indicates the power (P) and torque (M). Note that there are three groups of curves with different flow rates of the drilling fluid, and each group has an optimal RPM (n_0) associated with the maximum power.

The relationships among these parameters are expressed as follows.

- The RPM of the turbine motor is proportional to the flow rate of the drilling fluid, the torque and the pressure drop (Δp) are proportional to the square of the flow rate, and the power is proportional to the cube of the flow rate. The relationships can be mathematically expressed as follows.

$$\frac{n_2}{n_1} = \frac{Q_1}{Q_2} \tag{5.89}$$

$$\frac{M_1}{M_2} = \left(\frac{Q_1}{Q_2}\right)^2 \tag{5.90}$$

$$\frac{\Delta p_1}{\Delta p_2} = \left(\frac{Q_1}{Q_2}\right)^2 \tag{5.91}$$

$$\frac{P_1}{P_2} = \left(\frac{Q_1}{Q_2}\right)^3 \tag{5.92}$$

Obviously, sufficient flow rate of the drilling fluid can guarantee enough output of power and torque.

- At a certain flow rate, a lower torque causes a higher RPM of the turbine. The RPM reaches the maximum while idling. Accordingly, it is not appropriate to use the turbine motor for reaming.

5.5.1.3 Structures and Output Characteristics of PDM

Figure 5.25 illustrates a typical design of the PDM, including the stator integrated with the housing and the rotor integrated with the spindle. The upper section of stator connects the drill string, while the lower part of the rotor connects the drill bit via the cardan shaft (universal shaft). Obviously, the design of the PDM is much simpler than that of the turbine motor.

Figure 5.26 shows the output characteristic curve of the PDM, which can be explained as follows.

– The torque (M) of the PDM is proportional to the pressure drop (Δp). The RPM varies with different pressure drops, but it changes slightly within a certain range of the pressure drop. The pressure drop can be read from the pump pressure gauge, and the torque reflects the WOB. Accordingly, the operator can drill ahead according to the pump pressure gauge. For directional wells and horizontal wells, the friction on the drill string is too large, and it is very difficult to obtain the accurate WOB from the weight indicator. Therefore, it is of great advantage to drill ahead based on values read from the pump pressure gauge.

Fig. 5.25 Schematic of PDM

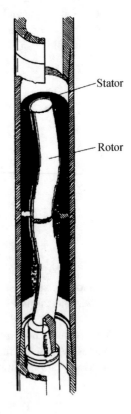

Stator

Rotor

Fig. 5.26 Schematic of PDM

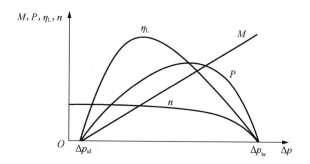

The pressure drop is also used to calculate the torque on the drill bit, which can help obtain an accurate reactive torque angle (see the specific calculation for azimuth correction run in Sect. 5.5). This is an outstanding advantage of the PDM in directional drilling.

5.5.2 Rotary Drilling Deflection Tools

Rotary drilling deflection tools include the whipstocks, the jetting bits, and the stabilizer BHAs. Whipstocks and jetting bits are only used to deflect a well and are briefly introduced here.

5.5.2.1 Whipstocks

As illustrated in Fig. 5.27, the whipstock was the first widely used deflection tool for directional drilling. Due to its complex techniques, the whipstock is now only used for the casing sidetrack or for the special conditions where the mud motor is not suitable.

5.5.2.2 Jetting Bits

As illustrated in Fig. 5.28, except for a large nozzle and two small nozzles, the appearance of the jetting bit has no distinctive differences from that of the ordinary drill bit. Its deflection operations are described as follows. First, orient the drill string and keep it from rotating. Then, move the drill string up and down within a narrow range, while the pump is on to circulate the drilling fluid until a powerful jetting from the large nozzle erodes a new deviated borehole along the predetermined direction. Next, start the rotary table to rotate the drill string, trim, and enlarge the deviated borehole. If more wellpath changes are required, the deflection operations described above must be repeated.

Jetting bits are merely used for soft formations. Now, they are only adopted if the mud motor is unavailable.

Fig. 5.27 Structure and
deflection principle of
whipstock

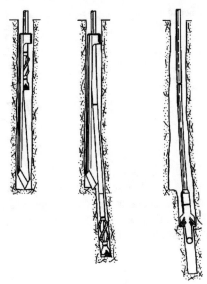

（a）Placement and   （b）Fix the whipstock and drill   （c）Borehole enlargement
orientation   an openhole with a small diameter   and drilling

5.5.2.3   Stabilizer BHA

Instead of deflecting the borehole, the stabilizer BHA merely aims to build up, drop
off, or hold on the inclination in a deviated wellbore. Based on the rotary drilling,
this method refers to a combination of BHAs with various deflection performances
by skillfully installing stabilizers on bottom drill collars.

Since the 1980s, more in-depth studies on the stabilizer BHAs have gradually
emerged. Utilizing mathematics, mechanics, and computers, several design methods have been put forward, including the differential equation method, the finite

Fig. 5.28 Structure and
deflection principle of jetting
bit

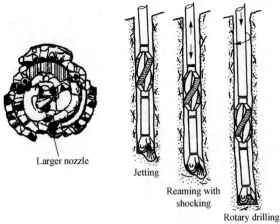

Larger nozzle

Jetting

Reaming with
shocking

Rotary drilling

（a）Jetting bit   （b）Deflection principle

element method, the continuous beam method, and the weighted residual method, all of which require complicated computer programs. In the absence of corresponding computer programs, the empirical data from Tables 5.4, 5.5, and 5.6 are widely used in the field. The data from the tables lie within a certain range and can be adjusted based on the operators' experience.

- **Build BHA**

Based on their ability to build up the inclination, the build BHAs are subdivided into strong, moderate and weak ones, and their structures and fit sizes are shown in Fig. 5.29 and Table 5.4. Note that a greater buildability results from a larger WOB, a smaller L_1 and a large diameter of the near-bit stabilizer. Also, low ROP should be kept while using the build BHAs.

- **Hold BHA**

Based on their ability to hold the inclination, the hold BHAs are subdivided into strong, moderate and weak ones, and their structures and fit sizes are shown in Fig. 5.30 and Table 5.5. It is essential to maintain a regular WOB and a relatively high ROP while using such BHAs. If the strong hold BHA is required, two stabilizers can be connected in series as a near-bit stabilizer.

Table 5.4 Fit sizes of build BHAs

Type	L_1 (m)	L_2 (m)	L_3 (m)
Strong build BHA	1.0–1.8	–	–
Moderate build BHA	1.0–1.8	18.0–27.0	–
Weak build BHA	1.0–1.8	9.0–18.0	9.0

Table 5.5 Fit sizes of hold BHAs

Type	L_1 (m)	L_2 (m)	L_3 (m)	L_4 (m)	L_5 (m)
Strong hold BHA	0.8–1.2	4.5–6.0	9.0	9.0	9.0
Moderate hold BHA	1.0–1.8	3.0–6.0	9.0–18.0	9.0–27.0	–
Weak hold BHA	1.0–1.8	4.5	9.0	–	–

Table 5.6 Fit sizes of drop BHAs

Type	L_1 (m)	L_2 (m)
Strong drop BHA	9.0–27.0	–
Weak drop BHA	0.8	18.0–27.0

Fig. 5.29 Build BHAs

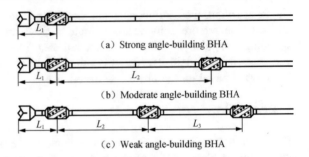

(a) Strong angle-building BHA

(b) Moderate angle-building BHA

(c) Weak angle-building BHA

Fig. 5.30 Hold BHAs

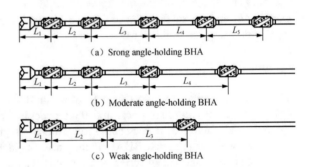

(a) Srong angle-holding BHA

(b) Moderate angle-holding BHA

(c) Weak angle-holding BHA

- **Drop BHA**

Based on their ability to drop off the inclination, drop BHAs are typically subdivided into strong and weak ones, and their structures and fit sizes are shown in Fig. 5.31 and Table 5.6. It is significant to keep a low WOB and ROP while using the drop BHAs. For the strong drop BHA, a longer L_1 generally indicates a greater drop ability, but a new tangency point between the stabilizer and the drill bit is quite undesirable.

Fig. 5.31 Drop BHAs

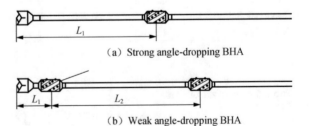

(a) Strong angle-dropping BHA

(b) Weak angle-dropping BHA

5.5.3 Steering Drilling System

The steering drilling system refers to a combination of deflection tools and measurement while drilling (MWD) tools. Combined with the highly efficient drill bit, this system can accomplish a series of wellpath control operations within one tripping, including the build section, the drop section, the hold section, and the azimuth correction run. MWD tools are used to measure the wellpath parameters and orientation parameters in real-time and transmit the measurement results to the surface. Operators can adjust the working modes and drilling parameters in real-time according to the wellpath tendency. As long as downhole tools and measuring instruments work normally, operators can continue the drilling process without tripping operations and replacing the deflection tool. Based on whether the drill string is allowed to rotate while deflecting, the steering drilling system is divided into the sliding steering drilling system and the rotary steering drilling system.

5.5.3.1 Sliding Steering Drilling System

Based on the mud motor deflection tool, the sliding steering drilling system, the most widely used steering drilling system at present, is a combination of the mud motor deflection tool and the MWD tool.

Specifically, there are two working modes in the sliding steering drilling system. The sliding drilling mode refers to the fact that when it is required for the build, drop, or azimuth correction run, the drill string is kept from rotating and can only slide into the new borehole along its axial direction while drilling. The rotary drilling mode indicates that when it is required for maintaining the borehole or slight changes in wellbore curvature, the rotary table drives the entire drill string and the housing of the mud motor to rotate while the mud motor rotates. Due to the rotary housing, only some special mud motors with the bent housing can be applied to the steering drilling system. Figure 5.32 illustrates two types of mud motors with the bent housing. One is known as a singly bent motor with a smaller bent angle

Fig. 5.32 Mud motor deflection tools in sliding steering drilling system

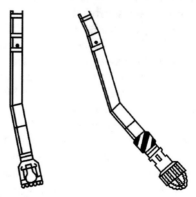

(a) Singly-bent motor   (b) Reverse doubly-bent motor

and a bending point near to its lower part. The other one is a double tilted universal joint motor with two bent angles in opposite directions. Their common characteristics are a greater buildability in the process of the sliding drilling and less resistance on the rotary bent housing in the process of the rotary drilling.

As one of the most important drilling systems at present, the sliding steering drilling system has greatly promoted the development of modern horizontal wells and ERWs.

5.5.3.2 Rotary Steering Drilling System

With the development of horizontal wells, ERWs, and three-dimensional multi-target wells, a variety of problems have emerged in the sliding steering drilling system. During the sliding drilling, the drill string does not rotate, and the frictional resistance between the drill string and the borehole is completely exerted on the drill string in its axial direction. This results in the drill string buckling or even self-locking. These serious problems may lead to issues in the bit feed and delivery of WOB and result in a relatively low ROP. Moreover, the differential pressure sticking is easy to occur, and the borehole distortion is easy to form since the reactive torque angle of the mud motor severely influences the tool face orientation. During the rotary drilling, the rotary bend housing may result in the borehole enlargement and the bit wear. Therefore, the rotary steering drilling system was developed to address these problems.

The rotary steering drilling system combines the rotary steering tool with the MWD tool, sometimes even with the logging while drilling (LWD) tool, or the formation evaluation while drilling (FEWD) tool for a more powerful drilling system.

Specifically, deflection tools in the rotary steering drilling system are divided into the push-the-bit steering tool and the point-the-bit steering tool.

- **Push-the-bit steering Tool**

According to whether the performing section of the steering tool rotates together with the drill string, the push-the-bit steering tool can be subdivided into two kinds: the static push-the-bit steering tool and the dynamic push-the-bit steering tool, respectively. Here, only the latter one is briefly introduced as follows.

As illustrated in Fig. 5.33, the structure of the dynamic push-the-bit steering tool comprises a controlling unit and a performing unit. The core of the controlling unit consists of an inertial platform, a power generator, and a computer control system. The housing of the controlling unit and the bit rotates together with the rotary drill string, but the inertial platform remains stationary.

Figure 5.34 shows the structure of the performing unit where a static disc valve and a movable disc valve are located. Controlled by the inertial platform, the static disc valve with a crescent-shaped hole remains stationary while the drill string rotates. The radial azimuth of the hole is fixed. It must be noted here that this azimuth can be adjusted by the inertial platform according to the wellpath tendency.

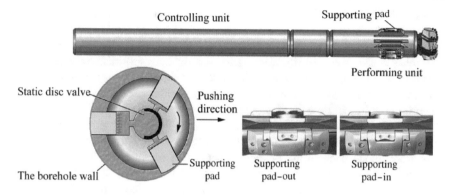

Fig. 5.33 Dynamic push-the-bit steering tool

Fig. 5.34 Principle of the
fluid control platform

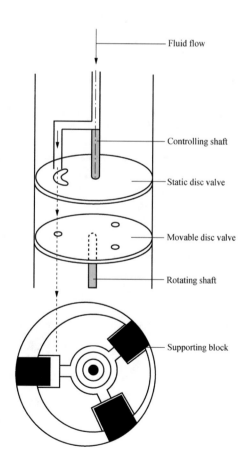

In each 120° circular direction, the movable disc valve has three separate round holes on it. Each one of these holes is connected with a hydraulic cylinder where a piston-equivalent support block is located. The movable disc valve is controlled by the rotating axis of the drill string, allowing the movable disc valve to rotate together with the drill string. Only when a certain round hole on the movable disc valve rotates to communicate with the crescent-shaped hole on the static disc valve, the drilling fluid will flow into the cylinder connected with the round hole, pushing the supporting block out to support the wellbore. When the round hole deviates from the crescent-shaped hole, the drilling fluid flow will be shut off, drawing the supporting block back to the cylinder. Then, other round holes follow the same pattern. With the fixed azimuth of the crescent-shaped hole, the resultant support force from each support block is always located on the same azimuth. The drill bit is controlled by the resultant support force on the wellbore and forced to another side, resulting in the side cutting. Thus, operators achieve the steering operation.

The drilling fluid flow pressure in the crescent hole depends on the support force from the support pad. The build-up rate can be adjusted. If the drilling fluid flow is completely shut off, the inclination can be maintained.

- **Point-the-bit steering tool**

Figure 5.35 shows the point-the-bit steering tool. A non-rotating sleeve is located near the bit, inside which two eccentric rings are present, namely the outer eccentric ring and the inner eccentric ring, respectively. In the drilling process, both the eccentric rings and the non-rotating sleeve do not rotate despite the rotary drill string. Due to the eccentric rings, the rotary shaft that drives the bit to rotate will be bent, producing a bias angle between the bit axis and the borehole axis and resulting in asymmetric cutting. Therefore, the steering operation is achieved.

The bending direction of the rotary axis varies according to different combinations of the inner and outer eccentric rings. Similar to the principle of the rudder, the steering tool leads the bit tilt. Hence, it is named as the point-the-bit steering tool.

Fig. 5.35 Point-the-bit steerable tool

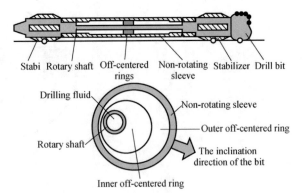

The adjustment of the eccentric rings can lead to alterations in bending directions of the rotary axis and different build-up rates by changing the bending degree. Furthermore, it can also achieve a straight rotary shaft to maintain the wellpath.

5.5.4 Basic Methods for Directional Wellpath Control

The predetermined trajectory of the 2D directional well generally includes four types of well sections: the vertical section, the build section, the hold section, and the drop section. When the steering drilling system is used, it is easier to control the wellpath and drilling modes (for the sliding steering drilling system) or drilling instructions (for the rotary steering drilling system). This can help realize the continuous wellpath control, making it easier to achieve wellpath control with better quality and higher efficiency. When the steering drilling system is not used, it is more complicated to control the wellpath, because the drilling tools vary according to different wellbore sections, leading to different methods for wellpath control. Here, only the scenario where the steering drilling system is not used is introduced. Overall, the process of a directional wellpath control is divided into the following three stages.

5.5.4.1 Drilling the Vertical Section as Vertically as Possible

The true wellpath of the vertical section should be as close as possible to the plumb, indicating that the inclination needs to be as small as possible. The vertical section of the directional well can be controlled in a similar manner as the drilling of vertical wells, but higher requirements are needed because the quality of the vertical section is essential for the following wellpath control for other sections. The issue of the wellpath control for vertical wells has been discussed in Sect. 5.3.

5.5.4.2 Strictly Control the Directional Deflection Operation

The directional deflection section is a part of the build section, but it starts from the end of the vertical section, which is known as *Kicking-off or deflection* because the inclination of the vertical section is equal to zero. As the azimuth of the vertical section is nonexistent, it needs to be oriented at the beginning of the build section. If the azimuth of the directional deflection section deviates from the pre-designed value, it will cause great difficulty in the following steps of the wellpath control. Accordingly, the directional deflection operation is critical for the overall wellpath control.

Mud motor deflection tools are widely used in almost all the processes of the modern directional kick-off operation, in addition to whipstocks which are still used as sidetracking tools for the casing window operation. The length of the kick-off section generally depends on the inclination (about 8–10°) to meet the requirement of the stabilizer BHAs by rotary drilling.

5.5.4.3 Track and Control the Wellpath Until the Target

The tracking control, generally known as the specific wellpath control, starts from the end of the kick-off section to the end of the drilling process. The main task of this stage is to understand the wellpath tendency in the drilling process and to maintain the wellpath less deviated from the predetermined one by using various deflection tools or BHAs. "Less Deviated" here means less error between the true wellpath and the pre-designed well trajectory and ensures that the actual wellpath precisely hits the target. Moreover, "Less Deviated" means more frequent wellpath surveys and replacements of deflection tool, which makes the drilling process more costly and time-consuming and results in more complicated conditions at the downhole. Accordingly, the following principle is applicable: To hit the target with a high ROP.

Another principle at this stage is to control the wellpath by using stabilizer BHAs in the rotary drilling. Due to its higher ROP, the rotary drilling is mainly used for the build, hold, and drop sections from the end of the kick-off section, except the following two scenarios where the downhole motor deflection BHAs are used instead.

- When the stabilizer BHAs in the rotary drilling cannot meet the requirements of the build or drop section, the mud motor deflection tool is applied instead.
- When the stabilizer BHAs in the rotary drilling cannot control the azimuth, it generally deviates from the predetermined value in the drilling process. When the true azimuth severely deviates from the predetermined value, the wellpath may miss the target. Therefore, the mud motor deflection tool is essential for the azimuth correction run.

In total, it is easier to control the inclination at the tracking control stage where more tools are also available. However, it is more difficult to control the azimuth, which involves many complicated calculations, especially when the azimuth correction run is achieved as the inclination is also required to change. Therefore, the main calculations for the azimuth correction run will be introduced.

5.5.5 Calculations for Azimuth Correction Run

The azimuth correction run refers to every change in wellbore azimuth, except the inclination which continues to build or drop along the original azimuth. Inclination changes typically occur in the process of azimuth correction run. For the directional wellpath control, the calculation for the azimuth correction run is essential, which is one of the basic techniques that directional drilling engineers must master. It includes the calculations for the tool face angle, the reactive torque angle of the mud motor, the inclination and azimuth at the new bottom hole with the given tool face angle, and the length of azimuth correction run.

5.5.5.1   Concept of Tool Face Angle

As mentioned above, there are two principles for the directional deflection. The first is that the bending axis of the drill bit generally results in asymmetric cutting at the downhole, and the second is that the side force on the drill bit causes side cutting. By either principle, the drill bit would lean toward the wellbore. The tool face direction line refers to the perpendicular line to the drill string axis and along the bending axis of the drill bit or the side force on the drill bit, and the tool face refers to the plane passing through the tool face direction line and the drill string axis, as illustrated in Fig. 5.36.

Due to the borehole deviation, the bottomhole surface (called bottomhole circular surface) is inclined as well. The high-side direction line typically indicates the direction from the center of the bottomhole circle to the highest point of the circle.

As the drill string lies at the downhole, the high-side tool face angle, generally symbolized by ω, represents the angle on bottomhole circular surface measured from the high-side direction line to the tool face direction line in the clockwise sense.

In addition, the tool face angle is also known as the magnetic-north tool face angle. The tool face angle, in general, refers to the high-side tool face angle.

On the horizontal plane passing through the center of bottomhole circular surface, the magnetic-north tool face angle, symbolized by ω_m, starts from the magnetic-north line to the tool face direction line in the clockwise sense. As the

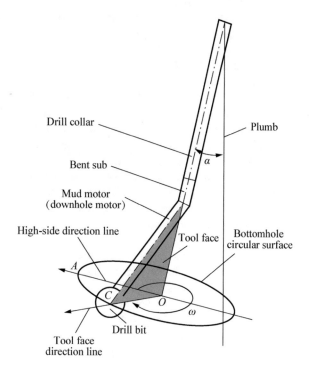

Fig. 5.36 Schematic of the tool face orientation

magnetic-north line does not lie on the bottomhole circular surface, the magnetic-north tool face angle is still obtained by the approximation of Eq. 5.7. Specifically, when the inclination is relatively small or the azimuth approximates $0°$ or $180°$, the degree of approximation is much higher. Moreover, the high-side tool face angle shows less accuracy when the inclination is small. Accordingly, in the practical process, the high-side tool face angle and the magnetic-north tool face angle are complementary to each other. Therefore, when the inclination is larger than $5°$, high-side tool face angle is typically used; otherwise, the magnetic-north tool face angle is used instead. The conversion is expressed as follows.

The magnetic−north tool face angle = the azimuth + high−side tool face angle

$$(5.93)$$

The tool face direction line rotates with the rotary drill string, leading to changes in the tool face angle in the range of $0°-360°$. In western countries, the tool face angle generally lies between $0°$ and $180°$ and indications, such as "turn right" (+) or "turn left" (−) are added before the tool face angle. The tool face angle typically starts from the high-side direction line or the magnetic-north azimuth to the tool face direction line in the clockwise sense (turn right) or in the counterclockwise sense (turn left).

The new borehole direction varies with different tool face angles.

− When the tool face angle is $0°$, the tool face direction line overlaps with the high-side direction line. In the new borehole, the inclination increases, while the azimuth remains unchanged.
− When the tool face angle is $180°$, the tool face direction line points in the opposite direction from the high-side direction line. In the new borehole, the inclination decreases, while the azimuth stays constant. When the inclination reaches $0°$, the unchanged tool face angle will result in the inclination in the opposite direction and a new azimuth at reverse $180°$.
− When the tool face angle lies between $0°$ and $180°$ or $180°-360°$, there must exist a special tool face angle within each range where the azimuth changes while the inclination remains unchanged.
− When the tool face angle equals a certain value other than the special values mentioned above, both the inclination and the azimuth in the new borehole will change. Thus, the wellbore direction changes, and its change rate generally depends on the tool face angle.

In general, when the dogleg is small at the kick-off section, both the inclination and azimuth increase in the new wellbore if the tool face angle is between $0°$ and $90°$; the inclination reduces, while the azimuth increases in the new wellbore if the tool face angle lies between $90°$ and $180°$; both the inclination and azimuth decrease in the new wellbore if the tool face angle is between $180°$ and $270°$; and the inclination increases, while the azimuth decreases if the tool face angle lies between $270°$ and $360°$. Accordingly, the tool face angle is critical for controlling the borehole direction after the capability of deflection tools is determined. In other words, when the borehole is required to be drilled along a given direction, the tool face angle should be calculated and controlled to achieve the required wellpath control.

5.5.5.2 Calculations for Tool Face Angle

The tool face angle ω can be calculated with the given data including the inclination α_1 and the azimuth ϕ_1 at current bottom; the build-up rate K of current deflection tool; the predicted length of new drilling section ΔL; the inclination α_2 and the azimuth ϕ_2 at new bottom. There are two calculation methods, specifically the graphical method and the analytical method. The graphical method was used widely in earlier days, while the analytical method is the dominant one now due to the advent and development of high-speed computational tools.

Here, only the analytical method is introduced by using the following equations.

$$\cos \gamma = \cos \alpha_1 \cos \alpha_2 + \sin \alpha_1 \sin \alpha_2 \cos \Delta \phi \qquad (5.94)$$

$$\cos \omega = \frac{\cos \alpha_1 \cos \gamma - \cos \alpha_2}{\sin \alpha_1 \sin \gamma} \qquad (5.95)$$

The dogleg γ can be obtained by Eq. 5.94. Then, the tool face angle ω can be determined by substituting the calculated dogleg into Eq. 5.95.

Note that the range of inverse cosine function is from $0°$ to $180°$. When $\Delta \phi$ is negative, operators need to pay attention to the range of ω. Specifically, we set $\cos \omega = C$. As a result, when $\Delta \phi$ is larger than zero, $\omega = \arccos C$; when $\Delta \phi$ is less than zero, $\omega = -\arccos C + 360°$.

The length of the azimuth correction run ΔL is

$$\Delta L = \frac{30\gamma}{K} \qquad (5.96)$$

There are seven parameters in the above three equations, namely $\alpha_1, \alpha_2, \Delta \phi, \gamma, K, \omega, \Delta L$. Knowing any four parameters, the rest can be obtained on the practical basis of the azimuth correction run.

Example 5.5.1 Determine ω and ΔL with the given data including $\alpha_1 = 15°$, $K = 3°/30$ m, $\Delta \phi = 22°$, and $\alpha_2 = 18°$.

Solution

By using Eq. 5.94, the dogleg $\gamma = 6.88°$. By using Eq. 5.95, the tool face angle $\omega = 75.19°$. Then using Eq. 5.96, it yields $\Delta L = 68.77$ m.

Example 5.5.2 Determine ω and ΔL with the given data including $\alpha_1 = 15°$, $K = 3°/30$ m, $\Delta \phi = -22°$, and $\alpha_2 = 18°$.

Solution

γ and ΔL are the same as those in Example 5.5.1. By using Eq. 5.95, the tool face orientation $\omega = -75.19° + 360° = 284.81°$.

5.5.5.3 Calculations for Reactive Torque Angle of Mud Motor

During the operation of the mud motor, the rotor is subject to the circulating drilling fluid, which produces a torque transmitted to the drill bit to break the rock. At the same time, the stator is subjected to a reverse torque caused by the circulating drilling fluid, which exerts a tendency of counterclockwise rotation on the drill string. However, at the wellhead, the drill string is locked by the rotary table, and only a small angle, known as the reactive torque angle symbolized by ϕ_n, is allowed in a counterclockwise sense. The existing reactive torque angle aims to reduce the determined tool face angle. Furthermore, to offset the effect caused by the reactive torque angle, it is necessary to add this reactive torque angle to the calculated tool face angle when orienting the deflection tool.

A number of factors affect the reactive torque angle, including the reactive torque, the drill string length, the polar inertia moment of the drill string, the friction between the drill string and the wellbore as well as the tool face angle. Due to some uncertain factors, it is difficult to establish a determined mode for calculating the reactive torque angle. In the engineering practice, the reactive torque angle is generally obtained by the empirical data method or the reverse calculation method.

The reverse calculation method is used to reversely compute the true reactive torque angle with the given parameters from a certain test drilling section. Before the test drilling, a tool face angle (ω_0) is first calculated based on the requirements of wellpath tendency and the capability of the current deflection tool. Then, an estimated reactive torque angle (ϕ_{n0}) is used to orient the deflection tool according to the expression that the tool face angle before adding the WOB is equal to $\omega_0 + \phi_{n0}$. During the test drilling, the true tool face angle (ω) and the reactive torque angle (ϕ_n) are unknowns because the estimated ϕ_{n0} deviates from the true one. However, on account of the locked drill string at the wellhead during the test drilling, the sum of ω and ϕ_n must be equal to that of ω_0 and ϕ_{n0} before adding the WOB, which can be mathematically expressed as $\omega_0 + \phi_{n0} = \omega + \phi_n$. Accordingly, after the test drilling, the ω and the ϕ_n can be obtained through the wellpath measurement and calculation.

With the measured data including the inclination α_1 and azimuth ϕ_1 before the test drilling and the inclination α_2 and azimuth ϕ_2 after the test drilling, the true reactive torque angle can be calculated as follows.

– The dogleg γ of the test drilling section is determined by Eq. 5.97:

$$\gamma = \arccos[\cos \alpha_1 \cos \alpha_2 + \sin \alpha_1 \sin \alpha_2 \cos(\phi_2 - \phi_1)] \qquad (5.97)$$

- The true tool face angle ω is computed by Eq. 5.98:

$$\omega = \pm \arccos \frac{\cos \alpha_1 \cos \gamma - \cos \alpha_2}{\sin \alpha_1 \sin \gamma} \qquad (5.98)$$

In Eq. 5.98, when $\phi_2 > \phi_1, \omega$ is positive; otherwise, ω is negative.

- The true reactive torque angle ϕ_n can be calculated by Eq. 5.99:

$$\phi_n = \omega_0 + \phi_{n0} - \omega \qquad (5.99)$$

5.5.6 Orientation Methods of Deflection Tool

The deflection tools include rotary drilling steering tools and mud motor deflection tools. As long as the directional drilling is carried out, the orientation for deflection tools is required for this process. The desired tool face angle (ω) should be calculated in advance according to the requirement of wellpath tendency. Operators need to know the tool face orientation at downhole and then adjust it at the predetermined site, which is known as the *orientation* for the deflection tool. A set of orientation techniques are required.

When the steering drilling system is adopted, the real-time data of the tool face orientation can be measured and transmitted to the surface by the MWD tool at the downhole and displayed on surface devices. If the tool face orientation does not match the desired one, the drill string can be rotated slowly to adjust the tool face orientation. However, rather than the steering drilling system, the single-shot inclinometer is mainly adopted for the orientation operation for a large number of directional wells. Due to its complicated working principle, the single-shot inclinometer is specifically introduced as follows.

The early orientation method is called the surface orientation method. Due to its outdated and complicated nature, the surface orientation method has been replaced by the bottomhole orientation method. According to this method, the deflection tool requires a mark on its tool face in advance. After the deflection tool is set at the downhole, the inclinometer is run into the borehole via the inner bore of drill string and located at the position of the mark by a steel line or a cable. In this way, the tool face orientation is measured by the inclinometer.

The popular notation of the mark is known as the *orientation key method*. As illustrated in Fig. 5.37, if the deflection tool is a combination of an orientation bent sub and a straight mud motor, the orientation key is installed inside the bent sub. If a bent housing mud motor is used, the orientation key, in the same direction of the tool face, is installed in a straight orientation sub above the mud motor.

Fig. 5.37 Notation and
working principle of the
orientation key

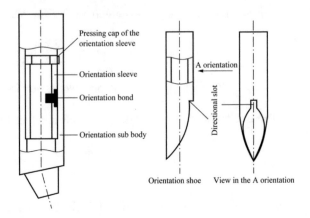

An orientation shoe is connected to the bottom of the inclinometer assembly. The lower half of the orientation shoe is a special curve, the top of which has an orientation slot. The extension of the generatrix of the cylinder where the orientation slot is located points to the negative X-axis direction, which is the reference direction of the probe tube of the inclinometer. When the inclinometer assembly is located at the downhole, its orientation slot automatically matches the orientation key due to the special shape of the orientation shoe. In this way, the tool face direction is transmitted to the inclinometer via the orientation key. Therefore, the tool face angle can be measured by the inclinometer.

According to the values measured from the gravity accelerator and the fluxgate, operators can obtain the inclination of the azimuth and the tool face angle(s). If the measured tool face angle does not meet the expectation, operators can slowly rotate the drill string to match the desired value.

Once the orientation operation is successful, it is also necessary to make marks on both the kelly and the outer edge of the rotary table, which can guarantee an unchanged tool face orientation after making a connection. When 1–2 singles are completed, a new wellpath survey and orientation operation are required until the predetermined requirements are met.

Exercises

1. What are the differences between directional wells and vertical wells? What is the right scenario for a directional well (please show it by graphical methods)?
2. What are the basic parameters of the wellpath? Why are they called basic parameters?
3. What are the two methods to represent the azimuth? What is the conversion between the two methods?
4. What are the differences between azimuth and direction? Please show them by example.

5. What are the differences between the horizontal projection length and the closure distance? What are the differences between the vertical section and the closure distance?
6. What are the differences between the vertical projection and the vertical profile?
7. How do the concepts of the dogleg, dogleg angle, and dogleg severity differ from one another? Why does the dogleg severity indicate the minimum wellbore curvature?
8. What is the meaning of the overall angle change? How is it calculated? Why does the overall angle change rate represent the borehole curvature?
9. How do the gravity accelerator and the fluxgate work? Why can they measure the inclination and the azimuth?
10. Why is a non-magnetic drill collar required for the electronic inclinometer? Is the electronic inclinometer installed on the upper or lower section of the non-magnetic drill collar? Does the length of the non-magnetic drill collar influence the inclination and the azimuth?
11. What is the magnetic declination? What is grid convergence or meridian convergence? What are the differences between the magnetic north, true north, and grid north? Why is the grid north regarded as the benchmark in drilling engineering?
12. Why is it necessary to correct the azimuth? How is it corrected?
13. Why is the absolute value of the azimuth change at a certain survey interval less than 180°? If it exceeds 180°, what should be done?
14. What parameters must be calculated for a survey interval? What are the relevant parameters for a survey station? What is the relationship between these two calculations?
15. What are the differences between the angle averaging method and the corrected angle averaging method? What are the differences in both results?
16. What are the hypotheses of the minimum curvature method and the cylindrical spiral method?
17. What are the main tasks in the wellpath control for vertical wells?
18. What are the two essential geological factors that cause the well deviation? How do they work?
19. What are the two essential factors of drilling tools that cause the well deflection? What factors are they concerned with?
20. How does the wellbore enlargement cause the well deflection? How can it be prevented?
21. How does a packed-hole BHA work? Is it used to drop the inclination?
22. How does a pendulum BHA work? Why is the ROP slow when using it?
23. What is the classification standard of wells? Is it a vertical well when the true wellpath is inclined?
24. What is the classification standard for directional wells? What are the types of conventional two-dimensional directional wells?
25. In the drilling technology, is a larger or smaller maximum inclination better?

26. What are the differences between the multi-target build–hold well trajectory and the build–hold well trajectory? What are the differences in the designs of these two well trajectories?
27. What are the final results of the well trajectory design?
28. What are the types of mud motor deflection tools? What principles do they have in common?
29. What are the advantages of PDM in the directional deflection? What are the characteristics of PDM used in the steering drilling system? Why are these characteristics essential?
30. What are the different principles between whipstocks and jetting bits? Can they continue to deflect?
31. What are the components of the rotary steering drilling system? How does it work at the downhole? What are the main advantages of the system?
32. What are the main tasks at the three stages of the wellpath control?
33. How are the high-side direction line and the tool face direction line generated?
34. What is the significance of the tool face angle? What is the wellpath tendency when the tool face angle equals 240°?
35. Why is the magnetic-north tool face angle required when the high-side tool face angle is available? What is the relationship between them?
36. How is the reactive torque angle of the mud motor generated? How does it always reduce the tool face angle?
37. What is the orientation operation? What are its purpose and significance? How does it work in the field?
38. Why can the orientation key on the deflection tool automatically fit into the orientation slot of the single-shot inclinometer assembly?
39. Perform the azimuth conversion between the degree scheme and the quadrant scheme.

   (1) Convert the following azimuth into the quadrant scheme: 50°, 90°, 175°, 200°, 315°, 0°.
   (2) Convert the following azimuth into the degree scheme: S13.5°E, S70°W, N50°E, N33°W.

40. Determine the borehole curvature of three survey intervals as follows by using the two formulae (Eqs. 5.1 and 5.3), respectively.

z	1	2	3
Length of the survey interval (m)	33	35	35
Inclination α_1 (°)	3	80	35
Inclination α_2 (°)	3	85	39
Azimuth ϕ_1 (°)	15	100	303
Azimuth ϕ_1 (°)	194	160	295
Dogleg severity [(°) $\cdot$ (30 m)$^{-1}$]			z
The overall angle change rate [(°) $\cdot$ (30 m)$^{-1}$]			

41. Determine the azimuth increment $\Delta\phi$ and the average azimuth angle ϕ_c of the following survey intervals.

Survey interval	1	2	3	4
ϕ_1 (°)	300	185	3	45
ϕ_2 (°)	10	25	355	230
$\Delta\phi$ (°)				
$\Delta\phi_c$ (°)				

42. Try to complete the following table by using the minimum curvature method and the cylindrical spiral method. The designed azimuth θ_0 is equal to 22°.

L (m)	α (°)	ϕ (°)	N (m)	E (m)	D (m)	V (m)	Borehole curvature [(°) · (30 m)$^{-1}$]
1524.24	16.11	1.06					
1532.24	17.34	0.89					
1542.89	19.00	358.2					
1551.94	20.12	0.99					

43. Please check the results of certain survey intervals and points in Table 5.1.
44. Calculation for the packed-hole BHA: Determine the optimum length L_p between the middle stabilizer and the drill bit with the given data including the wellbore diameter $d_h = 216$ mm, the stabilizer diameter $d_s = 215$ mm, outside diameter of the drill collar $d_{co} = 178$ mm, inside diameter of the drill collar $d_{ci} = 71.40$ mm, the elastic modulus of the drill collar $E = 20.594 \times 10^{10}$ Pa, the line weight of the drill collar $q_c = 1.606$ kN/m, the density of the drilling fluid $\rho_d = 1.33$ g/cm^3, and the tolerant maximum inclination $\alpha = 3°$.
45. Calculation for the Pendulum BHA: Determine the optimum length L_z between the stabilizer and the drill bit with the given data including the wellbore diameter $d_h = 216$ mm, the stabilizer diameter $d_s = 215$ mm, outside diameter of the drill collar $d_{co} = 178$ mm, inside diameter of the drill collar $d_{ci} = 71.40$ mm, the elastic modulus of the drill collar $E = 20.594 \times 10^{10}$ Pa, line weight of the drill collar $q_c = 1.606$ kN/m, density of the drilling fluid $\rho_d = 1.33$ g/cm^3, tolerant maximum inclination $\alpha = 3°$, and the WOB = 120 kN.
46. Design a build–hold well trajectory and fill out the following table with the given target point parameters (the TVD is 1200 m, and the closure distance is 200 m) and kick-off parameters (the TVD of the KOP is 500 m, and the build-up rate is 1.8°/30 m).

Node	Inclination (°)	TVD (°)	Closure distance (m)	Well depth (m)
0				
k				

(continued)

(continued)

Node	Inclination (°)	TVD (°)	Closure distance (m)	Well depth (m)
b				
t				

47. Design a build–hold–build well trajectory and fill out the following table with the given target parameters: the TVD (D_t) is 2930 m, the closure distance (C_t) is 1830 m, the length of target interval (ΔL_{td}) is 400 m, and the inclination angle (α_t) is 90°; kick-off parameters: the TVD (D_k) of the KOP is 500 m, the first build-up rate (K_z) is 3°/30 m, and the second build-up rate (K_{zz}) is 2.5°/30 m.

Node	Inclination (°)	TVD (°)	Closure distance (m)	Well depth (m)
o				
k				
b				
c				
t				
d				

48. Design a multi-target build–hold–build well trajectory and fill out the following table with the given target parameters: the TVD (D_t) is 2930 m, the length of target interval (ΔL_{td}) is 400 m, and the inclination angle (α_t) is 55°; kick-off parameters: the TVD (D_k) of the KOP is 500 m, and the build-up rate (K_z) is 3°/30 m.

Node	Inclination (°)	TVD (°)	Closure distance (m)	Well depth (m)
o				
k				
b				
t				
d				

49. A directional well is drilled to the well depth 2,000 m with the inclination 31°, the azimuth 102°. According to the requirement of wellpath control, additional drilling for 100 m is required, producing a new inclination 33° and a new azimuth 70°. Determine the tool face angle and the build-up rate of the deflection tool.

50. It is known that the current inclination is 22.5°, the azimuth is 205°, the build-up rate of the deflection tool is 4.5°/30 m, and the tool face angle is 85°. Determine the new inclination and azimuth after additional drilling for 120 m.

Well Control

6

Abstract

Balanced pressure drilling is an important technology developed in the 1960s. It plays an important role in improving rate of penetration, reducing drilling costs, protecting reservoirs and prolonging the life of wells. This chapter firstly introduces the basic concept of pressure control in oil and gas well drilling and the balance relationship of the wellbore pressure system. Then it introduces the invasion and detection methods of formation fluids and common well control equipment. After that shut-in procedure, the law of wellbore pressure change and its control method in the process of kick and well killing methods are introduced in detail. Finally, current controlled pressure drilling technologies are described.

6.1 Introduction

The implementation of well control technology must take into consideration the safety of oil and gas well drilling and completion, and also other factors such as the discovery of oil and gas reservoirs, and the protection of the formations. The main requirements of well control are different in terms of the different well types, and operation stages. For example, in the drilling phases of the exploratory wells and appraisal wells, the well control should focus on ensuring the safety of drilling process, and the effective drilling to the target zones since the staffs do not fully understand the formations. Therefore, the well control measures should take the relatively conservative strategy. As for the production wells, considering that the staffs have got rich cognition of the formation, the well control should pay more attention to the drilling safety, the effective discovery of the oil and gas formations, and reduction of the formation pollution. The well control measures should take the relatively open strategies of oil and gas well control technologies which can enhance the drilling speed, reduce the failure, and mitigate the pollution.

© China University of Petroleum Press and Springer Nature Singapore Pte Ltd. 2021
Z. Guan et al., *Theory and Technology of Drilling Engineering*,
https://doi.org/10.1007/978-981-15-9327-7_6

The general principle of conventional well control technology is to maintain the pressure balance within the wellbore to ensure the smooth progress of well drilling, well completion, and the normal production of the well. While modern well control technology has extended the concept and function of well control, which emphasize the integrity of the wellbore, focus on the cooperation of technology and equipment, coordinate the balance relationship of the wellbore pressure system and give consideration to the operation safety, discovering and protecting oil and gas formations, enhancing the drilling speed, reducing drilling failures and enhancing the comprehensive economic interest, and so on. This forms the new well control concept.

Well blowout brings enormous economic loss and painful lessons. According to incomplete statistics, during the period of 1994–2003, a total of 15 serious drilling blowout accidents occurred on the onshore in China. Among the 15 accidents, seven wells were caught on fire after the out-of-control blowout, which occupied 47% of the total wells. The blowout accident caused by the out-of-control pressure of the oil and gas well not only can bring the damage to the safety of wellbore, equipment, formation, resource and human beings but also can have bad effect on environmental and social safety.

In recent years, the states and oil companies have paid more attentions to the supervision and management of oil and gas well control issures. The awareness of well control has been continuously enhancing and the well control management has been improving day by day. The danger and harm caused by blowout have been effectively restrained. With the development of drilling and completion well toward deep wells, offshore wells and other high-risk drilling areas like deep, ultra-deep, high-pressure high-temperature, and high-sulfur well drilling have all raised higher requirements for well control. The blowout of the "Deepwater Horizon" and the sinking of platform in the Gulf of Mexico in the USA in 2010 and the Montara blowout in Timor Sea in Australia in 2009 are enough examples to show that there is still a long way to go in pressure control. Well control theory, equipment, technology, management, training, and procedures still need to be strengthened.

6.1.1   The Basic Conception of Well Control

Well control

The pressure control of oil and gas wells, or well control in short, refers to adopting certain technologies to control the pressure of wellhead and drilling fluid liquid column to maintain a certain balancing relationship with formation pore pressure, and ensure the safe and smooth implementation of drilling operation. The "certain technologies" and "certain balancing relationship" mentioned in the definition contain the following aspects:

- A certain well control equipment, including the casing string with enough strength, the wellhead blowout preventing system meeting the requirements, the piping system, and the effective blowout preventor (BOP) control system;
- Certain technologies, including reliable formation pressure prediction (monitoring) results, reasonable well plan design, and effective wellbore pressure control technology;
- A certain wellbore balancing relationship, including the selected technical programs, such as at-balanced drilling, under-balanced drilling, aerated drilling, and managed pressure drilling according to the needs of drilling;
- Certain well control management, including design, repair, and operation completed by the personnel with a certain qualification or skills.

Well invasion

The process of fluids in the formation pores (oil, gas, and water) invading into the well is called well invasion. The most common well invasions are gas, oil, and salt water.

Influx

Influx refers to that the formation fluid (oil, gas, and water) invades into the well, and the amount of drilling fluid coming out of the wellhead is greater than the drilling fluid pumped into the well, or the drilling fluid spillover from the wellhead automatically as follows stop.

Well kick

With the influx developing further, the drilling fluid flows out of the wellhead, which is called well kick.

Blowout

Blowout is that the uncontrolled formation fluid (oil, gas, and water) flows into the well, which blows from the wellhead or flows into a weak formation. It is a type of serious drilling accident and can appear as one or both of the following:

- Surface blowout, the formation fluid blows out to surface through the wellbore.
- Underground blowout, the formation fluid flows from the high-pressure formation uncontrolled into the lower pressure or weak formation.

Well control classification

- **Primary well control—drilling fluid**

Primary well control refers to the technique of balancing the formation pressure with the drilling liquid column pressure in the well during the process of normal drilling and drilling of high-pressure reservoirs.

The core of primary well control is to determine a reasonable drilling fluid density and a set of ground equipment that is compatible with the balance relation of wellbore pressure system. During primary well control, the wellbore pressure system can be maintained at a controlled, balanced relationship, and can ensure the safety of drilling and completion through drilling fluid hydraulic column pressure.

In the case of conventional drilling and completion, the primary well control requires that the hydraulic pressure of the drilling fluid must be equal to or greater than the formation pressure to ensure that the formation fluid does not invade into the wellbore when the wellhead is open. But special drilling and completion construction, such as underbalanced drilling (UBD), aerated drilling, and managed pressure drilling (MPD), is to maintain the wellbore pressure system balance relation within a specified range with the help of surface equipment, to ensure the safe and smooth implementation of drilling and completion.

- **Secondary well control—BOP stacks**

Secondary well control refers to the well control technology which can recover the wellbore pressure system to the primary well control status with the use of well control equipment after the emergence of influx or well kick.

If for some unintentional reasons, the hydraulic column pressure of the drilling fluid is less than the formation pressure, an undesirable influx occurs. Then, the well control technology and the ground equipment can be used to control the continuing of the influx and establish a new pressure balance. The key to smooth implementation of the secondary well control technology is to have a complete wellbore and to adopt a rational well control technology.

- **Tertiary well control**

The tertiary well control refers to the well control emergency measures and technologies when the well blowout is not under control due to the failure of the secondary well control.

Due to the excessive well flushing volume, the ground equipment or personnel may lose their control over the timing or ability of formation fluids to flow into the well. It can cause blowouts on the ground or underground, and even caused severe conditions such as fire and surface collapse, which require special techniques and equipment to resume wellbore pressure control system, and even need rescue, firefighting, drilling rescue wells, and other special measures to control well blowout [9].

Well control principle

The principle of well control work is focused on primary well control, improving secondary well control, and avoiding tertiary well control.

Efforts should be made to keep the well under control, and at the same time, we should be prepared for any emergency. Once the wellbore pressure is found to be beyond the control range of the design, it can respond promptly and take proper measures to restore the design-controllable-state of the wellbore pressure system as soon as possible.

The tertiary well control is the situation that is not expected to happen in oil and gas drilling operation; it is what the staffs in drilling process, management, researchers committed to avoid it.

The basic requirements of well control

The basic requirements of well control are:

- Control the formation/wellbore pressure effectively to prevent blowout.
- Assist in tackling the complicated situations of well loss, collapse, or wellbore shrinkage.
- Protect the oil and gas formation to the utmost on the condition of keeping the wellbore in a safe state.

The well control content

The well control content mainly includes:

- The prediction and monitoring of the formation pressure.
- The control of the drilling fluid density.
- Reasonable wellbore structure design.
- BOP equipment assembly and installment meeting the standard requirements.
- The control of the wellbore pressure profile.
- The prevention and management of influx, well kick, and blowout.

It should be noted that, due to the ever-changing underground conditions and the limitation of people's understanding of the objective world, uncontrolled blowout is still possible. However, it is a key part of the well control work to keep warning bells going and prevent the blowout accidents caused by ideological paralysis and lax management.

6.1.2 The Balance Relationship of Wellbore Pressure and Formation Pressure

The balance relationship of wellbore and formation pressure is affected by many factors. In addition to the pore pressure, fracture pressure, hydraulic static pressure, and anular circulating pressure, the relationship is also influenced by the movement and rotation of drilling string, the cuttings loaded in the wellbore, and the wellhead back pressure (if there is).

6.1.2.1   Fluctuating Pressure

The definition of fluctuating pressure

In-well fluctuating pressure refers to the value of the pressure at the bottom of the well that increases or decreases due to the up-or-down motion of the drilling string or fluid. It is a general term of the surging pressure and the swabbing pressure.

- **Surging pressure**

The drilling fluid in the wellbore flows upward when the drilling string moves downward, and the increased value of the bottomhole pressure affected by drilling fluid density, shearing stess, flowing speed, and so on is called surging pressure. Its calculation formula is:

$$p_{sg} = 0.00981S_gD \tag{6.1}$$

where

p_{sg}	Surging pressure, MPa;
S_g	Equivalent drilling fluid density (surging pressure coefficient), g/cm^3;
D	True vertical depth (TVD), m

Generally, S_g 0.015–0.040 g/cm^3.

- **Swabbing pressure**

The drilling fluid in the wellbore flows downwards when the drilling string moves upward, and the decreased value of the bottomhole pressure affected by drilling fluid density, shearing stress, flowing speed, and so on is called swabbing. Its calculation formula is:

$$p_{sb} = 0.00981S_bD \tag{6.2}$$

where

p_{sb}	Swabbing pressure, MPa;
S_b	Equivalent density of drilling fluid corresponding to the swabbing (swabbing pressure coefficient), g/cm^3.

Generally, S_b = 0.015–0.040 g/cm^3.

The influence of fluctuating pressure on drilling safety

The drilling fluid is of a certain viscosity and shearing force. When the drilling string is pulling out of hole, (especially when there is the hole shrinkage and bit balling, where the clearance between the bore and bit is smaller), it will result in an excessive swabbing, causing the decrease of the bottomhole pressure and thus the formation fluid invasion. While running in the hole the drilling tool too fast, it will lead to an excessive surging pressure, resulting in the increase of the bottomhole pressure and the wellbore leakage.

The main factors for fluctuating pressure

- **Shear strength of drilling fluid**
  The longer the drilling fluid be in static, the greater the strength of the molecular space lattice structure, the static shear force, the flow resistance to overcome for the drilling fluid from the static state to the flow state, and the pressure of fluctuation caused by the up-and-down movement of the drilling string will be.
- **Speed of tripping-out and tripping-in**
  When tripping out the drilling string, a negative pressure in the bottom of the drill bit will be generated and thus decrease the bottomhole pressure. When tripping in the drill string, the lower part of the drill bit will squeeze the drilling fluid to flow upside, increasing the bottomhole pressure.
- **Inertia force**
  When tripping in or tripping out the drill strings, the inertia force caused by the acceleration or deceleration of the drilling string will result in the fluctuating pressure. The greater the inertia force is, the greater the fluctuating pressure will be.

The measure of reducing the surge pressure

- Strictly control the drill string tripping speed, especially when the drill bit is near the bottomhole or in the open-hole section, it should be highly paid with attentions.
- The sudden lift or brake is forbidden when running down or out the drill string to avoid the emergence of the excessive inertial force of the drill string.
- Keep the drilling fluid in a good condition, and homogeneous. Avoid the emergence of excessive surge pressure caused by the enormous shearing force and viscosity.
- Keep the wellbore smooth. Avoid or mitigate the pressure surge caused by the hole shrinkage, bit balling, and so on.

6.1.2.2   The Pressure Increment with Cuttings Bearing in Drilling Fluid

The increased pressure value of drilling fluid containing cuttings can be expressed as:

$$\Delta p_r = 0.00981 \Delta \rho_r D \tag{6.3}$$

where

Δp_r The pressure increment with cuttings bearing in drilling fluid, MPa;
$\Delta \rho_r$ The density increment with cuttings bearing in drilling fluid, g/m^3.

6.1.2.3   The Effective Bottomhole Pressure p_{he}

Effective pressure at the bottom of the well refers to the sum of various pressures acting on the bottom of the well. The effective bottomhole pressure is different under different drilling conditions. The following calculation is based on rotary table drilling.

When the downhole drilling fluid is in static state, the bottomhole effective pressure p_{he} can be expressed as:

$$p_{he} = p_h = 0.00981 \rho_d h \tag{6.4}$$

where

p_h Drilling fluid column pressure, MPa;
ρ_d Drilling fluid density, g/cm^3;
h Vertical height of drilling fluid column, m.

Drilling fluid column pressure plays the main role in composing the bottomhole effective pressure and balancing the wellbore pressure, which is the key to the implementation of primary well control.

In the drilling process, the bottomhole effective pressure p_{he} can be expressed as:

$$p_{he} = p_h + \Delta p_a + \Delta p_r \tag{6.5}$$

where

Δp_a Annular circulating pressure drop, MPa.

The annular circulating pressure drop leads to the rise of the bottomhole pressure, which is beneficial for preventing the formation fluid from invading into the wellbore.

In the tripping-out process, the bottomhole effective pressure p_{he} can be expressed as:

$$p_{he} = p_h + \Delta p_r - p_{sb} \tag{6.6}$$

When tripping out the drilling string, the drilling fluid circulation generally stops. Therefore, we have $\Delta p_a = 0$. Considering that the drilling fluid will be circulated to sweep the rock cuttings, hence, Δp_r will decrease significantly. Besides, due to the suction effect, the effective bottomhole pressure will decrease. Exceptional attentions and cautions must be paid in the process of tripping-out. In addition, with the number of drilling strings in the well decreasing in the trip-out process, the height of fluid column in the wellbore will be dropped, resulting in the reduction of the effective bottomhole pressure. Therefore, the drilling process and the rest period after drilling are the key periods to induce the imbalance of bottomhole pressure. Special attentions should be paid to the observation of influx in the time when the drilling fluid is filled (fully filled) and the hole is empty. When the exploratory well, deep well, gas-containing well are in a long time of out of BHA state, the well head is suggested to shutoff, and casing pressure change should be overseen cautiously.

In the tripping-in process, the bottomhole effective pressure p_{he} can be expressed as:

$$p_{he} = p_h + \Delta p_r + p_{sg} \tag{6.7}$$

The surging pressure caused by the drilling string tripping in will increase the effective bottomhole pressure. If the speed of tripping in drilling string is too fast, it will generate an excessive surging pressure, which may cause the formation fracture and the drilling fluid loss. Therefore, the drilling fluid loss should be monitored.

The maximum and minimum effective bottomhole pressures

The effective bottomhole pressure reaches to the maximum during trip-in, which could be expressed as:

$$p_{he\ max} = p_h + \Delta p_r + \Delta p_a + p_{sg} \tag{6.8}$$

where

$p_{he\ max}$ the maximum effective bottomhole pressure, MPa.

When tripping out, the effective bottomhole pressure reaches to the minimum.

The effective bottomhole pressure reaches to the minimum (the influence of the drill cuttings on the well effective bottomhole pressure can be ignored since that the cuttings in the drilling fluid have been circulated clean before tripping out), which can be expressed as:

$$p_{he\ min} = p_h - p_{sb} \tag{6.9}$$

where

$p_{he\ min}$ The minimum effective bottomhole pressure, MPa.

It can be seen that the effective bottomhole pressure varies to a great extent in different operations, which is also the main reason for the occurrence of well kick and loss in the formation with narrow pressure window.

6.1.2.4   The Balance Relationship of Effective Bottomhole Pressure and Formation Pressure

The bottomhole pressure difference is the difference between the effective bottomhole pressure and the formation pore pressure, mathematically:

$$\Delta p = p_{he} - p_p \tag{6.10}$$

where

Δp Bottomhole pressure difference, MPa.

- When $p_{he} \gg p_p$, $\Delta p \gg 0$, the well bottom is in an overbalanced state.
- When p_{he} is slightly greater than p_p, and Δp is slightly greater than 0, the wellbore is in a near-balanced state, and the corresponding drilling strategy is balanced pressure drilling.
- When $p_{he} < p_p$, $\Delta p < 0$, the well is in an underbalanced state, the corresponding drilling strategy is underbalanced drilling.
- When $p_{he} > p_f$, there may be the formation fracturing;
- When $p_{he} < p_p$, the formation fluid may invade into the wellbore.

6.2   The Invasion and Detection of Formation Fluid

6.2.1   The Invasion of Formation Fluid

6.2.1.1   The Classification of Formation Fluid

The formation fluid includes oil, gas and water. Gas can be classified into natural gas, carbon dioxide, or hydrogen sulfide (H_2S); water can be classified as fresh or salt water.

As for permeable or fractured formation, the formation fluid will invade into the wellbore due to the negative pressure difference when the fluid column pressure in the wellbore is less than the formation pressure at this point. At any point in the borehole, near the borehole or at the imperfections of the wellbore casing, formation fluid may enter the wellbore as long as the annulus hydraulic static pressure is below the formation pressure. For multiple pressure series of strata in the same open hole, although the bottomhole pressure may be greater than the formation pressure, the formation fluids may still enter the wellbore if the pressure of the upper drilling fluid column is less than the formation pressure of the corresponding layer. Therefore, the annular drilling fluid column pressure must be greater than the formation pressure in the whole open hole under normal circumstances, so as to avoid formation fluid intrusion.

6.2.1.2 The Causes of Formation Fluid Invasion

Misunderstanding of the formation pressure

The problem of formation pressure prediction in new area and carbonate strata has not been well understood so far. The exact prediction of formation pressure of the old explored area is difficult to realize under the influence of oil and gas exploration, formation depletion, water injection, and so on. The uncertainty of the formation pressure may lead to the low value of designed drilling fluid density.

Formation fluid invasion

The invasion of oil, gas and water, especially for gas invasion, will result in the decrease of the drilling fluid density in the wellbore.

Violation of the regulations when filling the drilling fluid in the tripping out process

Due to the removal of drilling string in the trip-out process, the level/height of the drilling fluid in the wellbore decreases.

The well loss leads to the lowering of fluid level

The untimely detection of the drilling fluid loss in the well may result in the decrease of effective pressure in the bottom hole.

The suction in trip-out process

Suction caused by tripping-out may lead to the decrement of effective bottomhole pressure.

Circulation stop

When the drilling fluid is in circulation, the effective bottomhole pressure is relative greater due to the annular circulating pressure. When the circulation stops, the annular circulating pressure disappears, and the effective bottomhole pressure decreases.

6.2.1.3 The Formation Fluid Invasion Volume in Different Status

The influence of the difference between bottomhole pressure and formation pressure on the formation fluid invasion

The greater the negative pressure difference between the bottomhole pressure and the formation pressure is, the larger the invasion volume of the formation fluid under the same condition is.

The influence of drilling fluid circulation on formation fluid invasion

When the drilling fluid is in circulation, the effective bottomhole pressure is relatively high and the formation fluid invasion volume is relatively low. The invading fluid is rapidly carried by the circulating drilling fluid moving upward, which is not easy to generate aggregation; when the circulation stops, the effective bottomhole pressure is relatively low, the invasion volume is relatively high, and the invading fluid will produce aggregation.

The influence of shut-in on the formation fluid invasion

In the initial stage of shut-in, with the invasion of formation fluid and the gas slippage, the well head back pressure increases and the effective bottomhole pressure increases as well, and the formation fluid invasion volume will decrease. If the well is shut-in for a long time, the effective bottomhole pressure will increase due to the gas slippage and expansion, which may be greater than the fracture pressure of the weak formation, resulting in the loss of drilling fluid and the continued invasion of formation fluids, resulting in a serious imbalance between borehole pressure and formation pressure.

The influence of different types of strata on formation fluid invasion

As for dense formation, the formation fluid invasion rate is slow, and the invasion volume in the unit time is relatively fewer. The corresponding invasion volume is relatively small in formation of low pressure and permeability, and that is the opposite in the formation of high pressure and permeability. The invasion volume reaches to the maximum in the formation and cave of high pressure and developed fractures.

6.2.2   Gas Invasion

6.2.2.1   Kinds of Gas Invasion

There are three main means of gas invasion:

- The gas in the rock pores enters the wellbore with the drilling cuttings (background gas).
- The gas in the gas layer diffuses into the well through the mud cake due to the concentration difference.
- When the bottomhole pressure is less than the gas formation pressure, the gas permeates or flows into the well through formation pores or cracks.

6.2.2.2   Characteristics of Gas Invasion

Characteristics of gas invasion are:

- The gas density is far less than that of the drilling fluid; occupation of the same volume of the drilling fluid by the gas will reduce the bottomhole pressure.
- The gas is in a compression state when invading into the drilling fluid, resulting in the limited reduction of the drilling fluid pressure.
- The volume expands when the gas moves up from wellbore, the closer it is to the surface, the faster the expansion is and the greater it affects the bottomhole pressure.
- The high-pressure and high-yield gas formation will cause a large amount of invasion gas accumulation at the bottomhole to form gas column.
- Invasive gas can also be accumulated at the bottomhole to form gas column when stopping the circulation for a long time.
- Under the high pressure of the drilling fluid column, the volume of invasion gas at the bottomhole is limited.

6.2.2.3   Hazards of Gas Invasion

Hazards of gas invasion are as follows:

- Under the condition of not-shut-in, the gas column slips off due to the density difference or rises with the drilling fluid circulation. The pressure of the upper drilling fluid column gradually decreases, and the expansion of gas volume accelerates, and the closer it is to the wellhead, the faster the expansion is. The higher the volume of the drilling fluid replaced by the expanded gas is, the lower the equivalent density of the drilling fluid is, and the more the effective pressure at the bottomhole reduces. Reduction of the effective pressure at the bottomhole aggravates the imbalance of the wellbore pressure system, resulting in more gas flowing into the well faster, thus leading to blowout.
- When the well is shut-in for a long time, the gas holding formation pressure will slip off and rise. This not only increases the bottomhole pressure but it also increases the wellhead pressure, which results in the weak stratum rupture, wellhead overpressure, overcome internal resistance pressure of the casing, causing well leakage or underground blowout.
- High-pressure, high-yield formations, especially gas wells with high sulfide, the gas invasion is fiercer and develops faster, causing higher risks if not properly controlled.

6.2.2.4   Pressure Changes After the Gas Invasion

Calculation data: Hole depth 3000 m, borehole diameter $8^1/_2$ in, drill pipe size $4^1/_2$ in, drilling fluid density 1.2 g/cm^3.

Under conditions of no shut-in

It can be found from Fig. 6.1, with the slippage and rising of the gas, the pressure of the upper drilling fluid column gradually decreases, and the gas volume expands, and the wellbore pressure occupied by the gas increases, and the pressure of the drilling fluid column on the bottomhole is lower. In this example, the amount of invasion gas at the bottomhole is 0.26 m^3, the height of the gas column at the bottomhole is only 10 m. When the gas reaches to the wellhead, the gas volume is expanded to 8.35 m^3 and the gas column is 320 m, resulting in the reduction value of the bottomhole pressure of 3.8 MPa.

Under vonditions of shut-in

If well was shut-in after gas invasion, the gas will slips up, but cannot expand because the wellhead is closed. The gas pressure will almost keep the same as the bottomhole pressure (should be "formation Pressure"). If the shut-in period is too long, the gas will eventually migrate to the wellhead. The gas pressure will be superimposed on the pressure of the drilling fluid column, which forms an excessively high bottomhole pressure acting on the entire wellbore. This can easily fracture weak formation, exceed pressure rating of the well control equipment and

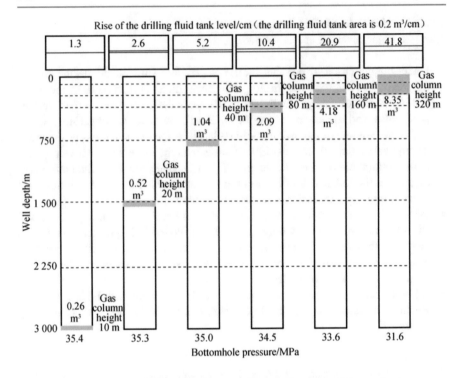

Fig. 6.1 Gas expansion and bottomhole pressure changes when the well is not shut-in

the internal pressure resistance strength of the casing (or casing burst pressure, causing) resulting in well leakage and blowout.

It can be seen from Fig. 6.2 that, during gas slippage and rising, the gas column pressure fails to be balanced due to the reduction of height of the upper drilling fluid, which results in the gradual increment of the wellhead pressure. When the gas slips to the wellhead, the wellhead pressure is equal to the original bottom-hole pressure. At this point, the entire wellbore will be increased by 35.4 MPa, and the bottomhole pressure will reach to 70.8 MPa. Therefore, in view of the bearing capacity of the wellhead, casing and bottomhole, the wellbore overpressure due to prolonged shut-in must be avoided.

6.2.3   Hazards of the H_2S Gas

6.2.3.1   Physico-chemical Properties of H_2S

H_2S is a colorless, virulent, strongly acidic gas with a relative density of 1.176. The H_2S with a low concentration has a smell of rotten eggs. It has a flashpoint of 250 ° C and shows a blue flame when burning, producing toxic sulfur dioxide. H_2S mixed with air can form an explosive mixture when the volume fraction reaches 4.3–46%.

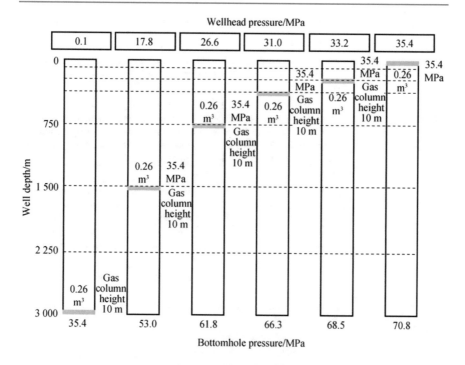

Fig. 6.2 Changes of the wellhead pressure and bottomhole pressure under the condition of shut-in

H$_2$S is 5–6 times more toxic than carbon monoxide, almost as toxic as cyanide which is a deadly gas. Under normal conditions, the critical concentration of H$_2$S for human safety is 14 PPM (Table 6.1).

6.2.3.2   Distribution of the H$_2$S Gas

H$_2$S mostly exists in carbonate rocks, especially in the sulfate sedimentary environment associated with carbonate rocks. The region with the highest content of H$_2$S gas is the South Texas gas field in the USA; H$_2$S gas has appeared in the carbonate gas reservoir of the Kongdian Formation in the Paleogene of the Zhaolanzhuang Gas Field in the Jizhong Depression of the Huabei Oil field in China, the Triassic Jialingjiang Limestone Gas Reservoir in the Wolonghe Gas Field in Sichuan Province the gas wells, the Ka-10 well, the Xi-4 well, the Dongwan-1 well and the An-4 well on the southern margin of the Tarim Basin and Xinjiang Karamay Oil fields. H$_2$S gas also appeared in the development of oil fields such as Hongshanzui, Baqu, heavy oil areas and other areas.

Table 6.1 Comparison of the hazard degree to human under different H_2S concentrations

H_2S concentration/PPM	Hazard degree
0.13–4.6	Can smell rotted egg flavour, do not produce harm to human body commonly
4.6–10	There is a heat sensation at beginning contact, but disappears soon
10–20	The safety critical concentration value. In this concentration, people can stay for eight hours under open conditions. Otherwise, people should wear a SCBA (Positive pressure, self-contained breathing apparatus) or positive pressure, supplied-air (airline) respirator
50	Only 10 min contact is permitted
100	After 3–10 min of contact, you may experience itching in your throat and cough, followed by damage to olfactory nerves and eyes, mild headache, nausea, a rapid pulse, and death for more than four hours
200	It can immediately destroy the olfactory system; the eyes and the throat have a sense of burning and will be burned. It will cause death when people do not leave immediately
500	Loss of consciousness, dyspnea, respiratory arrest within 2–15, if the rescue is not timely, will lead to death
700	Soon lose consciousness, stop breathing, if not immediately resuscitated, die immediately
1000	Loss of consciousness immediately, resulting in death or permanent brain damage or vegetative state
2000	Take a puff and die instantly, beyond rescue

6.2.4   Causes and Indicators of Kick

What needs to be explained is that, with the development of drilling technology, under-balanced drilling, aerated drilling, controlled pressure drilling and other drilling techniques are applied to reduce the damage to the formations, improve rate of penetration, reduce the occurrence of well leakage, and discover oil and gas earlier. When these drilling techniques are used, the drilling fluid density is often artificially reduced, making the drilling fluid column pressure slightly lower than the formation pressure, resulting in formations fluid entering the wellbore and causing a kick. This kind of kick is induced subjectively intentional under these circumstances. Unless otherwise stated, the kick discussed below is a non-subjectively intentional kick.

6.2.4.1   Cause of Kick

As mentioned in the drilling engineering, kick refers to the fact that the formation pressure is higher than the bottomhole pressure due to operation or strata, causing the formation fluid to enter the well and kick unexpected to the construction workers. Therefore, this kind of kick is induced by a non-subjective intention. Formation fluid may enter the wellbore at the time of kick, causing the drilling fluid to influx at the wellhead. At the same time, kick may be formed when it is discovered too late, and resulting in shutting the well in under pressure. In the normal drilling or tripping operations, the kick may occur under the following conditions:

- Hydrostatic pressure of in the wellbore is less than the formation pressure.
- The formation of kick occurrence has certain permeability, allowing the fluid to flow into the well.
- The formation contains a certain volume of fluid.

Although the formation pressure and formation permeability cannot be controlled, it is possible to maintain a proper drilling fluid column pressure. Drop of the drilling fluid column pressure due to any reason can result in invasion of the formation fluid into the well, and the most common causes are:

- During the tripping out, height of the liquid column in the wellbore drops due to not filling up properly.
- Negative pressure difference due to pulling out and suction.
- Height of liquid column decreased due to mud loss.
- Negative pressure difference due to the low density of the drilling fluid or cementation slurry fluid.
- Negative pressure difference due to the abnormal pressure over the effective pressure at the bottomhole.
- Too fast trip-in results in surging pressure and formation leakage, consequently the height of liquid column in the wellbore reduced.
- After the cement operation, due to the weight loss and channeling of the cement paste, gas invasion could be formed outside the casing, causing kick outside the casing.

6.2.4.2 Primary Indicators of a Kick

In all kinds of drilling operation, when gas, oil, or water invasion occurs, it will displace the drilling fluid out of the hole. Various displays of drilling fluid flow that can be found on the ground from the well are referred to as kick. By observing and analyzing these kicks, it is possible to judge whether the drilling fluid in the wellbore has been changed. Timely detection of kick and rapid control of the wellhead with correct operation are the keys to prevent blowout.

- Indicators of a kick while tripping in: Under normal circumstances, when running the drill string into the wellbore, the drilling fluid with a volume equivalent to the drilling string will be displaced. If the volume of the returned drilling fluid is more than the volume of the drilling string displaced then a certain amount of formation fluid has invaded into the well. If the drill strings are stopped to run in, and there is still drilling fluid flow at the wellhead, it means that there is kick. It is the most intuitive and effective way to stop going down and check whether there is fluid outflow in the wellhead.

- Indicators of a kick while drilling ahead:
  - Changed flow rate at the surface return;
  - Returned amount of the drilling fluid is over the pumped amount.
  - Level of the drilling fluid tank rises.
  - Influx after the pump stop.

- Changed in drilling fluid properties, such as gas measurement value, density, chloride ion concentration, resistivity, conductivity, temperature, and so on. Indicators of a kick during a tripping out: the volume of drilling fluid fillup is less than the volume of the drilling string displacement (depending on pulling dry or wet); and when stopping pulling, there is still influx at the wellhead.
- Indicators of a kick when BHA out of hole: There is influx of drilling fluid at the wellhead, and fluid level in a mud tank rises.

6.2.5   Detection of the Formation Fluid Invasion

6.2.5.1   Volume Detection Method for the Drilling Fluid

After invasion of formation fluids into the wellbore, the total volume of drilling fluid increases, and the level of the drilling fluid tank rises. For oil and water invasion, if we ignore the compressibility of the liquid in the annulus, the volume increment in the tanks is equal to that of the oil and water invasion into the wellbore. But for the gas invasion, due to its compressibility, the drilling fluid in the tanks will changed only on clear expansion of gas when it rises to the some part of the wellbore. Currently, the main instruments for monitoring the liquid level are float-level gauges (Fig. 6.3) and ultrasonic-level sensors. These sensors are installed in each drilling fluid tank to monitor its level, thus the cumulative volume changes of the drilling fluid can be obtained.

Fig. 6.3 Float-level gauge

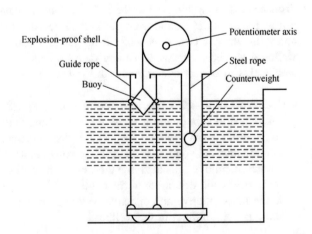

6.2.5.2   Flow Detection Method for Returning Drilling Fluid

Flow detection method for returning drilling fluid is also one of the means to detect the kick. The commonly used means of detection is the target-type flowmeter (Fig. 6.4), but also electromagnetic flowmeter and mass flowmeter detection are used.

6.2.5.3   Acoustic Wave Method of Kick Detection

At normal temperature and pressure, the sound wave propagation speed in water is about 1500 m/s, and about 340 m/s in air, but in two phases flow formed by air and water, could be as low as 10–100 m/s. Therefore, by using the appropriate detection equipment and tools, gas invasion along the wellbore can be detected.

6.2.5.4   Acoustic Wave Method of Total Loss Detection

If there is a total loss and the mud level cannot be seen at wellhead. Acoustic wave signals can be sent downwards along the wellbore, this signal will be reflected on the gas-liquid interface in the wellbore, and the reflected wave will reach the receiving equipment upwards along the wellbore. By measuring the time difference between the transmitting time of the acoustic wave and the receiving time, the drilling fluid in the wellbore (from surface/wellhead) can be calculated. This principle can be used to detect top level of mud if there is a total loss.

6.2.5.5   Other Detections

Mud logging techniques provide a set of means of detecting formation fluid invasion. These include

- The pump pressure or the hook load increases or decreases: When drilling into the high-pressure layer, the bottomhole pressure suddenly increases which causes the suspension weight to decrease and the pump pressure to rise; after the formation fluid invades into the drilling fluid, the drilling fluid density decreases, and the buoyancy decreases, and the suspended weight increases, pump pressure decreases.
- Temperature of drilling fluid in and out: Temperature of the abnormal formation is generally higher, so the temperature of the returned drilling fluid at the wellhead is higher when drilling into formation with abnormal high pressure.

Fig. 6.4 Target-type flowmeter

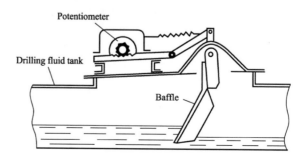

- Density of drilling fluid in and out: Invasion of oil, gas, and water from the formation into the well will lower the density of the drilling fluid.
- Conductivity (resistivity) of drilling fluid in and out: Invasion of oil, gas, and water from the formation into the well will change the physico-chemical properties of the drilling fluid. For example, invasion of formation brine into the drilling fluid will reduce the resistivity of the drilling fluid, and increase the conductivity of the drilling fluid.
- Viscosity of drilling fluid in and out: Invasion of oil, gas, and water from the formation into the well will have chemical reaction with the drilling fluid, and consequently viscosity of the drilling fluid changes.
- Changes in gas measurement values: When drilling into a hydrocarbon or gas stratum containing H_2S, some hydrocarbon or hydrogen sulfide will be dissolved or carried into the returned drilling fluid. The changes of hydrocarbon components and total hydrocarbons in the drilling fluid can be detected by a chromatograph. H_2S monitors can also be applied to detect whether H_2S is contained in the drilling fluid.

6.3 Well Control Equipment

6.3.1 Functions of Well Control Equipment

In drilling operation, the pressure system balance of oil and gas wells is often changed due to various factors, and a kick occurs. In this case, it is necessary to rely on the well control equipment to implement the shut-in operation and restore the control on the pressure system of the oil and gas wells. The well control equipment is a whole set of special equipment, instruments, and tools for implementation of the pressure control technology for oil and gas wells. The well control equipment has the following functions:

- Blowout prevention. Maintaining the sum of liquid column pressure and wellhead pressure in the wellbore to be always slightly over the formation pressure to prevent blowout conditions from forming.
- Discovering a kick in time. The implementation of monitoring the wellbore pressure balance, to timely detect the signs of a kick.
- Quick blowout control. Once a kick is detected, shut-in will be implemented quickly to stop further influx and blowout situations, to provide equipment for implementation of the killing operation and restore the pressure balance of the wellbore system.
- Handling complicated situations. In the case of uncontrolled blowout, the implementation of killing, firefighting, rescue, and other operations is initiated.

6.3.2 Basic Components of Well Control Equipment

Basic well control equipment is as follows:

6.3.2.1 Wellhead BOP (Blowout Preventor) Stacks

The wellhead BOP stacks mainly includes annular, single pipe ram, double rams, shear ram, casing head, drilling spoon, transition flange, HCR, and so on.

6.3.2.2 Hydraulic BOP Control System

Hydraulic BOP control system mainly includes the driller console, remote console, auxiliary remote console, and so on.

6.3.2.3 Well Control Manifolds

Well control manifolds mainly include choke manifold and hydraulic choke control box, relief pipeline, killing manifold, water injection pipeline, fire-fighting pipeline, reverse circulation pipeline, and so on.

6.3.2.4 Inside BOP Tools

Inside BOP tools (inside a drill string) mainly include check valve, upper kelly cock, lower kelly cock, drop-in check valve, bypass valve, and so on.

6.3.2.5 Well Control Instrumentation

Well control instrumentations include alarm instruments for monitoring the return temperature, density, drilling fluid return flow, drilling fluid tank liquid level, borehole liquid level on pulling out, pump stroke, and other parameters such as hydrogen sulfide and carbon dioxide monitoring alarm and similar ones.

6.3.2.6 Drilling Fluid Treatment Equipment

It mainly includes drilling fluid weighting equipment, Mud Gas Separator (MGS or Poor boy degass), vacuum degasser, and fill up/tripping pump on pulling-out, and so on.

6.3.2.7 Special Operation Equipment

It mainly includes snabbing unit (equipment for the pipelines without killing well), rotary BOP, firefighting equipment, disassembly and assembly equipment and tools of the wellhead, wellhead ventilation equipment, anti-poison gas equipment, and others.

A typical well control package is shown in Fig. 6.5

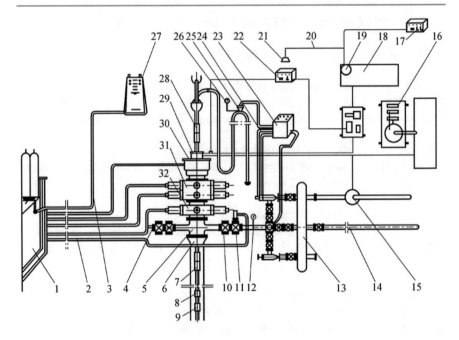

Fig. 6.5 A diagram for supporting well control equipment. 1—BOP remote console; 2—BOP hydraulic pipeline; 3—BOP air tube; 4—Kill manifold; 5—Drilling spoon; 6—Casing head; 7—Lower kelly cock; 8—Bypass valve; 9—Check valve for the drilling tools; 10—Manual gate valve; 11—Hydraulic gate valve; 12—Casing pressure gauge; 13—Choke manifold; 14—Relief pipeline; 15—Liquid–gas separator for the drilling fluid; 16—Vacuum degasser; 17—Drilling liquid tank-level monitor; 18—Drilling fluid tank; 19—Drilling fluid tank-level monitoring sensor; 20—Automatic drilling fluid charging device; 21—Drilling fluid tank-level alarm; 22—Alarm box for the automatic charging device; 23—Hydraulic choke valve control box; 24—Choke manifold control line; 25—Pressure transmitter; 26—Standpipe pressure gauge; 27—BOP driller console; 28—Upper kelly cock; 29—Anti-influx pipe; 30—Annular BOP; 31—Double ram BOP; 32—Single ram BOP

6.3.3   Common BOP Combination

6.3.3.1   BOP Combination with 14 MPa Pressure Rating

Generally according to the characteristics of the region, choose the combinations as follows:

- Single-ram BOP + drilling spoon;
- Double-ram BOP + drilling spoon;
- Blind-ram BOP + pipe-ram BOP + drilling spoon;
- Annular BOP + pipe-ram BOP + drilling spoon;
- Blind ram BOP + drilling spoon + pipe ram BOP.

6.3.3.2 BOP Combination with 21 MPa Pressure Rating

Generally according to the characteristics of the region, to choose from the following combinations:

- Annular BOP + double-ram BOP + drilling spoon;
- Single-ram BOP + drilling spoon;
- Double-ram BOP + drilling spoon.

6.3.3.3 BOP Combination with 35 MPa Pressure Rating

Generally according to the characteristics of the region, choose the combinations as follows:

- Annular BOP + pipe-ram BOP + drilling spoon;
- Annular BOP + double-ram BOP + drilling spoon;
- Double-ram BOP + drilling spoon;
- Annular BOP + ram-type BOP + drilling spoon + ram-type BOP.

6.3.3.4 BOP Combination with 70 and 105 MPa Pressure Rating

Generally according to the characteristics of the region, choose the combinations as follows:

- Annular BOP + single-ram BOP + double-ram BOP + drilling spoon;
- Annular BOP + double-ram BOP + drilling spoon + single-ram BOP;
- Annular BOP + double-ram BOP + double-ram BOP + drilling spoon, with the combination shown in Fig. 6.6.

If there is a special need, based on the above combination, shear RAM BOP, and rotary control head can be added. On selection of a higher grade well control equipment, for the combination of BOP, the combination of the previous pressure rating should be selected. On use of a composite drill, ram-type BOPs with a corresponding quantity should be provided in full, and ram cores of the corresponding sizes should be provided. Location for mounting of the pipe-ram BOP should ensure to seal the corresponding drilling rod body when it is closed. Generally, the pipe-ram core with a higher use percentage is mounted below, and the blind-ram core is installed on the top of the ram BOP. When it is needed to mount a shear BOP, the shear BOP should be mounted at the location for the blind ram of the previous combination, with the function of sealing an empty well.

6.3.4 Well Control Manifold

Well control manifolds include relief manifolds, choke manifolds and killing manifolds. Commonly used well control manifolds include double four-way

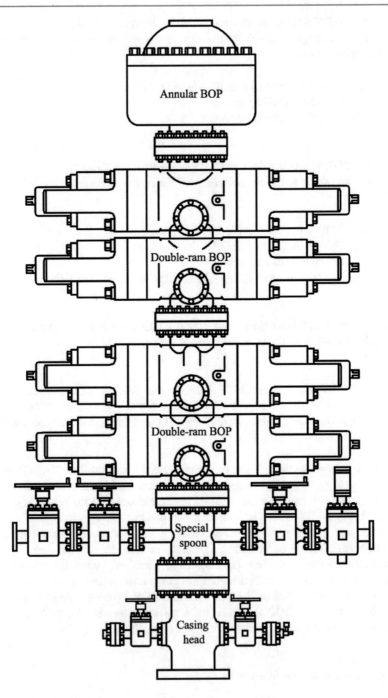

Fig. 6.6 BOP combination for the 70 and 105 MPa pressure rating

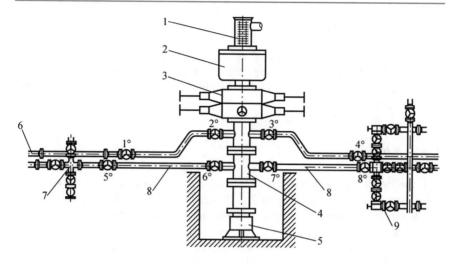

Fig. 6.7 Double four-way wellhead well control manifold. 1—Anti-influx pipe; 2—Annular BOP; 3—Ram BOP; 4—Drilling spoon; 5—Casing head; 6—Relief pipeline; 7—Killing manifold; 8—Blowout manifold; 9—Choke manifold

wellhead well control manifold (Fig. 6.7) and single four-way wellhead well control manifold (Fig. 6.8).

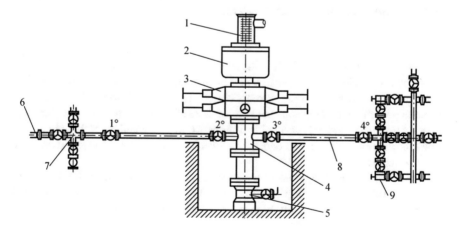

Fig. 6.8 Single four-way wellhead well control manifold. 1—Anti-influx pipe; 2—Annular BOP; 3—Ram BOP; 4—Drilling spoon; 5—Casing head; 6—Relief pipeline; 7—Killing manifold; 8—Blowout manifold; 9—Choke manifold

6.4   Well Shut-in

The key to controlling formation fluid invasion into the well is to prevent the formation fluid from flowing into the wellbore continuously. The most direct way to prevent the formation fluid from flowing into the wellbore is to shut-in the wellhead through the BOP equipment.

6.4.1   Shut-in

When an influx appears, the well should be shut-in instantly and the shut-in standpipe and casing pressures should be recorded. The earlier and faster the shut-in is, the less the formation fluid invasion volume is. Thus, the imbalance extent is lower, and it is easier to recover the pressure balance. The well control standards set the requirements of well kick detection and shut down procedures. When the volume of influx reaches 1 m^3, kick should be discovered. When the volume of influx reaches 2 m^3, the well should be shut-in. Moreover, the well should be shut-in immediately if there seems a well kick for exploratory well, high-pressure well, gas well and sour well, to ensure the wellbore and wellhead safety. As for familiar formations and wells with low pressure and low production, considering the issue of tool sticking and well wall collapsing, the drill bit can be uplifted to the casing shoe before shutting in the well under the condition of sufficient control of the situation. The rule is: ground safety is superior to the underground safety; and the staff safety is superior to the equipment safety.

6.4.1.1   Shut-in Method and Selection

Hard shut-in

Hard shut-in is to close the BOP with choke valve closed when influx is detected. The method shut in the well quickly, and the formation fluid invasion volume is the least, but when the blowout preventor is closed, the annulus fluid suddenly becomes stationary from the flowing state, and the wellhead device will have a water hammer effect. Strong water shock waves have influence on the annulus, casing shoes and open formations, which could damage the wellhead device, and cause the pressure leakage formation.

Hard shut-in can be applied to the following cases: Wellhead kicking speed is not high; salt water invasion affects the wellbore stability seriously; and when wellhead device is of relatively high-pressure withstanding capability.

Soft shut-in

Soft shut-in refers to that open the choke valve firstly after the detection of influx, and shutoff the blowout preventor secondly, and then closes the choke valve. This method takes a longer time but has the less impact on the wellhead device and the annulus. When the formation fluid invasion is fast, the invasion amount is large which may bring difficulties to the subsequent well control operations.

Soft shut-in can be applied to the following cases: when the wellhead kicking volume is relatively high and when wellhead device is of low-pressure bearing range.

Half-soft shut-in

Half-soft shut-in refers to open the choke valve with 50% opening extent first, and then close the blowout preventor, thereafter close the choke valve. The characteristic of the method is between hard shut-in and soft shut-in.

6.4.1.2 Shut-in Steps (Set Hard Shut-in as Example)

Kick during drilling

- Driller: Send an alarm signal, stop drilling, and pumping.
- Assistant driller: Go to the remote control station and get ready to receive the control command, to ensure the control system working normally.
- Driller: Uplift the kelly/tool joint to the position 500 mm away from the rotary stable surface.
- Driller: Open the hydraulic BOP valve.
- Driller: Shutoff the half-seal ram-type preventor fitting with the sizes of well-drilling tools.
- Driller, back-tong man, lead-tong man: Seat the drilling tool on the elevator slowly.
- Driller: Report to the drilling supervisor, drilling platform manager, or drilling engineer.
- Back-tong man: Record the standpipe pressure and casing pressure every 2 min.
- Derrick hand: Assist and reward the driller to take these steps correctly.

Kick occurs in trip-in or trip-out of the drilling pipes

- Driller: Send the alarm signal, and lower the drill collar to sit in the rotary stable.
- Assistant driller, back-tong man, lead-tong man: Grab the arrow check valve.
- Assistant driller: Go to the remote control station and get ready to receive the control commands, to ensure the control system working normally.
- Driller: Uplift the drilling tool for 100 mm away.
- Driller: Shutoff the half-seal ram-type preventor fitting with the well-drilling tools sizes.
- Driller: Seat the safety slip on the rotary stable surface slowly.
- Driller, back-tong man, lead-tong man: Connect the kelly.
- Driller: Report to the drilling supervisor, drilling platform manager, or drilling engineer.
- Back-tong man: Record the standpipe pressure and casing pressure every two minutes.
- Derrick hand: Assist and remind the driller to complete the steps correctly.

Kick occurs in trip-in or trip-out of the drill collar

- Driller: Send the alarm signal, and lower the drill collar to sit in the rotary table.
- Assistant driller, back-tong man, lead-tong man: Tighten the safety slip.
- Assistant driller, back-tong man, lead-tong man: Grab the blowout preventing riser (or single blowout preventing pipe).
- Assistant driller: Go to the remote control station and get ready to receive the control commands, to ensure the control system working normally.
- Driller: Lower the drilling tool to the position 100 mm away from the rotary table surface.
- Driller: Open the hydraulic blowout preventing valve.
- Driller: Shutoff the half-seal ram-type preventor fitting with the well-drilling tools sizes.
- Driller: Seat the drilling tool in the elevator slowly.
- Driller, back-tong man, lead-tong man: Connect the kelly.
- Driller: Report to the drilling supervisor, drilling platform manager or drilling engineer.
- Back-tong man: Record the standpipe pressure and casing pressure every 2 min.
- Derrick hand: Assist and remind the driller to complete the steps correctly.

Kick of empty well (BHA out of hole)

- Driller: Send the influx alarm signal.
- Assistant driller: Go to the remote control station and get ready to receive the control commands, to ensure the control system working normally.
- Driller: Open the hydraulic valve (or HCR).
- Driller: Shutoff the total-seal ram-type preventor fitting with the well-drilling tools sizes. Close blind RAM or shear RAM.
- Driller: Report to the drilling supervisor, drilling platform manager or drilling engineer.
- Back-tong man: Record the standpipe pressure and casing pressure every two minutes.
- Derrick hand: Assist and remind the driller to complete the steps correctly [5].

6.4.1.3 The Cautions During Well Shut-in

- Shutoff the annular firstly and the Ram-Type secondly when shutting off the BOP.
- The speed of shutting off the choke valve should not be too fast, and the casing pressure should not exceed the permissible value.
- When shutting off the Ram-Type blowout preventor, the drilling string should be in hanging state rather than seating on the rotary table, in case of that the drilling tool is not aligned to seal the well.

- If it is indeed necessary to run drilling string into the well, it is not possible to run in the open well as the drilling fluid will be spilling up. Instead, the upper and lower Ram blowout preventors should be used alternately to run in forcibly.
- The choke valve should not be shutoff totally when the shut-in casing pressure is greater than the allowable shut-in casing pressure, and the throttling circulation should be ensured under the condition of high-casing pressure [1].

6.4.2   The Calculation of Drilling Fluid Density for Well Killing

After shutting the well in, if the wellbore is intact, that is, the wellbore blowout preventor is in good condition and the casing internal pressure strength meets the requirements of well killing and the open-hole section is not lost, then the entire wellbore can form a closed space. Due to the formation fluid invasion or gas migration, the wellbore pressure and formation pressure will form a temporary balance after shutting in the well. Since the bottomhole pressure is less than the formation pressure, to re-establish the wellbore pressure system balance, it is necessary to increase the pressure of the drilling fluid column. Under conventional drilling methods, formation fluids that have entered the wellbore and low-density drilling fluids contaminated by formation fluids are typically circulated out of the well, usually by replenishing an appropriately weighted drilling fluid.

The density of heavyweight drilling fluid should be determined before configuring the heavyweight drilling fluid. It is necessary to know the formation pressure to determine the density of heavyweight drilling fluid, and we need to know the principle of U-tube effect and the recording method of the standpipe pressure during well shut-in.

6.4.2.1   The Mechanism of U-Tube Effect

Considering the drill string, annular space and the formation are connected, and the liquid pressure among them is transferred. Therefore, there is a certain balance among shut-in standpipe pressure, shut-in casing pressure, formation pressure, annular liquid column pressure, and the liquid pressure in the drill string. Because the size of the bit nozzle is very small, it is generally assumed that formation fluid does not enter the drill string during shut-in. With the invasion of formation fluid, the volume of fluid in annulus increases, the shut-in casing pressure increases resulting the increment of the bottomhole pressure and standpipe pressure. Thus, the pressure difference between bottomhole pressure and formation pressure decreases. Finally, the bottomhole pressure and formation pressure balance and formation fluid stops invasion.

The pressure balance relationship under the static condition

According to the working mechanism of U-tube effect, as shown in Fig. 6.9, the pressure balance relationship under the static condition (no circulation) can be explained as:

Fig. 6.9 The U-tube effect
mechanism in the shut-in
period

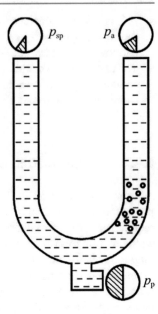

$$p_a + p_{ha} = p_p = p_{sp} + p_{hi} \tag{6.11}$$

where

p_a Shut-in casing pressure, MPa;

p_{ha} Hydrostatic pressure of the invaded drilling fluid in the annular space, MPa;

p_p Formation pressure, MPa;

p_{sp} Shut-in standpipe pressure, MPa;

p_{hi} The drilling fluid static hydraulic column pressure, MPa.

According to the assumption that the invading drilling fluid does not enter the drill string after the shut-in, the drilling fluid density in the drill string is the drilling fluid density used in the drilling operation, that is, the original drilling fluid density. The standpipe pressure and casing pressure should be read correctly, and the formation pressure can be calculated from Eq. 6.11, we can calculate the formation pressure, and figure out the new drilling fluid density needed to balance the formation pressure.

The pressure balance relationship under the throttling circulation condition

According to the U-tube effect mechanism, as shown in Fig. 6.10, the pressure balance relationship under the throttling circulation condition can be expressed as:

$$p_T + p_{hi} - \Delta p_{ld} = p_p + \Delta p_{la} = p_a + p_{ha} + \Delta p_{la} \tag{6.12}$$

Fig. 6.10 U-tube effect
mechanism under the
situation of throttling
circulation

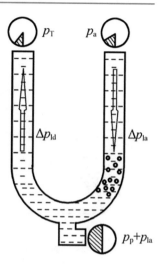

$$p_T = p_p - p_{hi} + \Delta p_{la} + \Delta p_{ld}$$
$$= p_{sp} + \Delta p_{la} + \Delta p_{ld} = p_{sp} + \Delta p_c \qquad (6.13)$$

where

p_T the standpipe pressure during circulation, MPa,

Δp_{ld} the circulation pressure loss in the drill pipe and drill bit, MPa;

Δp_{la} the circulation pressure loss in the annular space, MPa;

Δp_c the total pressure loss in the circulation, MPa.

Considering that the density of the drilling fluid in the annular space has changed due to the invasion of the formation fluid, and then the formation pressure cannot be calculated through the shut-in casing pressure. Apparently, the right reading and recording of the casing pressure are helpful in judging the imbalance status in the bottomhole and the characteristics of the formation fluid·

6.4.2.2 Method of Reading Standpipe Pressure and Casing Pressure of Shut-in Well Without Inside BOP Tool Installed Inside the Drill String

As for formations with good permeability, the balance between the formation and the wellbore can be established after 10–15 min of shutdown, but for the tight formation, it could takes a longer time to establish the balance (Fig. 6.11).

Oil and water invasion

Generally, the density of oil or water is lower than that of the drilling fluid. Though oil and water have the ascending motion to the drilling fluid, it can be ignored. Therefore, there is no trapped pressure caused by slippage expansion. When the shut-in pressure build-up is stable, the shut-in value can be obtained.

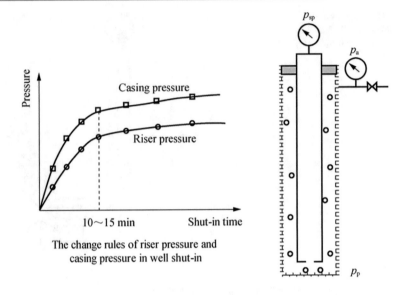

Fig. 6.11 Change of standpipe pressure and casing pressure after well shut-in

Gas invasion

As for gas invasion (oil–gas invasion, water–gas invasion), the drilling fluid contained gas exists in the annular space. When the standpipe pressure and casing pressure rise to certain values, the bottomhole pressure equals to the formation pressure, and the formation fluid no longer enters the wellbore. Theoretically, if the standpipe pressure is recorded at this time, then the formation pressure is relatively exact.

If the shut-in time is delayed, the gas in the wellbore will slip, rise and expand, leading to the rise of the wellhead pressure. Then the recorded standpipe pressure and casing pressure will greater than the actual pressure required to balance the formation, resulting in the formation pressure obtained from this will also be higher than the actual one.

The release of the trapped pressure

Trapped pressure refers to the value of recorded shut-in standpipe pressure or casing pressure which exceeds the needed value of shut-in standpipe pressure or casing pressure for balancing the formation. Trapped pressure is usually resulted by the slipping upward gas in the annular space and the restrictive wellbore space the late discovery of well influx, late shut-in, or the late record of the shut-in pressure after well shut-in. It limits the expansion of gas. If the well shut-in is earlier than the pump shut off during circulation process, then a certain volume of drilling fluid will be injected into the well, which will also lead to the pressure building in the wellbore and form the trapped pressure. `The trapped pressure will lead to the higher recorded values of shut-in standpipe pressure and casing pressure, and then the calculated drilling fluid density of well killing become higher. This may lead to the excessive well control, or even the formation leakage, and therefore the trapped pressure must be released firstly.

The steps of releasing the trapped pressure are as follows:

- Open the choke valve slowly to release a little amount of drilling fluid (40–80 L), and then close the valve.
- Observe the change of the pressures, if both the standpipe pressure and casing pressure decline, indicating trapped pressure, and then repeat above steps.
- When the standpipe pressure no longer declines, the shut-in standpipe pressure and casing pressure could be recorded.

If the standpipe pressure remains unchanged and the casing pressure rises after pressure relief, it means that there is no trapped pressure and drilling fluid should not be released continuously.

6.4.2.3 Method of Reading Standpipe Pressure and Casing Pressure of Shut-in Well with Inside BOP Tool

The well control standards stipulate that, each team and group must implement the slow pump rate (SCR) test to measure the total circulation pressure in well killing process. When the low-pump stroke test has been implemented, and the displacement rate of well killing and total circulation pressure loss Δp_c are already known, the steps for calculating the shut-in standpipe pressure are as follows:

- Record the shut-in casing pressure.
- Initiate the pump slowly and open the choke valve.
- Control the choke valve, so as to adjust the casing pressure equals to the shut-in casing pressure, and keep the casing pressure constant.
- When the pump rate reaches to the displacement rate of well killing, the casing pressure equals to the well shut-in pressure. Record the circulation standpipe pressure p_T, shutoff the pump and close the choke valve.
- Calculate the shut-in pressure (Δp_c is already known) according to Eq. 6.13.

6.5 Well Killing

Well killing refers to employ a reasonable killing method with the density obtained from the actual formation pressure to prepare the killing fluid, pump them into wellbore according to certain steps into the wellbore to re-establish the pressure balance of the wellbore system.

6.5.1   Purpose and Principle of Well Killing

6.5.1.1   The Purpose of Well Killing
The purpose of well killing is to restore the balance of pressure between the wellbore and formation, and to make hydrostatic pressure not lower than the formation pore pressure. On the one hand, the purpose of well killing is to make the bottomhole pressure slightly higher than the formation pressure after well killing; on the other hand, the purpose is to displace the invading fluid out of the wellbore, or to push the fluid back into the formation safely (Sulfide highly contained well).

6.5.1.2   The Basic Principle of Well Killing
The general method of keeping the bottomhole pressure constant is used in well killing. During the process of well killing, the bottomhole pressure is slightly higher than the formation pressure, and keeps the bottom hole pressure unchanged throughout killing process. At the same time, the formation fracturing pressure at the casing shoe, burst resistance of casing, and the pressure bearing capability of the wellhead blowout preventor should also be considered.

6.5.1.3   Maximum Allowable Shut-in Casing Pressure
The maximum allowable shut-in casing pressure should meet the following conditions:

- Not greater than rated working pressure of wellhead blowout preventor.
- The sum of hydrostatic pressure and maximum allowable shut-in casing pressure is not greater than formation fracture pressure.
- The sum of hydrostatic pressure and maximum allowable shut-in casing pressure is not greater than 80% of burst resistance in the weakest casing.

6.5.1.4   The Process of Well Killing
According to the determined killing method, the bottomhole pressure can remain unchanged through controlling choke valve to control the standpipe pressure and casing pressure. Through certain well-killing flow rate, the prepared well-killing fluid is circulated into the well to displace the contaminated drilling fluid in the well, and to rebuild the balance of pressure in the wellbore.

6.5.2   Calculation of Basic Parameters of Well Killing

6.5.2.1   Calculation of Well-Killing Fluid Density

Calculation of formation pressure

The formation pressure is calculated according to Eq. 6.11:

$$p_p = p_{hi} + p_{sp} = 0.00981(\rho_d + \Delta\rho_d)D \tag{6.14}$$

where

D True vertical depth of wellbore, m;
ρ_d Original drilling fluid density, g/cm^3;
$\Delta\rho_d$ The increment of drilling fluid density needed to balance formation pressure, g/cm^3.

Calculation of well-killing fluid density

$$
\begin{aligned}
p_{he} = p_p + \Delta p &= p_{sp} + p_{hi} + \Delta p \\
&= 0.00981(\rho_d + \Delta\rho_d)D + \Delta p \\
&= 0.00981\rho_{d1}D
\end{aligned}
\tag{6.15}
$$

$$
\rho_{d1} = \frac{p_p + \Delta p}{0.00981D}
\tag{6.16}
$$

where

ρ_{d1} killing fluid density, g/cm^3;
Δp additional safety value of bottomhole pressure, MPa.

Additional safety value by standards

Oil well: Additional safety value of bottomhole pressure is $\Delta p = 1.5 - 3.5$ MPa. Additional safety value of drilling fluid density is $\Delta p = 0.05 - 0.1$ g/cm^3;

Gas well: additional safety value of bottomhole pressure is $\Delta p = 1.5 - 3.5$ **MPa**. Additional safety value of drilling fluid density is $\Delta p = 0.07 - 0.15$ g/cm^3.

6.5.2.2 Calculation of Flow Rate for Well Killing

The low-pump strokes are used in the process of well killing. The flow rate is generally one-third to half the normal flow rate during drilling process. The purpose for low-pump strokes and small flow rate is to control the choke valve and the pressure easily in the process of well killing. At the same time, in order to reduce the difficulty of controlling and dealing, the expansion shouldn't be too fast, and the flow rate shouldn't be too high when the circulated gas arrived at wellhead.

$$
Q' = (1/3 - 1/2)Q
\tag{6.17}
$$

where

Q' Flow rate for well killing, L/s;
Q Normal flow rate during drilling process, L/s.

6.5.2.3 Variation Law of Standpipe Pressure During Well Killing

According to the relationship of pressure balance during well killing, it can be concluded based on Eqs. 6.11 and 6.13:

$$p_T + p_{hi} - \Delta p_c = p_p = p_a + p_{ha} \tag{6.18}$$

It can be seen from Eq. 6.18 that:

- As long as the value of the left side of the equation remains unchanged during the process of well killing, the bottomhole pressure can be maintained.
- Because the drilling fluid density of well killing ρ_{d1} is greater than original drilling fluid density ρ_d, p_{hi} gradually increases in the process of well killing. In order to maintain a constant bottomhole pressure, we must reduce the total standpipe pressure p_T.
- By increasing the opening of throttle, both the throttle pressure and the total standpipe pressure p_T can be reduced.
- The total standpipe pressure is called the initial circulating standpipe pressure (ICP) p_{Ti} when the killing fluid is pumped, and the total standpipe pressure is called the final circulating standpipe pressure p_{Tf}(FCP) when the killing fluid returns to the wellhead.
- When the killing fluid is circulated to the bottom of the well, the drilling fluid column pressure in the drill stem can balance the formation pressure, and the shut-in standpipe pressure p_{sp} is 0. The pressure loss of system circulation also changes from the initial circulating pressure loss Δp_{ci} to the final circulating pressure loss Δp_{cf} due to the change of drilling fluid density (as seen in Fig. 6.12).

$$p_{Ti} = p_{sp} + \Delta p_{ci} \tag{6.19}$$

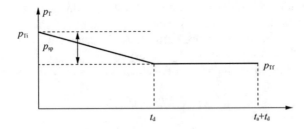

Fig. 6.12 Change of standpipe pressure during well killing

$$p_{Tf} = \Delta p_{cf} = \frac{\rho_{d1}}{\rho_d} \Delta p_{ci} \qquad (6.20)$$

6.5.2.4 Time of Well-Killing Fluid Travelling from the Surface to the Bit

The time required for well-killing fluid travelling from the surface to the drill bit refers to the time of pumping the drilling fluid from the surface to the drill bit through the drill string under the well-killing displacement rate, which may be expressed as:

$$t_d = \frac{V_d D}{60Q} \qquad (6.20)$$

where,

t_d The time when the killing fluid reaches the bit from the surface, min;
V_d Volume of drill string with unit length, L/m.

6.5.2.5 Time Required for Displacing the Annular Space with Killing Fluid

The time required for displacing the annular space with well-killing fluid refers to the full filling time of pumping the drilling fluid into the wellbore annular space through the drill bit, which may be expressed as:

$$t_a = \frac{V_a D}{60Q} \qquad (6.21)$$

where

t_a The time required for displacing the annular space with well killing fluid, min;
V_a Volume of annular space with unit length, L/m.

6.5.3 One-Circulation Well Killing Method

6.5.3.1 Characteristics

One-cycle well-killing method is also called Engineer's Method or Wait-and-Weight method. The basic method is as follows: shut-in the well as long as influx is observed, obtain the formation pressure data, and wait for the preparation of the killing fluid. After the preparation of the killing fluid is finished, start the pump, control the opening of throttle, and pump the killing fluid into wellbore; replace the original drilling fluid in a cycle, and restore the pressure balance of the formation hole system. One-cycle killing method is used for completing the operation of well killing. The advantages of this method are that the working time is short, the casing pressure is low during the well killing process, and it is not easy to break the weak formation. The drawback is that the waiting time for well killing is long and the possibility of pipe sticking increases in easy pipe-sticking formation. In the near stage, the exploration well, the remote well, and the main well will store a certain volume of heavyweight drilling fluid in the wellsite. The application of heavy killing fluid is more time-saving and labor-saving, and the advantages of one-cycle killing method are more obvious.

6.5.3.2   Operation Procedure

The pressure changes of one-cycle well-killing method are shown in Fig. 6.13. The procedure is as follows:

- Start the pump slowly, adjust the opening degree of choke valve, and keep the casing pressure equal to shut-in casing pressure.
- Adjust the flow rate to reach the flow rate of the well killing, keep the flow rate constant, and let the standpipe pressure be equal to the initial circulating standpipe pressure p_{Ti}.
- Pump killing fluid in the well, adjust the opening degree of choke valve and make the standpipe pressure decrease from the initial circulating standpipe pressure p_{Ti} to the final circulating standpipe pressure p_{Tf} within the time t_d when the killing fluid reaches to the bit from the ground.
- Adjust the opening degree of choke valve and keep the standpipe pressure equal to p_{Tf} within the time t_a when the killing fluid reaches to the surface from the bottom.
- When the killing fluid returns to the wellhead, stop the pump, casing pressure becomes 0, the well killing ends.

Fig. 6.13 Pressure change of one-cycle well-killing method

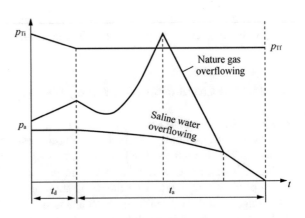

6.5.4 Two-Circulation Well-Killing Method

6.5.4.1 Characteristics

Two-cycle well-killing method is also called driller's method. The basic method is as follows: shut-in the well when discovering the influx and obtain the data of formation pressure. Firstly, make use of the original drilling fluid for the first-cycle of circulation to displace out the contaminated drilling fluid in the wellbore. Then, replace the original drilling fluid through the second-cycle of circulation with the prepared well-killing fluid, and restore the pressure balance of the formation wellbore system. The advantages of driller's method are that the contaminated killing fluid can be circulated as soon as possible, the waiting time of shut-in well is short, and the circulation can be restored as soon as possible. And the disadvantage is that the well-killing time is longer and the casing pressure of the first circulation cycle may even be higher.

6.5.4.2 Operation Procedure

Two-cycle well-killing method needs two full circulation cycles to complete the operation of well killing. The specific procedure is as follows (pressure change is shown in Fig. 6.14):

The first cycle

- Start the pump slowly, pump in the original drilling fluid, adjust the opening of choke valve, and keep the casing pressure equal to shut-in casing pressure.
- Adjust the flow rate of the pump to the required flow rate for well killing, and adjust the opening degree of choke valve to make the standpipe pressure equal to the initial circulating standpipe pressure p_{Ti}; make the standpipe pressure equal to p_{Ti}. Until the displacement of the drilling fluid in the well.

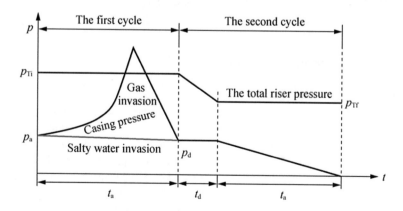

Fig. 6.14 Pressure change of secondary circulation well killing

- Stop the pump and close the choke valve; at this time, the casing pressure is equal to the shut-in standpipe pressure.

The second circulation cycle

- Start the pump slowly, adjust the opening of throttle, and keep the casing pressure equal to shut-in casing pressure of the first cycle.
- Adjust the pump speed and make the flow rate equal to that of well killing, and the standpipe pressure is close to the initial circulating standpipe pressure p_{Ti} at this time.
- Adjust the choke valve opening and make the standpipe pressure decrease from the initial circulating standpipe pressure p_{Ti} to the Final circulating standpipe pressure p_{Tf} within the time schedule t_d when the killing fluid reaches to the bit from the surface.
- Adjust the throttle opening and keep the standpipe pressure equal to p_{Tf} within the time t_a of the killing fluid reaching from bottom to surface. The casing pressure decreases to 0 and well killing stops when the killing fluid returns back to the wellhead.

6.5.4.3   Comparison of Two Killling Methods

- The time of waiting for well killing after shut-in process: Driller's Method < Engineer's Method.
- Well-killing operation time: Driller's Method > Engineer's Method.
- Peak value of casing pressure: Driller's Method > Engineer's Method.
- Operational difficulty: Driller's Method < Engineer's Method.
- The Engineer's Method is suitable for lower pressure bearing of wellhead installation and lower formation fracture pressure.
- Driller's Method is suitable for easy sticking cases.

6.5.5   Special Killing Method

6.5.5.1   Volumetric Method

This method is suitable for the well in which the drilling fluid has blown emptily or circulation cannot be established. The basic method is as follows: pump in a certain quantity of drilling fluid in wellhead, shut-in the well to let it fall down, and release a certain casing pressure. The decreased value of casing pressure should be equal to the column pressure of the drilling fluid pumped in. Repeat above steps, and reduce the casing pressure gradually. Trip in by snubbing when the casing pressure is decreased to certain extent, and then implement conventional well killing.

Volumetric method can also be used in this condition when circulation can't be in progress with no-drill tools in the well.

6.5.5.2 Bullhead Method

The applied conditions of this method are as follows: the well kick contains much hydrogen sulfide gas; the casing is set deeply, and the open-hole section is short, in which there is no drilling tools or only a small number of drilling tools; there is only one production layer with good permeability; the drill pipe is plugging and the killing fluid cannot reach the bottom of the well. An ultra-high-pressure formation that is unbearable for surface equipment; the well in which there is a leaky layer below the production layer, and a large amount of drilling fluid will leak into this layer during cycle of well killing. The basic practice is to pump the drilling fluid into the formation from the annulus, push the fluid from the wellbore back to formation and pump into the drilling fluid without exceeding the maximum allowable shut-in pressure.

6.5.5.3 Snubbing Method

When the wellhead is closed, snubbing the drill string the bottom of the hole, and then execute the kill operation according to conventional killing methods. When snubbing, the drilling fluid in the wellbore with the same volume should be discharged according to the volume of drilling tool. Snubbing could be completed by reducing the pressure on the annular BOP.

6.5.5.4 Dynamic Killing Method

Dynamic killing method is a method to kill well by increasing the flowrate of killing fluid and increasing pressure loss of circulating flow. This method is suitable for wells such as slim hole and ultra-deep well which are sensitive to return velocity and circulation friction.

6.5.5.5 Reverse Circulation Killing Method

After the influx occurs, the open-hole section is relatively long. In order to prevent the influx from entering into other strata, it is necessary to discharge the influx out of the well as soon as possible. The basic method is to pump the drilling fluid from the annulus into the well, circulate the influx from the bottom of the well into the drill string, and discharge it through the drill string, wellhead, reverse circulation manifold, and surface choke manifold. After the influx is discharged, the reverse circulation can be used to replace the slurry in the well, or the conventional circulation can be used to kill the well.

6.6 Controlled Pressure Drilling

Controlled pressure drilling (CPD) was regarded as a technology that includes aerated drilling, underbalanced drilling (UBD), and managed pressure drilling (MPD).

In fact, managed pressure drilling (MPD) is the new drilling technology developed to improve UBD and aerated technology. The common characteristic of these three technologies is to control the pressure in the wellhead or the pressure in certain depth of the wellbore by using dedicated equipment so as to control pressure distribution in the wellbore and to carry out safe and successful operation on special formations and areas such as low-pressure formation, narrow pressure window formation, formation with multi-pressure series of strata, extremely hard formation, and deepwater drilling.

6.6.1 Underbalanced Drilling (UBD)

6.6.1.1 Definition of UBD

Underbalanced drilling (UBD) refers to a drilling mode by which pressure of drilling fluid upon the bottom of the well is less than the pore pressure so as to allow the formation fluid to flow into wellbore and process the fluid.

UBD includes flow drilling and aerated fluid drilling. Flowing drilling refers to the UBD processing by using flow fluid while aerated fluid drilling is to inject gas into drilling fluids and take the mixture of drilling fluid and gas as drilling circulating medium.

6.6.1.2 Technical Characteristics of UBD

Advantages of UBD

Negative pressure difference can control the drilling fluid filtrate to flow into the formation as it seeks to reduce damage to formation and protect reservoir. Moreover, underbalanced drilling contributes to reduce drilling fluid density and resolve problems such as, lost circulation and differential sticking led by formation pressure exhaustion. Negative pressure difference can also lower or conquer chip hold effect to improve drilling rate. Besides, it is conducive to find hydrocarbon zone at early stage since the formation fluid flows into the wellbore under the negative pressure difference. Theoretically, real-time reservoir tests can be conducted by changing the wellhead back pressure to change the fluid production in the formation.

Disadvantages of UBD

Without mud cake as protection, improper pressure control may cause more damage to formation. What's more, it may cause well wall sloughing since open hole is in the negative pressure difference. Much more, formation fluid may exert higher demands to environment. Besides, man-made influx will multiply the risk of well blowout if the formation fluid contamination quantity is not properly controlled. Moreover, UBD requires special equipment and worker, sometimes one more casing may be run and the costs may be increased.

6.6.1.3 Major Equipment of UBD

Diagram of major flow drilling equipment is shown in Fig. 6.15.

- Rotating control head (rotating control device, normally called RCD): It is the key device for underbalanced drilling. It can make dynamic seal of wellhead annulus in the process of drilling and round trips.
- Liquid–gas separator or four phase separator: It can be used to separate liquid and gas from the output fluid, or separate oil, gas, water, and cuttings.
- Underbalanced choke manifold: It is used to choke control pressure and wellhead pressure so as to control the distribution of wellbore pressure based on the requests of UBD pressure.
- Torch and ignition device (flare stack): It is used for combustible gas separated by the liquid–gas separator to ignite and flare.
- Data acquisition system and pressure control system: They are used to collect the key parameters of underbalanced drilling, implement the lubrication and motion control of rotating blowout preventor, and complete the control action of throttling pressure.

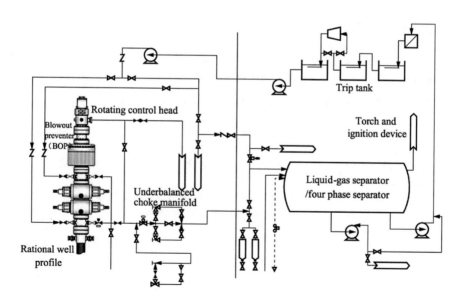

Fig. 6.15 Schematic diagram of fluid underbalanced drilling

6.6.1.4   Application Conditions of UBD

- Open holes should be drilled in the layer with single pressure. If open holes are in multiple pressure systems, the pressure difference between the layers should not exceed the allowable range of underbalanced drilling.
- Make borehole stability analysis before designing, and stratum stability should meet the requirements.
- The content of hydrogen sulfide discharged during construction should meet the following requirements: the concentration before exposure to the atmosphere is less than 50 PPM and the concentration at the outlet in contact with the atmosphere is less than 20 PPM.
- Ground equipment in line with the implementation of underbalanced drilling conditions. Besides, well profile should meet the requirements of the wellbore stability in underbalanced drilling so as to reduce the damage that production of formation fluids brings to the reservoir.
- Intermediate casing strings should block off possible broken belt, collapsible layer and outflow layer, and sealed to the top of the reservoir as far as possible.

6.6.1.5   Value of Underbalanced Pressure Differential

Definition of underbalanced pressure differential

Underbalanced pressure differential refers to the difference between bottomhole pressure and pore pressure, also known as underpressured value, especially refers to the value of bottomhole pressure that is less than the bottomhole pore pressure.

Principles of determining the underbalanced pressure differential

- Underbalanced pressure differential is less than the difference between pore pressure and formation collapse pressure;
- Underbalanced pressure differential design should be combined with ground equipment capabilities.
- Underbalanced pressure differential should be determined according to the reservoir type and lithology characteristics to avoid the stress damage of the reservoir.
- Underbalanced pressure differential should be determined on the basis of the formation of liquid (gas) volume.
- The higher formation pressure is, the smaller the liquid (gas) and underbalanced pressure differential should be as possible; on the contrary, it should be enlarged.
- If gas–oil ratio is high, the underbalanced pressure differential should be as small as possible; on the contrary, it should be enlarged.
- The maximum casing pressure value should not exceed 50% of the dynamic seal pressure of rotating control head.

6.6.2 Gas Underbalanced Drilling

6.6.2.1 Definition of Gas Underbalanced Drilling

Drilling with gas, mist, or foam as a circulating medium is called gas underbalanced drilling, or gas drilling for short.

Gas underbalanced drilling includes air drilling, nitrogen drilling, natural gas drilling, mist drilling, foam drilling, and so on. Air drilling refers to drilling with compressed air as a drilling fluid. Nitrogen drilling refers to drilling with nitrogen as a drilling fluid. Natural gas drilling refers to drilling with natural gas as a drilling fluid. Mist drilling refers to the drilling with gas and of small amounts of water, and foam injected into the gas stream before the compressed gas entering the drill string. Foam drilling refers to drilling with foam as a drilling fluid.

6.6.2.2 Technical Characteristics of Gas Underbalanced Drilling

Advantages of gas underbalanced drilling

- Drilling rate is many times higher than the normal drilling methods;
- Longer footage with single bit;
- The lower the bit weight, the straighter the well is.
- Water-sensitive formation shrinkage and collapse can be conquered.
- Differential sticking can be eliminated.
- Lost circulation in low-pressure formation can be resolved.
- The size of borehole could be more precise but borehole erosion may not be avoided.
- Less water in drilling;
- Maintenance costs on drilling fluid system can be lowered, but the drilling fluid must be reserved for well control.
- Shorter lag time of rock samples reaching the ground can reduce damage to the pay zone.

Disadvantages of gas underbalanced drilling

- It is difficult to cope with the formation with large quantities of liquid production.
- Low control over formation pressure and high risk of well control in unknown area (formation);
- High cost of auxiliary equipment, material, and operating;
- Risk of explosion exerted in the gas zone;
- Drilling pipe wear and tear without drilling fluids;
- Few buoyancy and high weight drill string;
- Sticking cannot be avoided when drilling in an unknown water layer.
- Measurement while drilling tool (MWD) cannot be used.
- Casing and corrosion of drilling string may not be avoided.
- Borehole washout may bring difficulties to cement well.

- The casing design method is different from normal drilling casing.
- Huge gas equipment needs larger wellsite.
- Small cuttings may take bad effect on cutting logging.

6.6.2.3 Major Equipment of Gas Underbalanced Drilling

The main equipment for gas underbalanced drilling is shown in Fig. 6.16. Gas underbalanced drilling equipment in Table 6.2. Besides major equipment required for flow drilling, gas underbalanced drilling needs equipment such as air compressors, superchargers, nitrogen equipment, booster, discharge pipes, combustion cells, and mist pumps.

- Compressor: It is used to provide pressure and compressed air for gas underbalanced drilling.
- Supercharger: It is to provide compressed air with higher pressure for gas underbalanced drilling based on the construction needs.
- Nitrogen Unit: According to the need to open the oil and gas formation, a certain amount of compressed nitrogen is supplied to the gas underbalanced drilling.
- Drainage pipes and combustion ponds: They are used for discharging cuttings and liquid discharged from wellbores and igniting and combustible venting of combustible gases.
- Mist pump: To inject small amount of water or chemical.

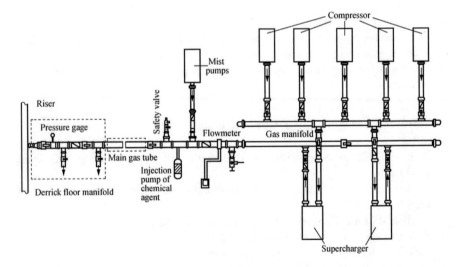

Fig. 6.16 Schematic diagram of gas injection equipment of gas underbalanced drilling

Table 6.2 Support equipment of gas underbalanced drilling

Corollary equipment	Equipment name	Method of underbalanced drilling		
		Air/atomization/foam	Nitrogen	Natural gas
Pressure control device	Rotating blowout preventor or rotating control device	✓	✓	✓
	Hydraulic valve		✓	✓
	Specific throttle manifold		✓	✓
Gas burning device	Equipment name	✓	✓	✓
	Automatic ignition device	✓	✓	✓
	Torch	✓	✓	✓
	Device for tempering	✓	✓	✓
Injection device	Vent line	✓	✓	
	Air compressor	✓	✓	✓
	Turbocharger		✓	
	Nitrogen device	✓	✓	
Non-snubbing trips device	Mist pump			
	Casing valve			

Note ①Rotating control head can be used; ②Natural gas drilling select special turbocharger; ③Select the device according to the actual situation of the site; ④The device was used for the whole process underbalanced drilling

6.6.2.4 Conditions for Gas Underbalanced Drilling

- Gas underbalanced drilling should be carried out at atmospheric pressure or low-pressure intervals, and gas drilling should not be carried out at high-pressure and high-production wells.
- The gas underbalanced drilling should be carried out in formation that is relatively stable and hard to collapse.
- The amount of liquid produced in the formation should not cause collapse of the shaft lining, the drilling mud packs, and the borehole plugging. Meanwhile, the mass concentration of hydrogen sulfide in the formation fluid should be lower than 50 PPM.
- On the basis of conventional well control equipment, it should be equipped with underbalanced drilling equipment and meet the relevant regulations.
- The wellsite has the installation location of aerated equipment to meet the relevant requirements of implementing the gas underbalanced drilling technology and drilling fluid treatment. The aerated equipment should be more than 15 m away from the wellhead with sufficient safe passage.
- The volume of foam drilling tank is not less than 1000 m^3. The location of cuttings pool should be convenient to install pipelines, and the capacity of cuttings pool should meet the requirements of sedimentation of sediment and

precipitation of dusting water. Moreover, it should meet the environmental protection requirements.

- Nitrogen gas drilling in the reservoir section and natural gas drilling should be completed. The combustion pond should be located in a safe zone 75 m downstream of the wellsite in the direction of the downwind, and it can be provided with the conditions for recovery and utilization of accumulated cuttings and dust-bearing water.
- The underbalanced technical service team has the corresponding qualifications, and the drilling team should be qualified above Class B (including Class B).
- During the implementation of gas drilling technology in the pay zone, the casing should be sealed to the top of the reservoir as much as possible.
- The setting depth of surface casing or deeper conductor will not result in formation leakage for non-reservoir gas drilling.
- Casing anti-extrusion strength should be checked by hollowing out the whole check and anti-extrusion safety factor should not be lower than 1.0.
- The upper casing cementing quality is qualified to meet the gas drilling operation requirements.

6.6.3   Managed Pressure Drilling (MPD)

Conventional drilling emphasizes that the bottomhole pressure is slightly higher than the formation pressure to ensure the well control safety. While underbalanced drilling emphasizes a certain underpressure value to protect the formation and increase rate of penetration. Gas drilling focuses on leak prevention in stubborn formations and increased drilling speed in hard formations. Practice has proved that both the conventional near-balanced drilling, underbalanced drilling or gas drilling, cannot fully solve some of the problems in the drilling; therefore, managed pressure drilling technology came into being.

6.6.3.1   Definition of MPD

MPD is a drilling program that can be applied to ensure safe and efficient drilling by precisely controlling the pressure profile of the entire borehole annulus based on the pressure range of the bottomhole (wellbore).

MPD includes constant bottomhole pressure drilling, mud cap drilling, pressurized mud cap drilling, and dual gradient drilling.

6.6.3.2   Technical Characteristics of MPD

- MPD combines tools and technologies to control the pressure profile of narrow liquid column in the ring. It is conducive to reduce the construction risk caused by drilling in formation with narrow pressure window.

- The purpose of precise control of annular pressure profile is achieved through comprehensive control of back pressure, fluid density, fluid rheology, annular fluid level, annular fluid friction, and borehole geometry.
- The pressure deviation can be corrected more quickly by changing the choke pressure (the way to change the annulus liquid level), effectively reducing the loss of manpower, money, material resources, and time due to the adjustment of drilling fluid density.
- It can control drilling fluid leakage or formation fluid intrusion more quickly, and control formation fluid flow in the construction process more quickly and safely in a safe and controllable range.
- The cost of equipment and service is higher, and the improper control will still bring the well control risk.

6.6.3.3 Major Equipment of MPD

The main equipment for MPD is shown in Fig. 6.17.

- MPD choke manifold: It is the key equipment for pressure drilling, mainly to achieve fine control of wellhead pressure during drilling and tripping.
- Flow meter: It is used to measure returned fluid flow.
- Back pressure pump: It is used to maintain the continuity of the circulation and well head back pressure for trips and making connection.

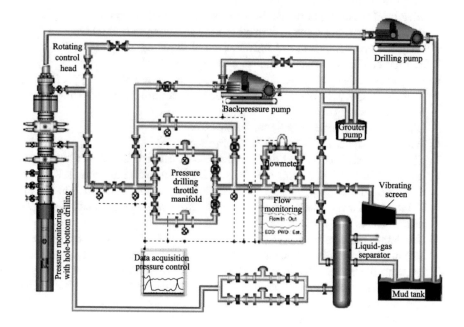

Fig. 6.17 Management pressure drilling device schematic

- Pressure while drilling (PWD) tool, monitoring of bottomhole pressure with drilling: It is used to measure annular bottomhole pressure at real time during operation, and to control the bottomhole pressure precisely.
- Data acquisition and pressure control system: It is used to collect the key parameters of pressure-controlled drilling and control the designed choke pressure.

6.6.3.4 Key Conditions of MPD

Three key conditions:

- A set of enclosed, pressured circulation system, and MPD control device;
- A complete set of drilling hydraulics design;
- A well-trained operation team acquainted with MPD concept.

Questions and Exercises

1. What is the main basis for determining the drilling fluid density? Why is important for drilling?
2. How to deal with the relationship between balanced pressure drilling and safety drilling?
3. Briefly, explain the characteristics, applicability, and limitations of underbalanced drilling.
4. Give brief reasons and precautions of formation fluid invasion.
5. Briefly, explain the signs of formation fluid intruding into borehole.
6. What are the methods to test formation fluid intruding into borehole?
7. In what conditions are hard shut-in should be applied?
8. How to correctly determine shut-in standpipe pressure and casing pressure?
9. How to judge well control principle of well control based on shut-in well pressure recovery?
10. What is the principle of bottomhole pressure?
11. What is the rationale for selecting killing methods?
12. The well depth is $D = 4000$ m, the diameter of drill bit is 215.9 mm, the size of drill pipe is 114.3 mm, unit weight is 24.72 kg/m, wall thickness is 8.6 mm; drilling fluid density $\rho_d = 1.50$ g/cm^3; technology casing string in $\Phi244.5$ mm is set to 3000 m, where the formation fracture pressure gradient $G_f = 19.9$ kPa/m. According to the record before influx, the pump rate is 30 strokes/min, circulation pressure is p_{ci}, and the pump type is 3NB-900. The liner diameter is 120 mm and the displacement is 30 r/min. The displacement rate $Q = 15.56$ L/s, and the maximum pump pressure is 33.27 MPa. After the wells shut in 10 min, the standpipe pressure is 4.67 MPa, the casing pressure is 6.65 MPa, and the drilling fluid volume in the drilling fluid tank increases to 4.5 m^3. Use Driller's Method and Engineer's Method to design well killing sheet.
13. What is the difference between conventional and unconventional well killing?

14. Briefly, explain the regular pattern of Engineer's Method, sketch the designs and make a brief description.
15. What is the reverse circulation well killing method? What are the conditions of reverse circulation method?

Well Cementing and Completion

7

Abstract

This chapter mainly introduces casing program design, casing strength design, cementing technologies, and well completion technologies. Casing program is the basis of drilling engineering which is to determine the casing levels, setting depth, top of cement, match of casing sizes, and hole section sizes. Casing strength design is mainly composed of the calculation method of external loading of casing, casing strength under combined external loadings and the conventional design method with safety factors for casing strings. The cementing technology includes types of oil well cement and additives, process of cementing, measures to improve displacement efficiency of cementing, weight loss of cement slurry and oil, gas and water channeling during wait on cement. The definition and application scope of perforation completion, open-hole completion, slotted liner completion, and gravel filling completion are introduced in completion technologies.

Cementing is an important link in the construction of oil and gas wells. Cementing engineering includes two processes: running casing and pump cement slurry. Running casing is to run a kind of casing string which is composed of some or several different steel grades and wall thicknesses in a specified depth in a drilled hole. Cementing is the process of injecting cement slurry through casing string into annular space between wellbore and casing string. Cement bonds the casing string firmly with the wellbore wall, which can seal oil, gas, water, and complex formations for further drilling or exploitation. The cementing quality not only affects the continuous drilling of the well, but also affects the smooth production in the future, the recovery capacity and life of the well. Therefore, the cementing quality of a well should be taken seriously from the beginning of design to the field operation. A lot of steel and cement are consumed in cementing engineering. According to statistics, the cementing cost of production wells accounts for

10–25% or more of the total cost of wells. Therefore, cost should be reduced as much as possible under the premise of guaranteeing quality.

Completion is also known as the completion of a well, including drilling production zones, determining the bottomhole structure of completion, connecting the wellbore and production zones, and installing the bottomhole equipment and wellhead. Completion is a key link between drilling and oil production. It is based on the geological structure of oil and gas reservoirs, rock mechanics properties, and reservoir physical properties to study the technological process of the best connection mode between reservoirs and wellbore. Therefore, the best completion method should be chosen to create the best conditions for stable and high production of oil and gas wells. The completion mode of a well directly affects the casing program, drilling method, cementing method, future production, and stimulation and downhole operation methods of the well.

7.1 Casing Program Design

Casing program is the basic spatial form of oil and gas well in the design process or the post well completion, which mainly includes casing levels and their setting depth, top of cement, the match of casing size, and drill bit size. Casing program design is the process of optimizing the above parameters, based on the information of the geological conditions of the region where the oil and gas well located, current technical equipment conditions, goal of drilling, safety requirements, engineering technological requirements, and so on. Casing program is the basis of drilling engineering design, and also the foundation of smooth drilling of a well. A reasonable casing program design can ensure the well-being drilled to the predetermined well depth, guarantee the safety in the drilling process, and prevent the reservoir pollution and maintain the subsequent productivity. The casing program design mainly relies on formation pore pressure and formation fraction pressure profiles.

7.1.1 Casing Types

Casing types can be divided into the following categories according to their functions (Fig. 7.1):

- Conductor;
- Surface casing;
- Intermediate casing, also called technical casing;
- Production casing.

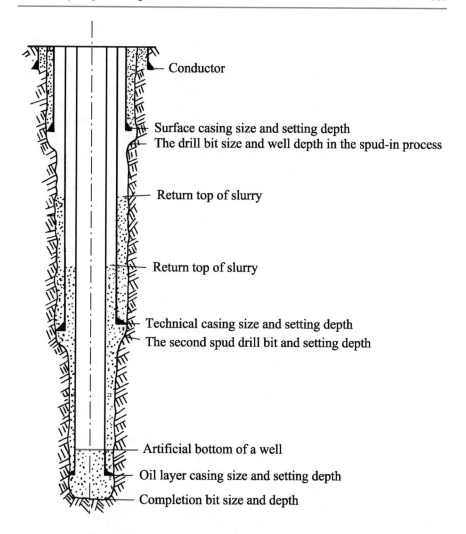

Fig. 7.1 Basic casing types

Conductor is the first, shortest and shallowest section of the steel pipe before the first spud-in, and the main role is to prevent the loose earth surface from collapsing. The surface casing is a layer of casing string that prevents the upper section of a wellbore from collapsing, the contamination of underground water, and the upper fluid intrusion. It is also used for installing a blowout preventer of the wellhead. It has two main functions: First, the top of surface casing is used for the installation of casing head suspending and supporting the subsequent levels of casing through the casing head; The second is to isolate the surface shallow water and shallow complex formation, so that the freshwater formation can be avoided from drilling fluid contamination. Surface casing shoes must be set in the formation with sufficient strength to avoid the occurrence of formation breakdown and underground blowout after shutting down for preventing the kick. The casing between surface casing and production casing is named as intermediate casing, which is mainly aimed at isolating formations of different formation pressure or the complex formations easy to collapse or leak. The intermediate casing can be one level, two levels, or several levels. Production casing is the last level of casing string run down after drilling to the target depth, and its role is to establish a solid channel for oil and gas production, to protect wellbore wall, and to meet the requirements of slicing, testing and transformation of reservoirs.

In order to save costs, liner is usually adopted after running down a level of intermediate casing, that is, running casing and pumping cement slurry only in the open-hole section, and the casing string does not extend to surface.

Typically, not every well contains the casing types described above, but surface casing and production casing are always necessary for a well completion.

7.1.2 The Principles of Casing Program Design

The main principles shown below should be followed in casing program design.

- Complying with the local laws and regulations, and meeting the safety, health, environmental system management requirements;
- Being helpful to identify, understand, and effectively protect the oil and gas layers, so that the oil and gas layers of different pressure gradients can avoid drilling fluid pollution damage;
- Avoiding the leakage, blow out, collapse, sticking, and other complex situations, for laying foundation of the full well drilling and ensuring that the drilling success rate, to reduce drilling costs as much as possible, and achieve the shortest drilling cycle;
- Having certain well control capability, the higher density drilling fluid used in drilling the lower part of the high-pressure formation or fluid column pressure generated by shutting down after well kick does not fracture a open-hole section;
- Under running casing and drilling process, the differential pressure between hydrostatic pressure of drilling fluid in the wellbore and the formation pressure won't result in differential pressure pipe sticking;
- The exploratory well design should take into consideration of the potential needs for deepening drilling and increasing intermediate casing running.

7.1.3 The Basis of Casing Program Design and the Relevant Basic Parameters

7.1.3.1 Basis of Casing Program Design

The basis of casing program design mainly includes the following aspects:

- Geological design for drilling;
- The profile of formation pore pressure, formation fracture pressure, and collapse pressure;
- Formation lithology profiles;
- Well completion methods and production casing size;
- Drilling datum of reference wells in adjacent blocks and adjacent wells in the same block;
- The level of drilling equipment and technology;
- The depth of riverbed bottom near the well, the depth of underground water, the distribution of nearby water source, the mining depth of the underground mining area, the water (steam) injection intake level of development adjustment wells;
- Drilling technology specifications and standards.

7.1.3.2 The Basic Parameters and the Range of Corresponding Value

- Swab pressure coefficient S_b: When the drill string moves upward, the deducted pressure value of the liquid column by the swabbing action is expressed in equivalent mud density. Typically, S_b is in the range of 0.015–0.040 g/cm^3.
- Surge pressure coefficient S_g: When the drill string moves downward, the increased value of fluid column in the well by the surge pressure generated by running down the drill string is expressed in equivalent mud density S_g, which typically has the range of 0.015–0.040 g/cm^3.
- Fracture pressure safety factor S_f: The increased value of formation fracture pressure to avoid fracturing of the open-hole formation near the upper casing shoe, which is expressed in equivalent density. The value of the safety factor S_f is related to the prediction accuracy of the formation fracture pressure. Typically, it is 0.03 g/cm^3.
- Well kick allowance S_k: The well fluid column pressure is increased due to the back pressure of the wellhead after well kick shut down. The well kick allowance means the allowed increased value of the well fluid column pressure equivalent mud density before and after shutting down the well. The value of well kick allowance S_k is related to the accuracy of the formation pore pressure prediction and the well control technology capability. Generally, the value of S_k is 0.05–0.10 g/cm^3.
- Differential pressure allowance: The maximum pressure difference between the well fluid column pressure and the formation pore pressure allowed in the

open-hole section. The differential pressure of the open-hole section is within the allowable range that can avoid problems of differential pressure pipe sticking. The value of differential pressure allowance is related to drilling technology and drilling fluid properties, and also the formation pore pressure and permeability of the open-hole section. If the normal formation pressure and the abnormal high pressure are in the same open-hole section, pipe sticking is easily to occur in the well section of normal pressure. So the differential pressure allowance can be divided into well section with normal pressure and abnormal pressure, which are expressed as Δp_N and Δp_A. The differential pressure allowance Δp_N of normal pressure is from 12 to 15 MPa, and the differential pressure allowance Δp_A of abnormal pressure is from 15 to 20 MPa.

The basic parameters data discussed above should be determined mainly by the available statistical data locally.

7.1.4 The Methods of Casing Program Design

One of the main contents of the casing program design is to determine the casing levels and their setting depth, and the key is to determine the difference of setting depths between two adjacent casings, that is, to determine the depth interval of the safe open-hole section. The so-called safe open-hole section implies that the open hole in the section, should avoid well kick, borehole wall collapse, differential pressure pipe sticking during the drilling process, well leakage occurrence during fracturing in drilling, well leakage occurrence of formation being fractured in well kick shut down or well killing, as well as the differential pressure casing sticking in running casing string and in other complex underground situation. As for the same well, the selection of different open-hole starting point and design order will deliver different designs of casing program and corresponding setting depth, when the casing levels and the setting depth are determined. Therefore, the determination of casing program can be divided into the bottom-up design and top-bottom design. Generally, the bottom-up design is adopted for the exploratory well in the proven block or well of explicit environmental and geological conditions. And the top-down and bottom-up designing methods are combined for exploratory well in new exploratory area or when the underground formation geological data are uncertain.

7.1.4.1 The Constraint Conditions of Safe Open-Hole Section

According to the principles of casing program design and the definition of the safe open-hole section, the requirements of preventing well kick, borehole wall collapse, fracturing formation in normal drilling, differential pressure sticking or casing sticking, and fracturing formation of shutting down after well kick, in open-hole section.

The constraints of preventing a well from kicking

During the routine drilling, the drilling fluid density should not be less than the equivalent mud density of the formation pore pressure when this formation drilled, which is the constraint condition of preventing well kick:

$$\rho_d \geq \rho_{p\,max} + \Delta\rho \tag{7.1}$$

where

ρ_d Drilling fluid density, g/cm^3;
$\rho_{p\,max}$ Equivalent mud density of maximum formation pore pressure, g/cm^3;
$\Delta\rho$ Additional value of drilling fluid density, g/cm^3.

The additional value of drilling fluid density is determined according to the industrial standard *SY/T 6426-2005*, sometimes $\Delta\rho$ is determined based on the value of S_b. As for oil wells, $\Delta\rho = 0.05$–0.10 g/cm^3. For gas wells, $\Delta\rho = 0.07$–0.15 g/cm^3.

The constraint condition of preventing wellbore from collapsing

Considering the effect of formation collapse pressure on the borehole stability, the maximum drilling fluid density in the open-hole section should also meet the following conditions:

$$\rho_{d\,max} \geq max\{\rho_{p\,max} + \Delta\rho,\ \rho_{c\,max}\} \tag{7.2}$$

where

$\rho_{d\,max}$ The maximum drilling fluid density adopted in the open-hole section during drilling, g/cm^3;
$\rho_{c\,max}$ The maximum formation collapsing pressure equivalent mud density in the open-hole section, g/cm^3.

The constraint condition of preventing fracturing the formation in normal drilling process

The maximum column pressures equivalent mud density that may occur at the well location of the weakest formation in the open hole is normally less than or equal to the equivalent mud density of the formation fracture pressure at that level, in the normal drilling or tripping operation, namely

$$\rho_{bn\,max} \leq \rho_{ff\,min} \tag{7.3}$$

where

$$\rho_{\text{bn max}} = \rho_{\text{d max}} + S_g \qquad (7.4)$$

where

$\rho_{\text{bn max}}$ The equivalent mud density of the maximum bottomhole pressure in normal drilling or tripping operation, g/cm^3;

$\rho_{\text{ff min}}$ The equivalent mud density of the formation fracture pressure, g/cm^3.

The equivalent mud density of the minimum safe formation fracture pressure is:

$$\rho_{\text{ff}} = \rho_{\text{f}} - S_{\text{f}} \qquad (7.5)$$

where

ρ_{ff} The equivalent mud density of safe formation fracture pressure, g/cm^3;

ρ_{f} The equivalent mud density of formation fracture pressure, g/cm^3.

The constraint condition of preventing differential pressure pipe sticking

During normal drilling or running casing operation, the maximum pressure difference Δp between the drilling fluid column pressure and the formation pore pressure in the open-hole section should not be greater than Δp_{N} or Δp_{A}:

$$\Delta p = 0.00981(\rho_{\text{d max}} - \rho_{\text{p min}})D_{\text{n}} \leq \Delta p_{\text{N}}(\Delta p_{\text{A}}) \qquad (7.6)$$

where

Δp The maximum pressure difference between drilling fluid column pressure and formation pore pressure, MPa;

$\rho_{\text{p min}}$ The corresponding formation pore pressure equivalent mud density of the level of the maximum pressure difference in the open hole, g/cm^3;

D_{n} The corresponding well depth of the level of the maximum pressure difference in the open hole, m.

D_{n} usually takes the maximum well depth in the normal pressure formation in terms of normal pore pressure formation; as for abnormal pressure formation, D_{n} the well depth corresponding to the minimum formation pore pressure equivalent density.

The constraint condition of preventing fracturing formation in well shut-in state

When the well is shut down after well kick, the equivalent mud density of the well jointly generated by the wellhead casing pressure and the drilling fluid column pressure in the wellbore changes with the depth of the well. The smaller the depth, the higher the equivalent mud density is. The formation at the top of the open-hole section (i.e., at the shoe of the upper casing) is easy to be fractured and can lead to leakage after well kick and shut down. Therefore, it should be ensured that the maximum pressure that may be generated in the wellbore at the casing shoe of the upper casing is not greater than the formation fracture pressure, and the other weak formation in the open-hole section should also be protected from being fractured. The constraint conditions are:

$$\rho_{ba\,max} \leq \rho_{ff} D_{min} \tag{7.7}$$

$$\rho_{ba\,max} = \rho_{d\,max} + \frac{D_m}{D_x} S_k \tag{7.8}$$

where

$\rho_{ba\,max}$	The equivalent mud density of maximum wellbore pressure when shut down after well kick, g/cm^3;
$\rho_{ff} D_{min}$	The equivalent mud density of the formation fracture pressure at the shallowest level in the open-hole section, g/cm^3;
D_m	The corresponding well depth of the maximum formation pore pressure in the open-hole section, m;
D_x	The corresponding well depth of the weakest formation in the open-hole section, and usually it is the shallowest well depth of the open-hole section, m.

7.1.4.2 The Bottom-up Method and Steps of Casing Program

Since the setting depth of casing is mainly determined by the completion method and the geological conditions of the oil and gas layer, the design step is to determine the depth of each casing from the bottom to the top starting from the intermediate casing. The design steps are as follows:

- The first thing is to obtain the profile of formation pore pressure, collapsing pressure and fracturing pressure of the design area, as shown in Fig. 7.2. The vertical coordinates stand for depth, the horizontal coordinates stand for the equivalent mud density (ρ_p, ρ_c, ρ_f) of formation pore pressure, collapsing pressure, and fracturing pressure.
- Determine the additional parameter of drilling fluid density—swab pressure coefficient S_b, surge pressure coefficient S_g, fracture pressure safe coefficient S_f, well kick capacity S_k, pressure difference capacity Δp_N and Δp_A according to the characteristics of the area and the properties of the designing well.
- Find the maximum pore pressure equivalent mud density $\rho_{p\,max}$ and the maximum formation collapse pressure equivalent mud density $\rho_{c\,max}$ in the three pressure profiles and record the corresponding well depth of the two values, respectively. Calculate the maximum drilling fluid density $\rho_{d\,max}$ of the

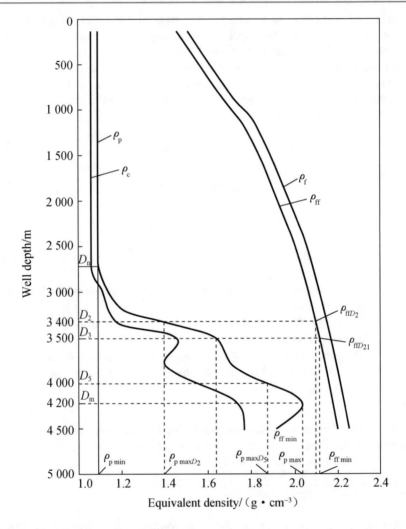

Fig. 7.2 Diagram of design workflow to determining the casing sequence and depth by bottom-up method

open-hole section based on Eq. 7.2. Use Eq. 7.4 to calculate the equivalent mud density of the maximum well pressure $\rho_{\text{bn max}}$ during normal drilling or tripping operation. Use Eq. 7.5 to calculate the equivalent mud density ρ_{ff} of safe formation fracture pressure at different well depth of the whole well and draw the equivalent mud density curve of the safe formation fracture pressure in the three pressure profile map, as the ρ_{ff} curve in Fig. 7.2.

- According to the constraint Eq. 7.3, considering $\rho_{\text{bn max}} \leq \rho_{\text{ff min}}$ and find the point $\rho_{\text{ff min}}$ in the horizontal coordinates of Fig. 7.2. Draw a vertical line upward from the point $\rho_{\text{ff min}}$ to cross with the curve of safe formation fracture pressure ρ_{ff}, the well depth of the cross-sectional point is the primary intermediate casing setting depth D_3.

- Find the equivalent mud density of maximum formation pore pressure $\rho_{\text{p max}}$ and the equivalent mud density of the maximum formation collapse pressure $\rho_{\text{c max}}$ in the three pressure curves in the well depth interval less than D_3 and calculate the maximum drilling fluid density $\rho_{\text{c max}}$ in the well depth interval based on Eq. 7.2. Meanwhile, scan and calculate the pressure difference Δp between the maximum wellbore pressure and the formation pore pressure, and record the corresponding equivalent mud density of formation pore pressure and the well depth D_n of the maximum pressure difference position in this well depth interval. Verify whether the primary intermediate casing setting depth D_3 if of the risk of differential pressure sticking according to the constraint condition equation of preventing differential pressure sticking.

  - If $\Delta p \leq \Delta p_N(\Delta p_A)$, then the primary selection of well depth D_3 is the check depth D_{21} of intermediate casing. Then check the risk of fracturing leakage of the intermediate casing at the check depth D_{21}, that is, use Eq. 7.8 to calculate the equivalent mud density of the maximum wellbore pressure $\rho_{\text{ba max}} D_{21}$ according to the maximum drilling fluid density $\rho_{\text{p max}}$ and the corresponding well depth D_m. The design requirement can be satisfied when $\rho_{\text{ba max}} D_{21}$ is less than or close to the equivalent mud density of the formation safe fracture pressure $\rho_{\text{ff}} D_{21}$ at the level of D_{21}. D_{21} is the setting depth D_2 of the intermediate casing. Otherwise, the setting depth value of intermediate casing should be increased and return to step (5) to recheck the risk of differential pressure sticking and finally determine the setting depth D_2 of the intermediate casing.
  - If $\Delta p > \Delta p_N(\Delta p_A)$, then the setting depth of intermediate casing should be less than the primary selection of D_3. According to Eq. 7.6, the allowed maximum drilling fluid density value $\rho_{\text{d max}} D_2$ at the level of depth of D_n when the pressure difference is $\Delta p_N(\Delta p_A)$ can be calculated. Calculate the allowed equivalent mud density of maximum formation pore pressure $\rho_{\text{p max}} D_2$ and find the corresponding points of $\rho_{\text{p max}} D_2$ in the horizontal coordinates. Draw a vertical line upward to cross with the equivalent mud density curve of formation pore pressure, and the well depth of the cross point is the setting depth D_2 of intermediate casing. Since the setting depth D_2 of the intermediate casing does not reach the primary selection of well depth D_3, so liner needs to be designed bellow depth D_2.

- Repeat step 3–5, and design the other layers of intermediate casing above the well depth D_2 until the surface casings setting depth are all determined.
- Liner casing design. Liner casing needs be ran down and its setting depth D_4 needs to be determined when the setting depth D_2 of intermediate casing is less than the pre-selection depth D_3.

- Firstly, determine the maximum running depth D_5 of liner casing. The equivalent mud density of safe fracturing pressure $\rho_{ff}D_2$ corresponding to well depth D_2 can be searched in the pressure profile. According to the constraint condition Eq. 7.3 of preventing fracturing the formation in normal drilling, consider $\rho_{bn\,max}D_2 = \rho_{ffD_2}$, $\rho_{bn\,max}D_2$ is the equivalent mud density of the allowed maximum wellbore pressure at the level of well depth D_2. Calculate the allowed equivalent mud density of the maximum formation pore pressure $\rho_{d\,max}D_5$ in the interval from D_2 to maximum linear running depth D_5 according to Eq. 7.4. Then use Eq. 7.1 to calculate the allowed equivalent mud density value of the maximum formation pore pressure $\rho_{p\,max}D_5$ of the well section interval from D_2 to D_5. Find the point of $\rho_{p\,max}D_5$ in the horizontal coordinates and draw a vertical line upward from this point to cross with the equivalent mud density curve of formation pore pressure. The crossing point closest to well depth D_2 (if there are several crossing points) is the maximum liner running depth D_5. After determining D_5, the differential pressure sticking check during drilling in the liner section and the formation fracturing and leakage check in well kick shut down.
- Investigate and check the risk of drill pipe sticking associated with the differential pressure when drilling in the liner section or during running liner casing. The check method is the same as step 5.
- Check whether there is a risk of fracturing the weak formation and leading to leakage after the well kick shutdown. According to the maximum formation pore equivalent pressure density $\rho_{p\,max}D_5$ and the corresponding well depth encountered in the liner running section, the equivalent mud density of the maximum well pressure $\rho_{ba\,max}D_2$ of D_2 is calculated by using Eq. 7.8. If $\rho_{ba\,max}D_2$ is less than the equivalent mud density of safe formation fracture pressure of D_2 $\rho_{ba\,max}D_2$, then the maximum liner running depth D_5 can meet the design requirements. Otherwise, the liner running depth should be reduced and calculated according to the constraint condition of preventing fracturing the formation after well kick shutdown again.
- After completing the check of differential pressure sticking during drilling in the liner running section and the formation fracturing leakage after well kick shutdown. If $D_5 > D_3$, then the final liner running depth $D_4 = D_5$; otherwise, another layer of liner should be designed according to step 7.

Example 7.1 The designed well depth is 4400 m, the equivalent mud density profile of formation pore pressure and formation fracture pressure is shown as Fig. 7.3. Suppose the design coefficients: $S_b = 0.036$ g/cm^3, $S_g = 0.04$ g/cm^3, $S_k = 0.06$ g/cm^3, $S_f = 0.03$ g/cm^3, $\Delta p_N = 12$ MPa, $\Delta p_A = 18$ MPa. Please try to design the structure of this well.

Solution Figure 7.3 shows that the equivalent drilling fluid density corresponding to the maximum formation pore pressure is 2.04 g/cm^3 at the depth of 4250 m.

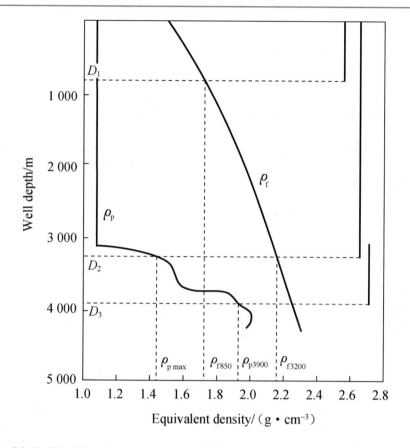

Fig. 7.3 Profile of formation pore pressure, and fracturing pressure

- Determine the initial selection point D_{21} of intermediate casing running depth.

From Eqs. 7.2, 7.5, 7.7, and 7.8, it can be obtained that if the well kick occurs when drilling the formation with the largest pore pressure, in order to ensure that the well leakage does not occur at the intermediate casing shoe, the equivalent mud density expression of the formation fracture pressure this layer should have is:

$$\rho_f \geq \rho_{p\,max} + S_b + S_f + \frac{4250}{D_{21}} S_k$$

Assuming $D_{21} = 3400\,\text{m}$ and putting the values into the equation above, it can be obtained:

$$\rho_f \geq 2.04 + 0.036 + 0.03 + \frac{4250}{3400} 0.06 = 2.181\,\text{g/cm}^3$$

It can be obtained from Fig. 7.3 that at depth of 3400 m, $\rho_{p3400} = 2.19\,\text{g/cm}^3$. Since $\rho_{p3400} > \rho_f$, the initial selection point of the intermediate casing running depth is determined to be $D_{21} = 3400\,\text{m}$.

- Check the possibility of drilling pipe sticking when running down the intermediate casing to the initial selection point of 3400 m.

It can be further obtained in Fig. 7.3 that at 3400 m, $\rho_{p3400} = 1.57\,\text{g/cm}^3$, $\rho_{p\min} = 1.07\,\text{g/cm}^3$, $D_n = 3050\,\text{m}$. Then, we can obtain the differential pressure using Eq. 7.6:

$$\Delta p = 0.00981\,D_n\left(\rho_{p3400} + S_b - \rho_{p\min}\right)$$
$$= 0.00981 \times 3050 \times 1.57 + 0.036 - 1.07$$
$$= 16.037\,\text{MPa}$$

Because of the condition $\Delta p > \Delta p_N$, the intermediate running depth should be shallower than the initial selection point.

When $\Delta p_N = 12\,\text{MPa}$, the allowed equivalent mud density of the maximum formation pressure can be obtained by coupling Eqs. 7.6 and 7.1:

$$\rho_{p\max} = \frac{12}{0.00981 \times 3050} + 1.07 - 0.036 = 1.435\,\text{g/cm}^3$$

Hence, the corresponding well depth at $\rho_{p\max} = 1.435\,\text{g/cm}^3$ is 3200 m by referring to the formation pressure equivalent mud density curve. The intermediate casing running depth $D_2 = 3200\,\text{m}$. A liner casing needs to be run down since $D_2 < D_{21}$.

- Determine the running depth of liner casing D_3

The liner casing running depth initial selection point is determined to be D_{31}. It can be obtained from Fig. 7.3, the formation fracture pressure gradient $\rho_{f3200} = 2.15$ g/cm^3 at depth of D_{31}. Put the value into Eqs. 7.2, 7.5, 7.7, and 7.8 and then the allowed equivalent mud density of the maximum formation pressure that does not fracture and leads to leakage at the intermediate casing shoe is:

$$\rho_{p\max} = 2.15 - 0.036 - 0.03 - \frac{D_{31}}{3200} \times 0.06$$

Supposing $D_{31} = 3900\,\text{m}$, and putting it into the equation above, we obtain

$$\rho_{p\max} = 2.01\,\text{g/cm}^3$$

In Fig. 7.3, the equivalent mud density of formation pressure at 3900 m is $\rho_{p3900} = 1.94$ g/cm^3. Since $\rho_{p3900} < \rho_{p\,max}$, and the difference is little, we obtained that the initial selection depth of liner casing is $D_{31} = 3900$ m.

- Verify the risk of differential pressure casing sticking in the process of running liner casing to the initial selection depth of 3900 m.

According to Eq. 7.6, it can be known that:

$$\Delta p = 0.00981 \times 3200 \times (1.94 + 0.036 - 1.435) = 16.983 \, \text{MPa}$$

Since we have found $\Delta p < \Delta pA$, the running depth of liner casing is determined to be $D_3 = D_{31} = 3900$ m, which can meet the design requirements.

- Determine the surface casing running depth D_1

According to Eqs. 7.2, 7.5, 7.7, and 7.8, it can be known that if well kick occurs when drilling to depth of D_2, then the equivalent mud density of the minimum formation fracture pressure at this position to prevent well leakage at the surface casing shoe is expressed as:

$$\rho_{f\,min} \geq \rho_{p\,max} + S_b + S_f + \frac{D_2}{D_1} S_k$$

Suppose $D_1 = 850$ m and put each value into the equation above:

$$\rho_{f\,min} \geq 1.435 + 0.036 + 0.03 + \frac{3200}{850} \times 0.06 = 1.727 \, \text{g/cm}^3$$

In Fig. 7.3, we found that that at well depth of 850 m, $\rho_{850} = 1.740 \, \text{g/cm}^3$. Because $\rho_{850} > \rho_{f\,min}$ and the two values are fairly close, the design requirement is satisfied. The Casing program design result is provided in Table 7.1.

Table 7.1 Casing program design result of a well

Casing program	Surface casing	Intermediate casing	Linear casing	Production casing
Setting depth (m)	850	3200	3900	4400

7.1.5 Size Selection and Coordination of Casing and Wellbore

In the design of a casing program, in addition to determining the setting depth of each level of casing, they should be matched reasonably to determine the casing size and wellbore dimension. The optimized selection and matching of casing and borehole (drill bit) dimensions result in a smooth progress and reduced cost in oil production, exploration, and drilling processes. In a casing program design, after determining the levels and their corresponding setting depths, the casing size and wellbore size should also be matched reasonably. The selection and matching of casing size and borehole (bit) size are related to the smooth operation and reasonable cost reductions in production, stimulation, and drilling phases.

7.1.5.1 Factors Considered in the Selection and Matching of Casing and Wellbore Size

- The determination of the size of casing and borehole generally starts from the internal to the outside level sequentially. Firstly, we determine the size of the production casing, secondly determine wellbore dimension corresponding to production casing, and then determine the sizes of each level of intermediate casing and the sizes of corresponding wellbore. By repeating of this workflow, we finally determine the wellbore size of surface casing and determine the size of conductor in the same manner.
- The dimension of production casing should satisfy the requirements of exploration and production engineering. For the production wells, the production casing size should be determined according to the productivity of the reservoir, the size of the tubing, the requirements of stimulation measures and the downhole operation. For the exploration wells, the requirements of drilling smoothly to the target layer and the requirements of exploration on the target formation wellbore size should be satisfied.
- For the exploration wells, it is necessary to consider whether the original target depth needs to be increased. For areas where there are uncertain geological conditions and geological data, the margin should be taken into account such that the number of layers of the technical casing can be increased during construction.
- The size of the casing and the wellbore (drill bit) should match to ensure running casing safely and meet the requirements of cementing, such as borehole conditions, curvature value, hole angle, the bottomhole pressure wave in running casing, geological complexity, and so on.

7.1.5.2 The Standard Combination of Casing and Wellbore Size

Figure 7.4 illustrates the selection map of matching of casing and wellbore (drill bit) size used in industry standard. We first determine the size of the last casing layer (or liner casing) by using this map. The solid arrow represents the typical

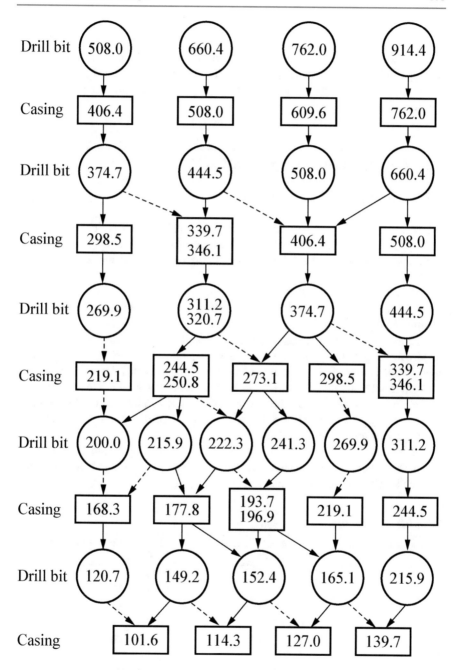

Fig. 7.4 Selection roadmap of matching casing and wellbore (drill bit) size (the unit of all data is mm)

match, which creates adequate gap for running casing and injecting cement slurry. Meanwhile, the dotted arrow represents non-typical match. When choosing the combination of sizes represented by dotted arrows, the impacts of casing joint, drilling fluid density, cement injection method, and the effect of wellbore curvature on running casing and well cementing quality should be fully taken into consideration.

7.2   Casing String Strength Design

7.2.1   Casing and Casing String

Oil well casing consists of seamless pipes or welded pipes which are made of high-quality steel, and both ends of pipes are processed into tapered threads. Most of the casings are connected with casing couplings to setup a casing string, which is used for cementing.

Oil well casing has its special standards, and each type of casing should comply with standards. China's current casing standards are similar to American Petroleum Institute (API) standards.

The API standard specifies that the steel of the casing body should meet the required strength, which is expressed in steel grade. The API standard does not specify the composition of casing steel, but it requires that the minimum yield strength of steel should be ensured. The number following steel grade represents the minimum yield strength, and the unit is kpsi (1 kpsi = 6.895 MPa). API standard divides casing pipe steel grades into eight categories of H, J, K, N, C, L, P, and Q, totaling to 10 grades, as shown in Table 7.2.

API standards require that the outside diameter of casing should be of standard dimensions. The common standard casing outside diameters range from 114.3 mm (4½ in) to 508 mm (20 in) which totally have 14 types:

Table 7.2 API standard casing strength

API standard casing steel grade	Minimum yield strength [MPa (kpsi)]	Minimum tensional strength [MPa (kpsi)]
H-40	275.79 (40)	413.69 (60)
J-55	379.21 (55)	517.11 (75)
K-55	379.21 (55)	655.00 (95)
C-55	517.11 (75)	655.00 (95)
L-80	551.58 (80)	655.00 (95)
N-80	551.58 (80)	689.48 (100)
C-90	620.53 (90)	689.48 (100)
C-95	655.00 (95)	723.95 (105)
P-110	758.42 (110)	861.84 (125)
Q-125	861.84 (125)	930.79 (135)

114.30 mm ($4\frac{1}{2}$ in)	244.47 mm ($9\frac{5}{8}$ in)
127.00 mm (5 in)	273.05 mm ($10\frac{3}{4}$ in)
139.70 mm ($5\frac{1}{2}$ in)	298.44 mm ($11\frac{3}{4}$ in)
168.27 mm ($6\frac{5}{8}$ in)	339.71 mm ($13\frac{3}{8}$ in)
177.80 mm (7 in)	406.40 mm (16 in)
193.67 mm ($7\frac{5}{8}$ in)	473.08 mm ($18\frac{5}{8}$ in)
219.07 mm ($8\frac{5}{8}$ in)	508.00 mm (20 in)

Casing strings have wall thicknesses ranging from 5.21 to 16.13 mm. Usually, the wall thickness of small-diameter casing is smaller than that of large-diameter casing.

In addition to standard steel grade and wall thickness, there are non-standard steel grades and wall thicknesses, which are also allowed by API standards.

The couplings of casings are all tapered threads. There are four types of standard casing connector: short round thread connector (STC), long round thread connecter (LTC), buttress-thread connector (BTC), extreme line thread (XL), and non-standard thread in addition.

The taper of the trapezoidal thread is 1:16, and the thread pitch is 3.175 mm (8 threads/in); the taper of tapered thread is 1:16 (external diameter < 16 in) and 1:12 (external diameter ≥ 16 in), and the pitch is 5.08 mm (5 threads/in).

The casing string usually consists of connecting casings of the same outside diameter, the same or different steel grades and the different wall thickness by couplings, and it should satisfy the requirements of strength and production.

7.2.2 The Calculation of Casing String Loadings and Strengths, and Casing Erosion

7.2.2.1 The Calculation of Casing String Loadings

In the strength design of casing string, the key problem is to determine the most dangerous loadings that casing bears in the whole operation process properly. At present, in the design of casing strings, the various loadings during drilling and production process are mainly divided into three types: axial force, internal pressure, and external force. The causes of these loadings vary. The purpose of load analysis is to find out the most dangerous combination of loads according to the actual working conditions in drilling, completion, and production, and to select the combination of casing to meet the strength requirements and production needs, taking into account the effects of wear, corrosion, and high temperature on casing strength, so as to maximize the service life and economic benefits of oil wells.

Axial force of casing

The possible axial force of casing includes gravity, buoyancy, inertia force, impact force, friction force, bending force, and additional axial force caused by temperature and pressure changes during well operation.

- The Axial Tension Generated by Casing Gravity

The axial tensional force generated by the gravity of casing strings increases gradually from the bottom to top in the casing string, and the maximum axial tension is sustained by casing at wellhead, the tension F_m is:

$$F_m = \sum qL_iK_B \times 10^{-3} \qquad (7.9)$$

$$K_B = 1 - \frac{\rho_d}{\rho_s}$$

where

F_m The axial tension on casing at the wellhead, kN;
q The casing weight of per unit of casing, N/m;
L_i The length of ith casing pipe (i = 1, 2, 3, …, n), m;
K_B Buoyancy coefficient;
ρ_d Drilling fluid density, g/cm^3;
ρ_s Casing steel density, g/cm^3.

In the field design of casing tensile strength, buoyancy reduction in drilling fluid is generally not considered. Axial tension is usually considered by the gravity of casing in the air. It is considered that drilling fluid leakage or gas drilling may occur under dangerous conditions.

- The additional axial force produced by casing bending

When the casing bends with the borehole, the tension load of the casing is increased by the bending of the casing. The additional axial force caused by bending can be calculated by the simplified empirical equation when the bending angle change and the bending change rate are little:

$$F_{bd} = 0.0733d_{co}\theta A_c \qquad (7.10)$$

where

F_{bd} The addition axial force generated by bending, kN;
d_{co} The casing external diameter, cm;
A_c Cross-sectional area, cm^2;
θ The deviation angle change per 25 m, (°)/25 m.

Additional axial forces due to casing bending should be considered in large inclination wells, horizontal wells, and drastic bending of the wellbore.

- The additional axial force of casing caused by the injection of cement slurry

During pumping cement slurry, if the amount of cement slurry is large, the density difference between cement slurry inside and the liquid outside is large, and when cement slurry does not return to the bottom of casing, the liquid inside the casing is heavy, which will cause a tensile stress on the casing string, which can be calculated approximately by the following equation:

$$F_c = \frac{\pi}{4} h \frac{\rho_c - \rho_d}{1000} d_{ci}^2 \qquad (7.11)$$

where

F_c The additional axial force generated by injecting cement slurry, kN;
h The cement slurry altitude inside the casing pipe, m;
ρ_c The density of cement slurry, g/cm^3;
ρ_d The density of drilling fluid, g/cm^3;
d_{ci} The casing internal diameter, cm.

- Other additional axial force

The dynamic load during running casing, such as the additional force of lifting casing or braking, or the change of pump pressure during injecting cement slurry can produce some additional force. These forces are difficult to calculate, usually buoyancy mitigation is used to offset or increase the safety coefficient.

In addition, the casing will be subject to the impact of temperature, causing the expansion of the casing in unconsolidated place which will also produce additional stress. If the temperature changes greatly, the additional temperature stress generated will be large, and this problem should be resolved.

External loadings on casing

Casing external loadings refer to the hydrostatic pressure of annular fluid and formation pressure on the outside wall of the casing in the well. The external loading exerted on a casing string comes mainly from liquid column outside the pipe, formation pressure, lateral extrusion force of high plastic rock, and the pressure generated by other operations. Casing strength design adopts the effective external force, that is, the value of the casing external loading minus internal pressure.

For non-plastic formation, the external loadings on casing are calculated according to the consideration of the drilling fluid column pressure in well cementing under the condition of a certain hollowing degree:

$$p_{ce} = 0.00981[\rho_d - (1 - k_m)\rho_{min}]D \tag{7.12}$$

where

p_{ce} The effective external collapse pressure, MPa;

k_m The hollowing coefficient (ranges from 0 to 1, 0 shows that the pipe is fully filled with fluid, and 1 indicates that the pipe is totally hollowed);

ρ_{min} The least fluid density, the minimum drilling fluid density of the next interval is adopted for surface casing and technical casing, and the well completion fluid density for production casing, g/cm^3;

D The vertical depth, m.

For plastic formation, under certain conditions, the lateral pressure generated by rock gravity in the vertical direction will be exerted on the casing, and the casing may be damaged if the maximum lateral extrusion force is all applied on the outside of casing. The effective external collapse loading can be expressed as:

$$p_{ce} = 0.00981\left[\frac{\mu}{1-\mu}G_o - (1 - k_m)\rho_{min}\right]D \tag{7.13}$$

where

μ Poisson's ratio of rock;

G_o Overburden pressure gradient, MPa/m.

Internal pressure

The internal pressure of casing refers to the pressure of the fluid in the pipe on the inner wall of the casing. The source of the internal pressure on the casing string comes from the pressure of formation fluid (oil, gas, and water) flowing into the casing and the external pressure of special operation (fracturing, acidification, and water injection) during the production process. The formation pressure is difficult to identify before drilling a formation in a new area; therefore the internal pressure of casing is uncertain. As for a known area, the datum of neighboring wells can be used as references. When the wellhead is open, the casing pressure equals to the pressure of the fluid in the pipe. When the wellhead is closed, the casing internal pressure is the sum of wellhead pressure and the fluid pressure. The effective internal pressure is used in designing the casing strength, that is, the actual pressure of the casing internal pressure offset with external pressure. The effective internal pressure can be ensured in the following way:

- Considering that the whole well is filled with natural gas, the internal pressure at any depth of the well is as follows:

$$p_{bh} = \frac{p_g}{e^{1.1155 \times 10^{-4}(D_s - h)\gamma_g}} \tag{7.14}$$

The effective casing internal pressure can be calculated through the following equation:

$$p_{be} = p_{bh} - 0.00981\rho_w D \tag{7.15}$$

where

p_{bh} The maximum casing internal pressure at any well depth, MPa;
p_g The natural gas pressure at the bottom, MPa;
p_{be} The effective casing internal pressure at any well depth, MPa;
D_s The setting depth of casing shoe, m;
h Depth for calculation, m;
γ_g The relative density of natural gas, which is 0.5–0.55;
ρ_w The formation water density, which is 1.03–1.06 g/cm^3.

- According to the maximum drilling fluid density in the next drilling process, the effective casing internal pressure at any well depth is:

$$p_{be} = 0.00981(\rho_{d\,max} - \rho_w)h \tag{7.16}$$

where

$\rho_{d\,max}$ The maximum drilling fluid density in the next drilling process, g/cm^3

- Regard the fracturing pressure at the casing shoe as the maximum wellhead pressure, that is, the sum of wellbore shut-in pressure and fluid column pressure in the pipe should be less than the formation fracture pressure at the casing shoe, then the effective casing internal pressure at any well depth is:

$$p_{be} = \rho_{head} + 0.00981(\rho_{d\,max} - \rho_w)h \tag{7.17}$$

where

ρ_{head} Wellhead pressure, g/cm^3

- The pressure-bearing capacity of the wellhead BOP is regarded as the wellhead pressure, and the calculation is the same with Eq. 7.17.

7.2.2.2 Casing Strength

The casing strength is the capacity value that the casing can bear internal and external loadings. The casing strength can be divided into tensile strength, collapse resistance, internal pressure resistance, and triaxial stress yield strength.

Tension strength

Tensile strength refers to the maximum tensile stress the casing can withstand under tensile action. The axial tension on the casing string peaks at the wellhead; therefore, the wellhead is the most dangerous section. Tensile stress can lead to two types of casing string damage: one is casing body being pulled off, and the other one is thread slipping. A large number of indoor experiments and field applications show that when the casing is loaded in tension, thread slipping occurs more often than casing body pulling off, especially when using the most common round thread casing.

The slipping load of round thread casing is usually less than the casing yield strength. The thread slipping load of each type of casing is given, and it is usually expressed in the total pulling force (kN) of screw thread slipping. In the design, it can be searched directly in the related casing manual.

Collapse resistance

Anti-collapse strength refers to the pressure value that the casing can withstand under the action of external collapse loads.

- The anti-collapse strength ignoring axial force

Originally, API did collapse tests on 2488 casings of three steel grades—K-55, N-80, and P-110. The result showed that there are four types of casing damage under single external collapse strength, that is, yield collapse, plastic collapse, transitional collapse, and elastic collapse. The collapse type is related to the radius-thickness ratio (the ratio of casing diameter to thickness). The anti-collapse strength calculation equation of the four types of damage can refer to the casing strength design standard of API BULL 5C3, ISO 10400, or SY/T 5724-2008. The anti-collapse strength provided in the casing strength design standard or manual takes no consideration of the axial force.

- The anti-collapse strength considering axial force

In practical applications, casing is under the action of biaxial stress. That is, the upper casing section is bearing the tension of the lower casing section in the axial direction, and the casing internal pressure and the collapse pressure outside the casing in the radial direction. Due to the presence of axial tension, the capability of the casing to withstand internal pressure or external collapse pressure will change.

Suppose the axial tension generated by the casing gravity is σ_z, the circumferential stress created by external collapse pressure or internal pressure is c, and the radial stress is σ_r. Since the fact that most casings are thin walled tube, σ_r is far less than σ_θ, which can be ignored. So there is need only to consider the two dimensional stress situation of axial tension σ_z and circumferential stress σ_θ. According to the fourth strength theory, the strength condition of casing damage is:

$$\sigma_z^2 + \sigma_\theta^2 - \sigma_\theta\sigma_z = \sigma_s^2$$

where

σ_s The yield strength of casing steel, MPa

This equation can be rewritten as:

$$\left(\frac{\sigma_z}{\sigma_s}\right)^2 - \frac{\sigma_\theta\sigma_z}{\sigma_s^2} + \left(\frac{\sigma_\theta}{\sigma_s}\right)^2 = 1 \tag{7.18}$$

This equation is elliptic equation, which regard $\left(\frac{\sigma_z}{\sigma_s}\right)$ as the horizontal coordinate, and $\left(\frac{\sigma_\theta}{\sigma_s}\right)$ as the vertical coordinate. The stress diagram as Fig. 7.5 can be drawn, which is called the bidirectional stress ellipse.

It can be obtained from Fig. 7.5:

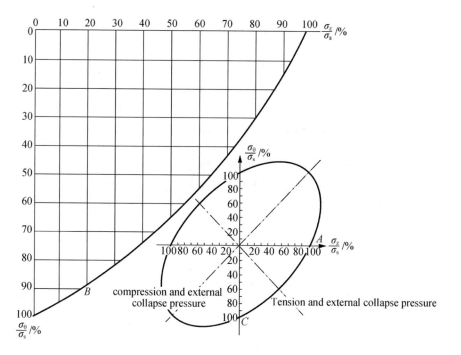

Fig. 7.5 Bidirectional stress ellipse

- The first quadrant is the combined effect of tension and internal pressure, which indicates that the internal pressure strength of the casing increases under the axial tension and the operation of the casing is safe. Therefore, there is not much to consider in this situation.
- The second quadrant is the combined effect of axial compression and casing internal pressure. It is rare for the casing to be subjected to compressive stress, so this situation is generally not considered.
- The third quadrant is the combined effect of axial compression stress and external collapse pressure. This situation will not be considered for the same reason as the second quadrant.
- The fourth quadrant is the combined effect of axial tension stress and the external collapse pressure, which occurs often. It can be known from the diagram that the axial tension stress reduces the collapse strength, which should be taken into consideration in the casing design.

When taking the axial tension stress into consideration, the casing collapse strength is calculated through Eq. 7.19 or the approximation Eq. 7.20.

$$p_{cc} = p_c \left[\sqrt{1 - 0.75 \left(\frac{\sigma_z}{\sigma_s} \right)^2} - \frac{\sigma_z}{2\sigma_s} \right] \tag{7.19}$$

$$p_{cc} = p_c \left(1.03 - 0.74 \frac{F_m}{F_s} \right) \tag{7.20}$$

where

p_{cc} The maximum casing anti-collapse strength taking the axial tension into consideration, MPa;

p_c The casing anti-collapse strength without the axial tension, MPa;

F_m The axial tension, kN;

F_s The tensile yield strength of casing, kN.

In the equation, p_c and F_s can be checked in the casing manual, the ratio of calculation error to theoretical calculation value of Eq. 7.20 is less than 2% in the range of $F_m/F_s \leq 0.5$.

Internal pressure resistance

Internal pressure resistance refers to the ability of the casing to withstand the internal pressure damage. Casing has three types of damage under the impact of internal pressure, that is, tube rupture, coupling leakage, and coupling cracking. Under normal circumstances, the coupling leakage pressure is less than the tube rupture and coupling crack pressure. Coupling leakage pressure is related to thread type, and it is not easy to calculate. For casing requiring high internal pressure resistance, the high-quality lubricating seal grease should be applied to the thread, and the thread should be tightened according to the specified torque.

The pipe rupture strength is generally obtained through *Barlow* equation:

$$p_{bo} = 0.875 \left(\frac{2\sigma_s t}{d_{ci}} \right) \tag{7.21}$$

where

p_{bo} The anti-internal pressure strength of casing tube, MPa;
T The casing wall thickness, mm;
d_{ci} The casing inside diameter, mm.

The permissible internal pressure values for the various casings are specified in the casing manual, and they can be found from the casing manual in the designing process.

Triaxial stress yield strength

The triaxial stress yield strength refers to the casing strength under the combined force of external pressure, collapse pressure, and axial tension. When the casing body is subjected to von Mises equivalent stress to the state of the minimum material yield strength, the casing begins to yield. Under the action of triaxial stress, the casing is subjected to axial stress σ_z, circumferential stress σ_θ, and radial stress σ_r. As shown in Fig. 7.6 σ_r, σ_θ, and σ_z, all take tensile stress as positive, and compression stress as negative.

The following assumptions are made to calculate the casing triaxial stress yield strength:

- The internal and external surfaces of the casing string are concentric cylinder surface;
- Isotropic yield;
- Ignoring the residual pressure;
- Ignoring elastic unstability of crossing section and the axial buckling of the casing string.

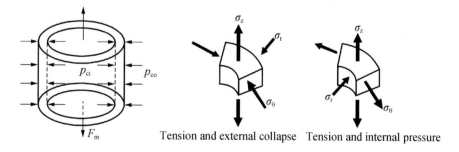

Fig. 7.6 Casing triaxial stress schematic diagram

The stress distribution of the casing string is given by the Lame equation of elasticity and considering the axial symmetry of the casing string and the radial distribution of the axial stress in the radial direction.

$$\sigma_r = \frac{p_{ci}r_{ci}^2 - p_{co}r_{co}^2}{r_{co}^2 - r_{ci}^2} - \frac{(p_{ci} - p_{co})r_{ci}^2 r_{co}^2}{r_{co}^2 - r_{ci}^2}\frac{1}{r^2} \tag{7.22}$$

$$\sigma_\theta = \frac{p_{ci}r_{ci}^2 - p_{co}r_{co}^2}{r_{co}^2 - r_{ci}^2} + \frac{(p_{ci} - p_{co})r_{ci}^2 r_{co}^2}{r_{co}^2 - r_{ci}^2}\frac{1}{r^2} \tag{7.23}$$

$$\sigma_z = \frac{F_t \times 10^3}{\pi\left(r_{co}^2 - r_{ci}^2\right)} \tag{7.24}$$

where

σ_r The radial stress, MPa;

σ_θ The circumferential stress, MPa;

σ_z The radial stress, MPa;

r_{ci} The casing inner radius, mm;

r_{co} The casing external radius, mm;

R The casing radius at any wall thickness, mm;

p_{ci} The casing internal pressure, MPa;

p_{co} The casing collapse pressure, MPa;

F_t The casing axial force at the calculated position, kN.

Equations 7.22 and 7.23 show that under the action of triaxial stress, the radial stress, and the magnitude of the circumferential stress are related to the internal and external pressure differential, which is also related to the casing radius r. Casing strength design is most concerned with maximum radial stress and circumferential stress. Theoretical deduction shows that when the casing is not subjected to bending stress, the inner wall of the casing yield first. According to the casing triaxial stress equation and the von Mises yield criterion, the von Mises equivalent stress σ_{VME} is equal to the yield stress of the casing material when the inner wall of the casing appears to yield, namely $\sigma_{VME} = \sigma_S$.

$$\sigma_{VME} = \frac{\sqrt{2}}{2}\sqrt{(\sigma_r - \sigma_\theta)^2 + (\sigma_r - \sigma_z)^2 + (\sigma_z - \sigma_\theta)^2} \tag{7.25}$$

Then the safety coefficient S_3—of casing triaxial stress yield strength is:

$$S_3 = \frac{\sqrt{2}\sigma_s}{\sqrt{(\sigma_r - \sigma_\theta)^2 + (\sigma_r - \sigma_z)^2 + (\sigma_z - \sigma_\theta)^2}} \tag{7.26}$$

7.2.2.3 Casing Erosion

Since the casing is to be used for a long time in the ground and will come into contact with various fluids which will cause corrosion of the casing material, as a result, the effective thickness of the tube is reduced and the casing-bearing capacity is reduced, or the properties of the steel are changed, resulting in decrease in casing-bearing capacity.

The main mediums that cause casing corrosion are gas and hydrogen sulfide (H_2S), dissolved oxygen, and carbon dioxide in fluid.

Hydrogen sulfide is mainly present in natural gas and is a chemical substance that has a strong destructive effect on casing steel. The hydrogen sulfide contact with casing steel will cause steel "hydrogen embrittlement" and fracture the casing. In environment of low pH, the corrosion of hydrogen sulfide is more intense.

As for the casing that may contact hydrogen sulfide, the anti-sulfide casing is an option. For example, the H-class, K-class, J-class, C-class, and L-class of the APT casing series can also be used in high alkaline environment.

As for other types of casing corrosion, the antiseptic substances, cathodic protection, and casing anticorrosive coating can be used for anti-corrosion.

7.2.3 The Design Principles of Casing Strength

The strength design of the casing string is based on the external load of the casing, and a safe balance relation is established according to the strength of the casing:

The casing strength > External load × Safety coefficient

The strength of the casing string is designed according to the requirements of the technical department. After determining the external diameter of the casing, the casing of different steel grades and wall thickness is selected according to the external load on the casing and the safety factor. The above safety balance is established in every dangerous section of the casing string. The casing string strength design must ensure that the maximum stress acting on the casing is within the allowable range during the entire life of the well.

The strength design principles should take the following aspects into consideration:

- Satisfy the need of drilling operation, reservoir development, and producing formation treatment;
- Having a certain storage capability when bearing the external load;
- Being economical.

In the design process of casing strength, the safety factor given in China's petroleum industry standard ranges from: anti-collapse safety factor $S_c = 1.00-1.125$; anti-internal pressure safety factor $S_c = 1.05-1.15$; anti-tension safety factor $S_t = 1.60-2.00$.

7.2.3.1   Common Design Methods of Casing String Strength

Many countries have specified their own casing string strength design method according to their own conditions. The most common methods are equal safety factor method, boundary load method, maximum load method, AMOCO method, BEB method, and Soviet method.

The casing string strength design is usually carried out from bottom to top. Conventionally, the casing at the bottom is called the first casing interval, the casing above the first interval is the second interval, and so on.

The equal safety factor method

The basic design idea of this method is to make the minimum safety factor on each dangerous cross section equal to or greater than the specified safety factor. In the design process, whether the anti-collapse strength can meet the demand is considered first. Usually the casing above the cemented section should also take into account the biaxial stress, the upper casing section should meet the requirements of anti-tension and anti-internal pressure.

The boundary load method

The boundary load method, also known as the overpull margin method, of which the anti-collapse strength design is the same as the safety factor method. The tension-resistant design is designed through the boundary load value determined by the casing tensile strength and safety factor of the interval of anti-tension design, and this value is regarded as the tensile strength standards designing of the upper casing. The design method is the available strength of the first casing interval based on the tensile strength design = tensile strength/tensile safety factor; boundary load (overpull margin) = tensile strength − available strength; the available strength of the second casing interval based on the tensile strength design = tensile strength − boundary load; the subsequent sections use the same boundary load to select the available strength.

The advantage of this design method is that the boundary loads of the intervals of the casing string are equal, so that the overpull margins of the casings under tension are the same, and the casing waste can be avoided.

The maximum load method

This method is proposed in USA, of which the design idea is to divide casings into the technical casing, surface casing, production casing, etc., and design each type of casing according to the properties and size of its external load. The design method is to select the casings based on the internal pressure and then design the strength according to the effective external collapse and tensile stress. The method is of meticulous consideration on the external precise design.

AMOCO design method

The design method unique features: consideration of the influence of tension in the anti-tension design; design according to the biaxial stress; taking the force at the coupling into consideration when calculating the external load; and the influence of the tensile stress is also taken into account when calculating the internal pressure. The analytic method and the graphic method are combined in the design.

BEB design method

This method is almost about graphic method, which mainly designs according to casing classifications. The impacts of tensile stress must be taken into consideration in designing anti-collapse and anti-internal pressure strength. The tensile stress is always calculated on the basis of the buoyancy in the drilling fluid, and considering that the impacts of buoyancy on the section of the bottom of the casing to make the bottom under positive pressure.

The Soviet design method

The design method is relatively cumbersome, and the design idea is to consider the external load according to different periods of change, taking into account the different anti-tension factors in different well sections, and do not consider the biaxial stress. However when the tensile stress reaches 50% of the tube yield strength, the anti-tension safety factor should be raised to 10%.

In the casing string strength design, the casing intervals should not be too complicated. In fact, the casings of 2–3 types of steel grade may be used. The wall thickness should only target on 2–3 species. Although the complicated casing intervals with many classifications are in line with economic principles, this will also cause more trouble in on-site management.

7.2.3.2 The Design Features of Each Casing Level

As for the design of the surface casing, technical casing and production casing, each has its design features and design focus.

The design features of surface casing

The surface casing is a layer of casing for the consolidation of the surface loose formation and the installation of the blowout preventer. It also bears the weight of the lower casing. Therefore, the surface casing is designed to withstand formation pressure when there is possible underground gas invasion or blowout. The anti-internal pressure is mainly taken into consideration to prevent collapsing the casing due to the high pressure when shut-in the well.

The design features of technical casing

The technical casing is a layer of casing that is run down to sequester the complex formation, which is subjected to the internal pressure and collision and wear of drilling tools in the blowout. Technical casing design is characterized by both high anti-internal pressure strength and the ability to resist the impact wear.

The design features of production casing

The production casing is a layer of casing finally run down in the oil and gas wells, and in which the oil tube is run down for the production. The anti-collapse capability of the oil layer casing should be taken into serious consideration since the running depth is relative large. The casing should be considered separately for the issue it may encounter in production. For example, some of the wells are used for water injection, and some wells need to be fractured or acidified in production, etc. The casing may also be subject to greater internal pressure, the production casing of this kind of well should be strictly checked for internal pressure strength. The production of some wells is delivered through thermal recovery such as hot steam injection, and the casing under long-term thermal effect will be inflated, resulting in greater compression stress. Therefore, the design should consider the tensile safety factor and other aspects of the initial tensile stress.

7.2.3.3 The Equal Safety Factor Method

The equal safety factor method is a relatively simple method; it has been applied for a long time, which is relatively safe in general wells. The force of the casing string varies with the well depth as shown in Fig. 7.7. It can be seen from the figure, the

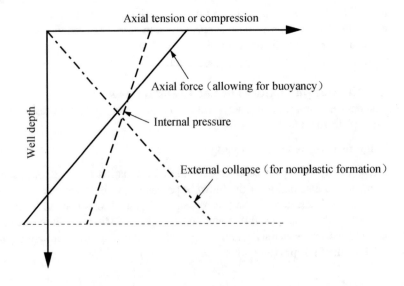

Fig. 7.7 Schematic diagram of casing string stress with the well depth change

axial tension, external collapse, and internal pressure vary in different casing string cross sections. The axial tension and external collapse decline from the bottom to top, and the effective stress of internal pressure inclines from the bottom to top. In order to maximize the strength of the casing string and to save the cost as much as possible in the design, the casing string should have different strength in different well sections. Therefore, the designed casing string is consisted of casings of different steel grades and wall thicknesses.

Firstly the design is usually based on the maximum internal pressure load to select the casing meeting the requirements of internal pressure strength and then continue according to the external collapse load from the bottom to the top. Finally, design and check the casing according to the casing axial tension.

The specific methods and steps of designing the safety factor method are:

- Calculate the maximum internal pressure may occur of this well and select the casings meeting the internal pressure strength. If it is the normal well, the internal pressure strength can be implemented after the completion of the whole casing string design.
- Primarily select the first interval of casing based on the maximum effective external collapse p_{ce} in the whole well. It can be calculated through Eqs. 7.12 and 7.13, and the allowed anti-collapse strength p_{c1} of the first casing interval must be greater than p_{ce}.
- Choose the casing of a lower level of thickness or a lower steel grade (or both lower) for selecting the second casing interval. The available running depth of the casing can be D_2, if Eq. 7.12 is adopted for calculating the effective external collapse pressure (the inner part of the tube is fully empty), then:

$$D_2 = \frac{p_{c2}}{9.81\rho_d} \tag{7.27}$$

where

p_{c2} The anti-collapse strength of the second casing interval, MPa

Based on this the allowed length L_1 of the first interval of casing is:

$$L_1 = D_1 - D_2 \tag{7.28}$$

In the equation,

D_1 The available running depth of the first casing interval, m

Calculate the casing gravity in the air according to the length L_1 of the first interval, that is $L_1 q_1$. Then check whether the tensile safety factor S_{t1} at the top of this casing interval is greater or equal to the designed tensile safety factor S_t.

$$S_{t1} = F_{s1}/(L_1 q_1) \geq S_t \tag{7.29}$$

In the equation,

q_1 The weight of casing per unit of length, kN/m;
F_{s1} The casing available tensile resistance, kN;
S_t The designed tensile safety factor.

- When the casing string designed according to anti-collapse strength exceeds the cement surface or the neutral point, consideration should be given to the reduction of the anti-collapse strength caused by the floating of the lower casing. Then the casing string can be designed based on the biaxial stress.

Calculate the reduced anti-collapse strength according to Eqs. 7.19 and 7.20 and check whether the anti-collapse safety factor can meet the requirements. If no, then extend the lower casing upward through the trial method until the anti-collapse safety factor of the biaxial stress meet the requirements.

Determine the lower casing intervals from the bottom to top likewise. The external collapse pressure on the casing becomes smaller as it climbs higher, so the casing of smaller anti-collapse strength can be taken into use. The tension load generated by casing gravity increases as the casing is at a certain depth, and the external collapse force decreases. Then the casing should be determined according to the tension-resistant design.

- Determine the upper intervals of the casing according to the tension-resistant design. Assume the total weight of the casings below the ith interval from the bottom to top is $\sum_{n=1}^{i-1} L_n q_{mn}$, and the tensile strength of the section is F_{si}, then the tensile safety factor S_t of the top section of the ith interval is:

$$S_t = F_{si} / \left(L_i q_{mi} + \sum_{n=1}^{i-1} L_n q_{mn} \right) \qquad (7.30)$$

where

c	The tensile safety factor;
L_i	The permissible length of the ith casing, m;
q_{mi}	The buoyant weight per unit length of the ith casing, kN/m;
F_{si}	The tensile strength of the ith casing, kN;
$\sum_{n=1}^{i-1} L_n q_{mn}$	The total weight of the casings below the ith interval from the bottom to top, kN.

The length of the ith casing based on the tensile strength design is:

$$L_i = (F_{si}/F_{si} - \sum_{n=1}^{i-1} L_n q_{mn})/q_{mi} \qquad (7.31)$$

Perform the casing design according to Eq. 7.31. If L_i cannot be extended to the wellhead, then we choose the casings of greater tensile strength above the ith casing to calculate and repeat this till the wellhead. With this, the whole casing string design will be completed.

- Verification of the anti-internal pressure safety factor. As for the regular well without selecting casing according to the internal pressure, the internal pressure can be verified through the following equation:

$$S_i = p_{ri}/p_i \qquad (7.32)$$

where

S_i The internal pressure safety factor;
p_{ri} The casing internal pressure safety factor at the wellhead, MPa;
p_i The internal pressure at wellhead, MPa.

The literature shows that, as for middle-deep well or deep well, the casing string designed utilizing the above steps can generally meet the anti-internal pressure requirements when the formation pressure is within the typical pressure gradient. If the actual internal pressure safety factor S_i is less than the specified internal pressure safety factor, then the wellhead pressure should be controlled during well control. The wellhead pressure should be limited to the maximum pressure allowed by the casing (or wellhead). Or the casing string design steps are changed to the internal pressure strength design firstly, then tension resistance design is applied to the casings after meeting the internal pressure strength requirements for tension-resistant design.

7.3 Cementing Technology

The cementing operation is placing an appropriate amount of cement slurry into the annulus between the walls of the hole and the casing. The purposes of this process are to secure the casing mechanically, selling off adjacent oil, gas and water layers and provide a leakproof base for following drilling and other operations. The most common cementing operation is pumping cement slurry into the well, flowing down through the casing and then flowing up through the annulus. Besides, other technologies such as stage cementing, inner string cementing, reverse-circulation cementing, delayed-set cementing, are also employed in some special conditions.

Key factors need to be considered in the cementing process including the choice of cement, cement slurry and cement additives, the well hole preparations, process design, and so on.

Basic standards need to be met including:

- The slurry height in the annulus and the height of cement plug have reached the predetermined standards;
- The drilling fluid in the annulus is replaced by the cement slurry without any residuals.
- The bonding between casing, formation, and cement should be strong enough to withstand acid fracturing and safe enough for running strings.
- The oil, gas, and water in adjacent zones are sealed off from influxes.
- The cement is strong enough to withstand long-term corrosion from formation liquids.

Modern cementing technologies based on above standards are involved in engineering disciplines of chemical, geology, mechanical, and other subjects and can be classified by cementing types, cement additives, cementing technologies. The knowledge and research in these aspects make sure cementing technologies satisfy cementing needs and requirements in conditions such as the complex well, the deep and ultra-deep well, and some special operation wells.

This chapter mainly discusses contents including the oil well cement, cement slurry characteristics and the operation requirements, cement additives and the mechanism of action, cementing technologies, special cement systems, and so on.

7.3.1 Oil Well Cement

7.3.1.1 Basic Requirements and Manufacturing Process of the Oil Well Cement

Portland cement is by far the most widely used oil well cement. It is also named as silicates cement as the main ingredients of Portland cement is silicates. Portland cement is known as hydraulic cement in fine powder. The cement becomes set and continues to develop comprehensive strength as a result of hydration. The setting and hardening occur when cement exits in air or mixed with water.

Portland cement was invented by William Aspdin, a British construction worker in 1824. By heating the mixture of limestone and clay, cementitious materials are produced and its name was derived from its similarity to the Portland stone which is quarried on the Isle of Portland.

The manufacturing process of Portland cement is listed as below (Fig. 7.8):

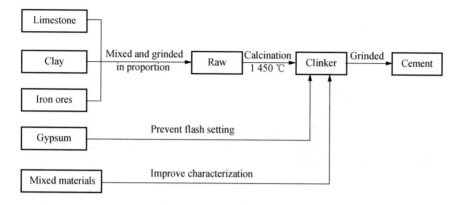

Fig. 7.8 Manufacturing process of oil well cements

The basic requirements for oil well cements are:

- The rheology of the cement slurry should be decent and is required to maintain throughout the process from pumping down to casing to being replaced in the annulus.
- The cement slurry is supposed to be stable in under hole temperature and pressure conditions.
- The cement slurry is required to set in a proper amount of time and develop required strength.
- The additives and cement slurry are compatible to improve the cement characterization.
- The cement stone is supposed to have low permeability.
- The oil well cement refers to special silicate cement that is suitable for oil well conditions based on above standards.

7.3.1.2 Main Compounds of the Oil Well Cement

The main compound composition of oil well cement include:

- Tricalcium silicate(C_3S) compromising 40–60% of the total material is the main compound of oil well cements. It has a significant influence on the cement strength especially at early stages. The content of C_3S for high-strength cement can be as high as 60–65%. In retarded cement, this percentage of C_3S is about 40–45%.
- Dicalium silicate (C_2S) has a proportion of 24–30%. The hydration process of C_2S is slow, and the strength development requires more time, thus having influence to the final cement strength. However, it will not affect the cement setting time.
- Tricalcium aluminate (C_3A) is the compound that can accelerate the hydration process, and it also plays an important role in deciding the cement presetting time and thickening time. C_3A has a significant influence upon the rheology of

the cement slurry and early strength development of the set cement, but it has few effects on the final cement strength.

It should be noticed that C_3A is very sensitive to sulfate, and thus, the quantity of C_3A in sulfate-resistant cement should below 3%. For some cement with high early strength, this amount could be 15%.

- Tetracalcium aluminoferrite (C_4AF), consisting 8–12% of the all compounds, has little influence on the cement strength. The hydration velocity is only lower than that of C_3A and early strength develops rapidly.

Besides four principal compounds discussed above, gypsum and alkali metal oxide also can be found in oil well cements. The typical properties of API cement are seen in Tables 7.3, and the influence of compositions exerts on cement is seen in Tables 7.4.

7.3.1.3 The Hydration of Oil Well Cement

When the cement comes into contact with water, the compounds dissolve and form hydration products with water. The cement slurry is then gradually depositing and setting their excess solids.

Cement hydration

The main chemical reactions are shown in equations below.

$$3CaO \cdot SiO_2 + H_2O- \rightarrow 2CaO \cdot SiO_2 \cdot H_2O + Ca(OH)_2$$
$$2CaO \cdot SiO_2 + H_2O- \rightarrow 2CaO \cdot SiO_2 \cdot H_2O$$
$$3CaO \cdot Al_2O_3 + 6H_2O- \rightarrow 3CaO \cdot Al_2O_3 \cdot 6H_2O$$
$$4CaO \cdot Al_2O_3 \cdot Fe_2O_3 + 7H_2O- \rightarrow 3CaO \cdot Al_2O_3 \cdot 6H_2O + CaO \cdot Fe_2O_3 \cdot H_2O$$

The hydration is a sequence of overlapping and secondary reactions among components, leading to continuous slurry cement slurry thickening and hardening.

The cement hydration is an exothermic reaction, and the heat of hydration reflects the degree of this reaction (Fig. 7.9). Therefore, the hydration rate can be followed by conduction calorimetry.

It can be observed from Fig. 7.9 that the induction time is located around point A, while the setting time is located around point B. The cement surface can be detected by analyzing the hydration heat characteristic. During cementing

Table 7.3 Typical properties of API cement

API class	Composition				Wagner fineness (cm²/g)
	C_3S	C_2S	C_3A	C_4AF	
A	53	24	8	8	1600–1800
B	47	32	5	12	1600–1800
C	58	16	8	8	1800–2200
G&H	50	30	5	12	1600–1800

Table 7.4 Influence of compositions exerts on cement

Composition		Early strength	Long-term strength	Hydration velocity	Heat of hydration	Shrinkage	Sulfate corrosion resistance
$3CaO \cdot SiO_2$	C_3S	Good	Good	Moderate	Moderate	Moderate	–
$2CaO \cdot SiO_2$	C_2S	Poor	Good	Slow	Low	Moderate	–
$3CaO \cdot Al_2O_3$	C_3A	Good	Poor	Rapid	High	High	Poor
$4CaO \cdot Al_2O_3 \cdot FeO$	C_4AF	Poor	Poor	Moderate	Low	Low	–

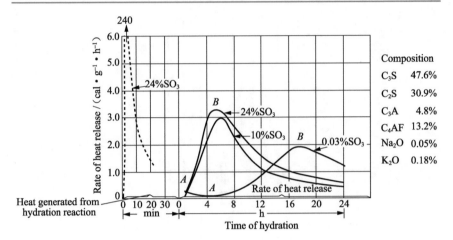

Fig. 7.9 Cement hydration rate

operations, it should be noticed that the hydrate heat may have an effect on casing extension and sealing.

Cement setting and hardening theory

The hardening process can be divided into three main periods:

- This period happens immediately after mixing when water and cement contact with each other. The increasing of hydrates is observed at this time, and when saturated condition is reached, some of the hydrates deposit in gel or micro-crystal forming the sol system with flowing abilities.
- Setting period: During this period, solid $Ca(OH_2)$ crystallize from solution and gel deposits into the water-filled space. The hydrates inter grow, and a cohesive network is formed. Consequently, the system begins to develop strength.
- Hardening period: The hydrates crystallization starts at this stage with the network of hydrate products becoming more and denser and strength increasing. The cement starts to set at the second peak period of the exothermal.
- C_3S is the controlling factor during the cement setting and hardening process while C_2S can extend the hardening time. At the early setting stage, aluminate is the important factor but it has little influence on the cement final strength.

There are three main compositions in the cement. The first composition is a non-qualitative substance, also known as cement gel, which has the crystalline structure and connecting with each other as a whole. The diameter of this particle is about 0.1 mm. The second composition is calcium hydroxide crystal produced from the hydration process. The third one is the unhydrated cement solids.

7.3.1.4 The API Class of the Oil Well Cement

The condition to which the oil well cement is exposed in wells can be quite complicated. The well depth varies from hundreds of meters to thousands of meters with more than 100 °C temperature difference and dozens of pressure difference sometimes. The cementing operation time also changes from tens of minutes to several hours. As a result, a single type of oil well cement cannot satisfy different under hole conditions. Based on different construction applications, oil well cements are classified into different types.

According to API classification requirements, oil well cements are classified as below:

Class A: Intended to for use from surface to a depth of 1828.8 m with temperature up to 76.7 °C, when special properties are not required.

Class B: Intended to for use from surface to a depth of 1 828.8 m with temperature up to 76.7 °C, when conditions require moderate to high sulfate resistance.

Class C: Intended to for use from surface to a depth of 1 828.8 m with temperature up to 76.7 °C, when conditions require high early strength. Class C is available in low, moderate, and high degree of sulfate resistance.

Class G and H: Intended to be used as basic cement from surface to the depth of 2440 m with temperature from 0 to 90 °C. They can be used for with additives to cover a wide range of well depths and temperatures. They are available in moderate and high degree of sulfate resistance.

Application range is shown in Table 7.5.

There are strict tests and evaluations on oil well cements before on site applications to ensure the quality. The tests and evaluations involve in testing on slurry thickening time, slurry fluid loss, rheological measurements, comprehensive strength and bending strength of set cement, the compatibility between slurry, preflush liquid and drilling fluid. For high-pressure and high-temperature wells, special tests are required to ensure the application under these conditions.

Table 7.5 Application range of API cements

API class	Application depth/m	Type			Comments
		Basic	Sulfate resistance		
			Moderate	High	
A	0–1828.8	√	–	–	Normal cement, no special requirements;
B		–	√	√	Moderate heat cement, moderate to high sulfate resistance;
C		√	√	√	High early strength cement, basic, moderate, and high degree of sulfate resistance;
G	0–2440.0	–	√	√	Basic cement, moderate to high sulfate resistance
H		–	√	√	

√ means the corresponding type of cement exists
– means the corresponding type of cement does no exists

7.3.2 Cement Properties and Performance Requirements

The slurry properties have significant influence on well cementing. Important slurry properties include density, thickening time, rheology, and stability. The set cement properties such as strength, permeability, and corrosion resistance should be considered as well.

7.3.2.1 Slurry Density

The density of cement powder varies from 3.05 to 3.20 g/cm^3 due to different composition proportion. The quantity ratio of water and cement ash is called water-to-cement ratio. For the sake of hydration, this ration is controlled between 0.2 and 0.25 for statistic cement slurry. To improve the flow ability, more water is added in. For different API class, the required water amount is different.

According to "The measurement of cement consistence, setting time, and stability" (GB/T 1346-2001), the water-to-cement ratio for class G is 0.44; for class A and B is 0.46; for class c is 0.56; and for class H is 0.38.

The standards for slurry density design are listed below:

- Satisfy the under downhole pressure limits. The hydraulic fluid column pressure must be higher than formation pore pressure. The sum of hydraulic fluid column pressure and the pressure caused by fluid resistance much not excess the formation fracture pressure.
- The density difference is suitable for displacement efficiency, tail slurry density > lead slurry density > preflush fluid density > drilling fluid density. In some cases, the density difference can change from 0.12 to 0.24 g/cm^3. The higher density, the larger flowing resistance is.
- Satisfy the set cement strength and bonding requirements. Basic density is preferred in sealing off oil and gas formations. It is recommended to reduce the usage of non-gel weighting additives and weight reduction additives.

The cement slurry with the water-to-cement ratio between 0.45 and 0.50 usually has the density of 1.8–1.9 g/cm^3. Additives may be introduced to adjust the slurry density as required. Weight reduction agents are added in when it is lower than normal density, while weighting agents are added when the slurry density is higher than normal density.

7.3.2.2 Thickening Time

When the cement powder is mixed with water, the hydration process going on the cement slurry is gradually setting. The flowing ability is constantly reducing during the cementing process, and it becomes more and more difficult to pump and replace the slurry.

The cementing operation has to be completed before the cement thickening; thus, the slurry thickening time has decided the time allowed for operation. For some deep wells with long operating time, the thickening time has to be long enough.

The thickening time of slurry cement is the time during which the cement slurry can be pumped and displaced into the annulus. According to the API standards, the thickening time is the time that takes the cement thickening to reach 100 Bc.

The thickening curve tested under a condition of the normal pressure and a given temperature is called the atmospheric thickening curve, while the thickening curve tested at a given temperature and pressure is called the pressurized thickening curve. Generally, the thickening time under atmospheric condition is longer than that under a given pressure condition as indicated in Fig. 7.10.

To avoid the gas migration during the cementing operation, the time takes to increase the slurry thickening from 30 to 100 Bc is supposed to be as short as possible (less than 30 min). The phenomena that the cement thickening increase from 30 to 100 Bc rapidly is called the right angle thickening.

The relationship between the pumpable and thickening is yet not fully understood, but generally it is agreed that:

- Easy to pump, 5–20 Bc;
- Not easy to pump, 20–30 Bc;
- Hard to pump, 30–40 Bc;
- Cannot be pumped, 40 Bc above (or 50 Bc);
- Easy to deposit, with free water, less than 5 Bc (or 2 Bc in some cases).

The pumpability of the cement slurry is a qualitative term. In practical, the rheology is the main factor in the quantitative slurry design. For the sake of

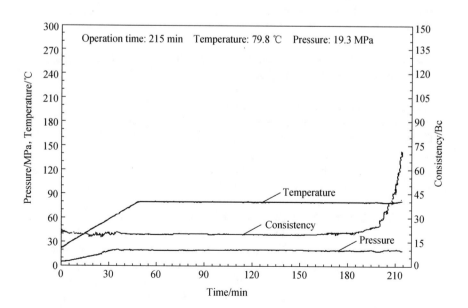

Fig. 7.10 Cement slurry thickening test profile at elevated pressure

pumping, according to the API standards, the maximum thickening of any slurry system during the first 15–30 min should be less than 30 Bc. A preferred situation is that during the operation time the slurry thickening is below 50 Bc.

The temperature has a significant influence on the thickening time. When the temperature increases, the hydration rate increases and the thickening time decreases. The thickening time is adjusted to by thickening time control agents coordinate with the cementing process. If the operation time is short, the thickening accelerator is introduced; if the operation time is long, then retarders are added. Moreover, when applying retarders, the effect of temperature should be considered.

7.3.2.3 The Cement Slurry Lost Circulation

In order to ensure the flowing ability, the water added in slurry is actually much more than the water required for complete hydration. When the cement solidifies, the excessive water will drop out. Lost circulation is defined as the total or partial loss of cement slurries into the highly permeable zones during the cementing operation. The drilling fluids will contaminate formations. If the dropping out water is trapped in cements rather than entering formations, it may create flowing channels for migrating liquids, leading to reduction of leakproof, and cement strength.

Based on lots of flied practices and researches, there are some widely accepted seminars of lost circulation:

- Well controlled, <200 mL;
- Moderate controlled, 200–500 mL;
- Poor controlled, 500–1000 mL;
- Out of control, >1000 mL.

According to API standards, at pressure of 6.9 MPa and a given temperature, the requirements of lost circulation of different slurries are listed below:

- Anti-gas-channeling, 30–50 mL/30 min;
- liner cementing and squeezing cementing, ≤ 50 mL/30 min;
- Casing cementing, ≤ 250 mL/30 min.

The lost circulation for untreated slurries can be as high as 1000 mL/30 min, and thus, fluids loss control agents should be used to control lost circulation.

7.3.2.4 The Setting Time of Cement Slurry

The setting time of slurry refers to the time at which cement paste loses its plasticity transforming from liquid to solid. It is also known as the initial time. The final setting time is the time period between the time water is added to cement and the time at which the slurry solidifies and strong enough to withstand pressure.

API requires that the initial setting time of slurries must be longer than 45 min while the final setting time is shorter than 12 h. A short gas between initial setting time and final setting time is desired in terms of preventing gas migration during the cementing operation.

A patent data from the Soviet Union shows that for HTHP wells, the thickening time of slurries is longer than the initial setting time by 15–30 min. Most of tests of the initial setting time are completed using Vicat apparatus under atmospheric temperature and various pressures. Thus, this relationship is not observed in these tests.

The time period between the time the slurry is pumped and setting into casing, and the time the set cement is strong enough. It has a significant influence on cementing operation and controls the time of the cementing process. For surface casing and intermediate casing, the cement is expected to have a higher early strength to shorten the operating time. Generally, the setting time is around 8 h. When the compressive strength reaches 2.3 MPa, the following drilling operation could start.

7.3.2.5 Rheology of Cement Slurries

The rheology as an important factor to describe the flow and deformation of materials in response to applied stresses. Accurate and justified rheology data can ensure the accurate calculations of flow friction.

The cement slurry is one of the non-Newtonian fluids, and its rheology characteristics have a close relationship with cementing operations. Some commonly used rheological models including two elements, three elements and four elements models. To be more specific, many models are frequently used for non-Newtonian fluids, such as Bingham model, power-law model, Carson model, Herschel–Bulkley model, Robertson–Stiff model, and Polynomial model.

Bingham model:

$$\tau = \tau_0 + \eta_{pv}\gamma. \tag{7.33}$$

Power-law model:

$$\tau = K\gamma^n \tag{7.34}$$

Carson model:

$$\tau^{\frac{1}{2}} = \tau_0^{\frac{1}{2}} + (\eta_C\gamma)^{\frac{1}{2}} \tag{7.35}$$

Herschel–Bulkley model:

$$\tau = \tau_s + K\gamma^n \tag{7.36}$$

Robertson–Stiff model:

$$\tau = A(\gamma + C)^B \tag{7.37}$$

Polynomial model:

$$\tau = a_0 + a_1\gamma + a_2\gamma^{1-m} \tag{7.38}$$

where

T	Shear stress, Pa;
τ_0	Dynamic shear, Pa;
τ_s	Herschel–Bulkley dynamic shear, Pa;
γ	Velocity gradient;
η_{pv}	Plastic viscosity;
K	Consistency index;
n	Flow index;
η_c	Structure viscosity under Carson model;
A, B, C, a_0, a_1, a_2	Experimental coefficients.

Among all models, the Herschel–Bulkley model has the least errors, while Bingham and power-law model tend to have larger errors usually. Sometimes, using Bingham model could lead to large error, but using power-law model could avoid this problem. These two models are easy to use as they have simple form with clearly defined coefficients and they have guiding significance to the controlling and adjusting of cement slurries. Bingham, power-law and Herschel–Bulkley model are usually recommended.

The research shows that the accuracy of Herschel–Bulkley model is higher than that of other models. If Herschel–Bulkley model is the priority when selecting flow models. When choosing flow models, it is important to select the model that matches the rate of shear and shear force by using the linear programming method or linear comparing method. F value is calculated by using the rotary viscometer. The equation is:

$$F = \frac{\phi_{200} - \phi_{100}}{\phi_{300} - \phi_{100}}$$

When $F = 0.5 \pm 0.03$, Bingham model is used; otherwise, power-law model is used.

Some common coefficients are: τ_0, η_{pv} (Bingham); n, K(H-B). Usually, 6 speed rotary viscometers are employed and rheological parameters are calculated:

Bingham model:

$$\left. \begin{aligned} \eta_{pv} &= 0.0015(\phi_{300} - \phi_{100}) \\ \tau_0 &= 0.511\phi_{300} - 511\,\eta_{pv} \end{aligned} \right\}$$

Power-law model:

$$\left.\begin{array}{c} n = 2.0921 \, \mathrm{g} \frac{\phi_{300}}{\phi_{100}} \\ K = \frac{0.511\,\phi_{300}}{511^n} \end{array}\right\}$$

H-B model:

$$\left.\begin{array}{c} \tau_s = \frac{\tau_x^2 - 0.261\phi_{300}\phi_3}{2\tau_x - 0.511(\phi_{300} + \phi_3)} \\ K = \frac{0.511(\phi_{300} - \phi_3)}{511^n - 5.11^n} \\ n = \lg \frac{0.511\phi_{300} - \tau_x}{\tau_x - 0.511\phi_3} \\ \tau_x = 0.511(0.255\phi_{100} + 0.745\phi_6) \end{array}\right\}$$

7.3.2.6 The Stability of Cement Slurries

The stability is an important factor of cement slurries. Poor stability will lead to non-uniform cement density, especially in high-deviated wells. From the hole bottom-up to the surface, the density and bonding strength are reducing, which will have a negative impact on cementing quality. Usually, unstable cement slurries have high proportions of free liquids. These liquids will form flowing channels and reduce the cement sealing quality.

Generally, the stability of cement slurries can be described by the concentration of free water and the vertical density distribution of cement column. Methods used to evaluate stability include: free water based on API standards (water absorption), longitudinal shrinkage of cement column, slurry deposition, and stability (longitudinal density gradient).

The minimum requirement of API standard for water release from conventional cement slurry systems is no more than 3.5 mL of 250 mL cement slurry, or 1.4 percent of the volume of cement slurry, and 0 mL in inclined or horizontal wells.

The sedimentary stability refers to the suspension stability of different components particles in a cement slurry system. The stability is described by the density difference between the surface and bottom of a cement column of a length longer than 20 cm. A cement column is prepared using the method that is used for testing deposition stability and maintained in ambient temperature for 12 h. It is then divided into several sections and the density of each section is measured and calculated. Based on the results, the biggest density difference is identified to estimate the deposition stability. According to SY/T 5480-2007 standard, the density difference between the low-density cement slurry and the normal slurry is less than 0.03 $\mathrm{g/cm^3}$; for directional wells, horizontal wells and highly deviated wells, the density difference should be less than 0.01 $\mathrm{g/cm^3}$. Besides, the density difference for ultra-high-density slurries and normal slurries is less than 0.05 $\mathrm{g/cm^3}$ and 0.03 $\mathrm{g/cm^3}$, respectively.

In practice, if any of above situation existed, the cement slurries are considered to be unstable, namely with the longitudinal density difference relatively huge and free water dropping out when the slurry is standing for 2 h. These phenomena could happen individually or happen at the same time.

7.3.2.7   Strength Characteristics of Set Cement

Sufficient mechanical strength is required to satisfy flowing conditions:

- Support and enhance casing. As research reveals that when the strength of set cement reaches 56 kPa, a cement mantle with a length of 10 m can support a casing with a diameter of 177.8 mm and a length of 94 m. It can be concluded that supporting casings do not require high mechanical strength.
- Could withstand the impact load put by the drill string;
- Could bear the pressure exerted by acid and fracturing operations.

Compressive Strength of Set Cements

The compressive strength represents the ability of the set cement to bear the pressure and support casings. The API standards and SY/T 6544-2003 have clear requirement of compressive strength. While using the ultra-low-density slurries to seal the production formations, it is required that the compressive strength is higher than 7 MPa in 24 h, and it is higher than or equal to 14 MPa in 72 h (Table 7.6).

Bonding strength of set cement

To ensure the efficient bonding between cements and casing and between casing and formation, two bonding characteristics—shear cementation force and hydraulic cementation force—should be taken into account. These two characteristics are important factors of bonding strength.

Table 7.6 Compressive strength requirements for sealing off different formations

Casing program	Slurry types	Density (g/cm^3)	Compressive strength in 8 h	Compressive strength in 24 h
Surface casing	Lead slurry	1.6	>1.8	>3.5
	Tail slurry	1.9	>3.5	>3.0
Immediate casing	Lead slurry	1.6	>2.1	>3.5
	Tail slurry	1.9	>5.0	>11.0
Liner		1.9	>5.0	>14.0
Production casing	Lead slurry	1.6	–	>8.0
	Tail slurry	1.9	–	>14.0
Production liner		1.9	–	>14.0

- Shear cementation force: Supporting casing. It can be identified by testing the force caused by shift between cement sheath and casing and describe by the force per unit area. Generally, shear cementation is 10–20% of the compressive strength.
- Hydraulic bonding cementation force: The ability to stop liquids from migrating in annulus. It is tested by the initial permeating pressure of the casing and cement sheath.

In terms of sealing off formations, hydraulic cementation force is more efficient than shear cementation force. There is no API standard specializing in requirements of cement boding strength.

Flexural strength of set cement

With the increasing of small interval well boreholes, the requirements of cementing tenacity have become higher. Main estimation methods include:

- Impact resistance tests. The impact resistance is the ability of the cement sheath to withstand a high force or shock. It is the consumed energy when the cement sheath fractures when high pressure applied on it.
- Flexural strength test. Flexural strength is also known as bend strength. This test is measuring the load that is applied to the cement sample constantly and uniformly until the sample fractures and then calculating the flexural strength.
- Elastic modulus test. Elastic modulus represents the ability of an object or substance's resistance to being deformed elastically when a stress is applied to it. A large elastic modulus means that the marital is hard to deform under stress and the rigidity of that material is large.
- Tensile strength test. This test is measuring the load applied to the tested sample constantly and uniformly until it fails.
- Perforation test. The perforation test is a method to estimate the plasticity of the cement sheath by evaluating the uniformity of perforation holes, fractures, and cracks under the simulated pressure and temperature similar to bottomhole environment.

These tests have guiding significance for cementing engineering, but in practice the difficulties of these tests vary. The flexural test and elastic modulus test are relatively easy to conduct without high requirements in terms of testing samples. Moreover, they can analysis the rigidity resistance of cement sheath quantitatively. The rigidity of set cement is evaluated by analyzing the data from flexural test and tensile strength test. There are no API standards for flexural resistance now.

7.3.2.8 The Permeability of Set Cement

One of the main purposes of cementing is to maintain the good sealing between casing and well walls and avoid formation liquids or drilling fluids migrating in annulus. The permeability of set cement is the ability of letting liquids to pass through it under pressure. According to SY/T 5480-2007 standards, the permeability of set cement should be less than 0.01×10^{-3} μm^2, and for formation contains corrosive fluids and shale gas formation, it is supposed to be as low as possible.

7.3.2.9   The Corrosive Resistance of Set Cement

The set cement should be resistant to corrosive fluids, especially to sulfate. To improve the sulfate resistance capability of cement, the concentration of C_3A should be less than 3%, and the total concentration of C_4AF and C_3A should be less than 24%. Sometimes, slag and silica sand should be added in the cement to improve its sulfate resistance.

7.3.3   Cement Additives and Mechanisms of Action

With the development of petroleum industry, the exploration and developing areas have been expanded and drilling technologies have been improved constantly. There are some challenging conditions encountered in deep wells, ultra-deep wells, and special wells which require high-performance cement slurries. Additives are needed to satisfy the special requirements in these complicated conditions.

7.3.3.1   Density Modifiers—Weighting Agents and Lightening
Agents

The formation pressure and formation fracture pressure are various and complicated in order to ensure the safety and quality of cementing, density modifiers are introduced. Generally, high-density slurries have a density of 2.0–2.3 g/cm^3 and that of ultra-high-density slurries are higher than 2.3 g/cm^3.

Weighting agents

When drilling through high-pressure formations or during cementing operation in adjustment wells, in order to prevent well blowout and gas channeling weighting agents are added into slurries to increase the slurry density. The most widely used weighting agents are barite, ilmenite, hematite, and superfine manganous manganic oxide.

- Barite

$BaSO_4$ is a white sheen powdery material with a density of 4.3–4.6 g/cm^3. Sometimes it shows in gray, light red, light yellow color because of impurities. The chemical characteristics of Barite are unstable. It is a nonmagnetic and non-toxic material and is insoluble in water. Slurries with densities up to 2.2 g/cm^3 can be prepared with Barite.

- Ilmenite

Ilmenite ($FeTiO_3$) is a black or gray granular material, with a density of 4.45 g/cm^3. This substance is metallic and chemically stable. By adding ilmenite in cement, the slurry density can go up to as high as 2.4 g/cm^3.

- Hematite

With the density of 5.0–5.3 g/cm^3, Fe_2O_3 occurs as red crystalline granules with high hardness and stable chemical characteristics. Hematite is effective weighting agent because of its high density. Slurries prepared with hematite can be as high as 2.6 g/cm^3.

- Superfine manganous manganic oxide

The density of it is 4.8 g/cm^3, and diameter of the particle is less than 10 μm. As a result of small diameter and high density, superfine manganous manganic oxide has functions of suspension stability and weighting. When it is mixed with hematite, they can form close packing which is beneficial for preparing ultra-high-density cement slurries.

Lightweight agents

Lightweight agents are used to reduce the slurry density, thus reducing the static pressure of cement slurries. These agents can prevent fractures in weak formations and improve the cementing quality. Generally, for low-density slurries, the densities are between 1.2 and 1.7 g/cm^3. And for slurries with density less than 1.2 g/cm^3, it is called ultra-low-density slurries. Some of the most common lighting agents are clay, coal ash, glass microsphere, microsilicon, and gilsonite.

- Clay

Densities of clay are 2.6–2.7 g/cm^3, and it is one of the most common lighting agents. Because of the high mud yield, clay can increase the content of free water, thus reducing the density of slurries. The proportion of Amargosite is about 20%, and the slurry densities can be reduced by increasing the proposition. However, the permeability of set cement will be reduced leading to a reduction in compressive strength and corrosion resistance.

- Sodium silicate

Sodium silicate can react with calcium oxide and calcium chloride in water and generate calcium silicate gel. This gel can provide a high viscosity which will dramatically increase the water content of slurries rather than increasing the content of free water. Sodium silicate can increase the water–ash ratio and reduce the density of slurry.

- Coal ash

Coal ash is also known as fly ash, which is ash collected from the boiler in the coal-driven power station. Coal ash as a lightweight agent has the advantages of high yield and wide supply. The mineral compounds are SiO_2 in glass state, Al_2O_3 in glass state, quartz, mullite, and carbon granules. The glass contents are the main compounds. The density of coal ash is usually between 2.0 and 2.5 g/cm^3; unit weight is about 1000–1250 g/cm^3; and the particle diameter is in the range of 0.5–300 μm. The density of slurries prepared with coal ash can be as low as 1.05–1.45 g/cm^3.

- Glass microsphere

Glass microsphere is hollow glass beads with thin walls selected from coal ash. Glass microspheres have a density of 0.7 g/cm^3 under atmospheric pressure and have diameters ranging from 30 to 1000 μm. The wall thickness of glass microsphere is only of 5–8% of the diameters. Main compounds are SiO_2, Al_2O_3, and Fe_2O_3 and so on.

Glass microsphere has been used with slurries to produce lightweight slurries which can reduce the density to 1.05–1.45 g/cm^3. The desired amount of mixture is 10–40%.

- Silica fume

Silica fume is also known as superfine silicon powder which is a byproduct of iron alloy production. The main compound of it is SiO_2 (with the proportion of 90–98%). The density of silica fume is about 2.6 g/cm^3. The particle size is very small with an average of 0.1 μm and ranging from 0.02 to 0.5 μm, and the size is much smaller than that of cement particle and glass microsphere particle. Microsilicon can effectively improve the stability of slurries and reduce the content of free water. However, it can only reduce the densities of slurries to a small range (1.35–1.68 g/cm^3) and it requires a large amount of water when mixing with slurries. The water–ash ratio is high, and it may have negative influence on the strength development of set cement. Because of the rheology and limitation in water using, microsilicon is not widely used in practical.

- Gilsonite

Gilsonite is a naturally occurring solid hydrocarbon, a form of asphalt (or bitumen) with a relatively high melting temperature. Its density ranges from 0.9 to

1.1 g/cm^3, which is close to the density of water and thus can be used to prepare low-density and high-strength cements. Gilsonite is usually applied together with water-absorbent materials. With adding percentage of 2.5–50%, the density could be reduced to 1.37–1.7 g/cm^3 at an applying temperature below 150 °C.

- Hollow glass microbeads

Hollow glass microbeads are made of borosilicate with wall thickness of 1 μm and diameters ranging from 10 to 200 μm. The density of hollow glass microbeads is between 0.2–0.6 g/cm^3, and the compressive strength can be as high as 82 MPa. With increasing diameter, both the density and the compressive strength of hollow glass microbeads are decreased. Slurry densities could be reduced to 1 g/cm^3 or below by adding hollow glass microbeads.

7.3.3.2 Thickening Time Modifiers—Accelerators and Retarders

Cementing operating time varies in different conditions, and the thickening time of cement slurries is effected significantly by temperatures. In order to meet the cementing requirements under complicated bottomhole conditions, accelerators, and retarders are introduced to adjust the thickening time of slurries.

Accelerators

During the cementing operations in shallow wells or deepwater wells, slurries often have relatively long thickening time and slow strength development although the pumpability has achieved requirements. These issues will lead to delay of drilling operations and poor cementing qualities. Consequently, accelerators are applied to shorten the setting time and to accelerate hardening process. Accelerators can be divided into two sections: inorganic salts and organic compounds.

- Inorganic salt accelerators

Many inorganic salts are used as accelerators, among which chloride salts are best known. Other salts such as carbonates, silicates, aluminates, nitrates, and sulfates can all be used as accelerators.

Calcium chloride is undoubtedly the most effective and economical of all accelerators.

Regardless of the retarding point, it always acts as an accelerator and is normally added at concentration between 2 and 4%. NaCl acts as an accelerator at concentration up to 10% by weight of mix water (BWOW). Between 10 and 18% (BWOW), NaCl is essentially neutral, and thickening times are similar to those obtained from freshwater. When the concentration is above 18%, NaCl can result in retardation.

- Organic accelerators

Organic accelerators of cements exist, including calcium formate, formamide, oxalic acid, and triethanolamine (TEA). TEA is an accelerator of the aluminate phases and a retarder of silicate phases. TEA is normally not used alone, but in combination with other additives to counteract excessive retardation caused by some dispersants.

- Mechanisms of action

For different accelerators, the mechanisms of action can be different. Currently, there are three main aspects:

- Common-ion effect (calcium chloride) and salt effect (NaCl) can change the solubility of gel materials and accelerate the hydration process.
- Accelerators can react with cement gels and then generate compound salts and clathrates that have smaller volume, which can lead to rapid hydration process.
- Accelerators can form crystal and accelerate setting and hardening process.

Retarders

During the cementing operation of mid-deep wells and ultra-deep wells, retarders are always added to increase the thickening time to meet the cementing requirements. There are organic retarders and polymer retarders available.

- Organic retarders

Currently, phosphoric acid salts and hydroxyl carboxylic acid slats are the best known organic retarders. There are various organic phosphoric acids used as retarders, and the most common ones include ethylenebis tetraphosphonic acid, etidronic acid, and so on. One of the advantages of these retarders is that they are not sensitive to minor change of cement compounds. The applicable temperature is between 40 and 170 °C. It should be noticed that the high concentration will have influences on slurry strength developments.

Hydroxyl carboxylic acid salts act as high-temperature retarders, including tartaric acid, gluconic acid, and technetium gluceptate, and so on. The performances of these retarders have been understood well. They are powerful retarders, but can easily cause over retardation. To address this problem, Chatterji used the ammonium salts generated by the reaction between alkanolamine and hydroxyl carboxylic acids to relieve the over retardation.

- Polymer retarders

In recent years, the polymer retarders have been studied and researched widely. Monomers with different functions can be assembled by aggregation technologies and the length of the molecular chain, relative molecular mass can be controlled. As a result, these retarder molecules can be designed to get comprehensive desired characteristics. At this stage, this kind of retarders is obtained by the copolymerization of AMPS, carboxylic acid monomers, and acrylamide.

AMPS and itaconic acid are representatives of the polymer retarders. They are more effective compared with the polymer of AMPS and acrylic acid. They can be used alone at the temperature of 260 °C without enhancers such as sodium borate and organic acid.

- Mechanisms of action

There are several different theories exist; however, it is commonly recognized that retarders can restrain the hydration process of cements minerals thus delay the crystallization process.

Four principal theories have been proposed and are summarized as below.

- Adsorption theory: Retardation is due to the adsorption of the retarder onto the surface of hydration products, thereby inhibiting contact with water.
- Complexation theory: Calcium ions are chelated by the retarder, preventing the formation of nuclei.
- Nucleation theory: The retarder adsorbs on the nuclei of hydration products, poisoning their future growth.
- Precipitation theory: The retarder reacts with calcium and hydroxyl ions in the aqueous phase, forming an insoluble and impermeable layer around the cement grains.

7.3.3.3 Fluid Loss Control Agents

When cement slurry is placed across a permeable formation under pressure, a filtration process occurs. This may lead to fluid loss of cement slurries, reduction of flowing abilities or even a job failure. If a large amount of slurries enters into formations, the formation may be damaged. To control the fluid loss, agents are added into slurries. Two principal classes of fluid loss agents exist: finely divided particulate materials and water-soluble polymers.

Particulate materials

These materials often have small diameters which enable them to enter into the filter cake and lodge between the cement particles. As a result, the permeability of filter cake decreases. This kind of material includes microsilicate, asphalts, latex, and thermoplastic resin.

Water-soluble polymers

Currently, most of the fluids loss agents are water-soluble polymers. The mechanism is that these polymers can help to generate stable gel aggregates by hydrogen-bond interaction, and they will enter into the filter cake, thereby reducing the permeability of filter cakes. Moreover, the polar groups of macromolecular chains will adsorb onto the surface of cements grains. Several classes of water-soluble polymers are identified as useful fluid control agents including chemically modified natural products and synthetic polymers.

- Chemically modified natural products

As a modified natural product, cellulose derivatives are used very often in the industry including carboxymethylcellulose (CMC), hydroxyethylcellulose (HEC), carboxymethyl-hydroxyethylcellulose (CMHEC), etc. All of them share similar disadvantages which are low water solubility, high viscosity, poor temperature resistance, cement strength development delay. To improve the performance, some functional monomers are introduced.

- Synthetic polymers

There are various synthetic polymers with high fluid loss controlling performance which are variable with some advantages that natural products cannot compete with.

- Non-ionic synthetic polymers: Polyvinyl alcohol (PVA) has been widely used in practices as a fluid loss control agent as it has some unique advantages. This system could be applied at temperatures up to 95 °C. When the temperature is higher than that, the cross-link is destroyed, thereby the fluid loss control performance is interfered. Besides, the system has a poor salt resistance and cannot be used when the salt concentration is higher than 5%. In order to address these issues, the performance of PVA can be improved by chemical cross-linking. After the modification, the temperature resistance can be as high as 120 °C, and the salt resistance is over 8%, and the fluid loss can be controlled under 50 mL.
- Anionic synthetic polymers: They are the most widely studied and used fluid loss control agents including AMPS, acrylic acid, maleic acid, and acrylamide. These fluid loss control agents have advantages of high temperature and high salt resistance.

7.3.3.4   Dispersants

The main mechanism of actions of dispersants is that they can release free water and destroy the gel construction between cement particles by weakening and breaking clouds links to obtain the desired rheological properties. Some of the most common dispersants are listed as below.

Sulfonate

Sulfonates are used frequently in practice. The preferred materials generally have 5–50 sulfonate groups attached to a highly branched polymer backbones. Other dispersants in this class include melamine sulfonate, polynaphthalene sulfonate, lignosulfonate, and polystyrene sulfonate.

Sulfonated aldosterone condensation polymers

This material contains –OH, –CH$_3$, –C– and –SO$_3$H groups with applicable temperature up to 150 °C. It is the most desired dispersant currently.

Hydroxyl polysaccharide with low relative molecular mass

Dispersants in this class include hydrolysis of starch, cellulose, hemicellulose, and other non-ionic polymers. They all have good dispersion.

Low molecular compound

An example in this class is hydroxycarboxylic acid with a high dispersant ability and it acts as a powerful retarder at the same time similar to citric acid.

7.3.3.5 Lost Circulation Prevention Agents

During the cementing operation, cement slurries often enter into permeable, fractured or porous formations. To solve these problems, lost circulation prevention agents are added. Commonly used lost circulation prevention agents are in schistose texture, threadiness texture or they are granular material and gel material. Xylonite, cellophane flakes, mica plates, and gilsonite are normally used. Currently, the most reliable lost circulation prevention agents are the silicate-based synthetic fibers (AFCS) invented by Schlumberger. They have average lengths of 12.7 mm and diameters of 20 μm. The shape and materials of AFCS are suitable for mixing with slurries. The cellulose in AFCS, different from normal cellulose, is easy to disperse in cement slurries. This characteristic enables it form a hard fiber-mesh that fixes the particles in the slurry and creates a bridge plug thus preventing lost circulation.

7.3.3.6 Antifoam Agents

During the preparation of cement slurries, especially when sulfonates are added, lots of foam will form. This will result in density changes and will reduce the height and strength of cement column. Usually, antifoam agents are required to address this issue.

Commonly used antifoam agents

There are two classes of these additives: polyglycol and silicones.

• Polyglycol

Ploy, including alcohol, ether and ester, is most frequently used because of its low cost. In most situations, polys are added and mixed with water. Materials such as glyceryl polyether, Span-80, tributyl phosphate, ethyl glycol, 1-butanol are used.

• Organic silicones

The silicones are highly effective antifoam agents, and they work by inferencing the formation of foams. Siloxane oil and oxosilane are main materials.

Siloxane is a transparent, odorless oil liquid. It has a low freezing point and stable chemical characteristics with low surface tension. Only small concentrations are necessary to achieve the performance usually less than 0.001% by weight of mixed water.

Oxosilane is suspension of finely divided particles of silica dispersed in polydimethylsiloxane or similar silicones. Oil-in-water emulsions at 10–30% activity also exist.

Mechanism of action

There are several theories proposed. Antifoam agents produce a shift in surface tension and alter the dispersibility of solids so that the conditions required to produce a foam are no longer present.

• Inhibition theory: When the foams are about to form, the antifoam agent particles will destroy the elastic membrane;
• Destroy theory: When the foams exist, antifoam agents reduce the surface tension and lead to reduction of the foam walls and finally destroy foams.
• Drainage theory: Antifoam agents encourage liquid film to drain leading to foam breaking.

7.3.3.7 Early Strength Agent

Early strength agents are used to reduce the thickening time and improve the early strength of cements. There are three classes: inorganic salts, organics, and compounds.

Commonly used early strength agents

- Inorganic salts

This class mainly includes chloride, sulfates, nitrates, and some hydroxides. The chlorides are the most commonly used early strength agents because of its low cost, high efficiency. Usually, 2–4% by weight is enough.

Sodium silicates as a cement filling material also shorten the thickening time. In cement slurries, sodium silicates can react with Ca^{2+} generating C–S–H gel and shorten the time of induction period.

- Organic early strength agents

This class mainly includes triethanolamine, triisopropanolamine, formic acid, and glycol.

The triethanolamine acts as accelerator for aluminates that can accelerate the hydration process of C_3A and accelerate the formation of ettringite. However, when it is applied to silicates, it may extend the hydration process and retard the whole process.

- High early strength agents

Generally, the initial liquid phase of cement slurries is made up of silicate liquid, calcium hydroxide, potassium hydroxide and sodium hydroxide. And they are all in a dynamic equilibrium:

$$CaSO_4 + 2(K, Na)OH = (K, Na)_2SO_4 + Ca(OH)_2$$

Any additives that may encourage this equilibrium move forwarding can accelerate the hardening and setting process. High early strength agents can do this.

There are three working theories of the high early strength agents, which are summarized below:

- Ion effect

Inorganic slats in slurries can have salt effect and common-ion effect, thereby changing the solubility of gel materials and encouraging hydration process. If there is no common ion electrolyte, early strength agents under the salt effect can increase ion concentrations of slurries and change the absorbed layer, hence improving the solubility of mineral materials and accelerating hydration process. If there is common ion electrolyte, early strength agents under the common-ion effect can reduce solubility of some mineral materials and one the other hand can encourage crystallization of hydration products. As a result, the early strength is enhanced.

The early strength effects of negative/positive ion are listed as below:

$$Ca^{2+} > Mg^{2+} > Li^+ > Na^+ > H_2O$$
$$OH^- > Cl^- > NO^{3-} > SO_4^{2-} > H_2O$$

- Generating double salts, clathrates, hard dissolve compounds

Some early strength agents can prepare grains of crystallization and then encourage hydration process. Some materials such as $NaAlO_2$ and Na_2SiO_3 present gel texture when dissolved in water. They can combine with calcium ion to generate crystallization center and improve the hardening and setting process.

7.3.3.8 Expansion Agent

Types of expansion agents

Two types of expansion agents exist. One of them is called "rigidity" expansion agent. Alkalis oxides such as CaO/MgO and $CaO/CaSO_4$ react with cement slurries and generate expandable crystal such as ettringite, calcium hydroxide, and magnesium hydroxide. These compounds can compensate volume shrinkage and expand slightly after cement is set. The other type is gas forming expansion agents such as aluminum powder. These additives are combined with foam stabilizer react with $Ca(OH)_2$ and generate H_2. This expansion can make up the volume loss after cement setting.

There are various expansion agents available in market and can be classified as below:

- Inorganic salts, such as $NaSO_4$, $NaCl$, $CaSO_4.0.5\,H_2O$, $4CaO.3Al_2O_3.SO_3$;
- Metallic oxide, such as $CaO + MgO$, CaO, MgO, $CaO + Al_2O_3$;
- Metal powder;
- Organics, such as rubber powder.

Under the bottomhole condition, because of the limitation of casings and formations, slurries with expansion agents expand slightly to compensate the volume shrinkage and block the microannulus. Moreover, it can improve the inner construction of cements and enhance the strength of bonding surfaces.

Mechanism of action

Commonly used rigidity expansion agents include calcium sulphoaluminate, vaterite, periclase, gas forming expansion agents, and aluminate powder. The expansion process of Calcium sulphoaluminate, vaterite, and periclase is similar. The cement expansion is resulted by the formation and development of hydration products. There hydration crystals form and develop in certain areas. When slurries are under limited conditions, the expansion could compact the cement structure and generate prestress.

- Mechanism of action of Ettringite

- Water swelling: AFt generated by dissolution–sedimentation is a gel material with colloidal size. They have negative charges and high specific surface. Ettringite particles in gel texture attract polarizes water molecule surrounding around ettringite crystals. The whole system expands due to the repulsive force. Besides, when they mix with water, diffused double layer is formed and present expansion macroscopically.
- Crystallization pressure: Ettringite crystals grow crossing each other and develop pressure that lead to cement expansion. The growth of crystals results in system volume increasing, and the expansion depends on number of crystal nucleus and growth development rate of crystals. Usually, a larger number of crystal nucleus and smaller size of crystals will result in larger expansion. Crystal nucleus can form in liquids or on the surface of cement particles. In fact, AFt (type 2) expand because of water swelling, while AFt (type 1) can generation crystallization pressure. Under limited conditions, AFt (type 1) acts as enhancement while AFt (type 2) acts as expansion agents.

- Mechanism of action of CaO

The volume increases when CaO transformed into $Ca(OH)_2$ after hydration, these expanses slurries. However, $Ca(OH)_2$ is not the specific product of CaO. Other materials such as C_3S and C_2S also generate $Ca(OH)_2$ during hydration with a weight percentage of 20%. They also produce CSH gel and $Ca(OH)_2$ phase, and these two materials combine closely thus do not cause expansion. When $Ca(OH)_2$ produced by CaO hydration, it mainly accumulated on the surface of CaO, namely the expansion is caused by part chemical reactions. When CaO hydrates and produces $Ca(OH)_2$, not only the volume increases but also the pore volume. It should be noticed that only when hydration products accumulated, the expansion will occur.

- Mechanism of action of MgO

The reaction between water and MgO that generates $Mg(OH)_2$ is a typical solid-phase reaction. The expansion is resulted by the crystal growth pressure. There are four steps of hydration process of MgO: Firstly, water molecule is physical and chemical absorption on the surface of MgO particles; secondly, Mg^{2+} and OH^- are absorbed on diffuse layer of water molecules; thirdly, crystal nucleus of $Mg(OH)_2$ form; fourthly, crystal nucleus of $Mg(OH)_2$ develops.

When $Mg(OH)_2$ crystals are small, the expansion force is mainly caused by water swelling. With the growth of $Mg(OH)_2$, the expansion force mainly comes from crystallization pressure.

- Mechanism of action of aluminum powder

As expansion agents, aluminum powder has to be smashed finely and coated by resin on surface to control the gas foaming and duration time. Under the action of bottomhole temperature and pressure, aluminum powder reacts with water and produces hydrogen bubbles. Under the action with foam stabilizer, hydrogen bubbles develop into individual bubbles and distributed in slurries evenly. These small bubbles are expanding under trap conditions and compensate the volume loss and pressure loss. On the macrolevel, slurry volume expands slightly and improves the bonding strength between casings and hole walls.

7.3.3.9   Strengthening Agents

The sealing ability of cement directly influences the efficiency of operation of drilling, completion, production, recovery enhancement. Usually, normal cements suffered from brittle failure thus cannot seal off formations effectively. Strengthening agents are introduced to improve the performance of cements. Commonly used agents include fiber, latex, and elastic particles.

Fibers

One of the best methods to improve the toughness is adding fibers. Frequently used fibers are steel fibers, polymer fibers, glass fibers, and carbon fibers.

- Steel fibers

Compared with other fibers, steel fiber has various advantages using with cements. It has a high elasticity modulus and compressive strength. Usually, five times higher than cements. Moreover, steel fiber can have various cross-sectional shapes thus the boding strength between base materials.

- Polymer fibers

Commonly used polymer fibers include polypropylene fiber and polyester fiber. Polypropylene is a crystalline polymer with uniform structure and is non-toxic and tasteless. It has a density of 0.90–0.91 g/L and is the lightest resin. Polypropylene is chemical stable, and it does not react with acid, alkali and organic solvents. The mechanic properties of polypropylene are decent with tensile strength of 33–11.4 MPa. In terms of compressive strength and bending strength, they are about 41.4–55.1 MPa.

- Glass fibers

Glass fibers are frequently used in practice due to the easy availability and high mechanical performance. They have stable sizes and high elasticity modulus. Besides, glass fibers are fire and mildew proof. Most importantly, they have high tensile strength. For example, the tensile strength of glass fiber with a diameter of 3–9 μm can be as high as 1500–4000 MPa. Glass fibers are the most widely used fiber and can be combined with carbon fiber to get a higher toughness economically.

- Carbon fibers

Carbon fibers have carbon mass fraction over 90% that are manufactured from polyacrylonitrile (PAN), rayon, and petroleum pitch. They have several advantages including high stiffness, high tensile strength, low weight, high chemical resistance, high-temperature tolerance, and low thermal expansion.

Latexes

Latexes are usually used as gas channeling prevention agents and corrosion resistance agents. The elastic reformation ability and impact resistance will be improved significantly after adding latex. Several latexes are available in practice, and most widely used latex is styrene–butadiene latex.

Elastic particles

The rubber powder is most frequently used. By adding rubber powders, deformation centers with certain strength are formed under bonding effects to restrain the growth and development of microfractures. The centers can reduce the rigidity of set cement so when it is impacted the centers can absorb vibration energy hence improving the impact resistance.

However, rubber powders are inert organics. The surfaces are not hydrophilic and are difficult to bond with cements. They have low densities and are easily float on the surface, which can lead to unstable cement systems. If rubber powders are applied, they have to be treated to increase the hydrophilia.

7.3.4 Cementing Design and Well Cementation Technology

7.3.4.1 Well Cementation String Structure

The cementation string is mainly composed of casing strings, also including the accessories of the casing strings, such as cementing head, cementing plug, guide shoe, float shoe, float collar, casing centralizer, cement basket, stage collar, and external packer. The structure of cementation string is shown in Fig. 7.11.

Fig. 7.11 Schematic
diagram of cementation string
structure

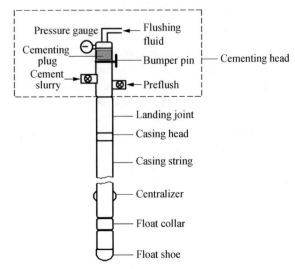

Landing joint
Casing head
Casing string
Centralizer
Float collar
Float shoe

Cementing head

Cementing head is located at the top of the casing string. It is the control head of the slurry injection during the cementing operation, and its lower part is connected to the landing joint. According to its connecting pipe fitting, cementing head can be divided into two categories: drill pipe cementing head and casing cementing head. Drill pipe cementing head can be used in the inner string cementing or liner cementing jobs. Casing cementing head is categorized as single plug casing cementing head and double plug casing cementing head. The double plug casing cementing head can be used in the double cementing plug well cementation and double stage cementing well cementation. The common types of cementing heads are shown in Fig. 7.12.

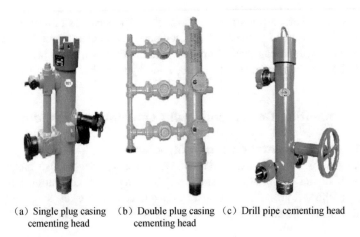

(a) Single plug casing   (b) Double plug casing   (c) Drill pipe cementing head
    cementing head   cementing head

Fig. 7.12 Commonly used well cementing head

Cementing plug

The cementing plug is used to separate the cement slurry and displacement fluid (washing fluid), to prevent two different fluids from mixing. The top cementing plug is a solid plug. In the process of displacement, the bump-pressure achieves when the top cementing plug descends to the choke ring position of the float collar, indicating the displacement process is complete. The upper part of the bottom cementing plug has a rupture disk, while the lower part is hollow. When the bottom cementing plug goes down to the choke ring of float collar and is blocked, the upper rupture disk is broken to ensure that the cement slurry continues to circulate. At present, only the top cementing plug is commonly used in the well cementation process. The cementing plugs used in conventional well cementation are shown in Fig. 7.13.

Guide shoe

Guide shoe is used to guide the casing smoothly down into the well and protect the lower casing. Most types of guide shoes have rounded convex heads, and the interior of the convex head is made of drillable materials (other materials can be used for the completion casing), and the coat is usually made of casing collar steel. At present, the commonly used guide shoes are cement guide shoes and aluminum guide shoes, as shown in Fig. 7.14. For the guide shoe of intermediate casing, after drilling out the guide part at the bottom, the lower part of the casing should form an inner slant, to guide the tripping-in and tripping-out of the subsequent drilling tools, to prevent the hanging of the casing. Thereby, the structure is called casing shoes.

Float shoe and float collar

Float shoe connects to the very bottom of the casing string, and the shape of its lower portion is similar to guide shoe (hence, it is no need to use guide shoe when the float shoe connects to the lower part of case string). The float shoe can guide the casing to smoothly tripping-in. There is a check value (float valve) inside the float shoe, whose function is to increase the buoyancy during the casing operation and to

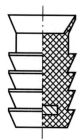 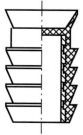 

(a) Top cementing plug   (b) Bottom cementing plug   (c) Picture of cementing plug

Fig. 7.13 Commonly used well cementing plug

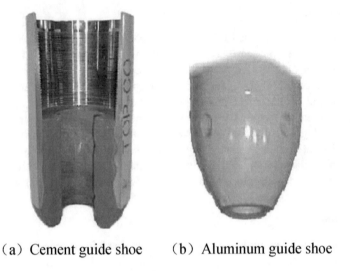

(a) Cement guide shoe   (b) Aluminum guide shoe

Fig. 7.14 Guide shoe structural diagram

reduce the hook load. Also in the meantime, the check value prevents the cement slurry pouring back into the casing after the completion of the displacement. In addition, the use of check value is easier to control the kick during casing operation and inject pre-filtered drilling fluid into the casing, to prepare for the cementing. Automatic cement grouting float shoe can support the automatic grouting during the casing. When the annulus pressure and casing internal pressure reaches a certain value, the float valve is automatically opened to perform automatic grouting.

Float collar is mainly used to prevent the cement slurry flow back into the casing and achieve bumping-pressure. In deepwater cementing or long cementing interval, for the sake of safety, two float collars are often being used. For intermediate casing, float shoes and float collars made of easy-to-drill materials should be utilized, as shown in Fig. 7.15.

Casing centralizer

Casing centralizer is a tool to centralize the casing. In order to improve the casing standoff in the borehole and improve the quality of well cementation, a centralizer is usually installed in the lower interval of each casing (or a certain length casing) during in the casing operation. The common types of centralizers include elastic centralizer (single bow or double bow), rigid centralizer, spiral centralizer, roller centralizer, and so on, as shown in Fig. 7.16. Especially in horizontal wells, it is very important to maintain a high casing standoff. Hence, the rigid centralizer is generally being used in horizontal wells.

Spiral centralizer allows the annular slurry to develop rotational flow, thereby increasing the efficiency of displacement. Roller centralizer has a drag reduction

(a) Conventional float shoe (b) Automatic cement grouting float shoe (c) Float collar

Fig. 7.15 Float shoe and float collar

effect, which can reduce the friction between the casing and borehole wall, and it is commonly used in horizontal wells.

Cement basket

Cement basket is made of a conical rubber basket and elastic support rods, as shown in Fig. 7.17. Cement basket is generally placed in the hydrocarbon reservoir or the upper part of the weak formation with a low fracture pressure. Its function is to support the cement slurry weight at the upper part and to prevent the leakage and contamination of the reservoir. It is directly set in the outer casing, similar to casing centralizer. Under the pressure generated by the weight of cement slurry, the outside diameter of cement basket increases, blocking the annular space.

7.3.4.2 Cementing Design

Cementing design process involves the cementing adhesion loss, accurate prediction of borehole diameter, drilling fluid compression and other factors. While considering extra margin of cement quantities, the amount of cement slurry, performance of cement slurry needs also to be designed appropriately.

The commonly used cement slurry design coefficients and the selection of cement additives

The common cement slurry design coefficients are as follows:

- Coefficient of borehole diameter enlargement: 1.05;
- Coefficient of extra margin for cement quantity:1.10;
- The rate of cement ground loss: 1.05;

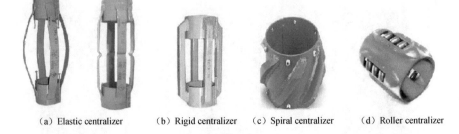

(a) Elastic centralizer   (b) Rigid centralizer   (c) Spiral centralizer   (d) Roller centralizer

Fig. 7.16 Casing centralizer

Fig. 7.17 Cementing basket

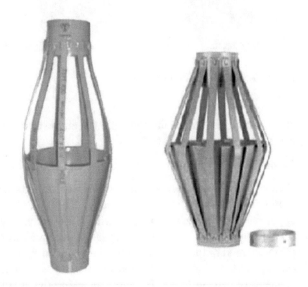

- Coefficient of drilling fluid compressibility: 1.05;
- Coefficient of the spacer fluid additional quantity in the annulus: 1.10;
- General cement slurry density: 1.85 g/cm^3;
- General G-grade cement water–cement ratio: 0.44.

In the process of well cementing, G-grade oil well cement is recommended. If the maximum temperature of the well cementation interval is greater than 120 °C, it is necessary to add high-temperature-resistant agent in the cement slurry and also to adjust the cement slurry additives in order to make the cement slurry system to attain good temperature resistance. Under special formation conditions, it is recommended to select the appropriate oil well cement base on API standard. Based on the requirements of the well depth, we need to properly select cement additives to adjust the cement slurry density, filter loss, thickening time, initial and final setting time, cement slurry stability, bleeding rate, and other properties.

Slurry column design

- Composition of cement slurry

The slurry used in the well cementing process includes preflush and cement slurry.

– Preflush

Preflush system is a specific fluid which is injected into the well before injecting the cement slurry. The purpose of preflush is to separate the cement slurry and drilling fluid. The preflush has the effects of isolating, buffering, and well cleaning and can improve the well cementing quality. Among the modern cementing technologies, these special fluids have become a specific fluid system.

Preflush can be divided into washing fluid and spacer fluid.

a. Washing fluid.

The purpose of utilizing washing fluid is to dilute and disperse drilling fluid, in order to prevent gelation and flocculation of the drilling fluid, to effectively flush the borehole wall and casing wall, and to wash the remaining drilling fluid and mud cake. The washing fluid is also able to buffer between cement slurry and drilling fluid and to improve the quality of consolidation. Washing fluid should not only have a low density around 1.03 g/cm^3 (close to water density), but also have very low plastic viscosity and good flow properties, such as low shear rate, low flow resistance, and the flow characteristic of developing turbulent flow under low flow rate. The critical velocity of developing turbulent flow is between 0.3 and 0.5 m/s. Washing fluid should also have good compatibility with cement slurry and drilling fluid.

Usually, washing fluid is prepared by adding surfactants in freshwater or diluting drilling fluid. The commonly used washing fluid formulations include CNIC aqueous solution, surfactant aqueous solution, seawater, etc.

b. Spacer fluid.

The spacer fluid is able to effectively separate drilling fluid and cement slurry, to generate the planar propellant displacement effect, and to buffer the low pressure and leakage zone. The spacer fluid has relatively high buoyancy and drag force, which can enhance the displacement effect. Generally, the spacer fluid is a viscous fluid, which has a larger viscosity than washing fluid, a slightly higher density and slightly larger gel strength. Spacer fluid is injected after the washing fluid, and then the cement slurry is injected after the spacer fluid.

Spacer fluid is generally formulated by adding viscous treating agent and barite into water. The performance requirement is that the density should be 0.06–

0.12 g/cm^3 larger than drilling fluid; high viscosity, shear stress should be 40–80 mPa s; water filtration loss is about 50 mL/30 min.

The common formulations of the spacer fluids are: adding guar gum or hydroxyethyl cellulose in aqueous solution, and adjust the density with barite.

– Cement slurry

Cement slurry is a mixture of dry cement (dry ash), water and various additives. According to the density requirement, water and dry cement are mixed proportionally.

The density of conventional cement slurry is generally between 1.80 and 1.95 g/cm^3. The solid phase of conventional cement slurry is basically cement particles; hence, the formed cement rock has a high strength, which can effectively carry the casing load or perforation pressure. In addition, the slurry density is relatively large; hence, the generated hydrostatic pressure is also large. Therefore, in the well cementing process, in order to effectively seal up the pay zone without causing pressure-induced leakage in formation, generally two or more densities of cement slurry are used. The reservoir area is sealed with higher density cement slurry (tail slurry), while the upper formation is sealed with lower density cement slurry (lead slurry).

In the process of cement slurry blending, we typically conduct the mixing of slurry at the mixing tank in the cementing truck in China. Due to the variations of the feeding rate of dry ash and water, the actual slurry density may deviate from the requirement of slurry design. During the operation process, the density deviation should be controlled in a reasonable range. We may effectively reduce the fluctuations in the cement slurry density using the automatic mixing equipment.

• The design requirements of cement slurry

– The length of washing fluid in the annulus usually takes 60–100 m, and the maximum may not be above 250 m, and the density usually takes 1.10–1.25 g/cm^3.
– The length of spacer fluid in the annulus usually takes 30–100 m, and the maximum length can be maintained at 200 m in the annulus. The fluid density should be selected properly based on the formation pressure in the well cementing intervals and generally takes 1.35–1.50 g/cm^3.
– Cement slurry consists of the lead slurry and tail slurry. The density of lead slurry is between 1.4 and 1.7 g/cm^3, and the tail slurry is between 1.85 and 1.95 g/cm^3. The cement slurry density can also be designed appropriately according to the length of cement interval, under the condition that the intensity of cement meets the well cementing requirement. Generally, the tail slurry outside of the casing should return at least 200 m above the top of oil formation.
– Pressure equilibrium verification. After completing the design of the slurry column, we should conduct the fracture pressure verification at the bottom of wellbore and upper casing shoes positions, in order to ensure that the formation

stays leakage-free by high pressure and the high-pressure zone is stabilized. If the pressure acting on the bottom of borehole does not meet the requirement, the cement slurry density, preflush density (washing fluid and spacer fluid) or length of intervals should be adjusted.

Calculation of annular fluid column pressure

- Calculation of hydrostatic fluid column pressure

In a well cementing operation, the hydrostatic pressure of fluid column generally refers to the annular fluid column pressure, when the cementing job completes. The calculation of hydrostatic pressure is shown below:

$$p_j = 0.00981(h_{c1}\rho_{c1} + h_{c2}\rho_{c2} + h_w\rho_w + h_s\rho_s + L_d\rho_d) + p_b \tag{7.39}$$

where

p_j Annular hydrostatic pressure of fluid column, MPa;
h_{c1} Length of tail slurry cementing interval, m;
ρ_{c1} Density of tail slurry cementing interval, g/cm^3;
h_{c2} Length of lead slurry cementing interval, m;
ρ_{c2} Density of lead slurry cementing interval, g/cm^3;
h_w Length of washing fluid, m;
ρ_w Density of washing fluid, g/cm^3;
h_s Length of spacer fluid, m;
ρ_s Density of spacer fluid, g/cm^3;
L_d Length of drilling fluid, m;
ρ_d Density of drilling fluid, g/cm^3;
p_b Annular build-up pressure, MPa.

- Determinationing the dynamic pressure of fluid column

In the well cementing process, the dynamic fluid column pressure generally refers to the column pressure, and its calculation equation is:

$$p_d = p_j + \Delta p_{la} \tag{7.40}$$

where

p_d Annular dynamic fluid column pressure, MPa;
Δp_{la} Pressure induced by annulus flow resistance, MPa.

The annular dynamic fluid column pressure p_d is a variable in the cementing operation, and it is related to the change of fluid column and its displacement. Generally, the dynamic fluid column pressure reaches the maximum value when the cement slurry bumps and exerts pressure at the bottom of the borehole. We typically calculate the dynamic pressure based on this condition.

- Basic conditions of well cementing pressure equilibrium design.

Based on the calculation of the annular pressure, we determine and compare the annular dynamic fluid column pressure p_d and annular hydrostatic fluid column pressure p_j at several critical places. Then we find the minimal annular hydrostatic fluid column pressure $p_{j\,min}$ and maximal annular hydrostatic fluid column pressure $p_{d\,max}$.

The well cementing design must satisfy the following conditions:

$$p_{j\,min} > p_p \text{ and } p_{d\,max} < p_f \tag{7.41}$$

where

$p_{j\,min}$	Minimal annular hydrostatic fluid column pressure, MPa;
$p_{d\,max}$	Maximal annular hydrostatic fluid column pressure, MPa;
p_p	Formation pore pressure, MPa;
p_f	Formation fracture pressure, MPa.

- Calculation of cement slurry volume

In the cement slurry volume design, the calculations of cement slurry volume, dry cement volume, and mixing water volume do not consider the weight and volume of all the additives (assuming in the form of powder).

- Density of cement slurry:

$$\rho_s = \frac{\rho_c \rho_w (1+m)}{\rho_w + m\rho_c}, \quad m = W_w/W_c \tag{7.42}$$

where

ρ_s	Density of cement slurry, g/cm^3;
ρ_c	Density of dry cement, g/cm^3;
ρ_w	Density of water (or mixing water), g/cm^3;
m	Water/cement ratio, dimensionless, generally the value is set based on the cement grade. For example, for G-grade cement, $m = 0.44$; for grade A and

B; $m = 0.46$; for grade C cement, $m = 0.56$; for D, E, F, and H grade cement, $m = 0.38$;

W_w Total water consumption for making cement slurry, kg;

W_c Total cement consumption for making cement slurry, kg.

– Cement slurry volume

Theoretical cement slurry volume:

$$V_c = \frac{\pi}{4} \left[\sum (d_{hi}^2 - d_{co}^2) h_i + d_{ci}^2 H \right], \ h_t = \Sigma h_i \tag{7.43}$$

where

V_c Theoretical cement slurry volume, m³;

d_{hi} Borehole diameter at varying intervals. If there is no measurement of borehole diameter, we assume d_{hi} to be the product of drill bit diameter and borehole expansion coefficient (1.05), m;

d_{co}, d_{ci} Outside and inside diameter of casings, m;

h_i Cement height of the outside of the casing corresponding to borehole diameter d_{hi}, m;

h_t Return height of external casing cement, m;

H Cement plug height of internal casing cement, m.

Actual cement slurry volume:

$$V_{actual} = K_1 V_c \quad \text{or} \quad V_{actual} = V_c + V_{add} \tag{7.44}$$

where

K_1 Additional coefficient of cement;

V_{add} Additional cementing volume, m³;

V_{actual} Actual cementing volume, m³;

– Dry cement volume:

$$W_c = V_{actual} \frac{\rho_c \rho_w}{\rho_w + m \rho_c} K_2 \tag{7.45}$$

where

K_2 Cement loss rate at surface

− Water consumption during slurry blending.

$$V_w = \frac{mW_c}{\rho_w} \qquad (7.46)$$

where

V_w Total water consumption volume during cement slurry blending, m^3.

− Preflush volume.

Washing fluid volume:

$$V_{pre} = \frac{\pi}{4} K_4 \sum \left(d_{hi}^2 - d_{co}^2\right) h_{pre\_i} \qquad (7.47)$$

$$h_{pre} = \sum h_{pre\_i}$$

Spacer fluid volume:

$$V_{spacer} = \frac{\pi}{4} K_4 \sum \left(d_{hi}^2 - d_{co}^2\right) h_{spacer\_i} \qquad (7.48)$$

$$h_{spacer} = \sum h_{spacer\_i}$$

where

V_{pre} Washing fluid volume, m^3;
K_4 Additional coefficient of spacer fluid in the annular space;
h_{pre_i} External washing fluid height of casing corresponding to borehole diameter d_{hi};
h_{pre} Total length of washing fluid at annular space, m;
V_{spacer} Spacer fluid volume, m^3;
h_{spacer_i} External spacer fluid height of casing corresponding to the borehole diameter d_{hi};
h_{spacer} Total length of spacer fluid at annular space, m.

– Displacement fluid volume:

$$V_{\text{dis}} = \frac{\pi}{4} K_3 \sum d_{\text{cij}}^2 h_j \qquad (7.49)$$

$$h_z = \sum_{j=1} h_j$$

where

V_{dis} Displacement fluid volume, m³;
K_3 Drilling fluid compressibility coefficient;
d_{cij}^2 Inside diameter of each casing intervals, m²;
h_j Casing length corresponding to each casing inside diameter, m;
h_z Depth of choke ring, m.

7.3.4.3 Conventional Well Cementation Work Flow Well Cementation Characteristics

- Well cementation operation is a one-time project. If the quality of work is not good, it is usually difficult to remedy.
- Well cementation operation is a concealed project, and the main process is in downhole. Hence, the operation cannot be directly observed. The cementation quality is impacted by numerous factors, including the accuracy of well cementation design and the quality control in the operation process.
- Well cementation operation impacts the development of oil and gas field and the follow-up projects. If the well cementation quality is poor, it may cause inter-zone channeling in the reservoir development process (especially in the water injection process) and significantly affect the routine oil and gas field development.
- Well cementation is a costly project.
- Well cementation project has short operation time and long working procedures.

Basic conditions of conventional well cementation

In a conventional well cementation operation, we must analyze carefully whether the operation conditions are satisfied before starting the cementation job on. It may cause engineering accident or quality issue, if we perform the cementing job without proper conditions. Typically, the basic conditions to satisfy the well cementation construction include:

- Unblocked borehole;
- Clean bottom of borehole;

- No leakage in downhole before cementing;
- No serious oil and gas cut, and the channeling velocity of oil and gas is less than 10 m/h.
- The casing standoff is not less than 67%.
- The drilling fluid is prepared to have the properties of low viscosity, low shear stress, low density, and good mobility, without affecting the borehole wall stability and downhole pressure stability.
- The physical properties such as cement slurry thickening time and slurry mobility should meet the operational requirement.
- The cement slurry and drilling fluid have a certain density difference, which is generally greater than 0.2 g/cm^3.
- The performance of cement equipment, water supply equipment, cementing equipment displaces drilling fluid equipment, and high- and low-pressure manifold satisfies the operational requirements.

Basic process of well cementing

The basic process of conventional well cementing is shown in Fig. 7.18. At the end of cementing operation, displacement fluid will push the top cementing plug to the choke ring position. When the top and bottom cementing plug meet, it will block the circulation channel, and the pump pressure rise rapidly, which is called bumping the plug.

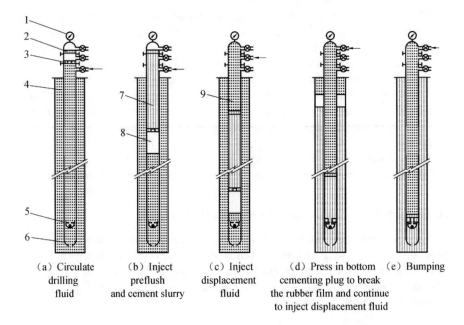

(a) Circulate	(b) Inject	(c) Inject	(d) Press in bottom	(e) Bumping
drilling	preflush	displacement	cementing plug to break	
fluid	and cement slurry	fluid	the rubber film and continue	
			to inject displacement fluid	

Fig. 7.18 Conventional well cementing basic process. 1—Pressure gauge; 2—Top cementing plug; 3—Bottom cementing plug; 4, 9—Drilling mud; 5—Float collar; 6—Guide shoe; 7—Cementing slurry; 8—Preflush

Detailed operation procedures are as follows:

- After drilling to the designed well depth, conduct electrical logging, then pigging, and then run casing.
- After running the casing, circulate drilling fluid and clean borehole. Generally two cycles of circulation are needed as requirement.
- Connect to the ground well cementation facility.
- Inject preflush (includes washing fluid and spacer fluid).
- Press in bottom cementing plug, and inject cement slurry, the rubber film will break when bottom cementing plug reaches the float collar.
- Inject flushing fluid, and press in top cementing plug.
- Pump in displacement fluid, and stop pumping when top cementing plug reaches the float collar and bumped.
- Shut-in the well and wait on the cement to set.

7.3.4.4 Commonly Used Well Cementation Technology

To improve cementing quality, reduce the damage to the pay zone in the cementing process, and meet the operation requirement, some advanced cementing technologies have been gradually developed in China and worldwide. The corresponding well cementation facilities and tools are also developed. Here, we mainly introduce the commonly used conventional (single-stage) cementing technology, multi-stage cementing technology, and liner cementing technology, inner string cementing technology, prestress cementing technology and reverse-circulation cementing technology.

Conventional (single-stage) cementing technology

Conventional (single-stage) cementing technology is to prepare the cement slurry using the cementing truck, bulk cement truck, and other ground facilities. This technology is to inject cement slurry through high-pressure manifold, cementing head, casing string into the wellbore all at once, by utilizing the preflush and bottom cementing plug (spacer plug) to isolate drilling fluid from cementing slurry. Then the slurry enters the annulus from the bottom of the casing string and reaches the designed position, in order to achieve the effective cementing at designed interval between casing and borehole wall (Fig. 7.19). Cementing operation process: Inject preflush → inject cement slurry → press in bump cementing plug (top cementing plug) → displace drilling fluid → bump the plug → wait on cement. In the conventional well cementation operation, we must carefully analyze whether the operation conditions are satisfied before starting the operation. It may cause engineering accident or quality issue, if we operate cementing without proper conditions. Please refer to the conventional cementing technology process for detailed operation process.

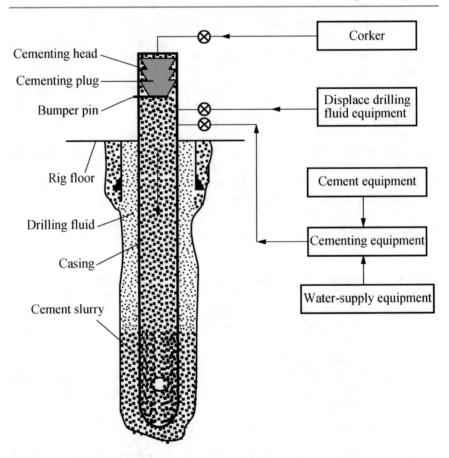

Fig. 7.19 Diagram of single-stage cementing

Multi-stage cementing technology

Multi-stage cementing technology is to place the stage collars in series at certain positions in the casing string, so that the cementing operation can be completed in two or more times during the cementing. Because of the difference of opening and closing mode in the stage collar circulation port, stage collar can be divided into mechanical stage collar, hydraulic stage collar and full-size stage collar.

Mechanical stage collar consists of opening sliding sleeve, closing sliding sleeve, shearing ball, anti-rotation device and collar body. Besides, it also includes opening plug and shutoff plug (Fig. 7.20). Mechanical stage collar is pressurized to open the circulation port by the gravity plug, or open the circulation port using the opening plug.

Hydraulic stage collar uses the hydraulic way instead of opening plug to open the circulation port, and its mechanism is: keep pressing after the primary

Fig. 7.20 Multi-stage
cementing and tools
(mechanical stage collar)

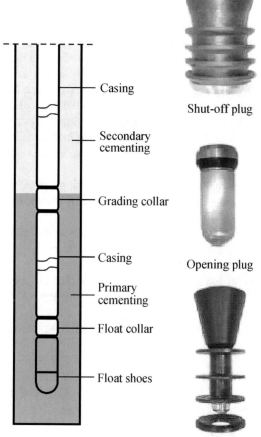

Casing

Shut-off plug

Secondary
cementing

Grading collar

Casing

Opening plug

Primary
cementing

Float collar

Float shoes

Primary cementing plug
and retainer

cementing plug bumping, and forming a downward force, to open the bottom sliding sleeve and expose the circulation port. Shutoff plug is still required to close the circulation port when the stage collar is close.

Full-size stage collar does not have opening plug and shutoff plug, and it uses the inner string and specific tool to open and close the circulation port. There are two types of open mode: by rotation or by lifting and lowering, to open the circulation port. Similar to the regular stage collars, the full-size stage collar connects to the designed position at casing string, and trips in specific tools after the casing string tripping-in. After finish injecting primary cement with inner pipe column, the secondary cementing operation is conducted using specific tools. After completion of cementing job, the circulation port is closed and the specific tools are taken out.

The mechanism of multi-stage well cementing is that for wells with very long cementing interval, we design in advance of the positions in need of stage cementing, based on the formation conditions, and we connect all safety-qualified stage collars and trip-in a wellbore with the casing string. Firstly, the primary cementing operation is conducted, which is similar to the single-stage cementing process. Then based on the different mechanisms of opening stage collars, open the circulation port of the stage collar and circulate the drilling fluid, until the primary cement slurry reaches certain intensity. Next, conduct the secondary cementing operation through the bypass hole of stage collar, and press in the shutoff plug after the secondary cementing, and shutoff the circulation port of the stage collar to achieve bumping. And then, shut-in the well and wait on cement to set. Lastly, drill out the shutoff plug and opening plug inside the stage collar.

Multi-stage well cementing application scenarios:

- Cementing interval is too long for single-stage cementing, and the displacement pump pressure is too high. As a result, cementing equipment can hardly meet the operational requirement.
- If the wells are low pressure and easy to leak, it is easy to cause leakage in cementing and contaminate the pay zone, due to the long cementing interval and high annulus pressure.
- In the case of a too long cementing interval for a single-stage cementing job, and also large temperature difference along the interval, the cement slurry properties cannot meet the cementing requirement.
- Oil and gas distribution is uneven, discontinuous, and the distance between formation layers is too large.

Multi-stage cementing is one of the effective methods for the protection of reservoir and long cementing interval. Stage collar is usually used in conjunction with external packer.

Liner cementing technology

Under deep well, ultra-deep well, horizontal well, small annular gap well, high-pressure gas well, and other complicated conditions, the liner is often used to meet the requirement of drilling and completion. Liner cementing (Fig. 7.21) not only saves the casing cost, reduces the load of the drilling rig, but also helps to protect the reservoirs, and to solve the technical challenges that conventional cementing cannot solve in a deep well or a complex well.

In liner cementing, the key component is liner hanger. Whether the liner cementing operation can succeed or not highly depends on the rationality and reliability of the design and usage of liner hanger. Liner hanger is a downhole tool, which is used to trip-in the liner into the well, and the liner hanger sits at the predetermined position on the lower part of the upper casing and then completes the cementing operation. Liner hanger can be divided into mechanical type (include orbit type and J-shape groove type), hydraulic type, hydraulic mechanical joint type, based on the hanging mode; it can also be divided into monotone single fluid tank, biconical single fluid tank, biconical double fluid tank, based on the conical structure. Based on its function, the liner hanger can be divided into regular type,

Fig. 7.21 Liner cementing

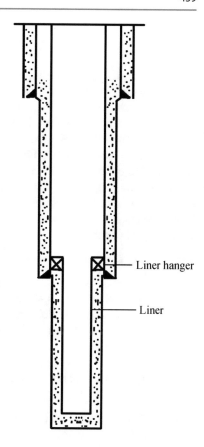

Liner hanger

Liner

inflatable type, spiral type, and corrosion control type. In addition, there are liner hanger with packers, and spiral liner hangers.

Liner cementing uses compound cementing plug, as shown in Fig. 7.22. The top cementing plug is drill pipe cementing plug, and it is installed in the cementing head of wellhead. The bottom cementing plug is liner cementing plug, and it is fixed to the lower part of central pipe for tripping-in tools, and liner cementing plug is hollow. In the displacement process, the drill pipe cementing plug goes down to the liner cementing plug, and stick in the inner hole of liner cementing plug, thereby

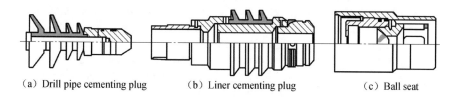

(a) Drill pipe cementing plug   (b) Liner cementing plug   (c) Ball seat

Fig. 7.22 Liner cementing compound cementing plug and ball seat

clip the fixed pin of liner cementing plug and goes down together. When compound cementing plug reaches the ball seat and achieve bumping, they lock to each other tightly.

After completing a conventional liner cementing job, the phenomenon of poor cementing quality or no cement at the overlapping interval might occur, which causes the formation of oil, gas, water cut into the casing. This is the main problem of liner cementing. At present, liner hanger with external packer is usually used to block the annulus channel at the overlapping intervals or use rotary liner cementing to improve the cementing quality.

Expandable liner hanger is a new type of liner hanger developed in other countries. The hanger clings at the external casing after expansion. It will form an effective seal in the annulus due to the pack-off element in the body. Because of its many advantages, the expandable liner hanger becomes a trend of domestic liner cementing in China.

Inner string cementing technology

In a surface casing or large-size casing cementing job, due to the large diameter of casing, low velocity of slurry, long operation time, and a serious mixing between cement slurry and drilling fluid, the cementing quality usually becomes very poor. For that reason, inner string cementing technology is usually utilized in large-size casings.

- The working mechanisms of inner string cementing.

Before run casing, install the inner string cementing jack at the bottom of the first casing. In the plug cementing, connect the inner string cementing plug to the bottom of drill string, trip-in the drill string, and insert the cementing plug into the jack, then use the sealing apparatus to achieve sealing between plug and jack. Start to inject cement from drill pipe after the drilling fluid starts to circulate well, and put in the drill pipe cementing plug after the cementing finish. Then pull out the drill string after cement displacement achieve bumping and finish the cementing operation.

- Inner string cementing tools.

The main tool of inner string cementing technology is the inner string cementer, which includes jack and plug. Cementing inner string is a drill pipe. When the plug is inserted into the jack, the seal at the connection spot can be achieved under an appropriate pressure, as shown in Fig. 7.23.

Inner string cementer has two connection modes, which are plug-in type and thread type. The commonly used mode at present is the plug-in connection. It can be pulled out after cementing, and no need to back-off.

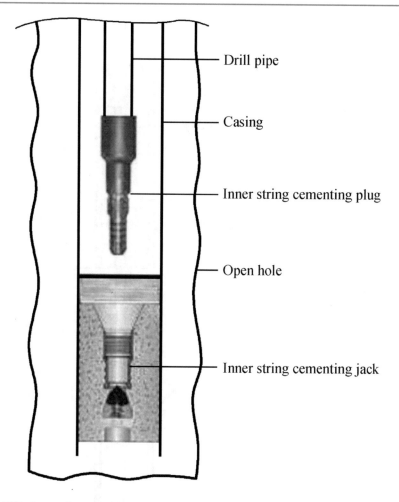

— Drill pipe

— Casing

— Inner string cementing plug

— Open hole

— Inner string cementing jack

Fig. 7.23 Inner string cementing

Prestress cementing technology

Prestress cementing is a type of cementing technology commonly used for the heavy oil thermal production well. In this process, before cementing or cement slurry sets, a tensile force is applied to the casing, in order to produce tensile stress at internal casing in advance. Thereby, it is to balance (reduce) the compression stress produced by the thermal expansion of casing, to prevent the damage of casing expansion in the process of thermal recovery.

At present, ground anchor cementing method (Fig. 7.24) is generally used to apply tension to the casing. Ground anchor is a type of slip, which connects to the bottom of the casing. Then cementing starts when casing runs into the predeter-mined depth, and keeps pressurizing after bumping, and clips the slip pin of ground

Fig. 7.24 Ground anchor
cementing diagram

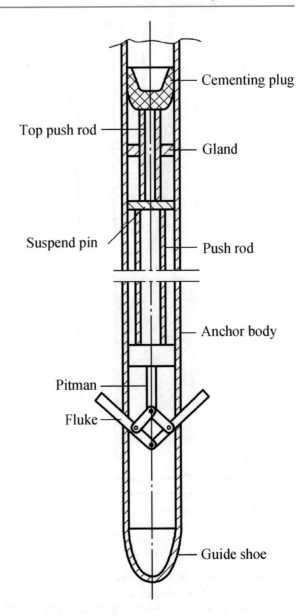

Fig. 7.24 Ground anchor
cementing diagram

anchor, to open the slip and support the borehole wall rocks, then the ground anchor
can bear the tension, and then raise the casing to predesigned tension and wait on
cement to be set.

Some ground anchors also use drilling tools to trip down. It connects to the
bottom of drilling tools. After the drilling tool tripping-in down to the bottom of
hole, we pressurize it at the ground, clip the pin of ground anchor, open it, back-off

and hook-off the drilling tools and then tripping-out. After run casing, casing and ground anchor are snap-connected, and cementing starts after the exerted tension is acceptable. Then, the tension to casing string is applied from the ground to the predetermined value, maintaining the tension, and waiting on cement. Lastly, unload the tension after the cement becomes set. To improve the tension resistance of the ground anchor, we can firstly use the cement to fix the ground anchor.

In addition, "double setting cement" method can be used to apply tension on the casing. The working principle of this method is: taking advantage of double setting cement with big difference in thickening time; after the bottom cement setting, pull the casing to build-up prestress, and wait on the top cement to set.

Reverse-circulation cementing technology

Reverse-circulation cementing is generally used for those wells with serious leakage problem at the bottom of the casing. The characteristics of this technology include: injecting the designed amount cement slurry from the external annulus of the casing to the wellbore, and the drilling fluid reversing out from the internal casing; When the bottom of hole has serious leakage, drilling fluid bleeds into the formation, and then drilling fluid displaces the cement slurry to the predetermined position, to achieve the purpose of cementing.

The advantage of reverse-circulation cementing is to reduce the cement slurry leakage created by the large back pressure at the bottom when the cement slurry returns. Thereby, it is to ensure the return height of cement slurry and reduce the contamination of cement slurry to the pay zone.

The disadvantage of reverse-circulation cementing is that the cement slurry displacement efficiency is low and easy to generate drilling fluid channeling; in order to ensure the cementing quality at the bottom of the casing, part of cement slurry needs to flow back into the casing, and the cement plug should be drilled out after cementing.

7.3.5 Procedures to Improve Well Cementing Quality

It is easy to bring in quality problems in the process of cementing. The main problem is nothing more than two categories: The incomplete filling of cement slurry in external string and the occurrence of oil, gas, water cut in the external tube after cementing.

7.3.5.1 Basic Requirements to Ensure Well Cementing Quality
Basic requirements of well cementing quality under various conditions

- According to the requirement of geological and engineering design, the tripping in depth of casing, return height of cement slurry, and the height of cement plug at inner-tube should meet the requirements.
- The drilling fluid in the annulus of cementing interval should be replaced by the cement slurry.

- The cement mantle is well-cemented between the casing string and the borehole rock.
- Cement can resist the long-term erosion of oil, gas, and water invasion.

Among all the indices of well cementing quality, the consolidation quality of cement mantle is the most important one. Its performance relies on the good and effective isolation on the both cement planes, between cement mantle and casing string, and between borehole rock and cement. The cement should bear two kinds of forces. One is the shear force of cement, which is used to support the weight of the casing in the well; the other is hydraulic cementing force, which can prevent the high-pressure oil, gas, and water from channeling through two cement planes from formation, resulting in oil, gas, and water occurrence at the wellhead.

Frequently occurring cementing quality problems in the well cementing

- The emergence of oil, gas, and water at wellhead;
- Failure to seal the various layer, forming pressure channels in the wellbore, which influences production of the well;
- Casing deformation due to poor bonding between cement sheath and casing, which may lead to well abandonment.

The most common quality problems are channeling and oil, gas, water burst out from the external tube, etc.

7.3.5.2 Prevent Channeling-Improve the Displacement Efficiency of Cement Slurry

In the process of well cementing, the drilling fluid inside the annulus may not be fully displaced by the cement slurry, which means the efficiency of the cement slurry to displace the drilling fluid is low. The degree of displacement of the drilling fluid by cement slurry in the annulus is usually referred as the displacement efficiency.

There are two ways to represent the displacement efficiency: the volumetric displacement efficiency and cross-sectional displacement efficiency.

The so-called volumetric displacement efficiency refers to the ratio of the volume occupied by the cement slurry in the cementing interval to the annulus volume in the same interval. When $\eta_v = 1$, the drilling fluid is fully displaced by cement slurry, and the quality of the cementing is excellent; when $\eta_v < 1$, the drilling fluid is partially displaced by the drilling fluid. The lower the η_v is, the worse the cementing quality is. The volumetric displacement efficiency cannot completely reflect the serious channeling scenarios of drilling fluid through the part of annular cross section. Sometimes, the volumetric displacement efficiency is better, while the displacement efficiency of the partial cross section is poor. One example is the narrow annulus caused by the casing eccentricity.

The so-called cross-sectional displacement efficiency η_{se} refers to the ratio of the cross-sectional area occupied by the cement slurry to the entire cross-sectional area.

The cross-sectional displacement efficiency η_{se} considers the displacement at each cross section, so the evaluation of the displacement quality becomes more realistic.

During the cementing process, if the cement slurry cannot fully displace the drilling fluid in the annulus, the partial annular space will not be sealed by the cement slurry. This phenomenon is called channeling.

Channeling can lead to the decline of cementing quality, cause of the cement rock to lose protection against the casing, which will damage the casing because of the squeeze by the lateral deformation of the rock. It may form a connected channel in the cement rock, and loss the capability of sealing the formation of the different pressure systems, and further cause the oil, gas, water leaks to the external tube or underground pressure channeling. Channeling is a common defect of cementing quality.

The reasons of the formation of channeling

The formation of channeling is related to the displacement efficiency of cement slurry in the annulus. When the displacement efficiency is poor, channeling happens. Low displcament efficiency can be caused by the following reasons:

- The casing is not well centered. When the casing is centered in the wellbore, the annular gap size in every direction is consistent. On a cross section of the annuli, the average flow rate in the annulus is the same, and the rise of the cement slurry in the displacement process is uniform. However, when the casing is not at the center, on one side the gap is big, the flow resistance is small, the flow rate is high; on the other side, the gap is small, the flow resistance is large, the flow rate is low. This result in the uneven displacement flow rate in all directions in the displacement process, and the rising height of the cement slurry is not consistent. It is easy to break through in the high-velocity side, while the low-velocity side has poor displacement, and not easy to displace the drilling fluid. The greater the eccentricity, the bigger the change of inclination angle, the more uneven displacement by cement slurry. If this phenomenon becomes more serious, even the drilling fluid at the small gap side could not be displaced by the cement slurry. It might form a dead zone and generate channeling in the annulus.
- Irregular wellbore. When the wellbore diameter is irregular, the flow rate is high at small wellbore diameter interval, and the flow rate is small at large wellbore diameter interval. A loss of flow resistance exists at the diameter changing zone, where the drilling fluid near the wall surface only produce vortex flow, but not axial flow, thereby form retention. Especially when the casing is not at the center, it is easy to find the drilling fluid to be left over at diameter changing zone and then induce the channeling problems.
- Low cement slurry performance and improper displacement measures. Poor rheology of the cement slurry and improper flow rate used in the displacement process can also aggravate the phenomenon of channeling. If the mobility of the cement slurry is poor, the displacement becomes difficult and the pump pressure becomes high, and it is difficult to reach the turbulent flow in the displacement process. When the laminar flow takes place, the flow velocity in the middle of

the annulus is large, resulting in a fingering effect. The wall of the external tube will remain in a layer of drilling fluid, which will aggravate the channeling.

Measures to improve the displacement efficiency

- Using the casing centralizer to improve the centralization condition of the casing. In the intervals of large variation in the inclination angle and azimuth angle, the use of a number of casing centralizer is an effective measure to improve the central degree of the casing. The centralizer can reduce the non-centered degree of the casing. Especially in the inclined section, we should use multiple casing centralizers.

In regular wells, since the deviation of the inclination angle and azimuth angle is not drastic, elastic centralizers can be used. In the inclined section and sections with a large change in inclination angle and azimuth angle, rigid centralizers should be used.

Where to place the centralizer depends on the support height of the centralizer, the casing performance parameters, the gap between the casing and the wellbore, etc. At least one centralizer is needed for a single casing in a horizontal well or a high angle well.

- Move the casing in the cementing process. In the displacement stage of cementing, the casing may be in motion (lift up, lower down or rotating the casing). This is helpful to improve the displacement efficiency and displace the drilling fluid in dead zone.

Lift up and lower down the casing can affect the surrounding drilling fluid, causing the drilling fluid to generate upward movement tendency. This is beneficial to the displacement of drilling fluid by the cement slurry, thereby improving the displacement efficiency.

When the casing is in motion, due to the existence of frictional resistance, the drilling fluid and casing will move together and obtain a traction force for the drilling fluid. This force gives a certain flow rate to the drilling fluid in dead zone and helps the cement slurry to displace the drilling fluid in that area. Usually, specialized tools are needed to rotate the casing, and the process is quite difficult.

- Adjust the properties of cement slurry to improve the displacement efficiency. It is advisable to use a method of increasing the density difference between the cement slurry and drilling fluid, so that the drilling fluid can obtain a buoyancy force, which causes it to "float" on the cement slurry and to be displaced. The use of robust preflush system strengthens the isolation of drilling fluid and cement slurry, and positively impact the flushing of the borehole wall. The use of washing fluid can dilute the drilling fluid and make it easier to be displaced.

- Adjust the flow velocity of the cementing column in the annulus to meet the requirement of turbulent displacement. In the displacement, the flow velocity of the cement slurry should be made to reach turbulent flow in the annulus. When the cement slurry in the turbulent flow has the same displacement rate in each cross section, the displacement effect will be the optimal. In general, the displacement time in turbulent flow at the focal point should be no less than a certain time, and this time is called turbulent contact time. The turbulent contact time should be no less than 7 min in the field operation.

If the flow of cement slurry is turbulent in the displacement process, then the cement slurry properties should be adjusted at this time. It is required that the cement slurry should have small flow resistance, small water loss, and good rheology properties, while the displacement of the turbulent flow has a large flow rate and high pump pressure. Sometime, the borehole, facilities, and other conditions are not allowed to achieve a turbulent displacement. For example, the allowable power of the facilities is not adequate to support the high-rate displacement, or high-rate displacement may cause a very high-pressure drop in the downhole, and may penetrate the formation. In these cases, a small flow rate of cement slurry should be used to form piston-like flow, which is a plane-movement displacement. We try not to use laminar flow, which is between piston-like flow and turbulent flow, to displace the drilling fluid. It is critical to use a very small flow rate to displace fluid, which results in a long cementing operation time. In this case, the cement slurry thickening performance is hard to control.

In actual operation, the critical flow velocity differentiating the turbulent flow from laminar flow can be obtained by studying the rheological properties of the fluid in the well.

- Injection of high-quality preflush system. Before cementing, it is advisable to inject a certain amount of spacer fluid or washing fluid, to effectively isolate the drilling fluid, and increase the density difference between these two fluids. It may generate the beneficial "floating" effect to drilling fluid and also meet the requirement of turbulent contact time. Generally, in the design of cement slurry, it is required that the density of cement slurry should be greater than drilling fluid by 0.2–0.4 g/cm^3. The density of washing fluid approximately equals to 1 g/cm^3, thus increasing the density difference between drilling fluid and cementing slurry, and making the drilling fluid easier to be displaced from the annulus.

In the actual field cementing process, since it is difficult to have the tail slurry to achieve turbulent flow displacement or piston-like flow displacement, the washing fluid or spacer fluid is usually pumped to turbulent flow state to meet the need of turbulent contact time. Washing fluid has the properties of low density, low viscosity and low flow resistance, and it can achieve the turbulent flow characteristics under low flow rate and its critical flow velocity of turbulent flow is 0.3–0.5 m/s. It is required that the turbulent flow contact time of the preflush system is greater than

7 min, thus to ensure a good displacement efficiency. Once the filling volume of the preflush system cannot meet the requirement of turbulent flow contact time, some of the lead slurries can be used to obtain the turbulent flow displacement, in order to improve the displacement efficiency and achieve the requirement of turbulent flow contact time.

7.3.5.3 Oil, Gas, and Water Channeling Problem in the setting Process of the Cement Slurry

In the setting process of the cement slurry, many factors contribute to oil, gas, water channeling into the annular space, causing the emergence of oil, gas, and water in the external tube. This is an equilibrium problem of the pressure system. The method to prevent the channeling is to always maintain the hydrostatic fluid column pressure of cement slurry greater than the formation pressure; or both the primary cementing interface (cement mantle and casing cementing interface) and secondary cementing interface (cement mantle and borehole wall cementing interface) have good hydraulic cementing bond strength obtained from the setting process of cement, in order to achieve the sealing of oil, gas, water inside the formation.

The reasons of oil, gas and water channeling

- Weight loss during cement slurry gelation. The cement slurry will gradually change from liquid phase to solid phase after mixing with water. When the cement slurry is in liquid phase, it has hydrostatic fluid column pressure. In general, the density of cement slurry is greater than the density of drilling fluid. In cementing conditions, the fluid column pressure in the annulus is usually greater than the formation pressure. After the cement slurry is converted to solid phase, it has a relatively high bond strength with casing and rock, which can prevent the formation pressure from breaking through its cementing surface and channeling. However, in the process of cement slurry turning from liquid to solid phase, as the hydration and gelation of the cement, cement particles in the cement slurry form various types of network structure with borehole wall and casing. The gel strength keeps increasing, and some of the weight of the cement column is suspended on the borehole wall and casing. As a result, the effective hydrostatic pressure against the lower formation acted by cement slurry is reduced, and even more, the equilibration between the hydrostatic fluid column pressure and formation pressure is lost. This phenomenon is called the cement slurry gelling weight loss, as shown in Fig. 7.25.

Although most of the weight of the cement slurry is suspended on the cementing surface, the cement slurry is not completely set, and the cementing strength is very low on the cementing surface. When the effective hydrostatic fluid column pressure of cement slurry is reduced after cement setting, the formation pressure cannot be balanced off. If the cementing strength of the primary interface and secondary interface are relatively low, the formation pressure may break through the

Fig. 7.25 Cement slurry gelling weight loss diagram

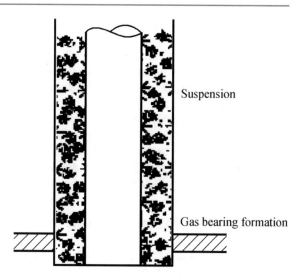

Suspension

Gas bearing formation

cementing surface and cause channeling. If the wellhead happens to be open, it can possibly result in the emergence of oil, gas, water in the surface. This phenomenon is caused by cement slurry weight loss.

- The weight loss caused by bridge blocking. During the process of cement injection and after the cement return to the designed height, a blockage (i.e., bridge blocking) might be formed in the permeable formation or small annular space intervals, due to the some instances such as filter cake formed by the cement slurry fluid loss, leftover drilling cuttings not carried out during drilling, corroded rocks formed during cementing, and settled cement particles, as shown in Fig. 7.26. Bridge blocking prevents the cement slurry pressure system from passing down, and the hydrostatic fluid column pressure decreases below the bridge blocking point. The formation fluid may be penetrating into the annulus

Fig. 7.26 Bridge blinding weight loss diagram

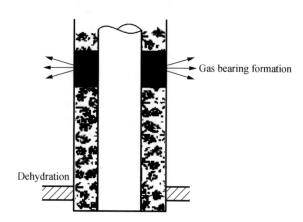

Gas bearing formation

Dehydration

below the bridge blocking point, and break through the bridge blocking, leading to leakage of oil, gas, and water.

- Cement volume contraction leads to oil, gas, and water channeling. The contraction of cement volume may cause the cracks in the cementing surface, which possibly producing the channels for oil, gas,and water. The volume of conventional silicate cement shrinks slightly when solidified, and the shrinkage factor is below 0.2%. This little contraction is very risky for high-pressure formations.
- Building pressure in the casing. During wait-on-cement period, if the pressure is maintained in the casing, the casing is in its expansion state. When the wait-on-cement is complete, the internal tube pressure is released, causing the casing to contract, which may also form channels of oil, gas, and water flow.

Methods to prevent oil, gas, water from channeling

- Release the internal tube pressure immediately after the cementing, and then pressurize the annulus. This can prevent oil, gas, water from channeling. Thus, in the process of cementing, the back pressure at the bottom of the casing should be kept well.
- Use expansive cement to prevent shrinkage of cement rock.
- Adopt multi-stage cementing technology or adopt two types of cement with a different rate of setting. In the process of cementing, multi-stage cementing technology will decrease the length of cementing interval, and reduce the oil, gas, water channeling possibility caused by cement slurry gelling weight loss. The other strategy is to use two types of cement with different rate of setting and make the rate of setting basically the same all around the annulus, thus extending the fluid pressure transmitting time. The rapid cement set also makes the oil and gas difficult to enter the cement rocks to form a channel.
- Use a scraper to remove mud cake at borehole wall. The removal of mud cake can make the cement and borehole wall contact directly, thus forming a good cementing bond at the secondary interface and avoiding the existence of microcracks at the secondary interface. This is to achieve the purpose of effectively sealing oil, gas, and water in the formation.

7.3.6   Special Cement Slurry System

Because of the continuous advancement in the oilfield exploratory development and drilling technology, more and more deep well, ultra-deep well, exceptional well, and problem well becomes the subjects of well cementing operation. The cementing operation in these wells has some special problems, and it is difficult to solve by cement slurry system formulated by the basic cement additives. Therefore, the special cement slurry system used to solve the special problems has shown great progress. Following are the introduction of several main special cement slurry systems.

7.3.6.1 High-Density Cement Slurry System

When drilling in an abnormal high-pressure formation, in order to balance the formation pressure in the cementing process, high density or even ultra-high-density cement slurry should be in cementing. The high-density cement slurry system generally refers to the cement slurry system with a density greater than 2.0 g/cm^3. At present, there are two composition methods for high-density cement slurry: one is to use a single solid weight additive to raise the density based on the principle of the grain size distribution; the other one is to increase the density of slurry.

In the high-density cement slurry, since the weight additive has greater density than cement slurry and less cement particles in the slurry, the suspension stability and strength of cement slurry are highly affected. Therefore, when conducting high-density cement slurry design, especially for ultra-high-density salt-resisting cement slurry, we must meet the prerequisite of flow property of slurry for the processes of mixing, injecting and displacement cement, and also to ensure the stability of precipitation and coagulation volume at pump-off moment in the injecting process, including drainage rate, cement rock density distribution and setting volume shrinkage. And we also need to meet the requirement of density to ensure the strength of cement rock. High-density cement slurry presents a higher requirement for the preparation of cement slurry, the development of cement slurry, the stability of settlement, and so on, especially when the density is higher than 2.5 g/cm^3. High-density cement slurry should have such characteristics as small liquid–solid ratio, good flow property, and good stability.

At present, in China, we mainly use hematite, ilmenite to make high-density cement slurry. The weighting material of DensCRETE high-density cement produced by Schlumberger is made of superfine manganese and moderate ilmenite and hematite, and the cement slurry has the relatively high compressive strength and good rheology and stability, and the density can be up to 2.9 g/cm^3.

7.3.6.2 Low-Density Cement Slurry System

In the cementing process, it is generally required that the annular fluid column pressure does not exceed the fracture pressure of the open-hole formation. Otherwise, it will fracture the formation, and result in cement slurry leakage in the formation, and contaminate the pay zone. With the development of underbalanced drilling technology, the demand for underbalanced cementing is increased. To adapt to the demand of hydrocarbon reservoir protection, in recent years, low-density cement slurry has been widely used. Low-density cement slurry system generally refers to the cement slurry system whose density is less than 1.7 g/cm^3.

The reduction of the density of cement slurry is generally achieved by the addition of lightweight additive, and the selection of lightweight additive should be considered thoroughly based on the density requirement and application condition of cement slurry system. The cement slurry lightweight additive can be divided into two categories based on its working principle: one is water-absorbent or viscosifying agents, such as bentonite, diatomite, coal ash, hard asphalt, expanded perlite, volcanic ash, sodium silicate, and other superfine powders. For this type of agents, the density of cement slurry mainly depends on the water–cement ratio. The other type mainly relies on the lightweight additive itself to reduce the density of cement slurry. For this type, the density of cement slurry mainly depends on the density of lightweight additive itself and the volume of additives.

In order to reduce the density, the low-density cement slurry needs to add a lot of lightweight admixtures, and increase the water–cement ratio, resulting in poor stability of cement slurry system, and difficult to control water loss. The strength of cement rock is also developed slowly and has low intensity. At present, the close packing theory is generally used to resolve the contradiction between the cement slurry low density and its rheology, stability, compressive resistance. The solid content of low-density cement slurry system is designed to be as high as possible and reduce the absorption of water (or water film) on the surface of the material, and make it with a better degree of sphericity.

7.3.6.3 High-Temperature-Resistant Cement Slurry System
The influence of high temperature on well cementing

In the high-temperature environment, the physical and chemical properties of the cement slurry and set cement will change obviously. This change is mainly indicated by the following: the thickening time of cement slurry is shortened rapidly, and the strength of set cement declines, cement permeability increases. This indicates that the hydration products and structures of cement have changed accordingly. With the increase of temperature, the strength curve of set cement rock has two distinct critical points. The cement strength gradually increases before reaching 110 °C, and the strength starts to decline rapidly as the temperature continues to increase after 110 °C. When the temperature reaches 150 °C, the strength shows another rapid decline. Hence, 110 and 150 °C can be considered as two critical temperature points for the strength decline.

The risks of high-temperature cementing are mainly related to the following aspects:

- The high temperature can shorten the thickening time of the cement slurry and make the cement slurry in the casing lose the fluidity without being displaced to the annulus, resulting in cementing failure.
- In the process of liner cementing or inner string cementing, drilling tools, or inner string can be set by the cement before they can be backed off or tripped out from the wellbore, result in cementing failure.
- The strength of set cement can decline severely. When the strength is low, the sealing property of the formation or the supporting capacity of the casing is lost, and the life of the oil and gas well is shortened.
- The high temperature causes the additives in the cement slurry to fail, resulting in poor rheological properties of the cement slurry, which increases the flow resistance in the cementing process and causes circulation losses, reducing the top of cement below design depth.

- The high temperature makes the water loss of cement slurry increase sharply, which can easily cause the "flash setting" phenomenon of the cement slurry, and generate the bridge blinding weight loss, and result in the gas channeling in the process of waiting on cement.

Commonly used high-temperature-resistant cement slurry system

The commonly used high-temperature-resistant cement slurry system mainly includes the following categories:

- Silica-enriched high-temperature-resistant cement slurry system.

The silica-enriched high-temperature-resistant cement slurry system consists of H grade and G-grade oil well cement, a certain number of admixtures (such as silica sand, microsilia, density regulator, etc.) and cement additives. Its performance is mainly controlled by the admixtures and additives. For instance, the strength of set cement after adding silica sand under different temperature conditions can be improved.

- Emulsion-like high-temperature-resistant cement slurry system.

Emulsion is a general name to describe an emulsified polymer, and the range of particle size of polymeric colloid in the emulsion is 0.05–0.5 µm. Most emulsion suspensions contain about 50% of the solid phase, and the small size of the emulsion particles is filled in the gap between the cement particles, blocking the channel, and reducing the permeability of cement rock. The latest product of emulsion is a copolymer of styrene–butadiene and other derivatives developed since the 1980s, and its temperature tolerance range can reach 176 °C. The emulsion products developed worldwide include: Latex 2000 emulsion by Halliburton, BA-86L emulsion by BJ Company, D500, D600, and D700 series by Schlumberger. In China, Jiang Hongtu from Daqing Oil Field has developed a high-temperature anti-channeling cement slurry system with DHL butyl benzene latex, based on the characteristics of the deeply buried oil reservoir, high well temperature, multi-target formation layers in Daqing Oil Field. This system has the characteristics of robust high-temperature resistance (circulating temperature reaches 170 °C, the static temperature reaches 220 °C), low permeability, right angle thickening, prevention of water and gas channeling and robust rheological properties, and the amount of emulsion admixture is 5–20%.

- High-alumina cement slurry system.

Li Zaoyuan, Guo Xiaoyang from the Southwest Petroleum University has explored the feasibility of using high alumina cement instead of ordinary silica cement for thermal recovery well cementing.

By selecting the CF additives, the high-alumina cement was solidified under low temperature (35 °C) and then cured for 7d at 150 and 320 °C, respectively, under water wet environment. Its compressive strength increases instead of declining with time; in the temperature range of 350–450°C, the high-alumina cement shows good high-temperature resistance performance. However, the field application of such cement slurry system was not found in the literature.

7.3.6.4 Low-Temperature Cement Slurry System

The hydration rate of various minerals in cement is greatly affected by temperature. The temperature of deepwater surface mud line is only 4 °C. At this temperature, the hydration of tricalcium silicate is very slow, and dicalcium silicate is hardly hydrated, causing the slow development of cement strength in a conventional oil well.

The main characteristic of low-temperature cement slurry system is to improve the early phase cement strength, and the main methods are as follows:

- Use early strength accelerant.
- Use ultra-fine cement and particle grading technology.
- Use high early strength high-alumina cement.
- Use sulphoaluminate cement.

At present, the commonly used low-temperature cement slurry system is fast-setting gypsum cement slurry system, Schlumberger's DeepCRETE cement slurry system. In China, we developed antifreeze latex cement slurry system and deepwater cementing cement slurry system.

7.3.6.5 Thixotropic Cement Slurry System

In the process of injection and displacement, the thixotropic cement slurry is a thin fluid. When the pump is stopped, it quickly forms a rigid, self-supporting gel structure. When the pump is resumed, the gel structure can be destroyed and the flow can be restored, so that the leakage problem can be effectively solved. Therefore, the thixotropic cement slurry is considered as an important technical tool to solve the problem of severe circulation loss.

At present, the applicable conditions of thixotropic cement slurry are:

- Suitable for the circulation loss during the cementing work and drilling process in leaking formation.
- Can prevent the occurrence of gas channeling under certain conditions.

- In the remediation of the cement in the permeable formations, the thixotropic cement slurry can be used as the lead slurry, to increase the pressure of squeezing and improve the success rate of cementing.
- Applicable to cementing operation in weak formation.
- Repairing the broken or corroded casing.

The thixotropic cement slurry system include the clay cement slurry system, sulfate cement slurry system, etc. Clay cement slurry system is to add absorbent expandable clay in the sulfate cement. The amount of clay can be up to 2%, which can be used effectively to prevent the gas channeling and sealing of annulus. Sulfate cement slurry system is to add calcium sulfate, aluminum sulfate or ferrous sulfate in the silicate, and the sulfate forms a gelatinous mass with thixotropy. The content of sulfate is generally less than 10%.

7.3.6.6 Anti-corrosion Cement Slurry System

The cement mantle can be corroded by the formation fluid during the long process of oil and gas well development. The acidic medium and sulfate can severely corrode the cement. In water wet environment, H_2S and CO_2 will cause severe corrosion to the hydration products, resulting in the failure of the sealing performance of the cement sheath, inter-zone communication, and gas emerging from the wellhead in a severe case. The key factors controlling the corrosion of the cement by acidic agents are the composition of the cement hydration product and the microstructure of set cement. The denser of the cement rock, the better the corrosion resistance.

The method of improving the corrosion resistance of the cement mainly includes:

- Mix the cement with certain grades of coal ash, then utilize the further action of coal ash activation vitreous body and cement hydration product, and the filling effect of the fine particles of coal ash on the internal pores of cement rock to improve the corrosion resistance of cement rock.
- Reduce the water–cement ratio, adding activated siliceous materials to react with free $Ca(OH)_2$ in the cement, or fill the small pore space in the cement, to reduce the permeability of the cement rock, and to help to reduce the corrosion speed of cement from CO_2.
- Adding styrene–butadiene latex into cement can obviously improve the anti-corrosion ability of the cement rock, and the ability of anti-sodium bicarbonate corrosion is stronger than the original slurry.
- Phosphate cement is used to produce calcium phosphate through an acid-base reaction, and it has high strength, low permeability and excellent capability to anti-CO_2 corrosion.

The typical products of the anti-corrosion cement slurry system include anti-CO_2 corrosion cement slurry system, latex system and Thermalock cement from Halliburton, anti-CO_2 corrosion cement EverCRETE from Schlumberger. EverCRETE is a technology based on the Cem-CRETE gradation. It reduces the amount of regular Portland cement, does not include $Ca(OH)_2$ system, and it can limit the performance degradation of cement. It can limit the crystallization reaction of calcium carbonate and make the cement rock to maintain a better mechanical property.

7.4 Well Completion Technique

Well completion is the process to connect wellbore to oil and gas reservoir (also known as production layer or paying zone). The technical procedures of oil–gas well completion include drilling through reservoir, determining the bottomhole structure for well completion, installing at bottom hole (running casing for well cementation or inserting screen), connecting wellbore to production layer, and installing wellhead equipment and so on. Well completion plays a key role in the stable and high production of well.

7.4.1 Drilling Through Reservoir

After drilling through reservoir rocks, the initial stress state of reservoir will be altered. Rocks will gain new force balance under new stress state; thus, their mechanical properties and oil-storage properties will change. Reservoir rocks are in contact with drilling fluid and completion fluid, and some chemical and mechanical reactions take place between well fluids and reservoir rocks, which will change the properties of reservoir layer. These changes may result in the following: contamination to the formation rocks; reservoir rock being instable, wellbore deformation, porous medium, and flow channels deformation. It will also lead to poor reservoir permeability, lower productivity, and reducing production life. Meanwhile, well kick may happen due to the disruption of pressure balance.

Well completion can make a good connection between formation layer and wellbore and can reduce the unfavorable effects reservoir rocks to minimum level. The ultimate purpose of well drilling is to exploit oil–gas resource rapidly and effectively. Hence during the well drilling and completion processes, we try to protect reservoir layers as much as possible and avoid any more damage to them. Therefore, drilling through reservoir is a crucial step in well completion.

7.4.1.1 Change of Oil-Storage Properties When Drilling Through Reservoir

While drilling through reservoir rocks, drilling fluid contacts with rocks. Since they are in different pressure system, chemical substance system, and concentration gradient system, drilling fluid will contaminate formation rocks.

Effects of pressure system

Typically, when drilling a reservoir, the column pressure of drilling fluid is larger than formation pressure. Under this fluid column pressure, the liquid and solid phases of drilling fluid will enter rock pores and fractures, causing plugging in porous flow path, and contaminating the reservoir rock.

If the column pressure of wellbore drilling fluid is smaller than the formation pressure, it will lead to well kick, which is not allowed.

Effects of chemical substance system

The chemical composition of drilling fluid is undoubtedly not the same as reservoir rock and also the fluid in reservoir rocks. Sometimes, these two types of fluids react and produce some water-insoluble deposits, which will accumulate on pore walls and then plug the porous channels. In other cases, some chemical compositions in drilling fluid will corrode and decompose rock components, damage cementing agents, cause rock expansion and collapse, which also lead to permeability changes in reservoir rocks.

Effects of concentration imbalance

Due to the concentration difference of chemical compositions in drilling fluid system and formation fluid system, chemical components will transport, and generate certain osmotic pressure. Under osmotic pressure, rock cementing agents might be damaged, or oil–gas flow will be retarded.

Because of the interaction between liquid/solid substances and reservoir rocks, rock cementing agents may be damaged, porous channels may be blocked, rock wettability may be changed, and flow channels may be water-blocked. Thus, it may lead to a permanent damage to reservoir structures, as shown in the followings:

- Solid or liquid phase particles enter into reservoir channels, clogging flow channels. It further causes oil incapable to be displaced, decreases porosity, reduces permeability, lowers down recoverable reserve, and diminishes cumulative oil production.
- Permeability changes due to the clogs caused by solid or liquid particle penetration, which are shown as water-locking, scaling, cementing agent detachment in porous channel.
- After solid and liquid particles entering into reservoir rocks, the rock matrix may be damaged, resulting in the sand production and other problems.

7.4.1.2   Changes of Rock Mechanical Properties When Drilling Through Reservoir

Downhole rocks were in a balanced state before drilling activity. When a borehole is drilled in the rock formation, their stress state will redistribute. It will not only cause shrinkage, collapse, and other problems for non-reservoir rocks, but also cause side effects on reservoir layers.

When drilling through reservoir, borehole rocks are drilled out and turn into an empty hole. Rocks near sidewall lose support, and the stress forces are redistributed. Thus, it will result in a tendency of rocks crushing to borehole center. Since drilling fluid exists at borehole, some stress can be compensated.

The degree of rock deformation is related to its lateral deformation ability. When lateral deformation ability is strong (appears as large Poisson's ratio, such as shale, halite), rock will protrude more to borehole center. For high-strength rocks such as limestone, the lateral deformation ability is small. Rock deformation ability is also related to the density of drilling fluids. Low-density drilling fluids have poor performance in compensating lateral stress and deformation; thus, borehole deformation phenomenon is severe. High-density fluids have strong compensation ability. But if it exceeds the compressive strength of rock, it will then lead to rock fracture.

After reservoir being drilled through, the lateral deformation of rocks will affect oil-storage structure. For porous sandstone reservoir, the influence is not drastic. While for a fractured reservoir, the influence is rather serious. Under lateral deformation, the opening of some tiny cracks will apparently decrease and even closes which will result in lower permeability of fracture reservoir, and will offset the lateral deformation of rocks.

In oil production process, the wellbore pressure drops due to the lateral compression force on sandstone reservoir. Meanwhile, by the washing effect of oil–gas flow, sand grains are under dragging force, resulting in sand production. During long-term oil production, formation pressure decreases, and sandstone matrix is under larger forces; thus, sandstone will be crushed, resulting in sand production.

7.4.1.3 Methods for Drilling Reservoir

The effective way to prevent contamination while drilling a reservoir is to use an appropriate drilling fluid system, balanced drilling technique, proper wellbore configuration, and fast drilling to protect production layer from long-time soaking in drilling fluids.

Use of an appropriate drilling fluid system

It is important to select proper drilling fluid system when drilling through reservoir. The properties of drilling fluid should be designed based on the reservoir characteristics.

The chemical system of drilling fluid should match with reservoir system as much as possible. The selection of the chemical system should be based on chemical properties of reservoir rocks and in-reservoir fluids, so as to prevent from adverse effects such as sedimentation or dissolution due to the mismatching of two chemical systems. Normally, we try to use drilling fluids with low solid phase or free of solid phase and reasonably increase the mineralization degree of drilling fluids, and treat drilling fluid with some surfactant.

Adopt proper density of drilling fluids, realize balanced drilling

The density of drilling fluids is high, which results in the column pressure of drilling fluid higher than stratum pressure. This pressure difference is the main reason for reservoir contamination. Meanwhile, the cutting-hold effect increases and penetration rate decreases, due to this pressure difference. Properly reducing the pressure difference between drilling fluid and stratum, and conducting balanced drilling are effective ways to prevent from reservoir contamination, which can also help to improve penetration rate and shorten well construction time window. We should determining reservoir pressure accurately, and adjust the density of drilling fluid to match with it, and properly increase some additional density if needed. When drilling through reservoir, the column pressure of drilling fluid should roughly equal to reservoir pressure. Under this pressure, reservoir contamination level can be minimized. The prerequisite of balanced drilling is to estimate reservoir pressure accurately, meanwhile, to cooperate with good well-control technology and solid-phase control technology.

In some special reservoir layers, underbalanced drilling technology can be applied, such as air drilling, mist drilling, foam drilling, and so on.

Adopt good wellbore configuration, reduce reservoir soaking in time

The longer the reservoir soaking in drilling fluids, the more serious contamination will be. And the solid and liquid phases will invade deeper into reservoir layers as well. Therefore, short drilling time is favorable for protecting reservoir. In addition to the increase of drilling rate, there are other methods to reduce reservoir soaking time. One good strategy is to use favorable well configuration. Set a casing to seal the drilled reservoir is a good strategy to stop the upper reservoir from soaking into drilling fluids. Also before drilling, seal the upper stratum by casing, and utilize high-quality drilling fluid when drill through reservoir in order to achieve balanced drilling. Meanwhile, using a good well configuration can reduce complex conditions and drilling accidents and shorten well construction time.

Prevent contamination in other production cycles

In well cementation, the cement can cause very serious contamination. Using low-water-loss and low-density cements can effectively prevent contamination in well cementation process. Reduce shut-in frequency or well-kicking time in formation testing and other borehole operations can also prevent contamination.

Apart from taking different measures to prevent reservoir contamination, there are also reservoir improvement measures, such as acidizing and fracturing. All of these can be applied to recover some oil-storage properties in contaminated reservoirs.

7.4.2 Principle of Well Completion and Types of Bottomhole Structure in Oil–Gas Well

7.4.2.1 Principles of Well Completion

Well completion is operated based on reservoir properties, production situation, and other conditions, under the premise of stable and high well productivity. For different reservoir rocks and different production modes, there are different completion methods.

Completion requirements

The main purpose of well completion is to establish a good connection between borehole and reservoir, so that well can have high productivity, and the borehole can maintain long-time stability for stable productivity.

For well completion, basic requirements are:

- Protect reservoir stratum to the maximum and prevent reservoir from any damage.
- Reduce flow resistance when oil–gas stream flowing in well hole.
- Able to effectively seal apart oil, gas, and water layer, prevent from crossflow and interference in each layer.
- Prevent from well collapse or sand production, assure the long-time stable production, extend well life.
- Able to operate water-flooding, fracturing, acidification, and other well stimulation measures.
- Simple process and low cost.

Completion design

Well completion is crucial to the whole production process. Thus, before drilling a well, there should be a comprehensive design for well completion. After determining the reservoir properties and oil–gas field development plan, completion design is conducted to determine the way to open reservoir and the bottomhole structure in completion, clarify the stratum layer and depth to set oil string, and determine the connection way of reservoir and well hole. Normally, completion design is conducted before the design of drilling engineering, or conducted simultaneously with the design of drilling engineering.

The process of completion design is: first prepare the reservoir analysis report based on laboratory analysis and electrical logging analysis of reservoir rocks, then the oilfield development department propose the completion plans, and well drilling department draw the detailed design for completion operation per well according to stratum conditions (oil companies are typically in charge of designing completion operation).

Completion design includes:

- Propose the type of bottomhole structure according to reservoir characteristics.
- Propose the borehole size of completion section, such as borehole diameter, length of drilled reservoir, length of pocket, and so on.
- Design the completion pipe string, including diameter of oil-string casing, downward depth, return top depth of cement slurry, perforation parameter of oil-string casing, size of screen pipe and liner pipe, and so on.

- Design completion fluid, and propose the type, parameter, using and adjusting method of completion fluid.

7.4.2.2 Types of Bottomhole Structure

When selecting the bottomhole structure, the following factors should be considered: reservoir type, reservoir lithology and permeability, oil–gas distribution, degree of formation stability in completion section, the existence of high-pressure layer, bottom water, or gas cap, and so on. For example, for homogeneous hard formation, open-hole well completion can be applied. While, for non-homogeneous hard formation, cased hole completion can be used. For non-stable stratum, we use non-fixed screen completion method, and for poor cementing production layer that produces sand, we use sand screen completion.

According to different reservoir conditions, the selection of bottomhole structure is listed in Fig. 7.27.

Bottomhole structure can be divided into four categories:

The first type is a closed bottom hole. When drilling at target stratum, after set production string or tail pipe, cement well to block reservoir, then perforate to connect reservoir with borehole, as shown in Fig. 7.27a and b.

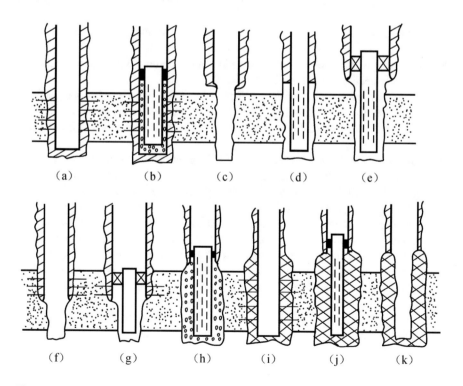

(a)   (b)   (c)   (d)   (e)

(f)   (g)   (h)   (i)   (j)   (k)

Fig. 7.27 Scheme of bottomhole structure in well completion

The second type is open bottom hole. The bottom hole is not closed after drilling reservoir and directly connects the exposed reservoir rock with borehole; or set perforated screens at producing interval to support formation, but no well cementing is applied, as shown in Fig. 7.27c, d, and e.

The third type is a mixed bottom hole. At bottom reservoir is open hole so that reservoir can be directly connected with borehole; at upper side, set casing to enclose then perforate to connect with borehole, as shown in Fig. 7.27f and g.

The fourth type is sand control completion, which is mainly for poor-cemented sandstone formation. And the production layer can be closed or open. It is used for sand control, as shown in Fig. 7.27h, i, j, and k. For the screen tube sand control completion method, it requires additional gravel to fill in between production layer and screen tube or other production string.

These four categories can be further divided into 11 commonly used methods.

- Single pipe perforation completion: It is for a typical closed bottomhole structure. We insert casing in drilled wellbore to cement well, as shown in Fig. 7.27a. In addition to a single pipe perforation completion, there are other methods such as multiple pipe perforation completion and permanent packer completion.
- Early stage open-hole completion: It is for a typical open bottomhole structure, as shown in Fig. 7.27c. Besides, there is the late-stage open-hole completion method.
- Slotted pipe completion: It is a type of open bottomhole structure. We insert screen pipe in open-hole section, as shown in Fig. 7.27d.
- Liner completion: It is a type of open bottom structure. We insert liner pipe in open-hole section, as shown in Fig. 7.27e.
- Semi-closed open-hole completion: The lower part of reservoir is open hole that directly connects to borehole. And at the upper reservoir, we insert casing, cement and then perforate, as shown in Fig. 7.27f. This is a mixed bottomhole structure.
- Semi-closed liner completion: Liner pipe is inserted in the open hole at lower reservoir, and casing is inserted and perforated at upper reservoir, as shown in Fig. 7.27g. This belongs to semi-closed bottomhole structure.
- Internal gravel packing sand control completion: After perforating sandstone formation, insert sand control screen pipes and pack gravel in the annual space between casing and screen, as shown in Fig. 7.27h. It is a closed bottomhole structure and is also a sand control type. It belongs to secondary completion.
- Open-hole gravel packing completion: It is a type of sand control completion method, as shown in Fig. 7.27h. This method is to insert screen pipe in the exposed sandstone layer and pack gravels in annual space.
- Permeable artificial borehole perforation completion: Insert permeable and solidifiable material in the space between casing and sandstone layer, use low-power perforating bullet to perforate casing but not break the inserted permeable zone, as shown in Fig. 7.27i. It belongs to sand control completion.

- Permeable artificial borehole screen pipe completion: Set screen pipe at sandstone layer and insert permeable and solidifiable material in the space between screen pipe and sandstone layer, as shown in Fig. 7.27j.
- Permeable artificial open borehole completion: Insert permeable and solidifiable material in open-hole section to form permeable artificial borehole, as shown in Fig. 7.27k.

In addition to the above common structures, there are also the variants of each structure type. With the advancement of drilling and production technologies, the bottomhole structure of completion will also continue to be developed.

7.4.3 Common Methods for Well Completion

7.4.3.1 Open-Hole Completion Method

Open-hole completion method means in the completion stage, reservoir at bottom hole is exposed; only upper part is cemented. Open-hole completion can be divided into pre-open-hole completion and final open-hole completion. Pre-open-hole completion is to insert production string casing to cement well before drilling to the target production layer, then to use the prequalified high-performance drilling fluid and drill bit to broach the production layer, so as to exploit the oil–gas interval in open hole (Fig. 7.28). Final open-hole completion method is to insert casing to the top boundary of oil layer and cement well after drilling to the target position below bottom boundary of oil layer, so as to keep the production layer exposed.

Open-hole completion method is only applicable to homogeneous hard reservoir including pore-type, fracture-type, fracture-pore-type or pore-fracture-type reservoirs. The homogeneous reservoir indicates that production layer has roughly the same permeability; the hard reservoir means that reservoir rocks are strong enough to be unbroken undertaking the pressure difference of overburden rocks and flowing fluids. The permeability of homogeneous reservoir ranges between 0.01 and 0.1 μm^2.

This method is suitable for wells in a single-layer reservoir, which does not need to produce by layers and has no water or gas-containing interlayers. Applicable reservoir rocks are limestone and solid sandstone, mudstone or shale. Due to the direct connection between reservoir and borehole, open-hole completion method has advantage that oil–gas stream has the lowest flow resistance when flowing into borehole. The advantage is more obvious especially in the pre-open-hole completion method. Surely, this method also has disadvantages.

Advantages of open-hole completion method

- It can eliminate interference from upper stratum and provide the most sufficient conditions for selecting drilling fluids compatible with the characteristics of opened production layer, so as to drill reservoir with least contamination.

Fig. 7.28 Scheme of
pre-open-hole completion
method

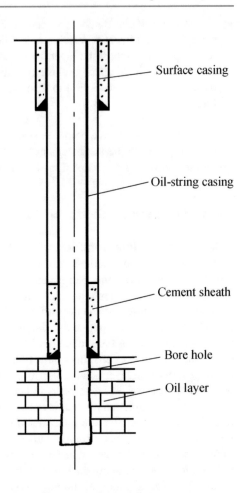

- In drilling process, it can lift drilling tool into casing when the subsurface condition is complicate, so as to avoid more complex accidents.
- It can shorten reservoir soaking time in drilling fluid, so the degree of reservoir damage can be reduced.
- Because well cementing is operated above the production layers, so it can eliminate the effects of high-pressure oil and gas flow on cementing, improve cementing quality, and avoid cementing contamination in reservoir section.

Disadvantages of open-hole completion method

- Only suitable for limited types of reservoirs, and not appropriate for non-homogeneous and weak cemented production layer. And it cannot avoid the unfavorable effects on oil production, such as wellbore collapse and sand production.

- It cannot avoid interference to production layer, such as interactions among oil, gas, and water layers, and the interference between different pressure systems.
- After putting into production, some well stimulation operations such as acidizing and fracturing are hardly to be conducted.
- In the initial open-hole completion method, upper stratum is cemented before opening the production layer. The actual information of production layer has not been investigated, and hence, the formation properties have not been determined. If any unexpected conditions are occurred in the reservoir-opening stage, it may cause a big impact on the further production.
- In the final open-hole completion method, it cannot avoid contamination to the production layer and unfavorable impacts caused by drilling fluid and cement slurry.

Initial open-hole completion

The typical configuration of an initial open-hole well is to select solid stratum that is 20 m distance from production layer, and stop drilling and set casing to cement the well. Before cementing, log the well and determine the depth of reservoir, in order to avoid drilling complex strata such as high pressure or unconsolidated formation before opening the production layer. Casing shoe should be set at solid stratum. The length of open-hole section and the thickness of production layer are related to the strength of rock formation. The length can range from several meters to more than a hundred meters. For high-strength rock formation, the length of open hole can be longer. The production layer can be all opened at one time or be partly drilled through in a thicker section. After all production layers are drilled through, the additional length should be drilled in order to save a pocket to stop drilling. The length of pocket is determined according to well conditions, which is at least 5 m, and above 10 m in most cases.

For each pore-fracture type reservoir, there is an allowable column pressure difference. Beyond this limit, pores may be plugged, fractures may be closed, and oil well will no longer produce oil. In the same way, for each pore-fracture type, there is certain rock strength. If oil flow rate is smaller than one critical limit, rocks will not be broken. For weak cemented production layer, the allowable production column pressure is small. Therefore, in drilling and oil production process, we should be cautious and avoid any crushing and damage to production layer.

7.4.3.2 Perforation Completion

Perforation completion refers to the following process: in completion stage, insert oil-string casing to seal production layer, then perforate casing, cement sheath and part of production layer by perforating bullet, so as to form the flow channels for oil–gas stream. After perforating the production layer, the production capacity of oil–gas well is affected by pressure and properties of production layer, and perforation parameters and quality as well. In petroleum exploration and development, perforation completion is the primary method, accounting for about 80–85% of the total completion cases.

Applicability of perforation completion

Perforation is the most commonly used completion method, which can be applied to almost all reservoirs. However, it is a misunderstanding that perforation is the best completion method. As research shows, the perforation completion is not always the best option for all reservoirs.

Perforation completion is applicable to all types of reservoirs, no matter the reservoir is pore type, fracture type, pore-fracture type or fracture-pore type, no matter reservoir strength is strong or small, no matter reservoir is homogeneous or not, and the pressure system is the same or not. That is to say, perforation completion is applicable for most reservoirs. However, only for heterogeneous reservoirs, perforation completion is the best suitable option. In a heterogeneous reservoir, the stable stratum interlocks with non-stable stratum, and strata under different pressure systems mutually interlock, and there are sometimes aquifers and gas interlayers, or with bottom water and gas cap. For homogeneous reservoirs, other completion methods might be more applicable.

Perforation completion method can be divided into single pipe perforation completion, multiple pipe perforation completion, tail pipe perforation completion, and perforation completion with packer, etc., where single pipe perforation is most common. According to perforation tools, it can be divided into cable-conveyed casing perforation and through tubing perforation, and tubing conveyed perforation. By perforation technique, it can be divided into overbalanced perforation and underbalanced perforation. No matter what perforation methods, there are specific requirements on casing, cementing and other aspects. In Fig. 7.29, it is the configuration of the most commonly used single well perforation completion.

In perforation completion, the technical conditions for opening reservoir are also very strict. Casing shoe of oil layer should be set at 1–3 m from bottom hole, the restrictive ring in the casing should be placed minimum 15 m below the bottom of production layer, and the height of in-casing cement slug below restrictive ring should be no less than 20 m. Other accessories on casing string such as centralizer and scraper should be placed in a depth to avoid production layer. There should be short casing on casing string, in order to log well by magnetic locator and correct perforation depth. The position of short casing should be 20–30 m above oil layer.

Perforation parameters include perforation density, diameter, and depth of perforation channel, perforating phase angle, perforated length in oil layer, and so on. These parameters are determined by strength of rock formation, properties of production layer and development plan for oil reservoir. Perforation density is 10–20 hole/m, and the maximum can reach 36 hole/m. The diameter of perforation channel is 10–16 mm, the maximum can reach 25 mm, and the shape of perforation channels is near cylinder. Phase angle of perforation is $72°$ to $180°$ and distributed in a spiral form. At the same cross section, only one perforation channel is permitted. The depth of perforation should be through casing, cement sheath, and should try to surpass reservoir damage zone.

Fig. 7.29 Scheme of casing perforation completion method

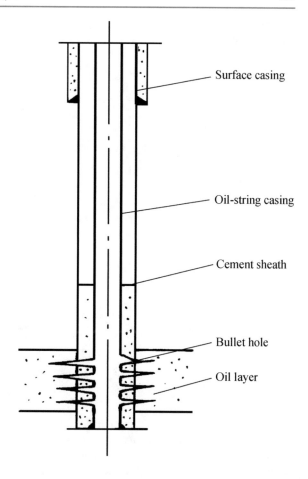

Surface casing

Oil-string casing

Cement sheath

Bullet hole

Oil layer

Perforation process

Perforation is a process to send the perforating gun to oil–gas layer, ignite it on surface to detonate perforating bullets, then shoot through casing, cement sheath and production layer all together, to form channels that connect borehole and reservoir. The scheme of a single pipe perforation method is shown in Fig. 7.30.

The tool for perforation is perforating gun, which can be divided into cable gun, and non-cable gun. Cable gun can be further divided into pipe-type gun and rope-type gun, such as through tubing perforation gun, steel wire perforation gun, steel pipe perforation gun. Cable gun is conveyed to downhole by cable or steel wire and is electrically ignited. Non-cable gun is conveyed to downhole by oil tube, so it is also known as tubing conveyed perforation gun. The commonly used perforation bullet is jet perforator, sometimes we also use bullet.

Perforation process can be divided into overbalanced perforation and under-balanced perforation. Overbalanced perforation is to perforate when wellbore column pressure is larger than reservoir pressure. In perforation stage, cable gun is

Fig. 7.30 Scheme of single
pipe perforation

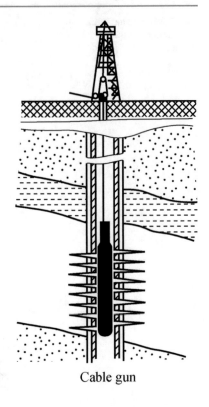

Cable gun

released to desired depth by winch, then the gun is ignited on surface through cable
to shoot bullet and perforate reservoir. Because the column pressure of wellbore
liquid is larger than pressure of production layer, formation fluid will not flow into
well immediately. Hence, well blowing will not occur. After all production layers
are perforated, we should pull out perforation tool, insert oil tubing, and discharge
perforating fluid, so wellbore pressure will decrease, making high-permeable layers
produce fluid. In overbalanced perforation method, if channels are not cleaned in
time, perforation residue will stay in pore rather than flowing out with fluid. So
pores in production layer will be plugged. The perforating fluid will intrude into
channels under positive pressure, causing reservoir contamination. Although there
are so many plug removal methods, the damage to reservoir is hardly eliminated
completely. In overbalanced perforation method, operational tools are simple and
there is also no blowout risk, but it may cause serious contamination to reservoir
layer. This method is almost completely abandoned in foreign countries, and we
will stop using it in China in near future.

The so-called underbalanced perforation refers to a process to perforate when
wellbore column pressure is lower than reservoir pressure by utilizing the low-density
perforation fluids or reducing the column height of fluids. Underbalanced perforation
has advantage in reducing reservoir contamination during perforation process.
Through tubing perforation is one of underbalanced perforation methods.

Fig. 7.31 Scheme of tubing
conveyed perforation

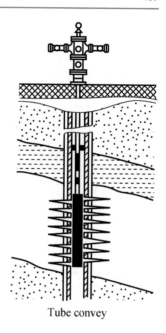

Tube convey

The process of through tubing perforation is that, first inserting oil tube 10–20 m above perforation layer, installing the wellhead, conveying perforation gun by cable through brace blowout preventer, turning off preventer, precisely locating through magnetic locator, and igniting on ground to perforate. In through tubing perforation, the degree of negative pressure should be determined appropriately. If it is too large, tools are hard to be conveyed; if it is too small, the desired effects will not be achieved. In high-permeable zone, the value of negative pressure can be 1.378–3.477 MPa (in case of producing liquid) and 6.89–13.78 MPa (in case of producing gas), and in low-permeable zone, the value can be twice of the above.

Underbalanced perforation can also be achieved by tubing conveyed perforation. Tubing conveyed perforation is also called non-cable perforation, as shown in Fig. 7.31. In this method, perforation gun is directly installed on oil tubing. By fitting with tubing size, insert tubing into well, calibrating the perforation depth, installing oil producing wellhead, shutting down wellhead, and shooting bullets by caber or annular pressure to perforate.

Perforation fluid is filled in wellbore during perforation. There were no specialized perforation fluids in old days, so displacement fluid which is used for cementing was also used for perforation, causing serious contamination to reservoir.

Basically, a special perforation fluid is formulated by adding solid weighting agents and chemical treating agents into water-based completion fluid, by which the quality perforation fluid can be prepared. The pH value of general perforation fluid is 9.5–10, and the density is 1.05–2.03 g/cm^3.

7.4.3.3 Slotted Liner Completion

There are two types of slotted liner completion methods. The first method is that, after drilling oil–gas formation layer, inserting casing string whose bottom is connected with liner at the position of oil–gas layer, and then cementing through external casing packer and sealing annular space above oil–gas layer. For this completion method, if liner is broken, repairing work would be very difficult. Therefore, the second completion method is more appropriate: When drill bit drills at the top of oil–gas layer, insert casing to the top of oil–gas layer, cement well, trip-in bit from casing shoe and continue drilling through oil—gas layer until reaching completion depth, then insert oil string with slotted liner, and seal the annular space between liner and casing by packer. For this method, if liner is broken, it can be tripped out and changed, as shown in Fig. 7.32.

Fig. 7.32 Scheme of slotted liner completion

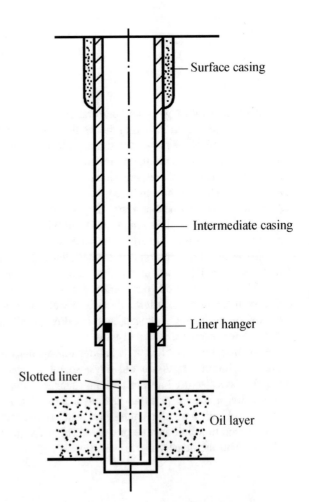

If we use slotted liner, some fine sands may flow into borehole with oil stream, and larger sand grains will block outside liner, forming "sand bridge" or "sand arch." Later, smaller sand grains will be blocked, and much smaller grains will also be blocked. Therefore, near borehole wall, it may form a sand filter blanket with grain size from large to small, which can prevent the vast sand production from oil layer, and have the effect of "sand control," as is shown in Fig. 7.33.

Parameters of slotted liner are:

- Slot shape. The profile of slot is trapezoidal shape (Fig. 7.34), and the intersection of bevel edge is 6°. This slot shape can prevent from slot-stuck by sand grains and the plug of oil–gas flow channel.

Fig. 7.33 Scheme of sand bridge formed around liner

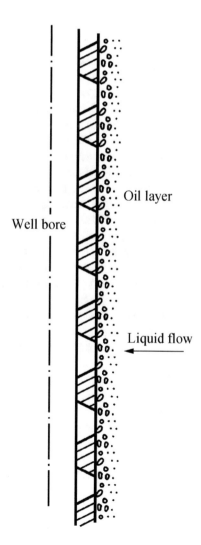

Fig. 7.34 Profiles of
different slots

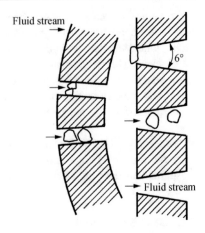

- Slot width. Slot width is the side length of smaller trapezoidal bottom. The determination of slot width in a correct way is key to sanding prevention. According to experimental results, the precondition for sand grains forming "sand bridge" or "sand arch" at slot places is slot width no more than twice of sand diameter, that is $e \leq 2d_{10}$ (e is slot width, d_{10} is the diameter of sand grains that take 10% of cumulative mass on the cumulative production curve formed by production sand). Thereby, large-diameter sand grains that account for 10% of total mass cannot pass through slots, but are blocked at outside liner pipe and form "sand bridge" or "sand arch."
- Slot length. Slot length is related to pipe diameter and slot arrangement. The strength of horizontal screen pipe is low; thus, the slot is short, normally 20–50 mm; for vertical slotted pipe, the slot is normally 50–300 mm. For large-diameter screen pipe, the slot should be short; for small-diameter screen pipe, the slot should be long and recommend taking the maximum value.
- Slot arrangement. There are two types of slot arrangement: one is horizontal to pipe axes, namely end-to-end arrangement, which can also be divided into streamline type, slot-wedge type, and combined slot-wedge type. The other is the side-to-side set up that is vertical to axes, also known as horizontal type, as shown in Fig. 7.35. Generally, the vertical slotted liner is utilized, because the strength of vertical slotted liner is higher than horizontal slotted liner.
- Slot number. The number of slots is related to flow area of liner pipes. Thus, in the premise of adequate slot strength, the flow area should be as large as possible. The total area of slots accounts for 2% of total outside surface of liner pipe. In the slotted completion method, slotted liner is applied to open-hole section; therefore, it cannot only play the role of open-hole completion, but also prevent from wellbore wall collapse in open hole. What is more, the method can also be used in coarse sand oil layer that sand production is not severe. This completion method is not complicated, easy to operate, and the cost is low. In slotted completion method, the sizes of casing, drill bit, and liner are listed in Table 7.7.

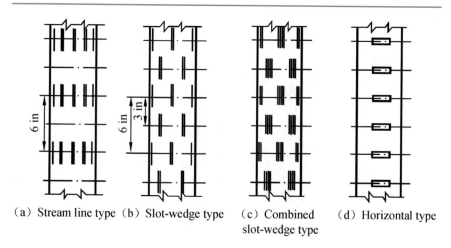

(a) Stream line type (b) Slot-wedge type (c) Combined (d) Horizontal type
                                              slot-wedge type

Fig. 7.35 Schemes of slotted liner

Table 7.7 Size chart of casing, drill bit, and liner in slotted completion

Size of top casing (mm)	Size of second drilling bit (mm)	Size of liner (mm)
177.8	152	127–140
219.1	190	140–168
244.5	215	168–194
273.1	244	194–219

7.4.3.4 Well Completion with Packer

Well completion with packer

In drilling process, bottom water may rise into the pay formation, or the rock stratum around open-hole stratum may change. For example, the collapse of rock formation above production layer may lead to abnormal operation of open-hole wells; or the rise of bottom water may lead to the increase of water cut in production layer, and the decrease of rock strength at wellbore bottom. Under this condition, the combination of liner (screen pipe) and open-hole packer can prevent complex situations from happening in an open-hole section. Packer completion can be used at the initial completion stage. Also, it can be used in the case of open-hole well production for a period of time, and then there are some minor complication occurred. If severe complication happened, packers may be invalid, then a casing should be inserted to seal the defective well section.

This process is attaching the open-hole packer to liner at a proper position, sending it to underground open-hole section, opening open-hole packer when hanging liner tube, isolating the defective upper rock layer and exposing the lower layer, so that the well can remain normal production. If bottom water arises under

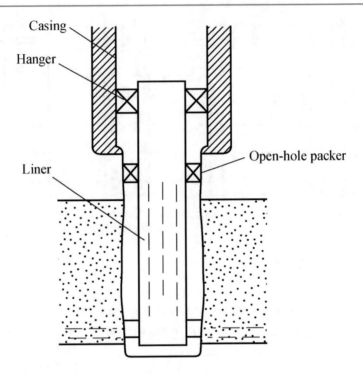

Fig. 7.36 Open-hole completion with packer

production layer, then open-hole packer should be set under liner tube. If rocks collapse above production layer, and then packer should be set under collapsed layer. In long open-hole section, the installation of packer at upper or lower places is quite flexible. This completion method also names as open-hole packer completion, as is shown in Fig. 7.36.

The applicable condition for open-hole packer is harsher than the in-casing type, because borehole wall is coarse and the diameter is uneven. In order to seal rocks, the in-packer rubber seals will carry more stress load. Therefore, the design standards for open-hole packer are stricter than casing packer.

The application of setting packer in open-hole layer is also commonly seen in horizontal wells.

Packer

Packer is a frequently used downhole tool in well completion, formation test, and oil production. In oil production process, packer is more commonly used inside casing. In well completion or other tests, open-hole packer is more frequently used. Packer is used to seal open-hole section or other in-casing layers. For example, in well completion, packer is used to seal complex stratum rock. In oil test or oil production process, packer is used to seal some particular production layers. The main component of packer is rubber drum, which can open under liquid pressure or mechanical force, seal the rock layer in open hole or the internal wall of casing, so as to obtain construction goal.

- Mechanisms of packer.

Packer is used to seal borehole rocks or casing wall by rubber expansion. The mechanisms of packer are: to push slips by mechanical force or liquid pressure, to make slip open and closely against borehole wall or casing wall, to anchor packer, then to immediately compress sealing components by all forces to open rubber sleeve and achieve sealing. Some packers can only be used for one time, since their slips and sealing components are unrecyclable once opened, and they are permanent type. For other packers, the opened slips and sealing components can be recycled and the packer can be ran and pulled for multiple times in well. Thus, they are recyclable type. Permanent packer has reliable sealing properties; thus, this type is often used under fixed conditions, such as well completion and well cementation, while recyclable packer type is often used inside casing.

As the main component in packer, the sealing component is contracted under normal conditions; and its size is smaller than borehole diameter or casing internal diameter. Thus, it can be opened by forces after being set to proper position. Therefore, sealing components are mainly composed by rubber and have extremely strong deformability. The sealing material is basically rubber. And to make material resistant to high temperature, some modifying agents are normally added into rubber. The hardness of material depends on the pressure difference at sealing component and the roughness level of sealing component. Normally, hardness is divided into high hardness and low hardness. High-hardness components are used in high pressure and not so rough casing wall, low-hardness components are used in rough borehole wall, and sometimes sealing components under two hardness levels are jointly used.

Packer should be resistant to corrosion. Sealing components should be anti-corrosion under the presence of H_2S and CO_2. Normally, Teflon, nylon, and other materials are added in rubber, and Teflon material can bear 450 °C high temperature at corrosive conditions. Packers should also be salvageable.

- Packer types.

By working principles, packers can be divided into gravity-set type, tension-set type, mechanical-set type, liquid pressure set type, and so on. By usage, packers can be divided into permanent type or recyclable type.

Gravity-set packer uses friction block to fix slips in usage. The accelerated setting of oil tube at target position can open slips, so packer rubber is sealed under the gravity of oil tube. Some other packers are used at bottom hole by cutting through plug under the gravity of oil tube, so slips are opened.

Tension-set packer is actually the overturn usage of gravity-set packer. Slips are open by applying tension to oil tube. Tension-set packers are normally used in shallow shaft.

Mechanical-set packer is set or pulled out by the rotation of oil tube. When rotating oil tube, slips are open, and sealing components are compressed. When rotating oppositely, we can pull out the packer.

Liquid-pressure set packer is to open slips by piston propelled in hydraulic cylinder, and open sealing rubber.

One feature of permanent packer is that it contains a devise to avoid loosening slips. Permanent packers are often used in well completion.

Downhole control unit and surface control unit are used with packers collaboratively, such as downhole safety valve and safety joint, and liquid-pressure pump unit on ground, and so on.

7.4.3.5 Sand Control Completion

During production process, some sandstone reservoirs may produce sand grains due to poor cementation of sandstone, high production output, or the reservoir is contaminated. Sand production will affect production output, or even make the well abandoned if sand production is severe. Therefore, this sand production phenomenon must be prevented. Sand control completion method is normally used in well completion stage. The common completion methods are open-hole gravel packing completion, in-casing gravel packing completion, and artificial borehole sand control completion.

Open-hole gravel packing completion

Open-hole gravel packing completion method is to set casing to seal reservoir before the production layer is drilled through, then drill through production layer, expand borehole in production section, insert screen pipe, and pack gravels inside annular space between borehole and screen pipe. Gravels and screen pipe can prevent the sand production from stratum.

In-casing gravel packing completion

After inserting the casing and perforation, if well produces sand, inserting screen pipe in sand production interval, and packing gravels in the annular space between screen pipe and oil-string casing. This sand control technique utilizes in-casing gravel packing completion method, which belongs to secondary completion.

The two key elements in gravel packing completion are the selection of gravel and the assurance of packing thickness.

The selection of gravel diameter depends on the diameter of produced sand. The diameter of formation-produced sand is unequally distributed. So after sieving sand grains, we plot the distribution graph of sand diameter accumulate mass. The diameter corresponding to 50% of accumulated mass is the medium diameter of sand grain. The diameter of gravels is normally 6–8 times of the medium diameter of produced sand grains, and gravel thickness is at least 8 times of grain diameter.

For the open-hole gravel packing, gravel layer thickness is no less than 30 mm; for the in-casing gravel packing, gravel layer thickness is no less than 15 mm.

The types of screen pipes include wire-wounded type, slotted type, porous material sintering type, and so on, in which the wire-wounded type is most commonly used. Wire-wounded screen pipe utilizes the wrap slits to prevent from sand grains. Distance between wire wrap is twice of the minimal diameter of produced sand grains.

Artificial borehole sand control completion

Artificial borehole sand control completion method is to insert permeable and solidifiable material in sand-producing layer, forming the artificial borehole to prevent from sand grains production.

This completion technique includes: (a) permeable artificial borehole perforation completion method, that is to insert good permeable material inside the gap between casing and formation layer, then use low-power perforation bullets to perforate casing by avoiding damage to inserted permeable layer; (b) permeable artificial borehole liner completion method, that is to insert permeable material inside liner and open hole; (c) permeable artificial borehole open-hole completion method, that is to insert permeable material in open-hole section and form artificial borehole.

The key element in this completion method is the selection of permeable solidifiable material. This type of material includes permeable material formed by adding silica in cement, resin-bonded mortar material, and so on.

7.4.4   Special Well Completion

Previously, we have discussed well completion methods using four types of wellbore structures. These methods are also applicable to horizontal wells and wells in fractured reservoirs and low permeability reservoirs. However, since each reservoir type has its specific features, there are special requirements for these complex reservoirs.

7.4.4.1   Completion Method for Horizontal Wells

Based on the statistical data from previously drilled horizontal wells, a single well production of horizontal wells is 2–5 times as much as that of vertical wells, whereas the cost of a single horizontal well is only 1.5 or 2 times that of vertical wells. Horizontal wells have gradually become an effective technique to improve single well production and enhance the oil and gas recovery factor, especially for fractured reservoirs, heavy oil reservoirs, low permeability reservoirs, and bottom water drive reservoirs.

There are five types of horizontal wells based on buildup rates, including long, medium, medium-short, short, and ultra-short radius. Different types of radius' horizontal wells have different build rate on the interval of each 30 m. The build rate of long radius well is around 2–6°, the build rate of medium radius well is around 6–20°, the build rate of medium-short radius well is around 20–80°, and the build rate of short radius well is around 30–150°. For the ultra-short radius horizontal well, the transition from vertical to horizontal section can be achieved within 0.3 m by special steering equipment.

Common completion methods for horizontal wells

The principles of horizontal wells' completion are as follows: obtaining maximum oil and gas production from wells while reducing the flow of other fluids (such as water), preventing wellbore instability and controlling effectively sand production, favorably reduce the potential workover frequency, prolonging the service time of all kinds of downhole strings without deformation, corrosion or scaling, maintaining the well condition for the potential of second or third round development, and most importantly staying cost-effective.

The technical requirements which should be satisfied in the horizontal wells' completion process are as follows: preventing water or gas invasion, being capable of performing production testing, well stimulation and workover operation, and maintaining well safety.

For horizontal well completion, the impact factors not only include all the factors related to vertical wells, but also consist of the impact of wellbore curvature on strings, and the impact of horizontal interval length on wellbore stability. These two factors are very important to horizontal wells' completion.

There are many techniques for horizontal well completion, in which open-hole completion, cased hole or liner completion, pre-slotted liner completion, screen completion, gravel packing completion and packer completion are most common measures, as shown in Fig. 7.37. Horizontal well completion techniques can be

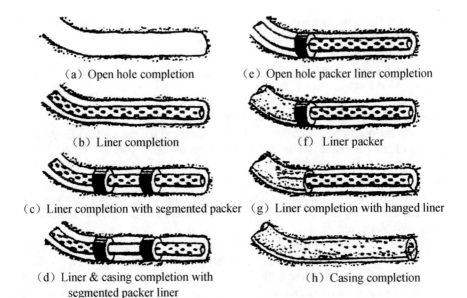

(a) Open hole completion   (e) Open hole packer liner completion

(b) Liner completion   (f) Liner packer

(c) Liner completion with segmented packer   (g) Liner completion with hanged liner

(d) Liner & casing completion with segmented packer liner   (h) Casing completion

Fig. 7.37 Diagram of horizontal well completion techniques

divided into two main categories: One category is a selective completion mainly used for cement-sealed reservoirs. Perforations at selectively zones or sections can be performed after sealing or cementing. The other category is a non-selective completion, which is mainly used for open-hole completion or its derivatives.

The characteristics of each different completion types are described as follows:

- Open-hole completion is the simplest well completion method, which can only be used for non-breaking and non-collapse solid rock formation. It is normally applied in these horizontal wells with medium, short, and ultra-short radius. Since there may be strong stress occurred while the casing goes through the bending section, it may have a significant effect on the casing strength. The advantages of this completion method include higher improvement coefficient, higher production, and easier elimination of contamination. In contrast, the disadvantage of this completion method is that, especially in the weak formation, the wellbore collapse is likely to happen if rock strength is inadequate. The wellbore condition limits the applications of stimulation in this regard.
- Casing or liner completion is a traditional completion method which has been widely utilized in vertical wells and was also frequently used in long radius horizontal wells. The advantage of this completion method is able to effectively seal various formations downhole. It can also overcome the challenges in side wall's complicated situations, make use of various stimulation measures, and selectively perforate sections or intervals required. In contrast, its application may be restricted in medium, short radius' horizontal wells with high bending degree.
- Pre-slotted liner completion is a relatively simple method, which can be used in wells with various radii and has already gained a widely application. Running pre-slotted liner in horizontal wells can be achieved by hanging the top of liner against casing or open-hole formations using open-hole packer or casing packer. The purpose of liners is to provide a support to weak rock formation. When using liners, there is no need to cement or only apply cementing on the partial well sections.
- Screen completion is a method of sand control completions, and the screens can eliminate cementing process. In addition, screen with pre-packed gravel may also be applied.
- Packer completion is widely used in horizontal wells. Open-hole packer should be utilized before and after horizontal sections where potential problems may occur. This completion method is usually applied with slotted liners or screens.

Based on the statistical data for various horizontal well completion methods, wells with cementing completion account for around 10%, wells with open-hole completion account for around 10%, wells with gravel-packed completion account for around 17%, wells with slotted liner completion account for around 40%, wells with screen completion account for around 3%, and wells with any other completions account for remaining 20%. From the perspective of technology development, the technologies including cementing completion, screen completion, and open-hole gravel-packed completion have shown most promising trend.

There are four steps in the selection of completion methods for horizontal wells. Firstly, based on reservoir types, fluid types, and rock stabilities, an initial proposal can be proposed by reservoir simulation studies; secondly, completion proposal should be evaluated by production technology since the aim of completion is to achieve the production targets; thirdly, the detailed completion operation procedures need to be proposed according to the current drilling technological level; fourthly, a technical and economic combined evaluation would be finally completed by technology and financing department. As a summary, similar to vertical well completion, horizontal well completion should be guided by the completion theories, based on the technical experience and technological conditions of reservoir, production, and drilling engineering, to gain the maximum economic results in current and future production.

Completion of long radius horizontal wells

Drilling methods of long radius horizontal wells are basically the same as methods of conventional vertical wells. All the vertical wells' completion methods can be utilized in long radius horizontal wells. Cementing is usually applied firstly after running the casing, and then perforation is performed to expose the pay zones.

Once predetermined well depth is drilled and reached, logging activities will be performed. Since it is extremely difficult to directly run the logging instrument into relatively long horizontal interval, drill strings can be used as a delivery tool. Logging can be carried out while tripping-out the drilling strings.

It is relatively easy to run casing in the long radius interval of horizontal wells. Conventional casing cementing can be used to seal the bottom of wellbore, and pay zone can be exposed by perforation instruments. Delivery of the perforation gun to the horizontal interval is really a great challenge, and hence, the tubing conveyed perforation technology needs to be used to perforate the pay zone.

The stability of long horizontal open-hole intervals is the main challenge in long radius horizontal well completion. The horizontal section length can reach up to couple thousands of meters. If the stability of rock formation is great, drilling or running casing can be continued without problems. If the stability of rock formation is poor, we need to find solutions by analyzing the drilling fluid performance and casing program. Intermediate casing needs to be run to the top of horizontal section before drilling into horizontal section. And oil-based drilling mud can be used to stabilize the borehole while drilling horizontal sections.

Completion of medium radius horizontal wells

For medium radius horizontal wells, build-up section comes with small radius and a sharp bend. Build-up drilling tool assemblies are necessary to overcome this challenge, such as bending subs and dynamical drilling tools. Friction resistance would be increased while tripping drilling tools and the bending stress would be greatly increased as well. Completion methods should be determined according to stress distribution of casing bending section. If casing cementing is selected, bending stress and threads' sealing performance of casing should be verified and checked to ensure the casing can be run successfully. Perforating activities are very hard to be performed in such wellbore settings. If there is a potential risk for casing to run through the bending sections, cementing with casings shall not be applied.

If the casing completion method is not applicable, other types of completion may be selected, such as open-hole completion, liner completion, liner completion with packer. The selection of bottomhole structure's completion is the same as the selection of horizontal wells with short radius.

Completion of medium-short and short radius horizontal wells

Such wells are mostly side tracking wells and old wells, and the diameter of horizontal wellbore is restricted by the casing size of old wellbore. The diameter of such well is extremely small compared with diameters in horizontal wells with long or medium radius. For examples, the diameter of horizontal section does not exceed 140 mm if the casing diameter is 177.8 mm (7 in). Similarly, the diameter of side tracking horizontal well is only 110 mm if the casing diameter is 139.7 mm (5 in), which brings up a much sharper bend by 1–5° increment per meter in the build-up section. Being restricted by wellbore diameter and bending sections, it is impossible to run casing in horizontal wells with medium-short and short radius. In this case, only open-hole completion, liner/screen completion, packer completion, or others can be selected.

Completion of ultra-short radius horizontal wells

For ultra-short radius horizontal wells, a very slim steel pipe (the external diameter is less than 40 mm) with high-pressure strength shall be used. By injecting high-pressure fluid, pipe body becomes plastic or semi-plastic and can be changed into horizontal from vertical in an extremely short distance (0.3 m) by a special whipstock. High-pressure jet flow from front nozzles can break rocks while moving forward to form a horizontal section. In some ultra-short radius horizontal wells, high-pressure hose may also be utilized to run through bending sections. Since the bending section is very short, normally less than 1 m, the bending in build-up section is dramatic and the wellbore diameter can be very small, less than 100 mm. Hence, running other pipe string into horizontal wellbore section to finish completion is almost impossible. In this situation, the high-pressure hose shall be retrieved after horizontal wellbore section has been washed out, and only open-hole completion method can be applied. In case of using high-pressure steel pipe, it can be kept in wellbore if necessary, after the horizontal section has been washed out. Through applying electrochemical corrosion methods, nozzles can be corroded off, and the slots can be corroded from steel pipe in horizontal sections. By this way, liner completion can be completed after upper section has been cut off. In addition, gravel-packed completion can be also considered and applied, which is a relatively complex and costly technology.

In case of drilling multiple wells in different directions in one oil layer, only open-hole completion method can be considered to apply rather than running string completion method since there will be some interference between these wellbores.

7.4.4.2   Completion of Wells in Tight and Fractured Reservoirs

Tight reservoirs are normally referred to low permeability sandstone formation, where the permeability range is from 0.001 to 0.01 μm^2. Such reservoirs are mostly featured with porous-fracture type or fractured-porous type, and they rarely are presented in a pure porous type. Fractured reservoirs generally refer to the reservoirs where oil-storage space and higher permeability have been generated from rock fractures. It could be sedimentary rock consisting of limestone, mudstone, shale or silicate, etc. Permeability in reservoirs has a wide variation range controlled by fractures, and such fractures are mostly developed vertically in sedimentary rock. There are many differences in well completion between these two types of reservoirs and a pure porous type of reservoir, and the differences are determined by special characteristics of the reservoirs.

Completion of tight reservoirs

Tight reservoirs are characterized by high rock mechanical strength, low permeability, and low productivity in pay zone. The oil production from porous rock in reservoirs is difficult and results in extremely low recovery factor. In order to improve production, stimulation methods like hydraulic fracturing and acidizing should be carried out to increase reservoir porosity and production. The main concern in the completion of wells in these reservoirs is whether the casing program can be durable and stable after fracturing and acidizing treatments or not.

The principles of well completion in tight reservoirs include: high strength of casing, excellent cementing quality, satisfactory cement return height, and high strength of cement. The following measures should be taken:

- Casing centralizer, scraper, and high-performance preflush and displacement fluids should be used.
- Slurry should be displaced at high flow rate, and casing should be kept moving in the displacement process.
- The threads of casing should have good sealing performance, high quality of sealing oil, or thread adhesive need to be selected.
- Thick wall casing made of steel grade more than N80 should be used, and the internal compressive strength should be more than 1.2 times of formation fracture pressure. To prevent casing from deforming, wall thickness should be kept consistent without sharp change.
- Some operations such as casing milling should be tried to avoid, and the top and bottom packers need to be utilized to seal the fracturing section while performing the fracturing job.

Well Completion in fractured reservoirs

One key characteristic of fractured reservoirs is that the fracture distribution and its development are not homogeneous, and fractures show orientation spatially. Oil productions take places only if fractures intersect with wellbore. And the more intersecting, the higher oil production from the wellbore. The principle of drilling such wells is to make the wellbore intersecting with fractures as much as possible. Similarly, the aim of well completion is to build an effective connectivity between wellbore and fractures. In drilling and completion processes, fractures shall be prevented from closing or being blocked off.

The principle of fractured reservoirs' completion is to make the wellbore intersecting with fractures as much as possible. Horizontal well drilling is the most widely applied technology to achieve this goal. In the process of well completion of fractured reservoirs, open-hole completion is widely selected. In the case that a perforating completion is applied, cementing needs to be carried out first after running casing, and then perforation can be carried out. The slurry may block off fractures and lead to a great reduction on pay zone's permeability. Also, the possibilities for the perforated tunnels intersecting with unevenly distributed fractures will be reduced greatly. In order to build an effective connectivity between wellbore and fractures, open-hole completion or its varieties should be the top choices.

In the reservoirs with high rock strength, open-hole completion is advisable. In case of relative weak reservoirs, screen completion, liner completion, or gravel-packed completion would be recommended. And if necessary, packers can also be placed on suitable locations of liners to seal weak formation. In any event, perforating completion should be avoided.

7.4.5 Wellhead Equipment for Completion Operation

In order to control and schedule operation and production in wells, there should be one set of reliable wellhead equipment in testing and production of oil and gas wells. Wellhead equipment refers to surface devices including wellhead control valves and the devices used to suspend/place downhole strings. It mainly consists of casing head, tubing head, and Christmas tree.

7.4.5.1 Casing Head

If cement is not able to return to the surface after each set of intermediate casing and production casing are cemented, there would be a partial casing hanging without support, i.e., free casing. The weight of this partial casing is supported against by casing which is already cemented and fixed. During the oil production, the reservoir pressure and temperature could bring a great impact on free casing, to elongate the length, to change the distribution of casing stress. In this case, casing may be easily damaged. Therefore, after cement has been solidified, casing head should be installed to seal the annulus between two layers of casing and shift the weight of free casing section onto surface casing.

According to the connection structure difference, casing head can be divided into slip type and thread type. Three layers of casing including production casing, intermediate casing, and surface casing are involved. The detail structure of wellhead equipment with standard casing head is shown in Fig. 7.38.

The first layer of casing head is mounted on the top of surface casing, which is connected by its internal threads and external threads of surface casing. And flange

Fig. 7.38 Standard wellhead
equipment. 1—tubing hanger;
2—tubing head; 3—tubing; 4,
9—casing hanger; 5, 10—
casing head; 6—production
casing; 7—intermediate
casing; 8—seal ring; 11—
surface casing

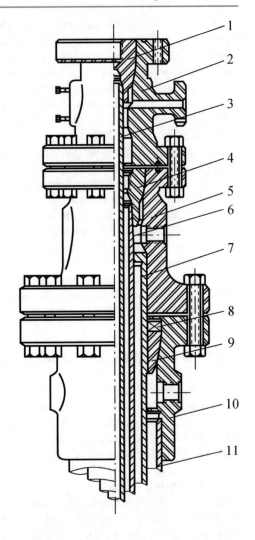

face is basically flat with ground surface. After cementing of intermediate casing is completed, the free part of intermediate casing is lifted and keeps the slips seated inside the cone of casing head. Then the suspended weight can be supported by the cone, and the sealing of the annulus between two layers of casing can be also achieved. The landing joint can be removed in the end. The second layer of casing head shall be seated on the flange of first layer casing head, and accordingly, tubing head shall be seated on the flange of second layer casing head.

If the formation pressure of oil and gas zone is relatively low and cement return level is approaching ground surface, a ring-shape steel plate can be directly used to seal the annulus between two layers of casing by welding. In this case, casing head is not required.

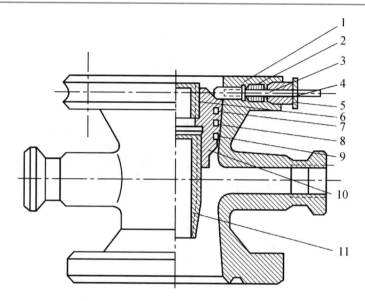

Fig. 7.39 CYB-250 tubing head assembly. 1—Special spool; 2—Packing seat; 3—Packing; 4—Jackscrew; 5—Packing gland nut; 6—Thread protector; 7—Upper seal ring; 8—O ring; 9—Lower seal ring; 10—Tapered tubing hanger; 11—Tubing pup joint

7.4.5.2 Tubing Head

Tubing head refers to a wellhead device used to hang tubing, seal tubing, and install Christmas tree. Tubing head is mounted by bolts and seated on the last layer of casing head. There is a cone inside tubing head. The top of tubing string can be suspended, and the annulus can be sealed by this cone, as shown in Fig. 7.39.

7.4.5.3 Christmas Tree

Christmas tree (Fig. 7.40) is composed of valves, chokes, seal box, and tees or spools. It is installed on the top of tubing head and can control the oil and gas flow, in order to make plans for oil production. By installing Christmas tree, many operations such as flowing production, rod production, fracturing, and acidizing can be proceeded.

Questions and Exercises

1. Please summarize casing types and their corresponding functions.
2. What are the working principles of casing program design?
3. A pay zone of one well is located around 2600 m, and the equivalent drilling fluid density of the predicted formation pressure is 1.30 g/cm^3. The surface casing has been run when drilling depth was reached at 200 m. Based on leakage testing results, the equivalent mud density of formation fracture pressure located in casing shoe is 1.85 g/cm^3. Please calculate and evaluate whether

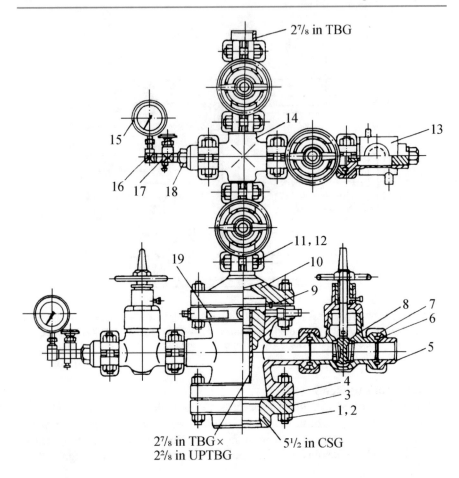

Fig. 7.40 CYb-250S723 type of Christmas tree. 1, 11—Nut; 2—Stud; 3—Casing flange; 4—Tapered tubing head; 5—Clamp pup joint; 6, 9—Steel ring; 8—Valve; 10—Tubing head flange; 12—Double end nuts; 13—Choke; 14—Small spool; 15—Pressure gauge; 16—Bend sub; 17—Pressure cut off valve; 18—Sub; 19—Nameplate

or not pay zone can be drilled in successfully without running intermediate casing? The key parameters are given as follows: $S_b = 0.038$ g/cm^3, $S_k = 0.05$ g/cm^3, $S_f = 0.036$ g/cm^3, $S_g = 0.04$ g/cm^3.

4. What does the casing string design consist of? And what is the design principle?
5. How many types of stresses may be subjected to the casing string in the downhole? Please summarize the main stresses.
6. Please summarize the main design methods of casing string currently widely used and describe the corresponding features.
7. Please describe the bidirectional stress elliptic theory, and when should the bidirectional stress be taken into consideration.

8. What are the main components and properties of well cement?
9. Please briefly describe the concept of water–cement ratio.
10. What is the principle of drilling fluid density design?
11. Please briefly describe the concept of cement slurry thickening time and its relation with the cementing construction.
12. Please briefly describe the characterization and significance of the stability of cement slurry.
13. Please briefly describe the characterization and significance of the cement bond strength.
14. What is preflush system and what is the purpose to use preflush?
15. What components are included in the lower section of casing string? Please introduce the functions of each component.
16. Please briefly describe the procedures and application areas of multi-stage cementing technology.
17. What are the main differences between liner cementing technology and single-stage cementing technology?
18. What are the main factors affecting the well cementing quality?
19. How to improve cementing quality?
20. Please describe the commonly used well completion methods and summarize the characteristics for each method.
21. Casing with 139.7 mm diameter, N-80 steel grade and 9.17 mm wall thickness has been used in a well, and its rated collapse strength is 60,881 kPa, tensile yield strength of pipe body is 2078 kN, and the gravity of casing suspended on the lower part is 194 kN. Please calculate the value of p_{cc}.
22. The allowable collapse strength of casing with 178 mm diameter, N-80 steel grade, and 9.19 mm wall thickness is 37,301 kPa. Suppose the safety factor of collapse resistance is 1.125, please evaluate and calculate the allowable running depth of casing assuming mud density is 1.23 g/cm^3.
23. Casing with 245 mm diameter has been run in a well depth of 2000 m. The density of drilling fluid is 1.20 g/cm^3, and the cement return depth is 1600 m. Please design the casing strength by computer programs (casing shall be selected among the given options below).

Steel grade	Wall thickness	Collapse strength	Yield strength	Thread slipping tension	Linear weight
	mm	kPa	kN	kN	kN/m
N-80	11.05	26,269	4470.50	3669.80	0.6384
N-80	10.03	21,305	4074.60	3278.30	0.5835
K-55	10.03	17,720	2802.40	2495.30	0.5835

Other Drilling Techniques and Operations

8

Abstract

This chapter introduces the procedures of handling some complicated downhole conditions and accidents, coring operations, and sidetrack drilling technologies. Classification, causes, prevention and handling of complicated downhole conditions, and accidents are described in detail. Furthermore, tools for fishing broken drill string, falling objects accidents, and core drilling are introduced. The main factors affecting core recovery and the measures to improve core recovery are summarized. Different types of sidetrack drilling methods, tools, and processes are presented.

8.1 Handling of Complicated Downhole Conditions and Accidents

During drilling, hazardous zones, improper selection of drilling fluids, poor wellbore quality, and other reasons may cause downhole tool stuck and blockage, serious bit jump, lost circulation, well kick, and other problematic conditions that can fail normal drilling activities and other operations. All these problematic conditions are generally referred as complicated downhole conditions. Drilling accidents such as drilling tool breakage, drilling stuck, blowout, and fire will occur due to faulty operations and improper handling of these complicated downhole conditions. Complicated downhole conditions and drilling accidents pose big challenges to drilling engineering, reduce drilling efficiency, and increase drilling costs. Severe drilling accidents will lead to abandonment of oil and gas wells, delay in exploration and development of oil and gas fields, and even severe damage to hydrocarbon resources. When drilling, if a complicated downhole condition cannot be diagnosed and fixed in time, it may further cause other complex situations or accidents.

© China University of Petroleum Press and Springer Nature Singapore Pte Ltd. 2021
Z. Guan et al., *Theory and Technology of Drilling Engineering*,
https://doi.org/10.1007/978-981-15-9327-7_8

To ensure smooth drilling operations, drilling operators must strictly comply to the operation instruction and other designed technical specifications and be aware of the possible causes and consequences of drilling accidents. They should also master the basic skills and abilities to deal with all drilling problems and accidents and adopt effective measures to prevent more damages from happening and solve the specific problems or accidents to ensure safe and successful drilling operations.

8.1.1 Uncontrolled Blowout and the Handling Procedures

Well kick refers to the phenomenon that formation fluids flow into a wellbore for various reasons during drilling and leads to continuous or intermittent pulse of drilling fluids. If a well kick is not controlled in time, it may cause well blowout. Uncontrolled blowout may become devastating, including not only waste of resources and pollution of environment, but also equipment damages, casualties, and well abandonment. Once a blowout wellsite is on fire, the damage could be more severe. Therefore, the greatest attention should be exercised on preventing wells from blowing out during both well design and operation, especially for a new exploration area with high-pressure structures.

8.1.1.1 Causes of Blowout

During drilling, when the hydraulic pressure of drilling fluid is lower than the formation pressure, the pressure balance of the formation–wellbore system will be lost, and well blowout may occur. It is more risky when a high-pressure oil and gas reservoir is drilled. There are two main reasons for the underbalance between wellbore and formation; one is the inaccurate prediction of formation pressure; the other is the decrease of hydraulic column pressure of drilling fluid. More specific reasons are as follows:

- Inaccurate evaluation of formation pressure. When designing the drilling and well construction plan, the depth of the abnormal high-pressure formation is not accurate, or the formation pressure is underestimated. So there will be a blowout when abnormal high-pressure oil zones are drilled out without adequate preparation. This is often the case especially when drilling in new exploration areas.
- Density of drilling fluids reduction. Formation fluids will invade into wellbore when drilling in high-pressure zones, which will decrease the density of drilling fluids and further decrease the fluid column pressure, resulting in the easier invasion of formation fluids into wellbore and further reduction of the density of the drilling fluid and the pressure of the fluid column, and leading to blowout at last.
- Column height of drilling fluids reduction. ① The drilling fluid is not filled up when hoisting: The trip-out of drilling tool will lead to column height decrease in wellbore. If no drilling fluids are supplemented in time to maintain the downhole column pressure, the formation fluids will invade into wellbore. According to statistics available, about 40% of blowout occurred for the reason that the

column height of drilling fluids was not properly maintained in pulling-out process. ② Drilling leakage zones: When drilling leakage zones, the height of the liquid column in the well will decrease due to leakage of drilling fluid. Therefore, special attentions should be paid to anti-leakage when adjusting the density of drilling fluids to balance high-pressure oil–gas–water zones.

- Swabbing effects during hoisting. The drilling fluids adhered on the inner and outer surface of drill string tend to move upward, leading to the instantaneous reduction of downhole pressure. This is the swabbing effect that downhole pressure decreases temporarily due to pulling string out of hole. Experimental results show that the faster the tripping-out speed and acceleration, the greater the swabbing effect. The resulted pressure reduction is generally equivalent to a decrease of density of drilling fluids by over 0.12 g/cm^3.

The most serious swabbing always occurs when bit is being pulled up from the bottom. Under the conditions of high viscosity and shear stress of drilling fluids, thick mud cake, bit balling, or backpressure valve installed in string, swabbing effects will be significantly amplified. It will lead to much sharper decrease in downhole pressure, resulting in the invasion of formation fluids into wellbore. And if not properly handled, it will cause well kick or even blowout.

8.1.1.2 Handling of Out-of-Control Blowout

The well-killing process after blowout has been introduced in the Chap. 6. Once blowout is out of control and causes fire, the conventional well-killing operation and purpose cannot be achieved. Effective measures must be taken to deal with it promptly; otherwise serious consequences will be caused. Fire extinguishing and well killing is the basic process to deal with fire blowout which can be handled with two main methods. One is surface fire extinguishing by circulating weighted drilling fluid through the original wellhead; the other is to drill a directional rescue well, which convey weighted drilling fluid from the rescue well to the on-fire well and extinguish the fire.

- Extinguish fire and kill well on ground. When well blowout is on fire, the fire will spread to almost the whole wellsite. Operators cannot get to wellhead to take any effective measures. If the fire time is too long, the derrick and other equipment as well as the wellhead will be burned down, causing huge losses, and bringing greater difficulties to the follow-up processing work. Therefore, the first step is to immediately manage to control the fire or put out the fire. The direct purpose of extinguishing fire on ground is to create conditions for workers to enter wellsite and control wellhead. If the fire can be put out in time, it is possible to directly control wellhead or install a new one to control the blowout. If the fire cannot be extinguished, obstructions must be cleared firstly, and then some protection measures must be taken to install a new wellhead in fire forcibly. There are many equipment and facilities in wellsite, which are unfavorable to the control of wellhead; thus these obstacles must be cleared as early as possible. Under the cooling protection by high-pressure water jet, oil tanks and other flammable and explosive materials should be removed from the wellsite. Move diesel engine, derrick, drill stand, and other facilities out of site to expose the whole wellhead. Change the new wellhead corresponding to the specific condition, and control wellhead pressure by its throttle manifolds and blowout preventer. Then circulate heavy drilling fluids into well through well-killing lines by designed killing method (see Chap. 6), and finally, new balance between column pressure in the wellbore and formation pressure will be established.

- Drill directional relief well to kill well. When it is extremely hard to put out fire on ground or the wellhead has already been burned down, and it's unable to extinguish fire and kill well on ground, the directional well drilling technology can be used. Choose an advantageous position on ground, design reasonable well track, and drill to the high-pressure layer of burning well with as little drilling footage as possible. And circulate heavy drilling fluids from the directional relief well to the high-pressure oil–gas layer according the designed killing methods, then the blowout can be stopped, and the pressure balance in wellbore can be maintained and the fire can be put out.

8.1.2 Lost Circulation and the Handling Procedures

Lost circulation happens if drilling fluids or cement slurry leaks into formation zones in drilling. It is one of the serious and common problems that happen in drilling. Lost circulation always lowers down the fluid column pressure in wellbore, leading to wellbore collapse or blow out. Thus, once lost circulation is found, effective measures must be taken in time to prevent worse problems.

8.1.2.1 Causes of Lost Circulation

During drilling, if hydraulic pressure of drilling fluid in the wellbore is greater than the formation fracture pressure, well leakage will occur. There are two main factors causing lost circulation: One is the abnormally low formation fracture pressure, large porosity, good permeability, or fractures, karst caves; the other is the inappropriate drilling parameters selection, such as excessive drilling fluid density, excessive pressure, sudden starting pumping, and too fast tripping-in speed, which cause pressure surge in the wellbore.

Lost circulation may occur in the following four types of formations:

- Unconsolidated formation or high-permeability formation;
- Natural fractured formation;
- Induced fractured formation;
- Cavernous formation.

Loss of circulation will result in less returned volume of drilling fluid and the fluid level decline in mud tanks. Serious loss of circulation of drilling fluid will result in the collapse of wellbore or blow out.

8.1.2.2 Types of Lost Circulation

Leakage rate is the leakage volume of drilling fluid per hour. According to the characteristics of a leakage zone, the leakage rate, the causes of leakage, and lost circulation can be divided into three types:

- Permeable leakage. This type of lost circulation often occurs in the shallow-hole section in poorly cemented gravel formations. Due to the good permeability in these formations, the drilling fluid will invade into pores under borehole pressure difference, but the formation of mud cake will then stop or weaken the loss degree. Thus, this type of lost circulation has normal loss velocity, which is often below 10 m^3/h.
- Fractured leakage. Natural fractures may exist in all types of rock formations in drilling, and fluid leakage to different degrees will occur in the formation of natural fractures. When drilling in the fractured zones, the correspondingly bit jumping, high rate of penetration, and other problems may cause lost circulation, whose leakage velocity is normally 20–1000 m^3/h.
- Cavernous leakage. In some limestone strata, large karst caves are formed by long-term erosion of groundwater. Drilling these large caves will lead to drill emptiness which can be 4 to 5 m long and lead to circulation loss that drilling fluids can only come in but cannot come out. The velocity of cavernous leakage is often larger than 100 m^3/h, and blowout or collapse often happens after lost circulation, which is the most severe leakage.

8.1.2.3 Handling of Lost Circulation

Handling of Permeable Lost Circulation

For minor lost circulation, the drilling can be restored to normal conditions by adjusting the property of drilling fluid, reducing downhole pressure difference, and plugging pore channels in formation, such as reducing the density of drilling fluid, adopting balanced drilling technique, simultaneously improving the viscosity and stress of drilling fluid to increase resistance of flowing into formation pore. The often applied treatment agents to increase viscosity and stress are CMC, bake glue, soda ash, econolite, mirabilite, and so on. If the effect is not obvious after increasing the viscosity, fine particle bridging agents can also be added, such as fine mica sheet and nutshell flour until the pores are plugged.

Handling of lost circulation in fractured zones

Since the specific causes of fractured lost circulation are different and the velocity of lost circulation also varies widely, the handling method should be correspondingly selected according to the specific circumstances when dealing with the specific case. The following three methods are often used in China now:

- Inject the hull flour and sawdust thick mud to plug. Hull flour and sawdust should be fine particles. The injection amount is 10–15%, and the viscosity should be appropriately high. The reverse circulation can also be applied since sawdust and hull flour may plug the bit nozzle.

- Cement or colloid cement can be injected when fractured lost circulation occurs in the non-reservoir. One formulation of colloid cement is 5 parts of cement, 1 part of clay, and 2% calcium chloride as catalyst. Now there are many cement-plugging methods used in place. The key point of cementing is to know the position of cement slurry in wellbore in order to ensure that most of cement slurries have entered formation before initial setting, leaving a little leftovers as plugs.
- Lime cream drilling fluid can be injected when fractured lost circulation occurs in the reservoir to temporarily plug the fractured layer. The formula is that the lime cream (the density is about 1.13 g/cm^3) and the drilling fluid (the density is larger than 1.25 g/cm^3) are blended at the ratio of 1:1 or 1:2, and the specification is determined by the downhole temperature and the required operation time. The other method is to inject the accelerated lime cream drilling fluid, which is prepared by adding water glass into normal lime cream drilling fluid, and the initial setting time can be adjusted by changing the volume of econolite. The larger the content of econolite is, the shorter the initial setting time is, which is suitable to large fractured lost circulation

Handling of lost circulation in cavernous zones

Cavernous lost circulation is the most severe one. Directly from the well-head, Stone balls and long fiber bundles (rice straw, branches,etc.) are often used to fill the underground caves or to set up "bridges", and then the cement slurry is used for plugging. Besides, a type of compound plugging fluid has also been developed, which is made by drilling fluid or cement slurry carrying with inert materials under different shapes, sizes, and amounts, such as multi-angular hard hull and mica, celluloid, and all kinds of plant fiber. These materials have sharp rims that can generate large friction, hanging and detention effects to the wall of caves, fractures, and holes and then form the bridge mesh. Mica and other stuff are thin but smooth and have bending deformation abilities; thus they can create the all-pervasive, smooth and flowing environment. The dense packing of plant fibers can seal joints and stop lost circulation, and then restore the circulation drilling. These practices have all obtained favorable effects.

8.1.3 Sticking and Its Handling

The phenomenon that drilling tools cannot move freely in wellbore due to some reasons in drilling process is called sticking. The reasons can come from the formation, the bad performance of drilling fluids, incorrect operation, and so on. It is necessary to analyze the specific situation to release the stuck effectively.

8.1.3.1 Types, Causes, and Prevention of Sticking

Sticking can be divided into the following types:

Sand settling sticking

In the drill process with clean water or low-viscosity and low-stress drilling fluid, the cuttings will settle down after stop pump due to their poor solids suspending ability. And the longer pump-off time is, the more settled the volume is. This is especially true when the speed of drilling is faster. When the situation is serious, the settled sands and cuttings may plug the annulus, bury the bit and part of the string, and lead to sticking, which is called as sand settling sticking. At this moment, if the pump is turned on too harsh, it may cause formation lost circulation under suppression, or stick the drilling tool much tighter.

The performances of sand settling sticking are: Drilling fluids will flow back even blowout seriously after connecting single or pull out and break up the standpipe; the pump will have high pressure or suffocation after restarting pump to circulate fluid; it gets stuck when pull out of hole and gets obstacle when run in hole, and it becomes more and more difficult to move the drill string upward or downward the drilling tools, and the resistance is great when rotating.

To prevent sand settling sticking, the performance of the drilling fluid should be ensured to meet the requirements to clean and suspense cuttings, and the equipment and circulation system should be checked and maintained at any time. Downhole circulation break should be avoided when the drilling process is stopped. The speed of making connection should be shortened. The speed of drilling should be controlled, and the flow rate should be increased to wash down the wellbore, if the pump pressure is found to increase and cutting return volume is small. Before stopping the pumps, the drill string should be pulled out from the bottom of wellbore and the string should be moved at all time.

Wellbore collapse sticking

Wellbore-collapse sticking is more likely to occur in formations such as water-swelling mudstone and shale, poorly cemented sandstone and conglomerate, or during drilling or reaming. The main reasons are large water loss of drilling fluid, long soaking time of formation, small density of drilling fluid, or late injection of drilling fluid when pulling bit out, and the swab effect which causes the collapse of wellbore.

The performances of wellbore-collapse sticking are: The large mud cake and small formation pieces usually fall off before the severe collapse, and it cannot reach the bottom after drilling bit changes; sometimes large pieces of uncut rocks from superior stratum will be carried out by the drilling fluid; there are sudden phenomenon such as unexpected bit bouncing, pulling-out resistance, pump pressure rising, pump built up, or even the non-rotation of drill bit in the drilling process.

The main measures for preventing the collapse sticking are using anti-collapse drilling fluid with low filter loss, high degree of mineralization and proper viscosity, increasing the density of drilling fluid properly in broken and fragile formations, ensuring the column height of drilling fluid, avoiding the wall collapse from bit balling and swab effect.

Differential pressure sticking

Because a wellbore cannot be perfectly vertical during drilling, when drilling tool is hold stationary in downhole, part of the drill string will stick to the wellbore and adhere to the mud cakes under the effect of differential pressure. The longer the resting time is, the larger the contact area between drill string and mud cakes is, and the resulted sticking is called differential pressure sticking, also known as wall sticking or mud cake sticking. The schematic diagrams are shown in Figs. 8.1 and 8.2.

The main causes of differential pressure sticking are poor performance of drilling fluid and high density, resulting in excessive pressure difference in well, large loss in drilling fluid, thick mud cake, and large viscosity coefficient. Once the circulation is stopped, if the drill string is not moved, a part of the drill string will be attached to the wall of the well and contact the mud cake. The increase of time will increase the contact area and depth, and increase the viscous force of mud cake on drilling tools, resulting in the drilling tool cannot move and rotate, but it can turn on the pump circulation and pump pressure is stable. The main preventive measures for differential pressure sticking are to adjust the performance of drilling fluid, reduce drilling fluid density as much as possible, improve the lubrication performance of drilling fluid, reduce the viscosity coefficient of mud cake, and enhance the activity of drilling tools or centralizer, so that the drilling tools are centered.

Key-seat sticking

Key-seat sticking often occurs in hard formation stratum, where well deviation and direction change abruptly, forming the sharp turn (dogleg). In drilling process, drill string rotates next to dogleg scrapes wellbore at dogleg section in the trip, forming a fine groove (like key seat). The groove is larger than tool joint, but smaller than drill bit. In trip-out, the drill bit will be pulled to bottom of key seat and get stuck, as shown in Fig. 8.3.

Before key-seat sticking occurs, the joint is eccentric worn. Tripping-in is not hard, drilling process and pump pressure are normal, but when tripping-out at dogleg position, there is resistance, and the phenomenon is severer with the increase of well depth or may lead to stuck.

Fig. 8.1 Diagram of differential pressure sticking. 1—Shale; 2—Sandstone; 3—Shale;4—Formation pressure; 5—Drilling fluid column pressure in wellbore

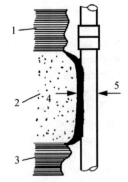

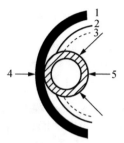

Fig. 8.2 Diagram that the contact area between drill string and mud cakes increases with still time. 1—The contact condition between drill string and mud cakes at one moment; 2—The contact condition some time later; 3—The contact condition after another time period; 4—Formation pressure; 5—Drilling fluid column pressure in wellbore

Fig. 8.3 Schematic diagram of key-seat sticking

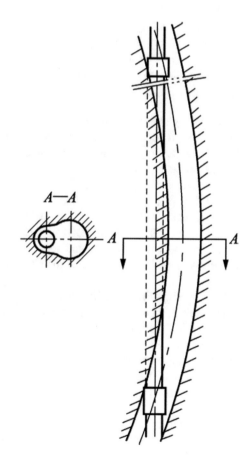

In order to prevent key-seat sticking, the quality of borehole must be guaranteed first to avoid forming dogleg section. Ream the hole again and again at key-seat section when trip-out the string or trip-in again to break the key-seat in time, and the speed must be slow at key-seat section. High-speed trip-out is forbidden.

Wellbore shrinkage sticking

Wellbore shrinkage sticking often occurs in swelling zones or in well section that has good porosity and permeability. Because of the poor performance of drilling fluid and excessive fluid loss, it is easy to form a loose colloidal mud cake on the wall of the well. When the flow rate of the pump is small and the up-return velocity on the drilling fluid is low, it is easy to deposit more clay particles, cuttings, and weighted agents on the mud cake, so that the diameter of the well is reduced or shrunk.

The main performance of the wellbore shrinkage sticking is that the shrinkage sticking position is fixed, and pump pressure increases in the circulation of drilling fluid; it is difficult to lift and easy to lower, and there are loose mud cakes on the top of tripped-out string joint.

Using high-quality drilling fluid of low density, low solidoid and low filter loss, or mixing oil in the former and reaming to expand the diameter of shrinkage can effectively prevent shrinkage sticking.

Junk sticking

Junk sticking may occur by careless operation. If the overshot, hinged jaw and other facilities are dropped into well, they will get stuck between the wellbore and drilling tool or the casing and drilling tool, causing junk sticking.

The causes are obvious, and this type of sticking can be avoided as long as we strictly follow the operation guidance and be more responsible.

Besides, there are many other types of sticking in fields, such as balling-up sticking, sand bridge sticking, and so on. Therefore, apart from the active prevention of sticking, we also need to accurately analyze the situation if sticking does occur, find out the real cause, and adopt effective measures to fix it in time, so as to avoid the further deterioration of the situation.

8.1.3.2   Handling of Sticking

Once sticking happens, the first thing is to analyze the operational conditions of trip-out, trip-in, rotation, pump circulation and borehole conditions and all events before the stuck to accurately judge the cause of sticking. The next is to take corresponding measures. Regardless of types of sticking, we all try to adjust the performance of the drilling fluid, clean rock cuttings, and clean borehole. The commonly used procedures are:

- Free stuck by lifting, lowering or rotating the drilling tool. By moving the drilling tool during the circulation of drilling fluid in well cleaning, we can unfreeze the tool if the sticking is not serious. But this method is only available to some types of sticking. For solids settling sticking or wellbore collapsing sticking, the lifting of drilling tool should be avoided in order to prevent from more serious stuck. But we can trip-in or rotate the drilling tool, and manage to start circulation under suppression, and then slowly lift the tool by back reaming method to releasing.

For resistance during trip-out (such as key-seat sticking, borehole shrinkage sticking, or balling-up sticking), we can abruptly trip-in the drilling tool at the original hanging load, but must not vigorously lift up; otherwise the sticking will be more serious. For resistance during trip-on or sticking due to over-large suppression, we should uplift the tool in large force, such as abruptly uplift and lower down, or rotate drilling tool, so as to releasing this type of sticking which has slight adherence.

- Spotting freeing stuck pipe. If the above methods are invalid, spotting freeing stuck pipe can be applied to pressure differential sticking, balling-up sticking, wellbore shrinkage block, and solenoid settling sticking. This method is to soak well by oil, salt water or acids, or circulate well by clean water to loosen the sticky mud cakes, decrease the viscosity coefficient, decrease its contact area with drilling tool, and decrease pressure difference, so that the drilling tool can be moved and the stick can be freed.

Before spotting freeing stuck pipe, the depth of stuck point must be figured out. The position of stuck point can be measured by calculating the relation between tensile force and elongation at the time of tension. Use a hook to pull out drilling tool with certain force, measure the elongation of drilling tool at this tension force, and then calculate the depth of stuck position by the following equation:

$$L = \frac{EA_p \Delta L}{10^3 F} \tag{8.1}$$

where

L	Length of the drill pipe above stuck point, m;
E	Elasticity modulus of steel, taken as 20.6×10^4 MPa;
A_p	Cross-sectional area of drill string, cm^2;
ΔL	Absolute elongation of drill string, cm;
F	Static tension force, kN.

After the depth of stuck point is determined, we then calculate the required volume of soaking oil (liquid) and inject it into stuck area to release the sticking. Generally, it requires the injected crude oil return to the position 100 m higher than stuck point that all drilling tool under stuck point is soaked in oil, and the oil level inside drill string is higher than the outside. The volume of soaking oil can be calculated by:

$$V_0 = K_{hD} \times 0.785 \left(d_b^2 - d_{po}^2 \right) H_1 + 0.785 d_{pi}^2 H_2 \qquad (8.2)$$

where

V_0 volume of soaking oil, m^3;
K_{hD} additional coefficient of borehole diameter, 1.2 to1.5 in most cases;
d_h drill bit diameter, m;
d_{po} external diameter of drill string, m;
H_1 height of soak oil in annual space, m;
H_2 height of injected oil inside drill string, m;
d_{pi} internal diameter of drill string, m.

After injecting crude oil, displace some drilling fluid every 10–14 min to return oil back and enhance its immersion effect. During wash down, keep moving the drilling tool. If fail to release sticking after soaking 6–12 h, replace the old crude oil and soak again, or add certain amount of diesel in crude oil to improve the effect.

Sometimes lye or clean water can be used as soaking agents, and the effect is good. In limestone formation, the low concentration of hydrochloric acid can be used to free sticking.

- Free stuck pipe by top jar and down jar.

A shock device can be attached between the drill rod and the drill collar or between collars if there are collapses, viscous, and expansible layers in drilling process. Once stuck, it strikes up and down immediately to free sticking.

In case of stuck during tripping out, such as pipe sticking caused by wellbore shrinkage or key-seating, which cannot be solved by moving drillstring up and down, the drilling tool can be reversed at the sticking point, and then the jar can be connected. After making up, bumping downward for loosing drilling tool and cleaning the wellbore circularly, the drill string can be lifted up slowly. If there is still sticking phenomenon, the drilling tool can be lifted gently by back-reaming.

In tripping-in process, if resistances are ignored and grow into well sticking, or wall sticking occurs due to the adherence of slight mud cakes, shock device can be used to free stuck.

- Free stuck by reverse milling. In serious sticking, if the above methods cannot release stuck and restart circulation, reverse or milling method are often used to pick up part of or all drilling tool. Backing-up means to reverse the rotatory table and back-up the right-hand string. The number of reversing strings is determined by the tightness degree of the screw head of stuck tool. Normally, the good backing-up is at stuck point. For drilling tool under stuck point, casing section mill is needed (Fig. 8.4), which is to wash down the cuttings and debris in outside space (i.e., the annular space between drilling tool and wellbore) and then backing-up drilling tool. This method is more complicated and time-lasting.

Fig. 8.4 Washover pipe

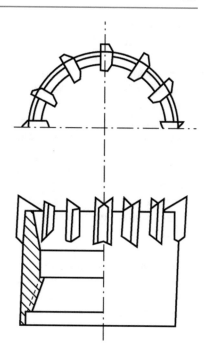

Processes of backing-up and milling are:

– Pull out all drilling tool above the stuck point (with large force) so that the drilling tool around stuck point undergoes no pull or push force. Then reverse rotatory table, back off screw head, and pull drilling tool out of well.
– If the pull-out tool does not reach the stuck point, use left-hand screw pipe ended with left-hand tap (lap safety joint on the top of tap if necessary). Run down to the drilling tool in the well to make cutting thread, and then lift it up to make the drilling tool free of force at the sticking point, and then reverse the rotary table to reverse the thread. This is repeated until all the drilling tool above the sticking point are pulled out, and then casing milling is carried out.
– Use mill sleeve to case section mill. The external diameter of mill sleeve or mill shoe is smaller than borehole, but its internal diameter is larger than the outer diameter of string joint. For drill pipe with diameter of ϕ119 mm ($4^1/_2$ in), the mill sleeve can be ϕ165 mm ($6^5/_6$ in) and the length is 58 m in general, or longer sometimes. Mill sleeve is jointed at the end of drill pipe, after it gets the string pipe case mill. Pull out the string pipe after casing at the length of a mill sleeve. Then insert left-hand screw pipe and left-hand tap to back off. Repeat the operations (case mill → back off → case mill), until all downhole drill tools are backed off.

• Explode to back off and case mill. This is a new method to deal with stuck problem. First, measure the stuck position. Then insert the cable through the inner hole of drill pipe to send the blasting fuse to the first joint thread above stuck position, align the middle of blasting fuse to the joint and meanwhile, pull out drill pipe to release all forces above stuck point and apply certain torque moment to drill pipe. Ignite blasting fuse to form explosion. The resulted strong blasting wave and shaking force will make part of the joint have elastic deformation and back off in time. It has the same principle with striking female joint of drill pipe by mallet if the joint of drill pipe cannot be dismounted. Meanwhile, the large amount of heat generated by explosion will heat the joint, melt the thread grease, and result plastic deformation, which is also helpful to remove thread.

This method is safe and fast. It will normally cause no damage to the drill tool and no need for the back-off of drill pipe and the use of fishing tool. The speed of free stuck pipe can be improved at the same time. But it demands the strict control on explosive charge and prudent operation. After back off, casing mill and fishing are required.

- Explode or sidetrack a new borehole. If the above methods are unsuccessful, or the stuck point is so deep that back-off method is time-consuming and may deteriorate the situation, pull out the free part of stuck tool after broken by explosion, plug cement on the top of remained tool, and sidetrack a new borehole.

8.1.4   Drilling Tool Accidents and the Handling Procedures

During drilling, especially in rotatory drilling, the drilling tool accidents are commonly encountered. Drilling tool accidents usually include drill pipe and drill collar breaking, slipping, tripping, and clasping. Drilling tools dropped into the wellbore are called "fish." Fishing is a very detailed work that needs a thorough analysis on downhole conditions. Proper tools and methods should be applied in time. The inappropriate handling or a long-time stay of fish in the wellbore may deteriorate the situation, making the accident more difficult to handle, or even leading to abandonment of the well.

8.1.4.1   Common Drilling Tool Accidents

Drill pipe or collar breaking

This type of accident is the most common one. When used with a thin thread drill pipe, it is prone to break at the end of its thread. The thin thread drill pipe is not used any more, but the relative weak point is transferred to the drill pipe itself because of the strength at the joint of the drill pipe. When drill pipe is under overlarge tensile force or twisting force (such as free stuck), or under large forces due to its defections such as cracks or corrosion, the wreckage is likely to happen.

Drill collars often break at the position of its thick thread. That is because the body of drill collar has large rigidity, but the thread part is relatively weak. The thread part also undertakes compound forces such as stress, torsion, or bending force. If the thread has bad quality or the operation is inappropriate, the drill collar may break at its thick thread.

Thread slipping or releasing

Thread slipping refers to the phenomenon that the jointed threads slip apart under some forces. Because the threads are badly worn, or the threads are not closely tighten that under long-time flushing with drill fluid, they slip apart. Sub-standard threads and threads that are difficult to be tightened or other factors may also lead to thread slipping.

Thread releasing is the automatic releasing of the undamaged thread from the joint under the abnormal reversion of drilling tool, which may be caused by bit bouncing, reverse rotation, or other situations.

8.1.4.2 Handling of Drilling Tool Accidents

When handling drilling tool accidents, appropriate fishing tools should be selected according to the actual conditions of the fish top and the borehole. Special fishing tools should be designed to quickly salvage the fish. The commonly used methods are listed as below:

Fishing by overshot

Overshot is the most commonly used fishing tool. It is salvaged from the outside of the fish, which is mainly used for fishing drill pipe, drill collar, oil pipe, and other smooth tube fish. After fishing, it can build the pressure to circulate and free stuck, and if stuck occurs, the overshot can be pulled out. The operations are easy and convenient. The basic configurations of an overshot are listed in Figs. 8.5 and 8.6.

Before fishing by overshot, slips, control ring and sealing components should be chosen according to the size of catching part of the fish. And the insert position of the overshot should be determined. When the overshot is reaching at the top of the fish, rotate it to the right slowly and move fishing pipe downward to catch fish into the guide shoe of slips. When the top of fish reaches basket control ring that has milling teeth on it, rotate fishing pipe slowly, mill off burrs on the fish head, keep

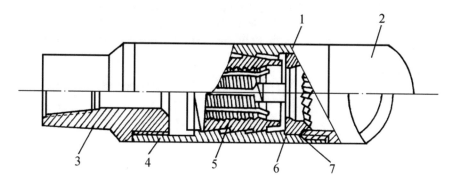

Fig. 8.5 Basket overshot. 1—Basket control ring; 2—Guide shoe; 3—Top joint; 4—Outer cylinder; 5—Basket slip; 6—R-shaped sealing ring; 7—O-shaped sealing ring

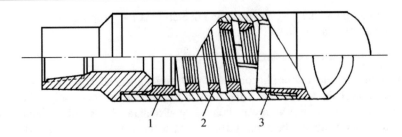

Fig. 8.6 Spiral overshot. 1—A-shaped sealing ring; 2—Spiral slip; 3—Spiral control ring

moving the overshot downward, and keep catching drill string. When the end of overshot reaches the fish head, compress with a force of 30–50 kN to make the fish head against the overshot. The overshot will open and allow the fish to come into it until it reaches the catching point of the fish. Pull the overshot out, and the overshot body will move upward relative to slips. The sawtooth slope of outer cylinder forces slip to shrink, making slips seize fish by the teeth. After catching fish, open the pump to circulate if necessary, but don't rotate drill string in case of fish slippery. Then trip-out the drill string and fish out the falling objects.

Fish by fishing spear

Fishing spear is a type of fishing tool with simple configuration and reliable effect. The basic structure is shown in Fig. 8.7. It is mainly made up of spindle, slips, release ring, and guide shoe.

Fishing spear is used to fish drill string and casing inside of the borehole. Because it seizes fish with large area, it will not damage fish. If fail to pull out fish, the spear is easy to get loose and retreat out, missing the fish. Inside slips is the left-hand saw-teeth indented jointing with spindle saw-teeth. The flange of release ring and the flange on the end face of guide shoe are safety devices that can resist from the locking, cementation, and stuck of fishing spear to ensure release easily. Forward rotation can make fishing spear out of fish.

When fishing by spear, the spear that matches with the inner diameter of the fish should be chosen first. Then push down it into well until the overshot reaches at the predetermined position inside fish. Rotate leftward for one circle or two and pull out

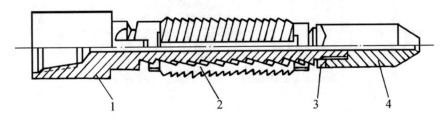

Fig. 8.7 Schematic diagram of fishing spear. 1—Spindle; 2—Slips; 3—Release ring; 4—Guide shoe

fishing pipe, and then, the slips will bulk up by the saw-tooth slope of spindle, making the external gear catch the inner wall of fish and catch the fish outside wellbore.

Fish by taper tap and box tap

Taper tap and box tap fishing were the most commonly used in oilfields in China. With the improvement of the strength of drill string material and the invention of new fishing tool, the method is no longer as popular as before, but it is still an effective and convenient fishing method.

The structure of a taper tap is shown in Fig. 8.8. It is a cone with nozzle inside and the thick thread on the top which can connect with drill tool. The fishing screw thread lathed on the surface of cone has been hardened by carburizing and quenching to make threads on the fish.

There are many types of taper tap, such as right-hand die tap, left-hand pin tap, and short pin tap. In fishing process, taper tap should be inserted into the nozzle of fish and be rotated by force to cut thread for fishing.

As long as the shape of nozzle is regular, it's able to cut thread and wall on the top of fish is thick enough such as the joint, the thickened part in drill pipe, and the

Fig. 8.8 Taper tap

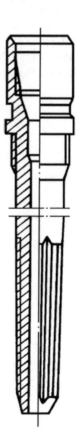

drill collar; taper tap can be used for fishing. But for the thin wall type, such as the drill pipe body and oil tube, taper tap is not appropriate.

The position of fish top should be well calculated when fishing. Insert drill to the depth of 0.5 m above fish top, and start pump to circulate for 10–20 min. Record the current pump pressure and hanging load, then push down TOF detector, and judge whether the taper tap gets in touch with fish top, whether it is inserted into the inner hole by the display of pump pressure and hanging load as well as the feeling of operators. Hanging load will decrease if the tap reaches the top of fish, and pump pressure will increase if the tap is inserted into the inner hole. If the tap is determined to enter into fish, pump can be shut down to cut thread. Load a stress of 20–40 kN and rotate for 3–4 rounds to cut threads. Turn on pump to circulate. If the pump pressure increases, then continue cutting threads. If the hanging load increases, it means that tap taper is cutting threads, and thus, we need to feed bit properly to keep up with the speed of cutting thread and record the exact thread circles at the same time. If threads are enough, turn on pump to circulate, wash over the deposited cuttings on bit, and then try to pull out the string. After the bit is pulled out 2–3 m away from downhole, drop and stop sharply three times to ensure the reliability of fishing and then pull out the drill to fish.

If the taper tap cannot reach the top of fish, we should check whether the depth of fish top and the pulling-down depth of fishing tool are correct. Or we can insert fishing tool a little bit deeper to reach fish top. If the top of fish deviates from the center of borehole, we can apply bent drill pipe with taper tap or hook with taper tap to fish, as shown in Fig. 8.9.

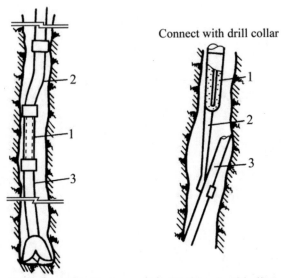

Connect with drill collar

(a) Bent drill pipe with die tap   (b) Wall hook with die tap

Fig. 8.9 Schematic diagrams of fishing by die tap. 1—Die tap; 2—Bent drill pipe or wall hook; 3 —Fish

8.1.5 Junk Accidents and Their Handling

8.1.5.1 Common Junk Accidents

During drilling or tripping-in, drilling tools or devices of poor quality may cause junk accidents if they are not thoroughly inspected, or wrongly operated or improperly handled. The common junk accidents include: cone lost (including cone, bearing pin, broken palm, and dropped pellet), blade breakage (drag bit), inclinometer falling into well, and drill floor tools falling into well (such as hammer, wrench, tong pin, and cable).

8.1.5.2 Handling of Junk Accidents

Side wall burying method

If the fallen objects are small (such as a pellet, a pin, and a blade), the borehole is shallow and the formation is soft, and the well construction is in its early stage; the objects can be left inside of the well. We only need to use the discarded drag bit to push the fallen objects into the side wall and press them into the wall and bury them. This method is simple and convenient, but not thorough. Sometimes the objects may fall out of the side wall again. For those large fallen objects, the hard formation and deep well, and the long openhole section, this method is not applicable.

Magnet fishing method

Magnet fishing method is widely applied to fish fallen cones, pellets, and small tools. In fishing magnet, there is a long-cylinder-shaped permanent magnetic core, which has magnetic poles at both ends. The upper inner joint and the shell are connected by screw threads and compressed against the top crown; the bottom shell and mill shoe are connected by screw threads and compressed against bottom crown. Between magnet and shell, there are insulated sleeves. The schematic diagram is shown in Fig. 8.10.

The fishing magnet is attached at the bottom of a drill pipe. Because drilling fluid has large magnetic reluctance, the magnetic core should be pushed as close as possible to the fish. Circulate fluid to flush the magnet core and the well bottom when pushing down the fishing magnet to a distance of 0.5–1 m from the well bottom. Then push down fishing magnet slowly until the weight indicator shows the magnet reaches bottom. The pressure should not be too high at this time to prevent the falling objects from being pressed into the formation. In pulling-out process, don't use rotatory table to shackle; otherwise the caught fish may fall down into the well again.

Reverse-cycling junk basket fishing method

The structure of reverse-cycling junk basket is shown in Fig. 8.11. The basket is a tool for fishing cones and small junks, and it has internal and external double-cylinder layers. After the injected valve balls fall onto the valve seat, drilling fluid will spray to borehole from a row of pinholes through the channel between internal and external cylinder and flush the junk into the basket cylinder. Drilling

Fig. 8.10 Magnet fishing
tool. 1—Joint; 2—Body; 3—
Upper magnet; 4—Cylinder;
5—Magnetic core; 6—
Bottom magnet; 7—Guide
shoe

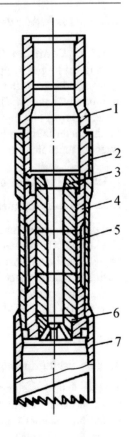

fluid will then reverse out through four holes B at the top of cylinder, forming a
back-cycling. The junk that failed to be moved by drilling fluid needs to be core
drilled by mill shoe and then be headed into cylinder by its core.

Before fishing, reverse-cycling junk basket needs to be attached with mill shoe
and core gripper at its bottom end. The operation procedures are: take out the valve
body first, start well flushing until the bit is about 1 m away from the well bottom,
inject valve body, open pump to start reverse circulation, rotate drill tool at low rate,
and trip-in the junk basket slowly to the bottom. If mill shoe is used, stress to 10–
20kN, core drill by 0.3–0.4 m, and then the core and junk can be taken out together
after cutting the core.

Fishing while drilling method

The tool used in this method is the drilling junk sub, as shown in Fig. 8.12. It trips in
with the bit, and it can be used to remove the drop teeth, broken teeth, marbles and
other granular object in the well. The drilling fish sub is attached between the drill bit
and the drill collar (be close enough to drill bit) in the well. When it reaches the
bottom of well, the grain-shaped metal fishes can be rushed to the upper of cylinder
by drilling fluid and settle down into cylinder due to smaller upward velocity.

Fig. 8.11 Reverse
circulation fishing basket.
1—Joint; 2—Valve ball;
3—Valve cup; 4—Valve seat;
5—Cylinder; 6—Mill shoe;
7—Core catcher

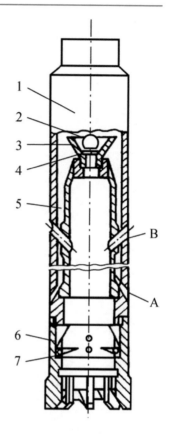

The usage of drilling fish sub is: Open the pump and start the circulation when the drill is down to about 0.5 meters from the bottom of the well. Meanwhile, rotate drilling tool at the first gear and slowly lower drill string to the bottom and flush for 5–10 min. Stop rotation and shut down the pump for 2–3 min, then pull out drill string, and repeat the operations for 2–3 times. After fishing out, we do not have to pull out the drill pipe and continue drilling.

Claw bit fishing method

Claw bit is a type of fishing tool that is simple in structure, easy to be manufactured, and convenient to use, which is often used to fish bit cone and other objects. The structure is shown in Fig. 8.13.

During fishing, lower the tool down to the bottom, flush well slightly, and then explore several positions. Find out the one with maximum kelly downs, press with 10–20 kN force, rotate the tool for 3–4 rounds, then press with 30–40 kN, and rotate for 5–6 circles. During the pressured rotation, if no stress bouncing is detected and the hanging load restores quickly, and it shows that the fish is caught. Pull it out of the hole.

Fig. 8.12 Drilling junk
basket. 1—Main part; 2—
Cylinder

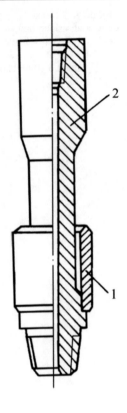

Fig. 8.13 Claw bit

If a claw bit is not used properly, the jaw of the drilling bit may be broken. Therefore, the method is not suitable for the wells that have resistance during trip-on and the wells that has large deviation or stuck during trip-in in case of holding the bit teeth.

Flat-bottom milling shoe method

If the fish is not easy to be caught out, flat-bottom milling shoe can be used. Figure 8.14 is the structure of a flat-bottom milling shoe. Its bottom end is radial-shaped teeth, which has cemented carbide welding layer on its surface. During the usage, load certain pressure and rotate at the first gear to maintain stable milling. Pull out and ream the hole once every 10–15 min to ream down the fragments squeezed into wellbore. In milling process, if the increase of pump pressure is detected, which may rise from the rub down of teeth, pull out the drill tool and change another milling shoe. The flat-bottom milling shoe can alternately work with magnet fishing, and the effect of milling shoe is better in hard formation.

There are many types of fishes and various fishing tools. Apart from the above mentioned tools that are commonly used, sometimes we need to design special tools according to the shape of fish, such as the fisher with wire line catcher.

Fig. 8.14 Flat-bottom milling shoe

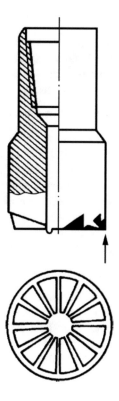

8.2 Coring Technique

8.2.1 Introduction

8.2.1.1 Purpose of Coring

Core is the only way to provide the primary rock sample of formation profile. Information unavailable from other methods can be obtained through the core samples. Various types of data can be collected by mud logging, geophysical logging, geochemical logging, and formation testing to evaluate the formation in oil exploration and development. However, all the above-mentioned methods have limitations. The complete first-hand data can only be obtained from core. Quantitative analysis and investigation of core allow us to make reasonable field development plans, calculate the reserve volume, and provide technical support for well stimulation. The main purposes of coring are:

Formation study

Comparison and analysis of subsurface formations can be achieved by investigating lithology, physical property, electrical property, mineral composition, and fossil profile from core. The strength, drillability, and grindability of formation can be known to guide drilling by measuring the mechanical properties of core.

Source rock study

To explore resources in a sedimentary basin, the first and foremost thing is to determine whether the basin has a good source rock. Therefore, various source indexes need to be analyzed by testing the core taken from the target reservoir. The target zones and potential areas can be identified according to the testing results.

Hydrocarbon-bearing reservoir study

After exploration and discovery, the distribution of oil, gas, and water in reservoir, the porosity, permeability and oil–gas saturation and the net pay thickness of oil–gas zones need to be further investigated to evaluate the field's commercial value.

All these parameters can be obtained by measuring core samples. The requirements for different tests vary significantly. For example, the routine core porosity and permeability can be measured by using conventional coring procedure, while a special low-invasion coring technique is necessary to preserve the original oil saturation of rock in order to minimize the invasion from drilling mud filtrate.

Oil–gas field exploitation guide

During the field development, fluids need to be injected into oil zones to maintain the energy to displace oil or natural gas in a secondary recovery process. The displacement of oil by water is a complex oil-water migration process. It is an important content to master the principle of water flooding and the law of oil and water migration under different conditions. People often use the core in the laboratory to prepare various oil-bearing models and perform water flooding experiments. The experiment can provide theoretical basis and practical experience for oilfield development.

The core-based study can also provide reliable reference for the acid fracturing of low permeability oil–gas reservoirs.

Field development check

During field development, in order to monitor the dynamical conditions of a field, the movement of injected fluid, and the efficiency of oil displacement, some inspection wells are often drilled to obtain the data of oil and water saturation, lithology, and physical properties. The relevant studies can be obtained to guide the practice and enhance recovery and exploit oil–gas field more reasonably.

Besides, well logs can be calibrated by using the related data from core, and relation between logs and core can be applied to wells without core.

Above all, in various stages of exploitation and development, in order to find out the nature of the reservoir or the comparison from the large area to the inspection of the development effect of the oil and gas field, evaluate and improve the development scheme, all research steps are inseparable from the observation and research of cores.

8.2.1.2 Processes of Core Drilling

- Crack bottomhole rock in ring shape to generate the drill core (cylinder).
- Protect core. In core drilling process, the formed cores need to be protected to avoid circulating drilling fluid to erosion the core and the mechanical collision of the drill string to damage the core.
- Take out core. After drilling to certain depth (normally at a single length or use long barrel to core, which is tens of meters or even hundreds of meters long), cut down the generated core from its bottom, clamp it tightly, and lift it to ground with drilling tool.

These above three steps are all very important in drilling core process. If any step is handled improperly, there might be no sufficient recovery of cores and sometimes it may cause an ultimate failure of the coring operation.

8.2.1.3 Evaluation Index of Core Drilling

Generally, drilling efficiency is evaluated by rate of penetration and total footage (each progression). The purpose of core drilling is to get core, and core with the same length of drilling footage should be taken out. But due to many other reasons (such as erosion and wear), we can't get the adequate length of core. Generally, core recovery rate is used to evaluate core drilling, and the calculation method is:

$$\text{Core recovery rate} = \frac{\text{Actual length of takenout core}}{\text{footage of core drilling}} \times 100\%$$

In core drilling, the first thing is to maintain the high core recovery rate. Under this premise, maximize the rate of penetration.

8.2.2 Coring Tools

Coring tool is the guarantee of drilling underground cores. To do coring work well, the composition, structure, application range, and operating requirements of coring tools should be known to get high core recovery rate and improve the speed of core drilling.

8.2.2.1 Compositions of Coring Tools

There are many types of coring tool to meet different situations. They all have three basic components: coring bit, which is used to drill core; core barrel and its suspension device, which is used to protect core; core gripper, which is used to cut and stuck core, and lift it with drilling tool to ground. Figure 8.15 is the schematic diagram of coring tool. The structures and applications of main components are described in the below.

Coring bit

Coring bit is the key tool to crush downhole rocks in ring and form core plug-in central area. The value of core recovery and the drilling speed are related to the

Fig. 8.15 Diagram of the components of coring tool. 1 —Coring bit; 2—Core gripper; 3—Inner core barrel; 4—Outer core barrel; 5—Centralizer; 6—Inside back pressure valve; 7—Hanger bearing; 8—Suspension gear

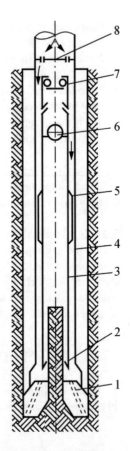

quality and selection of coring bit. The structure is designed to favor the formation of core and enhance the core recovery. The drilling should be stable to avoid the breakage of core caused by vibration. The bit edge should be concentric with center hole, and the nozzle should be at position where the jet flow can't directly shoot core and reduce the erosion of overflow. The bit cavity should be positioned where core gripper is close enough to the entry of core, therefore the core can be protected by entering into core barrel through core gripper as soon as possible after its formation, and it also can make core gripper close enough to the root of core in core cutting so that less core will be left in bottom.

According to the type of drilled stratum, coring bit can be divided into rock, drag, and diamond (Fig. 8.16).

- Rock coring bit has the type of four-cone bit and six-cone bit, which is suitable to medium-hard and hard formation. Due to the limitation of bit structure, the diameter of core should not be too large. Now, core cone bit is seldom used in China.
- Drag coring bit breaks rock by cutting. Same to the full face drilling drag bit, the cutting blade of drag coring bit is cemented hard alloy. The blades are evenly distributed in the concentric circular area, and drilling bit is stair-step shape to improve drilling efficiency.
- Diamond coring bit has many type of structures and can be applied to a large range of stratums. There are natural diamond coring bit, artificial diamond coring bit, and polycrystalline diamond compact drilling bit.

Core barrel and its suspension system

Core barrel is one important component of coring tools.

(a) Core cone bit

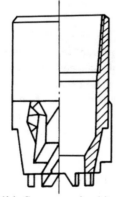
(b) Scraper coring bit

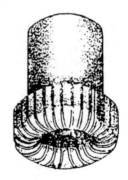
(c) Diamond coring bit

Fig. 8.16 Drilling bit

Coring tool can be divided into the monocular coring tool (which only has one core barrel) and the binocular coring tool (which has inner core barrel and outer core barrel). Currently, the mostly used tool is binocular coring tool. Binocular coring tool has the inner core barrel, outer core barrel, inside and outside centralizers, inside and outside back pressure valve, and hanger assembly, as shown in Fig. 8.15.

- Outer core barrel. Outer core barrel is made by high grade seamless steel pipe, and it is connected with drill string at the top and coring bit at bottom. In drilling process, the outer core barrel undertakes drill pressure and transmits torque to rotate drill bit. Centralizer is installed at the outer core barrel to avoid bending and swinging in drilling process.
- Inner core barrel. Inner core barrel is used to receive, store, and protect the core in drilling. To prevent circulated drilling fluid from entering into inner core barrel and to drain out the inside fluid (when core is entering inner core barrel), there are the water joint and back pressure valve (check valve) on the top of inner core barrel. Back pressure valve is made of 5 cm steel ball and ball seat. It doesn't have to install the steel ball before run in hole. After running in, the drilling fluid is circulated, and the fluid flows through the inner core barrel to wash away the accumulated sand in the barrel due to tripping in, and then the steel ball is put into the drill string from wellhead to the ball saet. Because ball valve is sealed, drilling fluid cannot enter inner core barrel but flows to watershed joint, passing through the annular space between inside and outer barrels and coming to drill bit. The core in inner core barrel will not be eroded by fluid. When core enters inner core barrel, the fluid opens the valve by pressure and expels the steel ball.

To make core successfully enter inner core barrel, the barrel should be no distortion, no crook, and no flattening and have smooth inside wall and uniform wall thickness.

- Inner core barrel hanger assembly. Generally, the inner core barrel is hanged on the top of outer core barrel to prevent from the core grinding. Hanger assembly includes hanger bearing unit and suspension system. The hanger bearing unit is used to prevent from the rotation and swing of inner core barrel. And the unit is composed by two or more than two pairs of bearings. The suspension system is used to suspend the inner core barrel to the top of outer core barrel by different methods. The method depends on the structure of coring tool. The most often used suspension systems are:

- Thread-connected suspension. This method is to use thread to hang the inner core barrel on the top of outer core barrel, which is most reliable method in suspension (Fig. 8.17).
- Pin-connected suspension. This method is done by inserting pin at symmetrical position to connect inside and outer core barrels. It has very strict requests on the

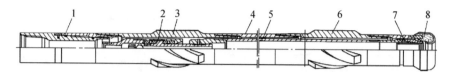

Fig. 8.17 Conventional coring tool by self-lockin. 1—Safety joint; 2—Hanger assembly; 3—Top centralizer; 4—Outer core barrel; 5—Inner core barrel; 6—Bottom centralizer; 7—Core gripper; 8 —Coring bit

material and size of pins. The pin should be cut off when cutting the core, but the pin should be intact when running in and drilling (Fig. 8.18).

– Ball and socket suspension. This method is done by using steel ball through hanger sleeve to connect the inside and outer core barrels together. The main components are steel ball, hanger joint, sliding sleeve, and release plug (Fig. 8.19).

Core gripper

Core gripper is used to cut off core and catch the cut core to prevent it from dropping. Core gripper should permit the successful entrance of core into inner core barrel without any damage.

Core gripper should be cooperatively used with tight hole set (core jaw seat). For some coring bit, the inner chamber is castled into jaw seat shape; thus the bit can

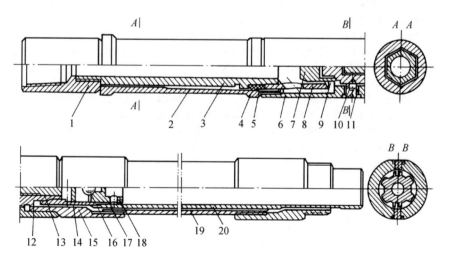

Fig. 8.18 Coring tool by mechanical loading. 1—Top joint; 2—Internal hexagonal hole; 3— External hexagonal hole; 4—Packing; 5—Pressure pad; 6—Locator sub; 7—Pressure head; 8— Pressured center rod; 9—Suspension; 10—Pin; 11—Plug; 12—Nut; 13—Bearing; 14—Back pressure valve cover; 15—Bearing seat; 16—Back pressure valve ball; 17—Outer barrel joint; 18 —Back pressure valve ball seat; 19—Outer core barrel; 20—Inner core barrel

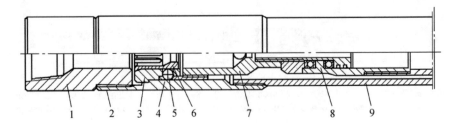

Fig. 8.19 Ball-socket hanging structure (conventional coring tool by loading). 1—Top joint of tool; 2—Support joint; 3—Suspension joint; 4—Sliding sleeve; 5—Hanging ball; 6—Release plug; 7—Water diversion joint; 8—Hanger assembly; 9—Outer core barrel

directly combine with core gripper. There are many types of core gripper which are designed to adapt to different stratums and coring tool structures.

- Core gripper with clamps: The shape is like circular clamps. It has a collar of cutout gaps which divides gripper into several sections, and in each section, several collars of teeth are castled (Fig. 8.20a, b). The outer wall of clamp is cone shape. It is cooperatively used with tight hole set, which has the same conical surface. When core gripper moves along jaw seat, the jaw will shrink to chuck the core. Core gripper with clamp is suitable to the soft and medium-hard formation.
- Core gripper with slips: It is suitable to medium-hard and hard formation. The components are hanger sleeve, pin roll, torsional spring, and slip slices (Fig. 8.20c). Slip slices can open under the force of torsional spring and cling to the inner wall of drill bit in drilling process. In core cutting process, core gripper moves downward along the inner wall of drill bit under external forces, and slip slices will constrict to wrap up the core.
- Core gripper with board: It is suitable to core in medium-hard and hard formations. The components are outside seat, torsional spring, and flake boards (Fig. 8.20d), and it is often used with other core grippers.
- Core gripper with slip-spring: It is suitable to hard formation, as shown in Fig. 8.20e. It is often used in geologic drilling, but seldom used in petroleum core drilling.

8.2.2.2   Classification of Coring Tools

We introduced the composition of a coring tool in the last section. However, in practice, coring tool is manufactured in an assembled unit. Coring tool is designed for different formations, different coring objectives, and requirements; therefore, it has different types and structures. The classifications of coring tools are:

Classification by tool structure

Single-barreled coring tool: There is no inner core barrel.

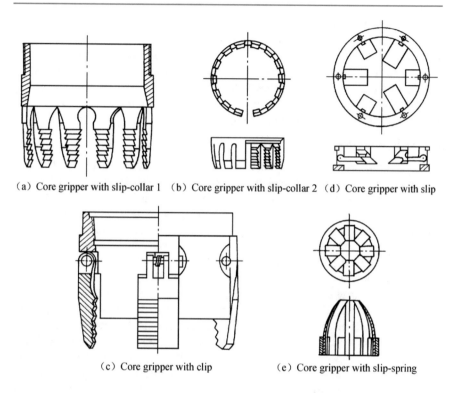

(a) Core gripper with slip-collar 1   (b) Core gripper with slip-collar 2   (d) Core gripper with slip

(c) Core gripper with clip   (e) Core gripper with slip-spring

Fig. 8.20 Core gripper

Double-barreled coring tool: There are both inner and outer core barrels, which can be classified into the double-action double-barreled coring tool (inner and outer core barrels both rotate in drilling process) and the single-action double-barreled coring tool (the inner core barrel keeps still in drilling process).

Classification by coring length

Short-barrel coring tool: Generally, the length of core drilling is within a single specified length, which means it can't make a joint pipe in core drilling.

Medium-long barrel coring tool: It can drill to tens and hundreds of meters.

Classification by core cutting method

Self-locking type, loading type, and differential control type.

Classification by coring method

Conventional coring tool: There are no special requirements for the core; thus most coring belongs to this type.

Special coring tool: There are certain requirements for the cut core, such as prevention from core contamination and maintaining its initial state at downhole conditions, or core orientation.

8.2.2.3 Conventional Coring Tool

If there are no special requirements on core, conventional coring tool is used to provide the first-hand data for the discovery of reservoir in exploitation and development process, and for the exploratory of sedimentary features, lithology, physical properties, electrical properties, electrical properties, and oil saturation of underground rocks. Conventional coring tools are widely used, and over 90% of the total coring operations are fitting into this category.

Self-lock-type coring tool

It is the basic tool for coring in medium-hard or hard formations and is also applicable to the formation that has good integrity. Self-locking coring tool is composed of coring bit, core gripper with slip collar, inner and outer core barrels, and safety joint (Fig. 8.17). The outer core barrel is connected with the box end; on the top of inner core barrel is the hanger assembly connected with the pin end; the bottom of the inner core barrel is connected with tight hole set (core gripper seat) where put the core gripper inside.

After tripping-in, large-displacement circulating drilling fluid is used to clean out the inner core barrel and the bottom hole; then the pump is shut down, and the steel ball (inside BOP steel ball) is injected in place and core drilling starts. Because the internal diameter of core gripper is a litter larger than the diameter of core, it will generate frictions when core is contacting with the inner edge of gripper. When drilling process is done and drilling tool is pulled out slowly, gripper will be relatively still to core due to the existed friction, and tight hole sleeve will also move up with drilling tool. The internal conical surface of tight hole sleeve and the external conical surface of core gripper tally with each other; thus the relative displacement will make core gripper contract, self-lock, clamp, and cut off core, making the core come up with drill tool to the ground.

Load-type coring tool

This type of coring tool is the basic tool for soft formation. When cut core by using this tool, external loads should be added to the inner core barrel to make inner core barrel and core gripper moves downwards relatively to the outer core barrel, pushing core gripper move along the inside conical surfaces of drill bit and making the gripper contract and clamp the core.

For the load-type coring tool, there are two loading methods; one is mechanical load, and the other is hydraulic load. The suspension methods for its inner core barrel are pin-type suspension, ball seat suspension, and so on. The next is the description of loading-type coring tool using mechanical loading.

The mechanical load coring tool features in its connection with mechanical loading joint on the head of the coring tool. Its components can move relatively, including the hexagonal bar (its top is connected with drill string), the hexagonal case (its bottom is connected with outer core barrel), and the central load rod (its bottom is connected with the hanger assembly of inner core barrel), as is shown in Fig. 8.18.

The working principle of mechanical load coring tool is shown as Fig. 8.21. In core drilling, drilling load and torsional moment are transmitted to drill bit through loading joint and outer core barrel, making core enter into inner core barrel by gripper (Fig. 8.21a). In core cutting process, pull up drill string to open the loading joint, inject the loading steel ball, and after it reaches the ball seat at the central load rod (Fig. 8.21b), lower down drill string and cut off the pin by loading some stress through steel ball, central load rod to loading joint. Therefore, the inner core barrel will slip down due to the loss of bearing (Fig. 8.21c), forcing the core gripper on the bottom of inner core barrel move along the colonial surface of drill bit, shrink and clamp the core, and then cut off the core under the torque of drill string.

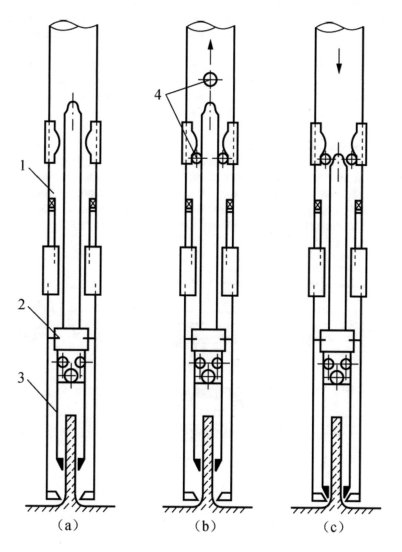

Fig. 8.21 Schematic diagrams of mechanical loading core cutting. 1—Loading joint; 2—Pin; 3—Coring tool; 4—Steel ball

Medium-long coring tool

Generally, the coring tool (short column) permits the coring length within a single length. But if the required core is long, many times of trip round are needed. Therefore, to improve drilling efficiency and decrease cost, medium-long coring tool is developed.

When drilling core by medium-long coring tool, the single is needed. But in the soft formation, drill bit is not allowed to be pulled up above bottom hole in drilling process; thus the special equipment should be added to single sliding joint (Fig. 8.22). Sliding joint is the differential device made of the sliding of inner and outer hexagons, whose effective sliding distance is around 11 m. It guarantees that the drill bit (including core barrel) is still contact with bottom hole when connecting stalk, (Fig. 8.23).

If the coring tool is very long and its weight exceeds the needed drilling pressure, the slippery will make the total weight of tool added to the drill bit. To adjust drill pressure flexibly, the adapter is assembled on the bottom of sub, which can pull and push coring tool, to adjust drilling pressure.

8.2.2.4   Special Coring Tool

The cores of special requirements are needed to assess oil and gas reservoir during exploitation and development. For example, the core needs to contain fluid and maintain its initial state. Special coring methods include sealed coring, pressure-maintaining sealed coring, directional coring, and so on. The tools used in these special coring methods are referred to as special coring tools.

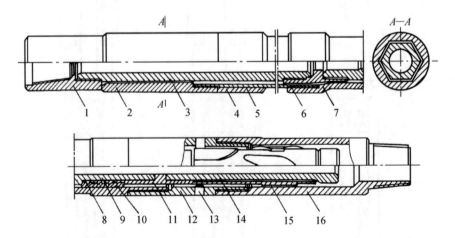

Fig. 8.22 Slip joint. 1—Top joint; 2—Inner kelly joint; 3—Kelly bar; 4—Loading case; 5—Case; 6—Riser; 7—Seal case; 8—Piston; 9—Central rod; 10—Seal ring; 11—Coupling; 12—Loading joint; 13—Pressure-bearing pin; 14—Sliding sleeve roller; 15—Sliding sleeve; 16—Lower joint

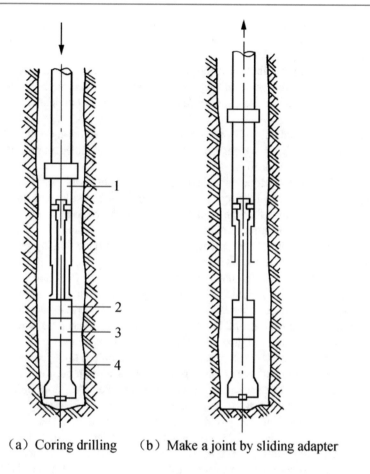

(a) Coring drilling   (b) Make a joint by sliding adapter

Fig. 8.23 Schematic diagrams of medium-long barrel coring connected with stalk. 1—Slip joint; 2—Lower joint; 3—Loading joint; 4—Coring tool

Sealed coring tool

In the conventional coring process, the filtrate of drilling mud will invade into cores, which will change the in situ oil or gas saturation, resulting in invalid and unreliable data. Sealed coring method is using sealing fluid (the polymer liquid preplaced in the inner core barrel that has high viscosity, good fluidity, good adhesive force and waterproof) to protect the drilled core timely. Sealed coring method can protect core from being polluted by the drilling fluid and can mitigate the following problems:

- Get reliable core data of the saturation of oil, gas, and water to provide guidance on calculating the volume of unexploited reservoir and making reasonable development plans;

- Know the saturation, distribution pattern, and dynamic flow properties of oil and gas reservoir at the current exploitation phase to determine reliable development method and adjust plan;
- Know the fluid distribution of reservoirs in a water-flooded field and the relations between lithology and physical properties of reservoir rocks to provide guidance for quantitative study on the oil displacement efficiency.

Sealed coring tools include pressure-maintaining sealed coring tool, self-locking sealed coring tool, and so on. Let's take self-locking sealed coring tool as an example to introduce its structure.

Figure 8.24 is the schematic diagram of the structure of self-locking sealed coring tool. It is the self-locking core cutting type and binocular and double-action type. It has the features that the sealed piston is on the bottom of inner core barrel unit and is fixed at the entrance of coring bit by pins. The sealing ring is fixed on the matching surface of inner core barrel and drilling bit; and the floating piston is fixed on the top of inner core barrel.

Before tripping-in, fill the inner core barrel with sealing fluid and place floating piston on the top of limit joint to form a sealing layer. When tripping-in, the liquid column pressure outside of the sealing area increases with drilling depth, so that the floating piston is moved up and the pressure maintains balanced between the inside and outside sealing area to guarantee the reliability of sealing.

After tripping-in, add some tracer material—ammonium thiocyanate (NH_4SCN) —into flushing fluid and start pumping to make tracer material evenly distributed and qualified to content regulation. Then trip tools into the bottom hole, add load increasingly, and cut off pins. In drilling process, the continuously formed core enters into the inner core barrel and moves the piston upwards. The sealing fluid in the inner core barrel is pushed downward and expelled outside. The discharged sealing fluid will form a protective area around the drilling bit, which will immediately adhere to core surface to form the protective film and avoid the contamination from drilling fluid to achieve sealed coring.

The sealing fluids used in sealed coring can be either oil-based or water-based. The raw material ratios are different in many fields. Taking Shengli oilfield as an example, the applied material ratio is (by mass) as follows:

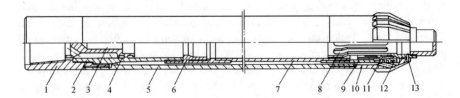

Fig. 8.24 Schematic diagram of a self-locking sealed coring tool. 1—Top joint; 2—Water diversion joint; 3—Floating piston; 4—Y-shaped seal ring; 5—Outer barrel assembly; 6—Limit joint; 7—Inner barrel assembly; 8—Sealed piston; 9—Neck sleeve; 10—Coring bit; 11—Core gripper; 12—O-shape seal ring; 13—Piston-fixed pin

Castor oil:	100;
Vinyl oxide resin:	12–14;
Zinc stearate:	0.84–1.68;
Bentonite or barite:	depends on the density of sealing fluid.

Pressure-maintaining sealed coring

The sealed coring can prevent contamination from the drilling fluid. However, the gas and light components of crude oil trapped in core will rapidly expand and dissipate due to the decrease of pressure and temperature during the trip-out process of core from the bottom to the ground, failing to preserve the initial state of core at the well bottom. Pressure-maintaining coring technology is an effective way to maintain core with its original reservoir fluids. By using this method, the obtained core can be used to calculate the exact parameters at the downhole conditions, such as the fluid saturation and reservoir pressure, the humidity of oil phase, and other physical properties of the reservoir.

These data are of great significance for understanding the geological conditions, calculating residual oil reserves, making reasonable development and adjustment plans and enhancing resources recovery.

Engineers in Daqing oilfield have successfully developed the pressure-maintaining sealed coring tool and completed the matching techniques, including pressure test, core frozen cutting, package and transportation and core analysis.

- Tool structures.

Figure 8.25 is the schematic diagram of pressure-maintaining sealed coring tool, which has the flowing parts:

– Drill bit: six-wing and three-step drag coring bit;
– Clamping structure: use the clam-hoop composite structure;
– Sealing head;
– Inner and outer core barrels assembly: the dual-barrel single-action type;

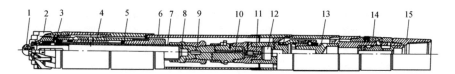

Fig. 8.25 Schematic diagram of a pressure-maintaining sealed coring tool. 1—Sealed head; 2—Coring bit; 3—Bearing; 4—Ball valve assembly; 5—Bottom joint of outer core barrel; 6—Outer core barrel; 7—Inner core barrel; 8—Needle valve; 9—Check valve assembly; 10—Gas chamber connection assembly; 11—Surge chamber assembly; 12—High-pressure gas chamber; 13—Suspension assembly; 14—Differential mechanism; 15—Joint

– Ball-valve structure: It is composed of ball, up-and-down valve seats, pre-loaded spring, valve, body valve, semi-slid ring, sealing packing, hinge pin and other components. Under the gravitational effect of the outer core barrel, the ball valve is forced to turn 90 degrees and closed, leaving the core in the sealed inner core barrel; therefore, it can maintain the formation pressure at the place of core.
– Air valve regulator: It is composed of high-pressure gas chamber, adjustable valve assembly, check valve assembly, and other components. High-pressure nitrogen (40 MPa) is stored in a high-pressure chamber; adjustable valve assembly is used to control the throughput of gas from the high-pressure chamber to realize the automatic pressure compensation and maintain the pressure balance of the inner core barrel in trip-in process.
– Gas chamber connection assembly: The connection joint is opened in trip-out process, and gas from the high-pressure chamber is compensated to inner core barrel through connection joint. Pressure joints are connected with pressure gauge to measure the pressure of the inner core barrel or the pressure of vent inner core barrel.
– Suspension assembly: The drop-hanger bearing is installed to keep the inner core barrel still. Bearing for hanging the inner core barrel is of ball seat type.
– Differential mechanism: It is composed of hexagonal rod, internal hexagonal sock, pipe reducer, pressure cap, packing, and other components. It is mainly used to transmit torque to drive the ball valve and seal the inner core barrel.

• Operation principle

The coring tool is dual-barrel and single-action type, which cuts off cores in clamp-hoop shape. The inner core barrel is the container for cores and also the sealed shell. Before tripping-in the tool, sealing fluid should be pre-filled from the sealed head. After finishing the core drilling, pull up the drilling tool to cut off cores, inject a steel ball of diameter 50 mm to set it onto the sleeve seat, flow back drilling fluid, and if the pump pressure is normal, it indicates the sleeve seat is well placed. At this time, under the gravity effect, the inner and outer hexagonal sleeves will rip out disengage, and it will make the outer core barrel move downward, adding gravitational effect on the semi-sliding ring of the ball valve. The added effect will generate a torque on the steel ball, rotate it in 90 degrees to make valve closed and seal the core into the inner barrel. The high-pressure gas chamber in pressure compensation system was pre-filled with high-pressure nitrogen, and the valve assembly was adjusted to set pressure. In trip-out process and the following operations, the pressure in the inner core barrel is maintained by pressure regulator.

• Core processing
• Take off the hexagonal joint and the hanging joint from the coring tool and connect the pressure testing device onto pressure testing valve to measure the pressure of the inner core barrel;

- Take off the inner core barrel; put the barrel with core inside into the cooling chamber and place adequate amount of dry ice around it; pressurize the cooling chamber until its pressure reaches the original value, and cool it for 6–8 h;
- Cut the frozen core into segments of certain lengths (cut together with the inner core barrel) to make it convenient for transportation and analysis;
- Seal the two ends of core segments; record the well number, date, well section, and other information on the drill card; put core segments into freezer and deliver them to laboratory;
- Mill off the inner core barrel before analysis. Clean out the sealing fluid, and prepare for analysis;

Directional coring

Directional coring is to drill a core and record its original position. Directional coring is required to know geologic structure in the secondary and tertiary oil recovery. The purpose of directional coring is to know the dipping angle, inclination, and trend of a stratum, the density and the distribution pattern of fractures, which will provide the basis for the development plan.

To achieve the objectives of directional coring and to understand the original position of cores, the directional coring tools need to:

- Mark lines on core during core drilling;
- Measure the position of the marking lines.

Directional coring tools are composed of two main parts: core barrel and devices (Fig. 8.26). On the bottom of the inner core barrel, there are three cutters to notch grooves on the core. Therefore, the core entering into core barrel will be vertically notched with marking lines during core drilling.

The measuring while drilling apparatus (multipoint inclinometer), whose bottom connects with extension elements, is connected to the top of core barrel by an orientation shoe. The extension elements can make apparatus in the place of non-magnetic drill collar so as to improve measuring accuracy. Cutter is in a line with the scratch line on the surface of instrument. The position recorded by compass table is the measured position by scratch line and also the position of core notch.

Fig. 8.26 Schematic diagram of a directional coring tool. 1—Guide vane; 2—Measuring instrument; 3—Non-magnetic drill collar; 4—Extension bar; 5—Bottom of instrument hole; 6—Outer core barrel; 7—Inner core barrel; 8—Directional shoe graver

For more details about the principles of instrument measurement and directional measurement, please refer to the chapter of directional drilling in this book or other reference books.

8.2.3 Enhance Core Recovery

Coring is to get the original data of subsurface rocks. Thus enhancing core recovery should be set as the main research goal under any circumstances. To improve core recovery, we must know the factors that influence core recovery and then take corresponding measures to ensure the core recovery and improve core drilling efficiency.

8.2.3.1 Factors on Core Recovery

During core drilling, there are many factors that will influence core recovery. These factors can be categorized into the following types:

Formation

Formation properties changes in many aspects, and it has various effects on core recovery.

Geological structure and stratigraphic structure (such as fault, high-dip angle formation, fracture developed formation, caves, etc.) will affect the core recovery rate.

Formation lithology is another influencing factor. For the consolidated, cemented, and tight formation that is easy to maintain its integrity during coring, the core recovery is high. But for the poorly cemented formation, the unconsolidated formation, the thin inter-bed formation, the hard and soft interbedded formation, the water-soaking and swelling formation, and the formation that has a lot of soluble salts, core recovery is low due to the negative effects of formation properties.

Core diameter

Core diameter is determined by the size of the coring tool. The core with larger diameter will have better strength, which is favorable for core recovery.

Core drilling parameters

If core drilling parameters are inappropriate, the core recovery will be adversely affected. Too large or too small drilling pressure is unfavorable to improve core recovery. If the drilling pressure is too large, the outer core barrel will have unstable buckling, leading to the unstable work of drill bit, and the inner and outer core barrels will clamp with each other, leading to swing and rotation of the inner barrel. If the drilling pressure is too small, the drilling rate will be very low correspondingly. The unstable drilling-in and vibration resulted from large drilling rate will lead to core damages. In terms of flow rate, too large flow rate can erode the core, and a small flow rate can cause a bit of mud bag, which will all affect the improvement of core recovery.

Complicated downhole conditions

For those complex circumstances occurring at downhole, such as lost circulation, collapse, shrinkage, junks, dogleg, key seat, and blowout, if they are not treated in time, it will directly lead to poor coring operations.

8.2.3.2 Measures to Enhance Core Recovery

There are many factors that influence on the core recovery. But for a certain place, the influence factors can be divided into several main aspects, and we should learn from these successes or failures of experiences. Generally speaking, to improve core recovery, we should pay attention to the following aspects.

- Execute proper core drilling plan;
- Select proper drilling bit and coring tool. The selected tools should not only satisfy the coring objectives (normal coring or special coring), but also fit for the drilling formations.
- Thoroughly inspect all tools before trip-in. For example, check any deformation of the core angle, the rotation of suspension bearing, the elasticity of core gripper and the matching of core gripper, and so on;
- Select proper parameters of core drilling;
- Follow the operation regulations, standards, and practices strictly;
- Summarize operation experiences and improve the core drilling operation constantly.

8.3 Casing Sidetracking Technique

Trip-in a special tool at the specified position of the installed casing, then mill a window (an oval hole), or mill off a section of casing to expose the formation. The technique above is called casing sidetracking, of which the purpose is to prepare for the sidetrack of directional well and horizontal well. The technique can significantly improve the success rate of exploratory drilling and oil recovery and improve productivity in the exploitation and development processes. Window-cutting technique and relative tools about the sidetracking of directional well and horizontal well will be introduced in this section.

Sidetracking through casing for sidetrack drilling technique is applied for the following scenarios:

- Drill multi-lateral wells;
- Sidetrack an accidental well;
- The drilled well that deviated from the target oil zones;
- Recompletion of depleted wells in a brownfield:

- If the casing at the bottom of a production well is damaged or the occurred accidents cannot be handled, the casing sidetrack technique can be used to drill to the target zone for recompletion and production.
- If the well section is seriously bridged or flooded, the technique can be used to drill to the production zone to fulfill recompletion.
- For a depleted well, residual oil areas can be exploited by sidetracking through casing for drilling directional wells or horizontal wells.

There are two methods for casing sidetracking: One is by using whipstock to cut an oriented window on casing, and the other is by using expanding tool to mill off a section of case pipe.

8.3.1 Sidetracking Through Casing by Whipstock

8.3.1.1 Construction Procedures of Simple Window-Cutting Tools

The construction procedures of sidetracking by simple tools are shown as below (Fig. 8.27):

- Confirm the position of sidetracking (depth, method);
- Cement plug. Cement an 80–100 m plug at the position prepared for sidetracking;

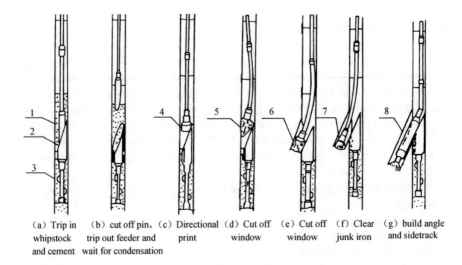

(a) Trip in   (b) cut off pin, (c) Directional (d) Cut off   (e) Cut off   (f) Clear   (g) build angle
whipstock   trip out feeder and   print   window   window   junk iron   and sidetrack
and cement   wait for condensation

Fig. 8.27 Schematic diagrams on operation procedures of window cutting. 1—Feeder; 2—Whipstock; 3—Tail pipe; 4—Lead print; 5—Compound mill cone; 6—Simplex mill cone; 7—Internal drilling shaft; 8—Turbine

- Control trips and wash down. Path by the gauge that has both larger external diameter and length than the whipstock, and wash down thoroughly;
- Trip-in the whipstock. The instrument is connected as: drill string + feeder + whipstock + liner. Use drill string to send the tool to the preset location (the top of cement plug), then cement until the slurry flows back about 10 m above the whipstock, and increase load to cut off the connection pin between the feeder and the whipstock. Pull out the drill string and the feeder, wash down extra cement slurry, and wait 2–5 days for solidification.
- Sidetracking through casing. Drill out the cement above the whipstock, and use the initial mill shoe to mill one side of casing; then use the cone-shaped mill shoe to trim window;
- Sidetrack. The mill shoes used are shown in Fig. 8.28.

If there are certain requirements on the position of sidetracking, the directional drilling is needed. This method is of low efficiency and low accuracy in orientation. If there are no special requirements on sidetracking position and orientation, the ordinary drilling method is applicable.

8.3.1.2 Construction Procedures of Fixed-Anchor Sidetracking Tools

The construction procedures of this method are similar to the above one, but additional tools (gyro inclinometer) can be applied in this method. The procedures are as follows:

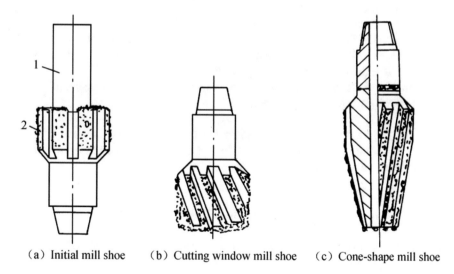

（a）Initial mill shoe   （b）Cutting window mill shoe   （c）Cone-shape mill shoe

Fig. 8.28 Mill shoes for cutting window on casing. 1—Guide rod; 2—Cutter blade

- Cement plug at proper positions;
- Trip-in the whipstock and ground anchor (drill string + feeder + whip-stock + ground anchor) to the top of cement plug;
- Insert measuring instrument through drill string to measure the working orientation of the whipstock, adjust the whipstock to the prescribed orientation, and keep drill string still;
- Inject cement slurry to seal ground anchor and increase load to cut off connection pins between the feeder and the whipstock. Pull up drill string, wash down extra cement slurry, and wait for solidification;
- Trip-in mill shoes to sidetracking on casing.

8.3.1.3 Construction Procedures of Ground Anchor with Slip Packer Sidetracking Tools

The procedures of methods described above are complex, which have to cement plug, or inject cement slurry to seal ground anchor. Two cement injection operations can be dismissed by using slip-packer-type ground anchor (Fig. 8.29) so as to simplify procedure. The used tools are: slip packer (there is a directional shoe pub joint on its top) and fix anchor (the top is connected with whipstock by thread, and the bottom has a directional shoe that can be matched with packer). The operation procedures are as follows:

- Set packer at preset position (replace cementing plug);
- Insert measuring tool to measure the orientation of the pub joint on packer's directional shoe;
- Adjust the relative position between the tool face of whipstock and the directional shoe of fixing anchor to ensure the working orientation of whipstock is at the preset position after the matching between the orientation shoe of fixing anchor and the directional shoe's pub joint of packer;
- Insert whipstock and fixing anchor. The combination is: drill string + initial mill shoe + whipstock + fixing anchor.

Drill to the position of packer to connect the directional shoe of the fixing anchor and the pub joint of directional shoe on the top of packer, then add load to cut off pins, and start the sidetracking of casing.

8.3.2 Casing Sidetracking by Expanded Casing Mill Shoe

This method is using the hydraulic expanded casing mill shoe to mill off a section of casing (30–50 m) and sidetrack by the conventional directional deflecting method.

Expanded casing mill shoe is composed of blade, piston, and mill shoe body. Insert casing mill shoe to the preset position, open pump, and start the circulation. The piston will compress the spring to drop below the circulation pressure, making

Fig. 8.29 Combination of slip packer window-cutting tools. 1—Initial shoe; 2—Whipstock; 3—Fixing anchor; 4—Packer

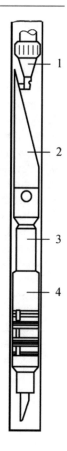

the cradle head against blade and expand outward to rotate drill string to cut off casing. Blade gradually expands after cutting off casing pipe, and pump pressure starts decreasing; press the milled casing slightly until it reaches the length required.

Exercises and Questions

1. What are complicated downhole conditions? What are drilling accidents?
2. What may cause well blowout?
3. What are the procedures of handling a wellsite catching fire due to a blowout out of control? What steps does each procedure have?
4. What are the reasons for lost circulation? List different types of lost circulation.
5. What are the commonly used plugging methods and typical lost circulation additives?
6. What is sticking? How many types of sticking?
7. What causes differential pressure sticking? How to identify and prevent?
8. What is key-seat sticking? How to identify and prevent?
9. How many ways of dealing with sticking? What type of sticking are they used for?

10. What are the characteristics of back-off case-milling free-sticking method and explosion back-off milling free-sticking method? What are the processes?
11. How many possible drilling accidents are there? What are the common fishing tools?
12. What are the working principles for pin tap and box tap in fishing? What circumstances are they applicable to?
13. What tools are commonly used in fishing? What features do they have?
14. What are the purposes of coring?
15. How many processes for drill coring? What are they?
16. What are the evaluation indexes for drill coring?
17. What are the main parts of the coring tool?
18. What are the types of coring bit? And what is the structure for each one?
19. What is the structure of a core barrel?
20. What is the structure of the suspension device of a core barrel?
21. What are the conventional core handling tools? And what formations are they used for?
22. What scenarios are sealed coring used for? What are the features of the sealed coring tools?
23. What are the characteristics of a pressure-maintaining coring tool?
24. Why do we need directional coring?
25. What are the purposes of sidetracking in casing?
26. What are the processes of operating simple sidetracking tools?

References

1. Liu XS (1992) Principles of drilling technology. Petroleum Industry Press, Beijing, China
2. Chen TG, Guan ZC (2000) Theory and technology of drilling engineering. Petroleum University Press, Dongying, China
3. The Writing Group of Drilling Manual (the first Party) (2013) Drilling manual (the first Party), 2nd edn. Petroleum Industry Press, Beijing, China
4. Tu HZ, Gao S (1990) Rock breaking mechanism. Geology Press, Beijing, China
5. Xu XH, Yu J (1984) Rock breaking mechanism. Coal Industry Press, Beijing, China
6. Shen ZH (1988) Basis and calculation of oil well design. Petroleum Industry Press, Beijing
7. Sun GZ (1993) Engineering geology and geological engineering. Seismology Press, Beijing, China
8. Li ZY, Wang ZJ, Yang YY (1990) Foundation of engineering geology. China University of Geosciences Press, Wuhan, China
9. Wang RH, Zhang WD (2016) Fundamentals of drilling technology, 2nd edn. China University of Petroleum Press, Dongying, China
10. Bit Plant of Jianghan Petroleum Administration (1992) Manual of Jianghan drill bit. Petroleum Industry Press, Beijing, China
11. Luo ZF (1984) Drilling technical manual. Petroleum Industry Press, Beijing, China
12. Zhao GZ, Gong WA (1988) Fundamentals of drilling mechanics. Petroleum Industry Press, Beijing, China
13. Han ZY (1995) Study on axial force calculating and strength testing for drilling in vertical holes. Pet Drilling Tech 23(s1):8–13
14. Huang HR, Yang KP, Luo PY (1981) Principle of drilling fluid technology. Petroleum Industry Press, Beijing
15. Li KX (1993) Drilling and completion technology for reservoir protection. Petroleum University Press, Dongying, China
16. Zhao FL (1997) Chemical agent for oil recovery. Petroleum University Press, Dongying, China
17. Sun SY (2006) Drilling machinery. Petroleum Industry Press, Beijing, China
18. Li JZ, Chen RZ (2001) Introduction to oil drilling and production Machinery. Petroleum University Press, Dongying, China
19. Ma YF, Kang T (2005) Manual for rig operation and maintenance. Petroleum Industry Press, Beijing China
20. Han ZY (2011) Calculation and design of directional well, 2nd edn. China University of Petroleum press, Dongying, China
21. Cha YJ, Guan ZC, Rong KS et al (2014) Drilling design. Petroleum Industry Press, Beijing, China
22. GB/T 28911-2012 (2013) Vocabulary of oil and gas drilling. Chinese Standard
23. SY/T 5435-2012 (2012) Wellpath planning & trajectory calculation for directional wells. Chinese Petroleum Industry Standard

© China University of Petroleum Press and Springer Nature Singapore Pte Ltd. 2021   555
Z. Guan et al., *Theory and Technology of Drilling Engineering*,
https://doi.org/10.1007/978-981-15-9327-7

24. SY/T 5619-2009 (2009) Methods of bottom hole assembly for planning directional wells. Chinese Petroleum Industry Standard
25. SY/T 5724-2008 (2008) Design for casing string structure and strength. Chinese Petroleum Industry Standard
26. SY/T 6544-2017 (2017) Performance requirements for oil well cement slurry. Chinese Petroleum Industry Standard
27. Sun XZ (2012) Common well control hazards: identification and countermeasures. Gulf Professional Publishing
28. Hao JF (1992) Balanced drilling and well control. Petroleum Industry Press, Beijing, China
29. Jiang R (1990) Well control technology. Petroleum University Press, Dongying, China
30. Liu G, Jin YQ (2011) Risk analysis and control of well control. Petroleum Industry Press, Beijing China
31. Sun ZC, Xia YQ, Xu MH (1997) Well control technology. Petroleum Industry Press, Beijing, China
32. Zhang MC (2017) Cementing technology. Sinopec Press, Beijing, China
33. Liu CJ, Huang BZ, Xu TT et al (2001) Theory and application of oil and gas well cementing. Petroleum Industry Press, Beijing, China
34. Xiong DY, Shen W (2003) Concrete admixture. Shaanxi Science and Technology Press, Xi'an, China
35. Zhang DL, Zhang X (2002) Design and application of cement slurry. Petroleum Industry Press, Beijing, China
36. Xu TT, Liu YJ (1997) Leakage prevention and plugging technology in drilling engineering. Petroleum Industry Press, Beijing, China
37. Li HS, Fu GQ (1993) Drilling coring technology. Petroleum Industry Press, Beijing, China
38. Mei J, Yan SC (1993) Adjustment well drilling and completion technology. Petroleum Industry Press, Beijing, China
39. Du XR (2012) Drilling tools manual. Sinopec Press, Beijing, p 2012
40. Ma BG, Xu YH, Dong RZ (2006) Influence of triethanolamine on the initial structures and mechanical properties of cement. J Build Mater 9(1):6–9
41. Yao X (2004) Study on expandant for oil well cement (II) expanding mechanism. Drilling Fluid Completion Fluid 21(5):43–48
42. Yao X (1999) The expansion mechanism of magnesia compound for oil well cement-thermodynamic principle and validation. J Southwest Pet Inst 21(2):77–80
43. Zhang QW, Deng M (2009) Review on expansion mechanism of cement paste mixed with MgO type expansion agent. Sci Tech Rev 27(13):111–115
44. Chen ZY (2007) Engineering application of aluminium-powder expansion agent cement and micro-expansion cement. Sci Inf 30(13):127–128
45. Fang EL, Xiong M, Liu YJ et al (2010) Laboratory study on antifreeze latex cement slurry. J Oil Gas Technol 32(5):278–280
46. Bu YH, You J, Jiang LL et al (2009) Research and application of thixotropic cement slurry. Pet Drilling Tech 37(4):110–112
47. Wang JD, Qu JS, Gao YH (2005) The review of deepwater sea cementing slurry technology abroad. Drilling Fluid Completion Fluid 37(4):110–112
48. Sun FQ, Hou W, Duan JZ et al (2007) Design and study of ultra-low density cement slurries. Drilling Fluid Completion Fluid 24(3):31–34
49. Rokinson LH (1959) Effects of pore and confining pressure on failure characteristics of sedimentary rocks. Trans AIME 216(6):26–32
50. Black AD, Green SJ (1978) Laboratory simulation of deep well drilling. Pet Engineer 50 (3):41–48
51. Moore (1974) Drilling practices manual. Petroleum Pub. Co., Wyoming
52. Rehm B, Mcclendon RK (1971) Measurement of formation pressure from drilling data. https://doi.org/10.2118/3601-MS

53. Eaton BA (1972) The effect of overburden stress on geopressure prediction from well logs. J Petrol Technol 24:929–934
54. Chilingarian GV, Vorabutr P (1981) Drilling and drilling fluids. Elsevier Scientific Publishing Company, Amsterdam
55. Gray GR, Darley HC et al (1981) Composition and properties of oil well drilling fluids, 4th edn. Gulf Publishing Company, Texas
56. Fillippone WR (1982) Estimation of formation parameters and the prediction of overpressures from seismic data. https://doi.org/10.1190/1.1827121
57. Olphen HV (1963) An introduction to clay colloid chemistry: for clay technologists, geologists, and soil scientists. Interscience Publishers
58. Wilcox RD, Jarrett MA (1988) Polymer deflocculants: chemistry and application. https://doi.org/10.2523/17201-MS
59. Chesser BG (1987) Design consideration for an inhibitive and stable water-based mud system. SPE Drilling Eng 2:331–336
60. Vidrine DJ, Benit EJ (1968) Field verification of the effect of differential pressure on drilling rate. J Petrol Technol 7:676–680
61. Bourgoyne YFS (1974) A multiple regression approach to optimal drilling and abnormal pressure detection. SPE J 14(4):371–375
62. Eckel JR (1967) Microbit studies of the effect of fluid properties and hydraulics on drilling rate. J Petrol Technol 19(4):541–546
63. Lummus JL (1970) Drilling optimization. J Petrol Technol 22(11):1379–1388
64. Mclean RH (1964) Crossflow and impact under jet bits. J Petrol Technol 16(11):1299–1306
65. Grace RD, Shursen JL (1996) Field examples of gas migration rates. https://doi.org/10.2523/35119-MS
66. Inglis TA (1987) Directional drilling. In: Petroleum Engineering & Development Studies, vol 2
67. Adams N (1980) Well control problems and solutions. Penn Well Books, Tulsa
68. Nichens HV (1985) A dynamic computer model of a kicking well. SPE Drilling Eng 2(2):159–173
69. Choe J, Juvkam-Wold HC (1996) Well control model analyzes unsteady state, twophase flow. Oil Gas J 94(49):49–53
70. Schubert JJ, Juvkam-Wold HC, Weddle CE (2002) HAZOP of well control procedures provides assurance of the safety of the subsea mud Lift drilling system. Paper presented at IADC/SPE Drilling Conference, 26–28 February, Dallas, Texas
71. Casariego V, Bourgoyne AT (1988) Generation, migration, and transportation of gas-contaminated regions of drilling fluid. https://doi.org/10.2118/18020-MS
72. Adams AJ, Maceachran A (1994) Impact on casing design of thermal expansion of fluids in confined annuli. SPE Drilling Completion 9(3):210–216
73. Alun W (1985) Coring operation. In: Exlog series of petroleum geology and engineering handbooks. Springer, Netherlands
74. API Bulletin 5C3 (1994, 2010) Bulletin on formulas and calculations of casing, tubing, drill pipe and line properties, 6th edn. API Production Department
75. ISO/TR 10400 (2007) Petroleum and natural gas industries-equations and calculations for the properties of casing, tubing, drill pipe and line pipe used as casing or tubing
76. Nelson EB (1990) Well cementing. Schlumberger Educational Services, Houston
77. Bourgoyne AT, Millheim KK, Chenevert ME et al (1986) Applied drilling engineering. Society of Petroleum Engineers, Richardson
78. Mitchell RF, Miska SZ (2011) Fundamentals of drilling engineering. SPE Textbook Series, vol 12
79. John F (2012) Drilling engineering. Heriot-Watt University, Edinburgh, UK, p 2012
80. Hossain ME, Al-Majed AA (2015) Fundamentals of sustainable drilling engineering. Scriener Publishing, Wiley

Printed in the United States
by Baker & Taylor Publisher Services